Hearing

Handbook of Perception and Cognition
2nd Edition

Series Editors
Edward C. Carterette
and **Morton P. Friedman**

Hearing

Edited by
Brian C. J. Moore

Department of Experimental Psychology
University of Cambridge
Cambridge, England

Academic Press

San Diego New York Boston
London Sydney Tokyo Toronto

This book is printed on acid-free paper. ♾

Academic Press, Inc.
A Division of Harcourt Brace & Company
525 B Street, Suite 1900, San Diego, California 92101-4495

United Kingdom Edition published by
Academic Press Limited
24-28 Oval Road, London NW1 7DX

Library of Congress Cataloging-in-Publication Data

Moore, Brian C. J.
Hearing / by Brian C. J. Moore.
p. cm. -- (Handbook of perception and cognition (2nd ed.))
Includes bibliographical references.
ISBN 0-12-505626-5
1. Auditory perception. 2. Hearing. 3. Psychoacoustics.
I. Title. II. Series.
BF251.M65 1995
152.1'5--dc20 95-1082
CIP

PRINTED IN THE UNITED STATES OF AMERICA
95 96 97 98 99 00 BC 9 8 7 6 5 4 3 2 1

Contents

7 *Across-Channel Processes in Masking*

Joseph W. Hall III, John H. Grose, and Lee Mendoza

8 *Pitch Perception*

Adrianus J. M. Houtsma

9 *Spatial Hearing and Related Phenomena*

D. Wesley Grantham

10 *Models of Binaural Interaction*

Richard M. Stern and Constantine Trahiotis

Contributors

Numbers in parentheses indicate the pages on which the authors' contributions begin.

Robert P. Carlyon (123, 387)
MRC Applied Psychology Unit
Cambridge CB2 2EF, England

C. J. Darwin (387)
Laboratory of Experimental Psychology
University of Sussex
Brighton BN1 9QG, England

David A. Eddins (207)
Department of Speech and Hearing Sciences
Indiana University
Bloomington, Indiana 47405

D. Wesley Grantham (297)
The Bill Wilkerson Center
Vanderbilt University School of Medicine
Division of Hearing and Speech Sciences
Nashville, Tennessee 37212

David M. Green (207)
Psychoacoustics Laboratory
Department of Psychology
University of Florida
Gainesville, Florida 32611

John H. Grose (243)
Head and Neck Surgery
Division of Otolaryngology
The University of North Carolina at Chapel Hill
Chapel Hill, North Carolina 27599

Joseph W. Hall III (243)
Head and Neck Surgery
Division of Otolaryngology
The University of North Carolina at Chapel Hill
Chapel Hill, North Carolina 27599

Stephen Handel (425)
Department of Psychology
University of Tennessee
Knoxville, Tennessee 37996

William Morris Hartmann (1)
Physics Department
Michigan State University
East Lansing, Michigan 48824

Adrianus J. M. Houtsma (267)
Institute for Perception Research
Eindoven University of Technology
5600 MB Eindhoven, The Netherlands

Lee Mendoza (243)
Head and Neck Surgery
Division of Otolaryngology
The University of North Carolina at Chapel Hill
Chapel Hill, North Carolina 27599

Brian C. J. Moore (161)
Department of Experimental Psychology
University of Cambridge
Cambridge CB2 3EB, England

Alan R. Palmer (75)
MRC Institute of Hearing Research
University of Nottingham
Nottingham NG7 2RD, United Kingdom

Christopher J. Plack (123)
Department of Experimental Psychology
University of Sussex
Brighton BN1 9QG, United Kingdom

Richard M. Stern (347)
Department of Electrical and Computer Engineering
and Biomedical Engineering Program
Carnegie Mellon University
Pittsburgh, Pennsylvania 15213

Constantine Trahiotis (347)
Department of Surgery (Otolaryngology), Surgical Research Center, and Center for Neurological Sciences
University of Connecticut Health Center
Farmington, Connecticut 06032

Graeme K. Yates (41)
The Auditory Laboratory
Department of Physiology
The University of Western Ontario
Nedlands, 6009 West Australia Australia

Foreword

The problem of perception and cognition is in understanding how the organism transforms, organizes, stores, and uses information arising from the world in sense data or memory. With this definition of perception and cognition in mind, this handbook is designed to bring together the essential aspects of this very large, diverse, and scattered literature and to give a précis of the state of knowledge in every area of perception and cognition. The work is aimed at the psychologist and the cognitive scientist in particular, and at the natural scientist in general. Topics are covered in comprehensive surveys in which fundamental facts and concepts are presented, and important leads to journals and monographs of the specialized literature are provided. Perception and cognition are considered in the widest sense. Therefore, the work will treat a wide range of experimental and theoretical work.

The *Handbook of Perception and Cognition* should serve as a basic source and reference work for those in the arts or sciences, indeed for all who are interested in human perception, action, and cognition.

Edward C. Carterette and Morton P. Friedman

Preface

The aim in editing this volume was to cover all the major areas of hearing research with a series of coordinated and well-integrated chapters. Authors were asked particularly to emphasize concepts and mechanisms and to attempt wherever possible to explain not only what the empirical data show, but also why they show a particular pattern.

The volume begins with "The Physical Description of Signals" by William Hartmann. Many students (and even some researchers) have difficulty with the physical and mathematical concepts used in hearing research, and this chapter lays the essential groundwork for acquiring these concepts.

Chapter 2, "Cochlear Structure and Function" by Graeme Yates, describes the great advances in knowledge that have occurred in the past 15 years, especially the concept of the cochlear amplifier, a physiologically vulnerable mechanism that appears to be partly responsible for the high sensitivity, wide dynamic range, and good frequency selectivity of the auditory system.

Chapter 3, "Neural Signal Processing" by Alan Palmer, describes how different aspects of auditory signals are represented, processed, and analyzed at different stages of the auditory nervous system. The emphasis is on the functional role of the response properties of the neurons, and the organization is in terms of the type of stimulus being analyzed.

Chapter 4, "Loudness Perception and Intensity Coding" by Christopher Plack and Robert Carlyon, reviews both empirical data and theories concerning the way in which stimulus intensity is represented in the auditory system. The experimental data are used to test and evaluate theories of intensity coding.

Chapter 5, "Frequency Analysis and Masking" by Brian Moore, uses the concepts of the auditory filter and the excitation pattern to characterize the frequency selectivity of the auditory system. The chapter describes how the shape of the auditory filter can be estimated in masking experiments and summarizes how its shape varies with center frequency and level. The chapter then discusses various aspects of perception that are influenced by frequency selectivity.

Chapter 6, "Temporal Integration and Temporal Resolution" by David Eddins and David Green, describes the empirical data within the framework of a clear theoretical perspective. Many temporal phenomena are described, analyzed, and interpreted.

Chapter 7, "Across-Channel Processes in Masking" by Joseph Hall, John Grose, and Lee Mendoza, covers an area of research that has expanded rapidly over the past ten years. It is concerned with situations in which the "traditional" model of masking, based on the assumption that subjects make use of the single auditory filter giving the highest signal-to-masker ratio, clearly fails. Instead, performance appears to depend on the pattern of outputs across different auditory filters.

Chapter 8, "Pitch Perception" by Adrian Houtsma, describes how pitch theories have been developed and refined on the basis of experimental data. It includes a comprehensive description of a variety of pitch phenomena and describes both the clear tonal pitches evoked by sounds with discrete sinusoidal components and the less salient pitches that are sometimes evoked by noise-like sounds.

Chapter 9, "Spatial Hearing and Related Phenomena" by Wesley Grantham, gives a comprehensive description of the ability to localize sounds in space and to detect shifts in the positions of sound sources. The cues used for localization are described and compared in effectiveness. The chapter also covers binaural unmasking effects and the "precedence effect" that helps to reduce the influence of room echoes on localization.

Chapter 10, "Models of Binaural Interaction" by Richard Stern and Constantine Trahiotis, gives a comprehensive overview of how models of binaural processing have been developed and refined on the basis of experimental data. The chapter covers the perception of subjective position, discrimination of changes in position, and binaural unmasking.

Chapter 11, "Auditory Grouping" by Christopher Darwin and Robert Carlyon, describes an area of research that has expanded considerably over the past decade. It is concerned with the ability of the auditory system to analyze a complex mixture of sounds, arising from several sources, and to derive percepts corresponding to the individual sound sources. The chapter is particularly concerned with the information or "cues" used to achieve this.

The final chapter, "Timbre Perception and Auditory Object Identification" by Stephen Handel, relates the physical characteristics of sound sources to their perceived qualities and describes the ability of the auditory system to identify musical instruments, voices, and natural environmental events on the basis of their acoustic properties.

There are many links between the phenomena and the theories described in the different chapters. These links are pointed out in the extensive cross-references between chapters. This will help the reader who wants to find out as much as possible about a specific topic.

The contributors to this volume have done an excellent job, and I thank them for their patience and cooperation. I also thank Ian Cannell for his assistance with figures.

Brian C. J. Moore

CHAPTER 1

The Physical Description of Signals

William Morris Hartmann

I. INTRODUCTION

It is appropriate that the study of the perception of sound begins with a physical description of the sound itself. Indeed, the study of the perception of sound is usually an attempt to discover the relationship between the human response to a sound and a precise physical characterization of that sound.

In the usual physical acoustical situation, there are three steps: the generation of the sound by a source, the propagation of the sound by a medium from the source to a receiver, and the reception of sound. The generation of sound always originates in mechanical vibration. The first section describes the particular kind of vibration known as *simple harmonic*. Specializing the study to this kind of motion is not as restrictive as it might first appear, which becomes clear later when the Fourier transform is introduced. The next section describes the propagation of sound in air. It too is specialized to simple situations. Subsequent sections deal with mathematical aspects of signals that are particularly useful in psychoacoustics, the Fourier transform, spectral representations and autocorrelation, and the topic of modulation. The chapter ends with a section on filtering of signals.

Hearing

II. SIMPLE HARMONIC MOTION

Few concepts in physics have had such wide application or so many variations as the simple harmonic oscillator. The study of simple harmonic motion is a sensible starting point for the study of hearing as well.

The study of the harmonic oscillator begins by describing the motion of a mass m that is free to move in one dimension, x, about some equilibrium point, as shown in Figure 1. We might as well take the equilibrium point as the origin of the coordinate system; therefore, equilibrium is at $x = 0$. When the mass moves away from equilibrium by a displacement equal to x, there is a force on the mass tending to move it back to equilibrium. The force (and this is really important) is proportional to the displacement but in the opposite direction. Hence, the force is $F = -kx$. Parameter k is called the *spring constant*.

Newton's law of motion relates the force on the mass to the acceleration of the mass, d^2x/dt^2, by the law

$$m \frac{d^2x}{dt^2} = F, \tag{1}$$

and therefore, for the simple harmonic oscillator,

$$m \frac{d^2x}{dt^2} = -kx. \tag{2}$$

Equation (2) is a differential equation for displacement x that is second order in the derivative with respect to time t. It is linear in x, because x appears to only the first power on both sides of the equation.

The solutions to this equation of motion describe the displacement as a function of time, $x(t)$. There are three solutions. One of them is $x(t) = 0$, where the mass sits at its equilibrium point for all time and does not move.

The other two solutions are more interesting; they are the sine and cosine functions. To show that $x =$ sine and $x =$ cosine are solutions, we recall that the derivative of a sine function is a cosine and the derivative of a cosine function is the negative of the sine function. Therefore, the second deriva-

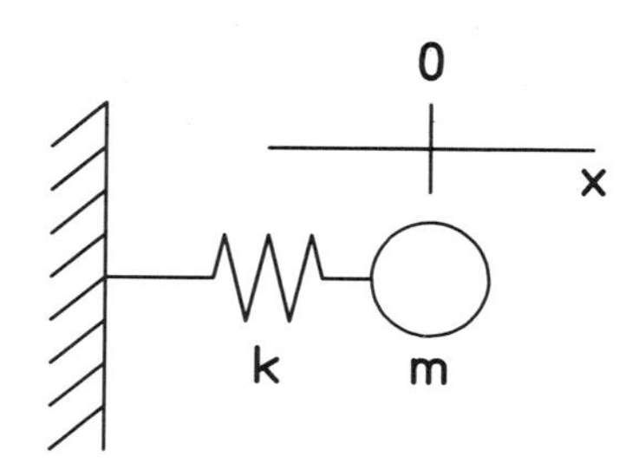

FIGURE 1 A mass and a spring make the simplest harmonic oscillator.

tive of sin $\omega_0 t$ is $-\omega_0^2$ sin $\omega_0 t$, and the second derivative of cos $\omega_0 t$ is $-\omega_0^2$ cos $\omega_0 t$. These are just what we need to solve Eq. (2), so long as the parameter ω_0 is related to the spring constant and mass by the equation

$$\omega_0^2 = k/m. \tag{3}$$

Because both the sine and cosine functions individually are solutions, it follows that the sum of the two is a solution. Further, we can multiply sine and cosine functions by constants, calling them A and B. Finally, the general solution to the simple harmonic oscillator differential equation is

$$x(t) = A\cos(\omega_0 t) + B\sin(\omega_0 t). \tag{4}$$

The solution $x(t)$ is well known in acoustics as the *sine wave* and in auditory science as the *pure tone.* The sine and cosine functions depend upon an angle that increases linearly with time. As the angle, sometimes called the *instantaneous phase,* grows, the equation traces out oscillations with angular frequency ω_0 radians per second.

The sine and cosine functions are periodic. Every time that the instantaneous phase $\omega_0 t$ increases by 2π rad (or 360°) the function $x(t)$ starts over again to trace out another cycle. Every cycle is the same as any other cycle.

Because there are 2π rad in a complete cycle, the relationship between angular frequency ω_0 in rad/sec and frequency f_0 in cycles/s [or Hertz (Hz), to use modern units] is

$$\omega_0 = 2\pi f_0. \tag{5}$$

For example, if f_0 is 1000 Hz then ω_0 is 6283 rad/s.

The periodicity can also be represented by a repetition duration called the *period,* symbolized T. The period of a sine is related to the frequency by the simple reciprocal relation

$$T = 1/f_0. \tag{6}$$

For example, if f_0 is 1000 Hz then T is 0.001 s or 1 ms.

Equation (4) is not the most convenient form for the solution. More convenient for psychoacoustical purposes is the amplitude and phase form,

$$x(t) = C\cos(\omega_0 t + \phi), \tag{7}$$

where C is the amplitude and ϕ is the phase.

The key to transforming from Eq. (4) to Eq. (7) is the trigonometric identity

$$\cos(\psi + \phi) = \cos\phi\cos\psi - \sin\phi\sin\psi, \tag{8}$$

for any angles ψ and ϕ.

We have only to associate ψ with $\omega_0 t$ to compare Eqs. (4) and (8) and arrive at the following correspondences:

$$A = C \cos \phi \tag{9}$$

$$B = -C \sin \phi. \tag{10}$$

The amplitude C is given by

$$C = \sqrt{A^2 + B^2}, \tag{11}$$

and the phase is given by the equation

$$\tan \phi = -B/A. \tag{12}$$

The solution for phase angle ϕ is

$$\begin{aligned} \phi &= \arctan(-B/A) & A < 0 \\ \phi &= \arctan(-B/A) + \pi & A \geq 0. \end{aligned} \tag{13}$$

An example of simple harmonic motion is shown in Figure 2. The amplitude is $C = 5$. The frequency is 1000 Hz, and so the waveform repeats itself three times during the 3 ms interval shown. For this waveform to be a correct representation of a 1000 Hz sine wave, however, the duration must be infinite. The initial phase of the cosine pattern is given by the angle $-\pi/4$ rad ($-45°$).

A. RMS Value

Figure 2 shows one other measure of the wave namely the rms (root mean square) value. The rms value is computed by performing the named steps in reverse order: first one squares the waveform instantaneous values (s), then one finds the mean over a cycle (m), and finally one takes the square root (r).

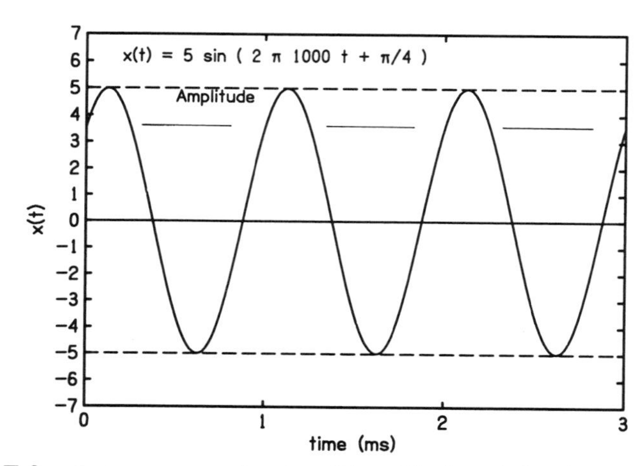

FIGURE 2 A sine wave as a function of time. The long dashes show the rms value.

Because operations of square and square root both occur, the rms value is on the same scale as the amplitude. If the amplitude is in millimeters of displacement then the rms value is also in millimeters of displacement. For a sine waveform, the rms value is related to the amplitude by a simple scale factor:

$$x_{\mathrm{rms}} = C/\sqrt{2} = 0.707C. \tag{14}$$

This completes the description of the basic solution to the oscillator equation consisting of the sine and cosine trigonometric functions and the concepts of amplitude, frequency, period, and phase. It is important to examine the physical situation that led to the original equation, Eq. (2), to determine what it is and, particularly, what it is not. This is done in the following section.

B. Other Oscillators

The situation of Eq. (2) is somewhat unrealistic in that it describes perpetual motion. Once the system is set in motion it oscillates forever. There is no dissipative element that would extract energy from the vibrations. Equation (2) can be modified to include dissipation, the most convenient form being viscous damping. Viscous damping is a resistive force proportional to the velocity. It is not the same as ordinary friction, which is independent of velocity, but it does resemble what is popularly known as *wind resistance.* With viscous damping, the differential equation becomes

$$m\,\frac{d^2x}{dt^2} = -kx - r\,\frac{dx}{dt}. \tag{15}$$

The solution to this equation is a damped sine, a pure tone that decays exponentially in time. If the damping is not too large, so that there are more than ten cycles of oscillation, then the natural frequency of the system is still given by Eq. (3) to a good approximation.

The oscillators described in Eqs. (2) and (15) are free vibrators, unaffected by anything in the outside world. The next step in the evolution of vibration dynamics is to include influences from the outside. These are represented as forces, so that Eq. (15) becomes

$$m\,\frac{d^2x}{dt^2} = -kx - r\,\frac{dx}{dt} + F_{\mathrm{ext}}(t). \tag{16}$$

The amplitude of oscillation is no longer arbitrary. It is proportional to the magnitude of the driving force F_{ext}. If the external force has a sinusoidal time dependence we have a driven oscillator. This mechanical system and its electrical analog, made with a capacitor and inductor (tuning), and a resistor (damping), are the prototypical tuned system. The natural frequency of vibration is approximately given by Eq. (3). In the steady state this system oscillates at the frequency of the external driver. It exhibits a resonance behavior, where the displacement becomes particularly large, when the frequency of the external driver agrees with the natural frequency of vibration of the system, ω_0. The sharpness of the tuning is described by a parameter Q, which is inversely proportional to the damping. Quantity Q will be described further in the section on filtering.

A final restriction of all the preceding systems is that they are linear. That means that the quantity x appears only to the first power in the equation of motion. The fact of linearity is of enormous importance. Only because the equations were linear could we begin with a solution to Eq. (2) like $\cos \omega_0 t$, multiply it by a constant A, and find that we still had a solution. Only because the equations were linear could we take two solutions, $\cos \omega_0 t$ and $\sin \omega_0 t$, and add them to get a more general solution. None of these features applies if the dynamical equation is nonlinear.

Simply to illustrate an example of a nonlinear dynamical equation, we write down the equation for the Van der Poll oscillator:

$$m \frac{d^2x}{dt^2} = -kx + (r_1 - r_2x^2) \frac{dx}{dt}. \tag{17}$$

What makes this equation nonlinear is the factor of x^2 that multiplies the otherwise linear term dx/dt. Whenever the dynamical variable x appears to a power higher than the first, whether differentiated or not, there is a nonlinearity. For example, a term like x^2, or $x\, dx/dt$, or $x\, d^2x/dt^2$, would make the equation nonlinear.

The solutions to nonlinear equations are usually difficult to find, and they are also rather specialized to particular circumstances. They may even be chaotic. Nonlinear mechanics is an important part of auditory physiology because the initial transduction in the cochlea is nonlinear. However, up to the present the implications of nonlinearity for perception have been handled phenomenologically with only sporadic attempts to relate results to first-principles dynamical equations.

III. PROPAGATION OF SOUND

If the mechanical vibrations discussed in Section II have frequencies in the audible range, they are heard as sound by a listener if they can propagate from the vibrating source to the listener's ear. The topic of sound propagation is a major part of the science of acoustics. The principles and applications are the subject of a vast literature, and the treatment in this section can deal with only the simplest concepts.

Sound is propagated in the form of waves. Unlike light, sound waves require a physical medium. Sound can be propagated through a gas or a liquid or (as in the case of bone-conduction audiometry) through a solid. The most familiar medium is gaseous; namely, air.

We who live on the surface of the earth are at the bottom of the atmosphere. Because of earth's gravity, the weight of the atmosphere exerts a pressure in all directions on everything down here, appropriately called *atmospheric pressure.* Its value is 10^5 Newtons/m^2; that is, 10^5 Pascals (Pa), equivalent to about 14.7 pounds per square inch.

Acoustical waves consist of variations in atmospheric pressure. Compared to the static pressure itself, the variations caused by speech, music, and the sounds of the environment are miniscule. A wave at the threshold of hearing for human listeners has an rms pressure of 2×10^{-5} Pa, almost ten orders of magnitude smaller than atmospheric pressure. A wave with an rms pressure of 10 Pa (one ten-thousandth of atmospheric pressure) is painful and dangerous to hearing.

The speed of sound waves depends somewhat on the temperature of the air. At room temperature (20°C) the speed is $v = 344$ meters per second (m/s), corresponding to 1238 km/hr, 1129 ft/s, or 770 mi/hr. Therefore, sound in air travels about a million times slower than light ($c = 3 \times 10^{10}$ m/s).

Accompanying the small pressure variation in a sound wave is a systematic variation in the average velocity of the air molecules, given the symbol u and measured in m/s. If the signal is propagating as a plane wave (more about which later) then the rms velocity is related to the pressure by a simple proportionality:

$$u = p/z, \tag{18}$$

where z is the specific acoustical impedance. This impedance is equal to the product of the air density and the speed of sound:

$$z = \rho v. \tag{19}$$

The density of air is $\rho = 1.21$ kg/m^3, and with a speed of sound of $v = 344$ m/s, the specific impedance is 415 kg/m^{2}s or 415 rayls. It follows that a plane wave at the threshold of hearing moves molecules with an rms velocity of $u = 4.8 \times 10^{-8}$ m/s. This is enormously slower than the rms velocity of air molecules due to the kinetic energy of thermal motion at room temperature, which is about 500 m/s. However, in the absence of a sound wave, the molecular velocities are equally often positive and negative, and they average to zero. As a result, a systematic oscillation with rms velocity that is one ten-billionth of the thermal velocity is audible.

A. Energy, Power, and Intensity

A sound wave carries *energy* as it propagates. Energy is measured in units of Joules (J), and the rate of transporting energy is known as *power,* measured in watts (W), 1 W = 1 J/s. As a plane wave travels, the power is distributed all along the surface of the wavefront so that the appropriate measure of the strength of the wave is power per unit area of wavefront, also known as *intensity* (I). Intensity, therefore, has units of W/m^2.

The intensity of a plane wave is simply the product of the rms pressure and velocity variations. Therefore, intensity is proportional to the square of the pressure:

$$I = pu = p^2/z. \tag{20}$$

For example, if p is the nominal rms pressure at the threshold of hearing then

$$I = \frac{(2 \times 10^{-5})^2}{415} \approx 10^{-12}\ W/m^2. \tag{21}$$

This value of intensity is another measure of the nominal threshold of hearing, $I_0 = 10^{-12}\ W/m^2$.

B. Radiation

The acoustical field of a sound source depends upon the geometry of the source and upon the environment. The simplest source is the monopole radiator, which is a symmetrically pulsating sphere. All other sources have some preferred direction(s) for radiating.

The environment affects the sound field because sound waves are reflected from surfaces where there is a discontinuity in specific acoustical impedance; for example, where air meets a solid or liquid or even air of a different temperature. The reflected waves add to the direct wave from the source and distort the shape of the radiation field. The simplest environment, called *free field,* is completely homogeneous, without surfaces, and can be obtained by standing in an open field and then removing the ground. A practical means of accomplishing this is bungee jumping. Free-field conditions are approximated in an anechoic room where the six surfaces of the room are made highly absorbing so that there are no reflections. From an acoustical point of view, it does not matter whether the outgoing sound waves are absorbed on the walls or whether there are no walls at all.

A monopole radiator expands and contracts, causing, respectively, overpressure and partial vacuum in the surrounding air. In the free-field environment the peaks and troughs of pressure form concentric spheres as they travel out from the source. The power in the field a distance r away from the source is spread over the surface of a sphere with area $4\pi r^2$. It follows that for a source radiating acoustical power P, the intensity is given by

$$I = P/4\pi r^2. \tag{22}$$

This equation expresses the "inverse square law" for the dependence of sound intensity on distance. If the source is not spherically symmetric (not a monopole) then, in free field, the intensity, measured in any direction with respect to the source, is still inversely proportional to the square of the distance; however, the constant of proportionality is not $1/4\pi$ but contains a directionality factor.

The inverse square law is a result of the fact that the world is three dimensional. In a two-dimensional world the sound power is spread over the perimeter of a circle so that the sound intensity decreases only as the inverse first power of the distance from the source. As a result, the inhabitants of Flatland are able to communicate over long distances. In a one-dimensional world the sound intensity does not decrease at all with distance from the source. The ship's first officer on the bridge can communicate with the engine room crew at the other end of the ship by means of the speaking tube, which has no amplification but only confines the sound waves to a one-dimensional channel.

C. Plane Waves

The pressure wavefronts that radiate from a monopole source are spherical in shape. However, a receiver far away from the source does not notice this curvature of the wavefront. The wavefront seems to be a flat surface, just as the surface of the earth seems to be flat because we are so far away from the center. The flat wavefront is known as a *plane wave.* A plane wave is characterized only by its propagation direction. (It is called the z direction in the following.) The pressure does not depend upon the other two directions of space. It follows that the pressure is constant everywhere in any plane that is perpendicular to the propagation direction, as shown by Figure 3.

The traveling plane wave is described by a cosine function (or sine) with an instantaneous phase that depends upon both space and time,

$$x(t) = C \cos(\omega t - kz + \phi), \tag{23}$$

where k is known as the *propagation constant.* Because of the negative sign in the instantaneous phase, we know that this wave is traveling in the positive z direction. The propagation constant is related to the periodicity in space in the same way that the angular frequency is related to the periodicity in time. We recall that $\omega = 2\pi/T$ and, therefore,

$$k = 2\pi/\lambda, \tag{24}$$

where λ, the periodicity in space, is called the *wavelength* and is measured in meters.

The wavelength and the period are related by the speed of sound,

$$\lambda = vT, \tag{25}$$

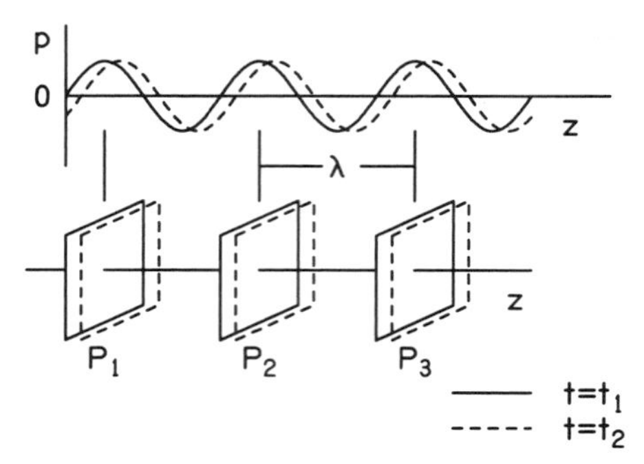

FIGURE 3 Two snapshots taken of a traveling wave: (top) The solid line shows the pressure as a function of spatial coordinate z seen in a snapshot taken at time $t = t_1$. The dashed line is a second snapshot taken at a later time, $t = t_2$ (where $t_2 - t_1$ is about one-tenth of a period). Evidently, the wave is propagating in the positive z direction. (bottom) An alternative representation of the two snapshots illustrates the wavefront concept. Everywhere in plane P_1 the pressure is at a maximum. Everywhere in planes P_2 and P_3 the pressure is also at a maximum because all the planes are mutually separated by the wavelength λ. The solid line shows a snapshot taken at time $t = t_1$. The dashed line shows a snapshot taken at the later time, $t = t_2$, when the wave has moved along the z axis. The wavefronts, shown by planes, have moved rigidly away from the source.

because when time has advanced by one period from time t_1 to $t_1 + T$, the plane at P_1 will have advanced to the original position of plane P_2, and the plane P_2 will have advanced to the original position of plane P_3. Equation (25) is usually written in terms of frequency:

$$v = f\lambda, \tag{26}$$

which says that wavelength and frequency are reciprocally related. For example, given that the speed of sound in air is 344 m/s, the wavelength of a 20 Hz tone is 17.2 m and the wavelength of a 20 kHz tone is 17.2 mm.

IV. MEASUREMENT OF ACOUSTICAL STRENGTH

The strength of an acoustical signal is measured electronically, beginning with a transducer that converts the acoustical signal into an electrical signal. The transducer is a pressure microphone, where the displacement of the diaphragm is proportional to the pressure in the sound field. The displacement generates a proportional electrical voltage that leads to the final reading on the meter, which is an rms voltmeter. In the end, therefore, the reading is a measure of rms sound pressure.

The sound pressure and sound intensity are universally quoted as "levels," which means that they are measured on a logarithmic scale. This scale is foremost a ratio scale, whereby the intensity is compared to a reference. The mathematical statement of this idea is

$$L_2 - L_1 = 10 \log(I_2/I_1). \tag{27}$$

The verbal translation is that the difference in levels, measured in decibels (dB), between sounds 2 and 1 is given by the common (base 10) logarithm of the ratio of the intensities. The factor of 10 is included to expand the scale.

A review of the log function makes a few features of the level (decibel) scale apparent. Because log(1) = 0, if sounds 1 and 2 have the same intensity then the difference in their levels is 0. Because the logarithm of a number less than 1 is negative, if $I_2 < I_1$ then $L_2 - L_1$ is negative. Because the log of a negative number is not defined, it is not possible to transform a negative intensity (whatever that might be) into decibels.

The function of the log scale is to transform ratios into differences. If I_2 is twice I_1 then $L_2 - L_1 = 3$ dB, no matter what the actual value of I_1 might be. That is because log(2) = 0.3. The log scale exhibits logarithmic additivity: if I_2 is twice I_1 and I_3 is twice I_2, then $L_2 - L_1 = 3$ dB, and $L_3 - L_2 = 3$ dB, so that $L_3 - L_1 = 6$ dB.

A. Levels and Pressure

Because the intensity of sound is proportional to the square of the rms pressure, it is easy to express level differences in terms of pressure ratios. From Eq. (27) we have

$$L_2 - L_1 = 10 \log[(p_2/p_1)^2], \tag{28}$$

or

$$L_2 - L_1 = 20 \log(p_2/p_1). \tag{29}$$

Level differences may be derived for other physical quantities, too.

There is sometimes uncertainty about whether the correct prefactor in a decibel calculation is 10 or 20. To resolve the uncertainty, we observe that there are two kinds of quantities for which a decibel scale is appropriate, energylike quantities and dynamical quantities. An energylike quantity or a fluxlike quantity, used in the description of a signal, is real and never negative. Such quantities are acoustical energy, intensity or power, electrical energy or power, optical luminance, or doses of ionizing radiation. When converting ratios of such quantities to a decibel scale the appropriate prefactor is 10.

Dynamical quantities may be positive or negative; usually they are positive and negative equally often. In some representations they may even be complex. Examples of such quantities are mechanical displacement or velocity, acoustical pressure, velocity or volume velocity, electrical voltage or current, or electric and magnetic fields. Dynamical quantities have the property that their squares are energylike or fluxlike quantities. Therefore, when putting these on a decibel scale, the appropriate prefactor is 20.

Other quantities are neither energylike nor dynamical, and for these quantities a decibel scale is inappropriate. For example, frequency is neither energylike or dynamical and it is inappropriate to measure frequency in terms of decibels, even though the logarithm of frequency is a well-motivated measure, by both psychoacoustic results and musical tradition. The log of frequency is given other names such as octave number or semitone number or cents.

B. Absolute dB

Although the decibel scale is a ratio scale in which a quantity is always compared with another quantity, it is common for individual sound levels to be expressed in decibels as though the measure were absolute. For example, sound level meters read in dB SPL (sound pressure level). This practice is possible because it has been agreed upon in advance that the reference quantity shall be a convenient fixed value corresponding roughly to the threshold of hearing, either an intensity of $I_0 = 10^{-12}$ W/m^2 or a pressure of $p_0 = 2 \times 10^{-5}$ Pa. With these references, the sound pressure level for a signal with intensity I or rms pressure variation p can be written as

$$L = 10 \log(I/I_0) \tag{30}$$

and

$$L = 20 \log(p/p_0). \tag{31}$$

The left-hand sides of these equations are not differences. They are simply levels, put on an absolute scale by convention.

There are absolute decibel units for signals in electrical form, too. The unit dBv uses a reference signal with an rms voltage of 1 volt. The unit dBm uses a reference signal with a power of 1 milliwatt. The former is dynamical, the latter is energylike.

1. Example: Addition of Intensities

The logarithmic transformation from intensity or power to decibels leads to some mathematical awkwardness. In the following, we give an example to illustrate techniques for coping with this.

We suppose that two sine waves x_1 and x_2 are sounding simultaneously, where x_1 has frequency 1000 Hz and level 70 dB SPL and x_2 has frequency 1720 Hz and level 67 dB SPL. Our task is to calculate the level of the combination.

Because the air is a linear medium at these sound pressure levels and the frequencies are different, the intensities of the two waves simply add so that the final intensity is $I_3 = I_1 + I_2$. The only problem is that I_1 and I_2 are given in dB as L_1 and L_2.

To find I_3 we first express I_1 and I_2 by inverting the dB formula,

$$I_3/I_0 = 10^{70/10} + 10^{67/10} \tag{32}$$

Next, we extract from all the exponents a large common factor:

$$I_3/I_0 = (1 + 10^{-3/10})10^{70/10}, \tag{33}$$

$$I_3/I_0 = 1.5 \times 10^7. \tag{34}$$

Then to find level L_3 we use the final formula $L_3 = 10 \log(I_3/I_0)$, whence

$$L_3 = 70 + 10 \log(1.5) = 71.8 \text{ dB SPL}. \tag{35}$$

Thus, the addition of a 67 dB signal to a 70 dB signal has led to an overall increase of 1.8 dB in the level.

V. THE SPECTRUM

The pure tone of Eq. (7) oscillates indefinitely, and it is not possible to draw a figure to represent its infinite extent in time. One can, however, represent this tone in a frequency-domain representation called the *power spectrum,* which plots the power, normally expressed on a decibel scale, against frequency. Figure 4(a) shows the power spectrum of the pure tone of Eq. (7). The single line at frequency f_0 means that all the power is concentrated at the single frequency.

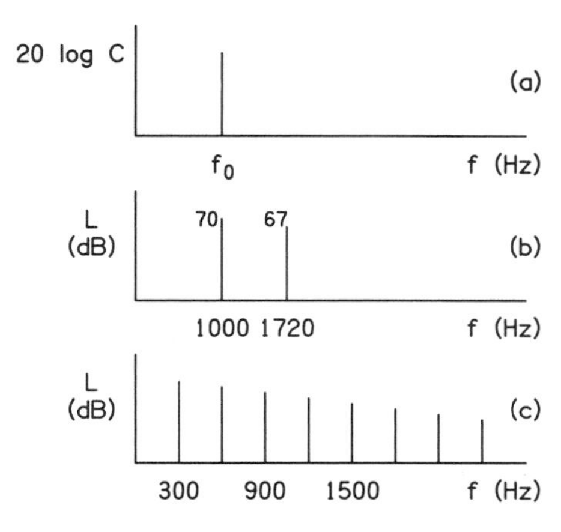

FIGURE 4 (a) The spectrum of a pure tone. (b) The spectrum of a complex tone made by adding two pure tones. (c) The spectrum of a complex periodic tone.

Figure 4(b) shows the power spectrum for the sum of two sines, the signal introduced at the end of the last section. Such a signal is a complex tone because there is more than a single sine component. Figure 4(c) shows the power spectrum of a complex periodic tone, with period of 1/300 s. All of its spectral components are harmonics of the fundamental frequency of 300 Hz.

The three spectra of Figure 4 are examples of line spectra. A noise signal, by contrast has a continuous spectrum, where there is power at every value of frequency. To deal with noise we need the concept of spectral density, $D(f)$, which gives intensity per unit frequency. Units of watts/(m²Hz) are appropriate, for example. To find the total intensity one integrates the spectral density,

$$I = \int_{-\infty}^{\infty} df\, D(f). \tag{36}$$

Function $D(f)$ might have any form $[D(f) > 0]$. In the simplest case it is a rectangular function, with constant value D over a frequency range Δf, as shown in Figure 5(a). Then the integral is trivial, and one may write

$$I = D\,\Delta f. \tag{37}$$

This equation is the basis for a logarithmic measure of spectral density called the *spectrum level*. Dividing by I_0 and taking the logs of both sides, we find

$$10 \log(I/I_0) = 10 \log(D/10^{-12}) + 10 \log \Delta f, \tag{38}$$

or

$$L = N_0 + 10 \log \Delta f, \tag{39}$$

where L is the level in dB SPL, and N_0 is the "spectrum level,"

$$N_0 = 10 \log\left(\frac{D}{10^{-12}\mathrm{W}/(\mathrm{m}^2\ \mathrm{Hz})}\right), \tag{40}$$

As an example, we consider Figure 5(b), where the spectrum level is 30 dB and the bandwidth is approximately 10 kHz. Such a sound would cause a flat-weighted sound level meter to read 30 + 40 = 70 dB.

It should be noted that the correct unit for spectrum level N_0 is dB because the spectrum level is the logarithm of a dimensionless ratio. Ten times the log of a ratio has units of dB. The literature sometimes gives spectrum level the units dB/Hz, which is completely understandable but also completely illogical. One can get units of dB/Hz only by making a mathematical error.

Except when a noise is digitally synthesized, actual noise bands are not rectangular. A shape like the solid line in Figure 5(c) is more typical of noise

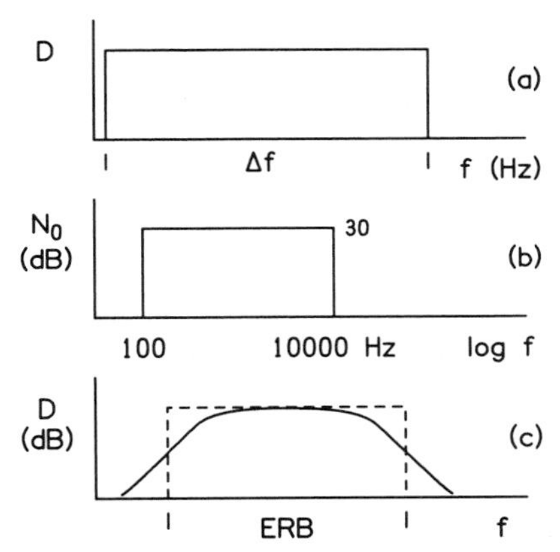

FIGURE 5 (a) The spectrum of a noise that is white, within limits, as a function of frequency. (b) The same noise, described by a constant spectral density as a function of log frequency. (c) A noise passed by a bandpass filter.

bands passed by a filter. One can, however, still use the concept of a rectangular band by defining an "equivalent rectangular band," as shown by the dashed line in Figure 5(c). The equivalent rectangular band has the same center frequency and the same maximum spectral density as the real band under discussion. The width of the rectangular band, the "equivalent rectangular bandwidth," or ERB, is adjusted so that the power in the rectangular band is equal to the power in the real band. In mathematical terms,

$$\Delta f_{\mathrm{ERB}} D_{\max} = \int_{-\infty}^{\infty} df\, D(f). \tag{41}$$

Still another concept used in the characterization of spectral density is that of effective noise voltage. The power in an electrical signal is proportional to the square of the voltage, $P \sim v_{\mathrm{rms}}^2$. In fact, by definition the rms voltage is simply proportional to the square root of the average power. Applying this idea to a density, we know that spectral density goes as $D \sim v_{\mathrm{rms}}^2/\Delta f$ and has units of volts squared per Hertz. The square root of D therefore has units of volts per root Hertz, $v/\sqrt{\mathrm{Hz}}$, which is a common unit of measure for the effective input noise of a device.

As a final word about spectral density, we note that dimensional arguments can sometimes be misleading. Spectral density has units of power per unit frequency. Because frequency has units of reciprocal time, spectral density has units of power multiplied by time, which are the units of energy. There is, however, no way in which spectral density is equivalent to energy; they both just have the same dimensions. [In fact, the mechanical quantity called *torque* (force times lever arm) also has units of energy, but torque is not equivalent to energy either.] Confusion begins with models of signal detection in noise, because the power in the signal multiplied by the signal duration is a legitimate energy. Models of detectability end up with detection performance given by a dimensionless quantity, and it is not really surprising that this quantity often turns out to be the ratio of signal energy to noise spectral density. There is no logical problem with this result. The problem arises only when it is assumed that there is a fundamental significance to the dimensionless ratio of signal energy to noise spectral density outside the context of the particular signal detection model. In fact, the ratio is still a comparison of apples and oranges; only the model makes this ratio a sensible quantity.

VI. THE FOURIER TRANSFORM

The Fourier transformation of a signal $x(t)$ transforms the functional dependence from time to frequency. Because the auditory system is tuned according to frequency, considerable insight into perception can often be gained by working with a mathematical representation in the frequency domain.

The Fourier transform of function $x(t)$ is $X(\omega)$, where ω is a running variable angular frequency. It is related to a frequency variable f measured in Hz by $\omega = 2\pi f$.

The Fourier transform of $x(t)$ is given by the Fourier integral,

$$X(\omega) = \int_{-\infty}^{\infty} dt\, e^{-i\omega t} x(t). \tag{42}$$

The inverse Fourier transform is defined by the integral,

$$x(t) = \frac{1}{2\pi} \int_{-\infty}^{\infty} d\omega\, e^{i\omega t} X(\omega), \tag{43}$$

which expresses the original function of time, $x(t)$, as an integral over angular frequency, or over frequency.

$$x(t) = \int_{-\infty}^{\infty} df\, e^{i2\pi f t} X(2\pi f). \tag{44}$$

A. Real and Complex Functions

Several points should be noted about the equations for the Fourier transform and its inverse. First, the Fourier transform is an integral over all values of time, positive and negative. It is necessary to have a definition of the signal for all time values before one can Fourier transform it. Second, the inverse

Fourier transform is an integral over both positive and negative frequencies. Negative frequencies play an important role, as will be evident soon.

The third point is that the Fourier transform $X(\omega)$ is generally a complex number, with real and imaginary parts. The factor $e^{-i\omega t}$ in Eq. (42) makes it so. In practical calculations, it is often helpful to replace this factor by the sum of real and imaginary functions, by using the Euler relation:

$$e^{-i\omega t} = \cos \omega t - i \sin \omega t. \tag{45}$$

The inverse Fourier transform $x(t)$ is also capable of being a complex number according to Eq. (43). But a complex value for $x(t)$ would contradict the fact that the signal is a real function of time. We know that it is real: We can display it on an oscilloscope, and we can hear it. What actually happens is that the negative frequency components arrange themselves to keep $x(t)$ a real function. Function $X(\omega)$ for negative values of ω must be the complex conjugate of $X(\omega)$ for positive values of ω. In symbols, this says that

$$X(-\omega) = X^*(\omega). \tag{46}$$

where X^* is the complex conjugate of X. (The conjugate of a complex number X has a real part that is the same as the real part of X and an imaginary part that is the negative of the imaginary part of X.) In other words, starting with function X, evaluated at a particular ω, changing the sign of ω leaves the real part of X unchanged while reversing the sign of the imaginary part of X. For a specific example, we suppose that signal x contains a component at a frequency of 100 Hz, and that $X(2\pi 100) = 3 + i4$. Then, because x is real, we must have $X(-2\pi 100) = 3 - i4$.

B. Transforms of Sine and Cosine

Because the sine and cosine functions represent pure tones with a single frequency, we expect that their Fourier transforms will be particularly simple.

If the signal is a cosine with angular frequency ω_0

$$x(t) = C \cos \omega_0 t, \tag{47}$$

then

$$X(\omega) = \pi C[\delta(\omega + \omega_0) + \delta(\omega - \omega_0)], \tag{48}$$

where function $\delta(\omega - \omega_0)$ is a function of ω that has a spike when ω equals ω_0, as described in the following.

For a sine function

$$x(t) = C \sin \omega_0 t, \tag{49}$$

the Fourier transform is

$$X(\omega) = i\pi C[\delta(\omega + \omega_0) - \delta(\omega - \omega_0)]. \tag{50}$$

The positive frequency component is given by $\delta(\omega - \omega_0)$, the negative by $\delta(\omega + \omega_0)$.

C. The Delta Function

The delta functions in Eqs. (48) and (50) are spikes of infinitesimal width and infinite height. A delta function has unit area

$$\int_{-\infty}^{\infty} d\omega\, \delta(\omega \pm \omega_0) = 1. \tag{51}$$

The most important feature of the delta function is its selection property: For any function $f(\omega)$,

$$\int_{-\infty}^{\infty} d\omega\, \delta(\omega - \omega_0) f(\omega) = f(\omega_0), \tag{52}$$

and similarly

$$\int_{-\infty}^{\infty} d\omega\, \delta(\omega + \omega_0) f(\omega) = f(-\omega_0). \tag{53}$$

The selection property means that $\delta(\omega - \omega_0)$ is able to suppress any contribution that function $f(\omega)$ tries to make to the integral except for the contribution at ω_0.

D. Amplitude and Phase Spectra

Suppose that $x(t)$ is a pure tone given by the general formula

$$x(t) = C \cos(\omega_0 t + \phi_0). \tag{54}$$

To find the Fourier transform, or spectrum, we begin by separating the time and phase dependences with a trigonometric identity,

$$x(t) = C \cos \phi_0 \cos \omega_0 t - C \sin \phi_0 \sin \omega_0 t, \tag{55}$$

and then use the transforms of sine and cosine functions together with the fact that the transformation is linear to find that

$$X(\omega) = \pi C(\cos \phi_0 - i \sin \phi_0)\delta(\omega + \omega_0) + \pi C(\cos \phi_0 + i \sin \phi_0)\delta(\omega - \omega_0). \tag{56}$$

Using Euler's relation, $e^{i\phi_0} = \cos\phi_0 + i\sin\phi_0$, we find

$$X(\omega) = \pi C\, e^{-i\phi_0}\delta(\omega + \omega_0) + \pi C\, e^{i\phi_0}\delta(\omega - \omega_0). \tag{57}$$

To find the spectrum in terms of frequency f we use the fact that $\delta(\omega) = \delta(2\pi f) = (1/2\pi)\delta(f)$. Then

$$X(f) = (C/2)\, e^{-i\phi_0}\delta(f + f_0) + (C/2)\, e^{i\phi_0}\delta(f - f_0). \tag{58}$$

Figure 6 shows the magnitude and phase angle of the components of this tone, as a function of frequency.

E. The Power Spectra of the Pure Tone and the Spike

The purpose of the power spectrum is to give an indication of the amount of power a signal has at each frequency. Because power is a real physical quantity, the power spectrum must be real and nonnegative. The power spectrum $P(f)$ is obtained from the Fourier transform $X(f)$ and its complex conjugate:

$$P(f) = |X(f)|^2 = X(f)X^*(f). \tag{59}$$

The amplitude spectrum is the square root; namely, the absolute value itself, $|X(f)|$.

The power spectrum is a density, with units of power per Hertz. There are logical difficulties with this definition for periodic signals, somewhat beyond the scope of this chapter. The solution is that the power spectrum can be found simply from the squared magnitudes in the components of the spectrum. For the pure tone of Eqs. (47) or (49), for example,

$$P(f) = \frac{C^2}{4}\,\delta(f + f_0) + \frac{C^2}{4}\,\delta(f - f_0). \tag{60}$$

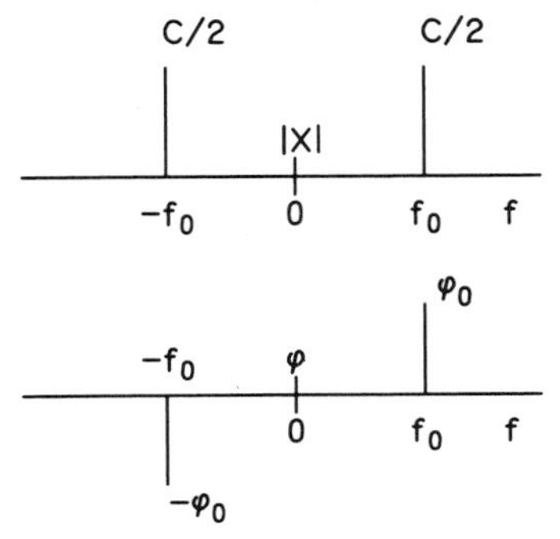

FIGURE 6 Spectral amplitudes and phases for a cosine tone with amplitude C and phase ϕ_0.

We now turn from the pure tone, which lasts forever, to the delta-function spike, which is of infinitesimal duration. A single spike at time t_0 is represented by

$$x(t) = A\,\delta(t - t_0), \tag{61}$$

where A is the strength of the spike. The Fourier transform is

$$X(\omega) = \int_{-\infty}^{\infty} dt\, e^{-i\omega t} A\delta(t - t_0), \tag{62}$$

By the selection property of the delta function, this is

$$X(\omega) = A\, e^{-i\omega t_0}. \tag{63}$$

Therefore, the Fourier transform is only a constant multiplied by a phase factor.

The amplitude spectrum is the absolute value of the Fourier transform, and the absolute value of a phase factor is simply unity,

$$|e^{-i\omega t_0}| = 1. \tag{64}$$

Therefore the spike has a "flat" amplitude spectrum, equal to A independent of frequency, and it has a flat power spectrum, with constant value A^2.

The cases of pure tone and spike are opposites. The pure tone lasts forever, but its spectrum is a delta function of frequency. The spike is a delta function of time, but its spectrum is a constant along the entire frequency axis.

F. The Lattice Function

The lattice is a periodic train of delta functions, a series of infinite spikes separated by a common period. The lattice function of time is

$$l(t) = \sum_{m=-\infty}^{\infty} \delta(t - mT), \tag{65}$$

where m is an integer index. The function is shown in Figure 7(a).

The Fourier transform of the lattice $l(t)$ is

$$L(\omega) = \sum_{m=-\infty}^{\infty} e^{-im\omega T}, \tag{66}$$

or

$$L(\omega) = \omega_0 \sum_{n=-\infty}^{\infty} \delta(\omega) - n\omega_0). \tag{67}$$

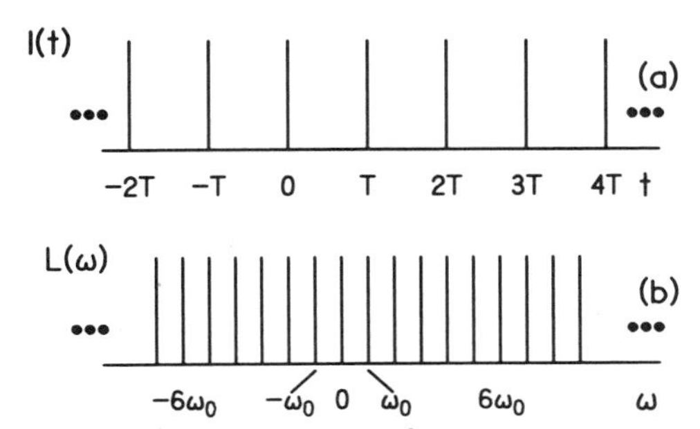

FIGURE 7 (a) The lattice function of time. (b) Its Fourier transform is a lattice function of frequency, with spacing proportional to the reciprocal of the period, $\omega_0 = 2\pi/T$. The functions continue to infinite time and infinite frequency, both positive and negative.

Thus, the Fourier transform of a lattice function of time is another lattice function, a lattice function of frequency, as shown in Figure 7(b). The fundamental angular frequency is ω_0, related to the period by $\omega_0 = 2\pi/T$. The fundamental frequency is f_0, $f_0 = 1/T$. The lattice function will play a role later in transforming all periodic functions, because all periodic functions of time can be written in terms of the lattice function. The preceding central result will therefore prove that all periodic functions of time have Fourier transforms consisting of harmonics. Before showing that, we need an important theorem, the convolution theorem.

G. The Convolution Theorem

The convolution theorem is a mathematical relationship of enormous power. It relates the Fourier transform of the product of two functions to an integral known as the *convolution* (or folding) *integral.*

If function z is a product of x and y, that is,

$$z(t) = x(t)y(t), \tag{68}$$

then the Fourier transform of $z(t)$ is given by the convolution integral:

$$Z(\omega) = \frac{1}{2\pi}\int_{-\infty}^{\infty} d\omega' X(\omega') Y(\omega - \omega'), \tag{69}$$

where $X(\omega)$ and $Y(\omega)$ are Fourier transforms of $x(t)$ and $y(t)$, respectively.

The relationship is symmetrical. If W is the product of X and Y, that is,

$$W(\omega) = X(\omega)Y(\omega), \tag{70}$$

then the inverse Fourier transform of $W(\omega)$ is given by the convolution in time,

$$w(t) = \int_{-\infty}^{\infty} dt' \; x(t')y(t - t'). \tag{71}$$

The convolution integral will be described in further detail in the section on filters.

H. Complex Periodic Signals

A periodic signal is a signal that repeats itself indefinitely into the future and also indefinitely back into the past. The periodic signal is clearly a mathematical abstraction, but its spectral properties are often an excellent approximation for the real-world signals that are periodic for only a finite time.

To create a periodic function $x(t)$ we begin with a function $s(t)$ that is a single cycle. This means

$$s(t) = x(t) \qquad -T/2 < t < T/2 \tag{72}$$
$$s(t) = 0 \qquad \text{otherwise}$$

The complete function $x(t)$ can be recovered from $s(t)$ by repeating $s(t)$ by convolving it with the lattice function,

$$x(t) = \int_{-\infty}^{\infty} dt' \; s(t')l(t - t'). \tag{73}$$

From the convolution theorem it is evident that the Fourier transform of the periodic function is

$$X(\omega) = L(\omega)S(\omega), \tag{74}$$

where $S(\omega)$ is the Fourier transform of $s(t)$. From Eq. (67) we find

$$X(\omega) = \omega_0 \sum_{n=-\infty}^{\infty} \delta(\omega - n\omega_0)S(\omega), \tag{75}$$

which is equivalent to

$$X(\omega) = \omega_0 \sum_{n=-\infty}^{\infty} \delta(\omega - n\omega_0)S(n\omega_0), \tag{76}$$

or, for frequency in Hertz,

$$X(f) = f_0 \sum_{n=-\infty}^{\infty} \delta(f - nf_0)S(2\pi nf_0). \tag{77}$$

Equation (77) says that the only spectral components of a periodic signal occur at frequencies that are integral multiples of the fundamental angular frequency ω_0. The component with angular frequency $n\omega_0$ (frequency nf_0) is called the nth harmonic.

The sum in Eq. (77) goes over both positive and negative values of n. However, $S(-n\omega_0) = S^*(n\omega_0)$, and therefore all the essential information about the spectrum is available in the half of the series with n positive.

1. Example: The Rectangular Pulse and the Rectangular Pulse Train

As an example of the techniques that have been developed in this section we find the Fourier transform of a rectangular pulse train. A rectangular pulse train with duty factor p is a periodic signal with period T that is in a high state for a fraction p ($p < 1$) out of every cycle and is in a low state for a fraction $1 - p$ of the cycle. Such a wave is shown in Figure 8(c) for the particular case where $p = 1/3$.

First, for the function of time, we begin with a lattice, $l(t)$, that establishes the periodicity, as shown in Figure 8(a). We next construct the element to be built on each of the lattice points, namely the single-cycle $s(t)$. Function $s(t)$ contains no information about the periodicity but does contain the information that the pulse has the value 1 for a time $\tau = pT$ and is zero for all other time, as shown in Figure 8(b). The final pulse train, given in Figure 8(c), is the convolution of the lattice function in Figure 8(a) and the single cycle in Figure 8(b).

For the Fourier transform of $x(t)$, the harmonic frequencies are given by the Fourier transformed lattice $L(\omega)$ shown in Figure 9(a). To find the harmonic amplitudes we use the Fourier transform of $s(t)$,

$$S(\omega) = \int_{-\tau/2}^{\tau/2} dt(1)\, e^{-i\omega t}, \tag{78}$$

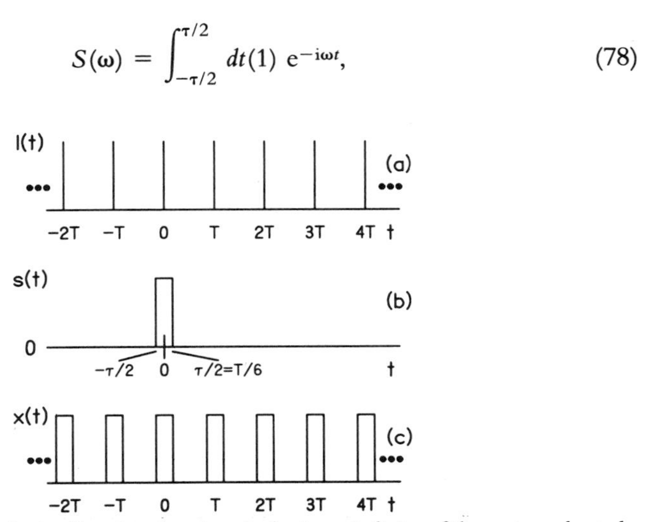

FIGURE 8 (a) The lattice function showing the basic periodicity of the rectangular pulse train. (b) The single pulse that is repeated to make the pulse train (a). (c) The pulse train is the convolution of (a) and (b).

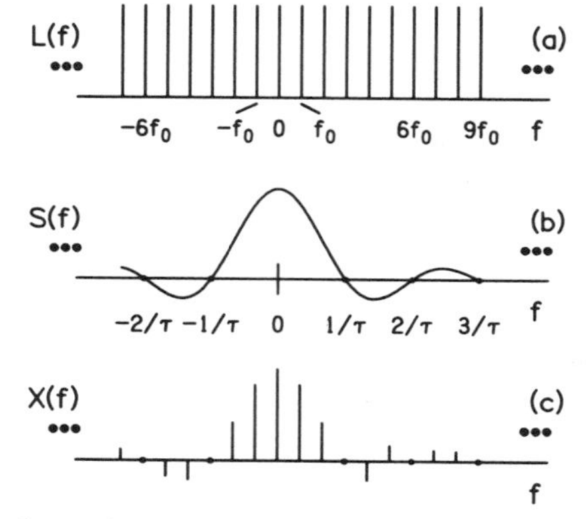

FIGURE 9 (a) The lattice function obtained by Fourier transforming Figure 8(a). (b) The spectral envelope obtained by Fourier transforming Figure 8(b). (c) The Fourier transform of the pulse train is the product of (a) and (b).

which can be solved to give

$$S(\omega) = \tau \frac{\sin(\omega\tau/2)}{\omega\tau/2}, \tag{79}$$

or

$$S(f) = \frac{\tau}{2\pi} \frac{\sin \pi f\tau}{\pi f\tau}. \tag{80}$$

Function $S(f)$ is shown in Figure 9(b).

Finally the Fourier transform of the pulse train, $x(t)$, is given by the product of $L(\omega)$ and $S(\omega)$, and so we substitute into Eq. (77). Using the facts that $pT = \tau$, and $f_0 T = 1$, we find

$$X(f) = p \sum_{n=-\infty}^{\infty} \delta(f - nf_0) \frac{\sin \pi pn}{\pi pn} \tag{81}$$

Function $X(f)$ is shown in Figure 9(c). It consists of a series of harmonics, with a spectral envelope given by the function $S(\omega)$, calculated from a single period of the wave. This kind of mathematical structure, a lattice function for harmonics multiplied by an envelope function determined by a single cycle applies to the Fourier transform of all periodic functions.

The power spectrum of the pulse train is given by squaring the strengths of the delta functions, as they are determined by the spectral envelope function. Therefore, the powers of the harmonics in the power spectrum are proportional to $|S(\omega)|^2$. They are never negative. For the spectrum of Eq. (81), the power spectrum can be computed, combining terms with positive and negative n and treating $n = 0$ separately,

$$P(f) = p^2\left\{\delta(f) + \sum_{n=1}^{\infty}\left|\frac{\sin \pi pn}{\pi pn}\right|^2 [\delta(f - nf_0) + \delta(f + nf_0)]\right\}. \quad (82)$$

It is common to write the amplitudes of the harmonics in terms of levels, so that the level of the nth harmonic is given by

$$L_n = 20 \log\left(\left|\frac{\sin \pi pn}{\pi pn}\right|\right). \quad (83)$$

An important attribute of the rectangular pulse train is that the harmonics decrease in level relatively slowly as the harmonic number n increases. Their amplitudes are given by the product of a rectified sine function and the factor $1/n$. As the harmonic number becomes large ($np >> 1$), the sine function oscillates perpetually and the asymptotic behavior of the amplitudes is proportional to $1/n$, and so (from Eq. 83) decreases at -6 dB per octave ($L_{2n} - L_n = -6$ dB). This $1/n$ dependence is the direct result of the fact that $x(t)$ has a discontinuity. It is not hard to generalize this result: If function $x(t)$ is continuous, but there is a discontinuity in the first derivative (e.g., a triangle wave) then the spectral envelope is asymptotically proportional to $1/n^2$, decreasing at a rate of 12 dB/octave. A discontinuity in the second derivative of $x(t)$ leads to a spectral envelope decreasing asymptotically as 18 dB/octave, and so forth.

Figure 9(c) shows that the third harmonic of the ($p = 1/3$) pulse train has zero amplitude. This is true as well of the 6th, 9th, 12th, and so on. In general, for a pulse train with duty factor p, the $1/p$th harmonic is missing, so is the $2/p$th, $3/p$th, and so on.

I. The Fourier Series

The Fourier series is a way of writing a periodic waveform $x(t)$ as a sum of sines and cosines. The general form of the Fourier series is

$$x(t) = DC + \sum_{n=1}^{\infty} A_n \cos(n\omega_0 t) + \sum_{n=1}^{\infty} B_n \sin(n\omega_0 t). \quad (84)$$

The DC is a constant term, independent of t. It is the average value of $x(t)$ over a cycle. Coefficients A_n and B_n are time independent; all the time dependence comes from the oscillating cosine and sine functions.

It may seems strange to represent functions such as the rectangular wave by sums of oscillating functions. In fact, it does not seem possible at first glance. However, Fourier's theorem says that, subject to rather few restrictions, with enough terms (enough harmonics) added together any periodic function can be represented by a Fourier series. More than that, Fourier's theorem says that, for a particular function $x(t)$, the series is unique; only one set of numbers $\{A_n\}$ and $\{B_n\}$ will give function $x(t)$.

It is a rather simple matter to find the Fourier series for a periodic function once one knows the Fourier transform $X(\omega)$. One simply has to plug $X(\omega)$ [or $X(f)$] into Eq. (43) or (44). For example, for the rectangular pulse train,

$$x(t) = p \int_{-\infty}^{\infty} df\, e^{i2\pi ft} \sum_{n=-\infty}^{\infty} \delta(f - nf_0) \frac{\sin \pi pn}{\pi pn}, \tag{85}$$

or

$$x(t) = p\left[1 + 2 \sum_{n=1}^{\infty} \frac{\sin \pi pn}{\pi pn} \cos(2\pi nf_0 t)\right]. \tag{86}$$

The symmetry of the original rectangular pulse train about zero time has resulted in a series with no sine terms (all the B_n are zero). There is a *DC* term given by p, and coefficients A_n are given by

$$A_n = 2p \frac{\sin \pi pn}{\pi pn}. \tag{87}$$

J. Temporal–Spectral Trading

The spectral envelope in Figure 9(b) or 9(c) illustrates an important fact about frequency and time representations of signals. The spectral envelope is an especially broad function of frequency f when the duration of the rectangle τ is especially short. In fact, the spectral width and the duration are reciprocally related. If, for example, we define the width of the spectral envelope to be the frequency difference between the first zero crossings at $\pm 1/\tau$ and give this width the symbol Δf, then the reciprocity between frequency and time is simply

$$\Delta f \tau = 2. \tag{88}$$

As an example of temporal–spectral trading, we consider another waveform that employs the rectangle function; namely, the rectangularly gated pure tone. It is described by the equation

$$x(t) = s(t) \cos(\omega_0 t), \tag{89}$$

and it looks like Figure 10(a).

Because the waveform $x(t)$ is the product of $s(t)$ and a cosine, the Fourier transform $X(\omega)$ is the convolution between the respective Fourier transforms:

$$X(\omega) = \frac{1}{2}\int_{-\infty}^{\infty} d\omega' S(\omega')[\delta(\omega - \omega' - \omega_0) + \delta(\omega - \omega' + \omega_0)]. \tag{90}$$

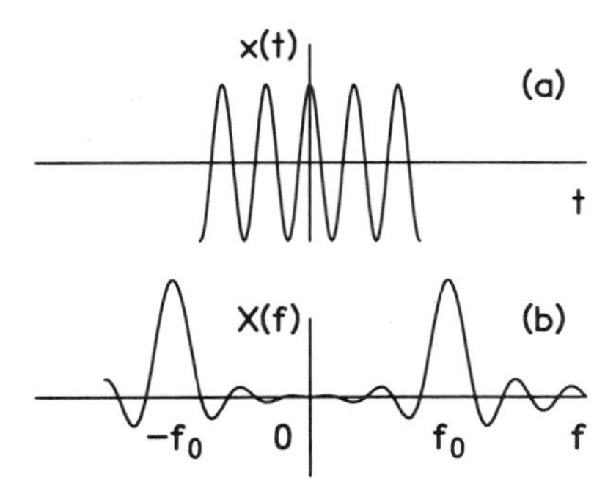

FIGURE 10 (a) A rectangularly gated sine tone with five cycles. (b) The Fourier transform of (a).

Convolving function $S(\omega)$ with the delta functions is easy to do. It has the effect of translating $S(\omega)$ so that it is centered on ω_0 (or $-\omega_0$). The result is

$$X(\omega) = \frac{\sin[(\omega - \omega_0)\tau/2]}{(\omega - \omega_0)} + \frac{\sin[(\omega + \omega_0)\tau/2]}{(\omega + \omega_0)}, \tag{91}$$

as shown for $X(f)$ in Figure 10(b).

The effect of gating the pure tone with a brief rectangular envelope is to take the sharp spectrum of the pure tone and broaden it so that it covers an appreciable part of the audible frequency axis, a result known as *spectral splatter.*

The reciprocal relation between time and frequency is sometimes called the *uncertainty principle,* in analogy to the uncertainty principle of wave mechanics. In that context, an appropriate interpretation of the temporal–spectral trading is that one has a choice to make about the precise point in a frequency–time space where a tone will fall. One can make the tone very brief so that one knows almost precisely when the tone occurs, but in that case one is somewhat uncertain about the frequency of the tone (the spectrum is broad). If one is willing to sacrifice a precise time for the tone and to give it a long duration, then the frequency can be established with greater accuracy.

The uncertainty principle is a fundamental limitation in the physics of waves, and yet, interpreting its correct role in different situations is a continuing challenge. The principle applies only to linear systems; nonlinear transformations can render it invalid. For example, a reciprocal-reading period-measuring device can measure frequency to arbitrary precision, given only a single cycle of the waveform.

The uncertainty principle is one of the cornerstones of information theory. However, such devices as the reciprocal-reading frequency counter do not necessarily violate information theory principles, because the measurement process includes assumed knowledge of the waveform. The counter would behave differently given a complex waveform that happened to have two positive-going zero crossings per cycle.

Correct interpretation of the uncertainty principle as it applies to audition is especially challenging. There are clearly demonstrable effects of spectral splatter manifested as off-frequency listening in narrowband masking experiments, but the uncertainty principle does not establish the limits of frequency discrimination (or pitch perception) for brief tones. Human listeners find it easy to beat the limits of the uncertainty principle by a factor of 5, and the dependence of frequency difference limens on duration follows the inverse first-power law for only a restricted range of durations. (See Chapter 8.)

The temporal–spectral trading has been illustrated here for a rectangular temporal envelope, but the concept applies for an envelope of any shape. A variety of envelopes has been considered by Ronken (1971). Different shapes lead to different distributions for the spectral splatter, but for all envelopes an uncertainty principle applies of the form

$$\Delta f \tau = k, \tag{92}$$

where Δf is some sensible measure of the spectral width and τ is some measure of the signal duration. With Ronken's definitions, constant k falls in the range from 0.4 to 1.2.

VII. AMPLITUDE MODULATION

A pure tone is described by the formula

$$x(t) = C \cos(\omega_0 t + \phi_0), \tag{93}$$

where C is a positive real constant called the *amplitude*. The amplitude determines the intensity of the tone and is the major element in establishing the loudness of the tone.

One can imagine relaxing the requirement that the amplitude be strictly constant and let it be a function of time $C = C(t)$ so long as the time dependence of $C(t)$ is slow compared to ω_0. This is the idea of amplitude modulation (AM). Typically one imagines that there are many cycles of the function $\cos(\omega_0 t + \phi_0)$ (called the *carrier*) in the time that it takes for $C(t)$ to change significantly. It is also typical to demand that $C(t)$ remain positive so that it remains a proper amplitude. It is further typical for the signal called the *carrier* to have no DC component. (In Eq. (93), the carrier is simply a cosine, and there is no DC component.)

The amplitude modulation of a cosine by another cosine serves as a useful example of AM. We let the amplitude be

$$C = C(t) = 1 + m \cos \omega_m t, \tag{94}$$

where ω_m is the modulation angular frequency and m is the "modulation depth," occasionally called the *modulation index*. So long as m is not greater than 1 (100% modulation) the amplitude remains positive.

The AM signal is, therefore,

$$x(t) = (1 + m \cos \omega_m t) \cos(\omega_0 t + \phi_0). \tag{95}$$

Using a simple trigonometric function we can rewrite this as

$$x(t) = \cos(\omega_0 t + \phi_0) + \frac{m}{2} \cos[(\omega_0 + \omega_m)t + \phi_0]$$

$$+ \frac{m}{2} \cos[(\omega_0 - \omega_m)t + \phi_0]. \tag{96}$$

Eq. (96) makes it evident that there are three components in the spectrum of the AM signal: one with the frequency of the carrier and two sidebands that are displaced in frequency from the carrier by $\pm\omega_m$. Because the amplitude of the carrier is unity, the amplitude of each sideband, relative to the carrier, is $m/2$. The level of each sideband, relative to the carrier is given by

$$L = 20 \log(m/2) \tag{97}$$

The maximum value occurs for 100% modulation ($m = 1$), so that the sideband levels are -6 dB with respect to the carrier.

A. Balanced Modulation

If the modulation is strong enough that $C(t)$ is driven to negative values (for example, if $m > 1$ for the cosine modulator) then one must use the idea of balanced modulation. Balanced modulation results from the multiplication of two signals, and both signals are allowed to be positive or negative. Sideband levels are still given by Eq. (97), and there is no limit on how large they may become.

In general, no distinction is made between carrier and modulator. If z is the balanced modulation product of x and y, then

$$z(t) = x(t)y(t). \tag{98}$$

In the simplest possible case, $x = \cos \omega_1 t$ and $y = \cos \omega_2 t$. From a trigonometric identity for the product of two cosines, we find that the spectrum of the balanced modulation signal has one component at the sum frequency $\omega_1 + \omega_2$ and another at the difference frequency, $|\omega_1 - \omega_2|$. There is no component at the carrier frequency.

The balanced modulation product of two multicomponent signals x and y has a spectrum that is easy to describe. There are components at frequencies given by the sum of each frequency in signal x with each frequency in signal y. There are components at frequencies given by the difference between each frequency in signal x and each frequency in signal y. Thus each component in the product has its origin in one component of x and one component of y. The amplitude of a component in the product is equal to one half of the product of the amplitudes of the original x and y components. An exception occurs for *DC* terms (frequency equal to 0), where the amplitude in the product is equal to the product of the amplitudes in the originals.

VIII. FREQUENCY MODULATION

In frequency modulation (FM), as in amplitude modulation, there is a carrier signal that is normally assumed to be a high-frequency sine wave and there is a lower frequency modulator. FM consists of modulating the frequency of the carrier.

The first step in a mathematical treatment of FM is to be quite clear about what is meant by *frequency*. We assert here that the frequency $\omega(t)$ is the time derivative of the instantaneous phase of a sine. If the signal is

$$x(t) = \sin[\Theta(t)], \tag{99}$$

then the instantaneous phase is Θ. (Here, and later, we set the amplitude equal to 1 for convenience.) It follows that the phase is the integral of the time-dependent frequency:

$$\Theta(t) = \int^{t} dt' \; \omega(t'). \tag{100}$$

In an FM signal, the instantaneous frequency consists of a constant carrier frequency, ω_c, plus a variation. If the variation is sinusoidal, with frequency ω_m, then

$$\omega(t) = \omega_c + \Delta\omega \cos(\omega_m t + \phi). \tag{101}$$

Therefore, the phase is given by

$$\Theta(t) = \omega_c t + \frac{\Delta\omega}{\omega_m} \sin(\omega_m t + \phi). \tag{102}$$

The quantity $\Delta\omega/\omega_m$, called the *modulation index*, is usually given the symbol β,

$$\beta = \frac{\Delta\omega}{\omega_m} = \frac{\Delta f}{f_m}. \tag{103}$$

Finally, substituting Eq. (103) into (102) and (102) into (99) we find the FM signal, given by

$$x(t) = \sin[\omega_c t + \beta \sin(\omega_m t + \phi)]. \tag{104}$$

The amplitude of the signal is constant, and therefore, unlike the case of AM, the power in the signal is not affected by modulation.

A useful alternative form for Eq. (104) is obtained by using the trigonometric identity:

$$\sin(\psi + \phi) = \cos\phi \sin\psi + \sin\phi \cos\psi. \tag{105}$$

Then the signal becomes

$$x(t) = \sin(\omega_c t) \cos[\beta \sin(\omega_m t + \phi)] + \cos(\omega_c t) \sin[\beta \sin(\omega_m t + \phi)]. \tag{106}$$

The next step in our treatment of FM is to calculate the spectrum. There are two approaches, depending upon whether β is small or not. If $\beta << \pi/2$, we have the case called *narrowband* FM (NBFM), and we can proceed by using an expansion that is a Taylor's series in β. If β is not small, we can deal with the spectrum only by using Bessel functions.

A. Narrowband FM

If the modulation index β is small ($\beta << \pi/2$) then we can simplify Eq. (106) by using the approximations that the cosine of a small angle is equal to 1 and the sine of a small angle is equal to the angle itself (expressed in radians).

Therefore, in this approximation,

$$x(t) = \sin(\omega_c t) + \beta \sin(\omega_m t + \phi) \cos(\omega_c t). \tag{107}$$

The product of sine and cosine can be expanded to obtain a sum of sine functions, a form that makes the spectrum apparent:

$$x(t) = \sin(\omega_c t) + \frac{\beta}{2} \sin[(\omega_c + \omega_m)t + \phi] - \frac{\beta}{2} \sin[(\omega_c - \omega_m)t - \phi]. \tag{108}$$

The NBFM spectrum has a carrier and two sidebands separated from the carrier by $\pm\omega_m$. This spectrum bears a great resemblance to the spectrum of an AM signal (see Eq. 96). In fact, the power spectrum of FM in the narrowband approximation is identical to the power spectrum of the AM signal, where β plays the role of m. Only in the matter of relative phase [the minus sign in front of the lower sideband in Eq. (108)] do AM and NBFM differ.

The signal in Eq. (108) can be more than just an approximation to a true FM signal. Digital techniques are often used to create that signal exactly in order to investigate the role that sideband phase plays in the perception of modulation. In that case, parameter β can have any value and the signal is known as quasi-FM (QFM).

B. Wideband FM

If the modulation index β is not small, the NBFM approximation will fail and we must calculate the FM spectrum exactly. To do this we begin with Eq. (106) and observe that the signal is composed of two terms: $\sin(\omega_c t)$ $\cos[\beta \sin(\omega_m t + \phi)]$ and $\cos(\omega_c t) \sin[\beta \sin(\omega_m t + \phi)]$. If we expand $\cos[\beta \sin(\omega_m t + \phi)]$ and $\sin[\beta \sin(\omega_m t + \phi)]$ in their Fourier series, then the two terms in the signal will be changed into two series in which each term is a product of two sines or cosines. By expanding these products into sums using trigonometric formulas, we get the spectrum.

Here we perform the steps just outlined for the case $\phi = 0$. The Fourier series for both $\cos[\beta \sin(\omega_m t + \phi)]$ and $\sin[\beta \sin(\omega_m t + \phi)]$ have component amplitudes that are Bessel functions, known as *cylindrical Bessel functions* or Bessel functions of the *first kind*, $J_n(\beta)$:

$$\cos(\beta \sin\omega_m t) = J_0(\beta) + 2[J_2(\beta) \cos 2\omega_m t + J_4(\beta) \cos 4\omega_m t + \ldots] \tag{109}$$

and

$$\sin(\beta \sin \omega_m t) = 2[J_1(\beta) \sin \omega_m t + J_3(\beta) \sin 3\omega_m t + \ldots]. \tag{110}$$

By substituting Eqs. (109) and (110) into (106) and then using trigonometric sum and difference formulas, we arrive at a form that shows all the components:

$$x(t) = J_0(\beta) \sin \omega_c t + J_1(\beta)[\sin(\omega_c - \omega_m)t - \sin(\omega_c + \omega_m)t] + J_2(\beta)[\sin(\omega_c - 2\omega_m)t + \sin(\omega_c + 2\omega_m)t] + J_3(\beta)[\sin(\omega_c - 3\omega_m)t - \sin(\omega_c + 3\omega_m)t] + \ldots \tag{111}$$

This is a time function consisting of a carrier, at ω_c, and an infinite number of sidebands. This is in contrast to the AM case, where there is only the carrier and a single set of sidebands. The first sidebands are separated from the carrier by a frequency separation of $\pm\omega_m$; the second sidebands by a separation of $\pm 2\omega_m \ldots$, and so forth. The Bessel functions give the amplitudes. The carrier has amplitude $J_0(\beta)$, the first sidebands have amplitude $J_1(\beta)$, the second sidebands amplitude $J_2(\beta)$, and so on. Functions J_0, J_1, and J_2 are shown in Figure 11.

It is evident that the amplitudes are not monotonic functions of the modulation index β. For $\beta = 3.8$ the second sidebands are strong whereas the first sidebands have virtually disappeared. There is an orderly behavior for small β; for $\beta << 1$, the sideband amplitudes grow as β^n.

It is also possible to generalize somewhat in the opposite case when β is large; for instance, for very low modulation frequencies, f_m, where sidebands are close together. Intuitively one feels that the bandwidth over which the sidebands are strong ought to agree with the overall frequency excursion

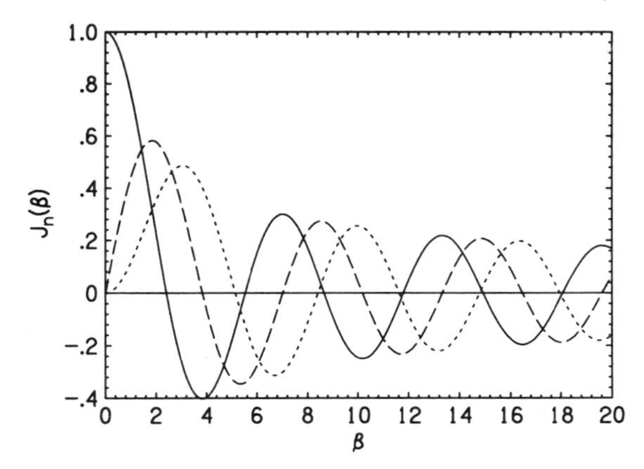

FIGURE 11 Bessel functions giving the amplitudes of the spectral components as a function of modulation index β for the carrier and two sets of sidebands; namely, $n = 0$ (solid line); $n = 1$ (long-dashed line); and $n = 2$ (short-dashed line).

of the tone, $2\Delta f$. In fact, that is what happens. For large values of n, the sideband amplitudes, $J_n(\beta)$, remain small as β increases, until β becomes approximately as large as n. Therefore, for $\beta \gg 1$ the number of significant sidebands is approximately equal to β. Since the sidebands are all f_m Hz apart and there is one set of sidebands on each side of the carrier, the bandwidth, $2B$, of the FM signal for $\beta \gg 1$ is approximately

$$2B \approx 2\beta f_m = 2 \frac{\Delta f}{f_m} f_m = 2\Delta f. \tag{112}$$

IX. FILTERS

A filter is a signal processing device with an input and an output. The filter causes the spectrum of the output signal to be different from the spectrum of the input signal. To quality for the category of "filter," the device must modify the spectrum of the input signal in a particular way; namely, by multiplication.

A. Frequency Domain

If the spectrum of the input signal is $X(\omega)$ then the spectrum of the output signal, $Y(\omega)$ is given by

$$Y(\omega) = H(\omega)X(\omega), \tag{113}$$

where $H(\omega)$ is a function that characterizes the filter, called the *transfer function*.

The simple multiplicative form of Eq. (113) is the only form that is allowed. It shows that the filter is a linear processor. If the input is doubled, then the output will also be doubled. If the input is the sum of two signals, then the output is the sum of the two inputs, as they would be filtered individually. In mathematical terms, if

$$X(\omega) = X_1(\omega) + X_2(\omega), \tag{114}$$

then

$$Y(\omega) = H(\omega)X(\omega) = H(\omega)[X_1(\omega) + X_2(\omega)], \tag{115}$$

or

$$Y(\omega) = H(\omega)X_1(\omega) + H(\omega)X_2(\omega). \tag{116}$$

The simplicity of Eq. (113) makes it natural to think of a filter as a device that operates in the frequency domain. Function $H(\omega)$ changes the spectral content of signals. For instance, if H is 0 at a particular frequency, then the output will have no power at that frequency regardless of how much power is put in. Accordingly, filters are classified by their frequency response. A low-pass filter allows low frequencies to pass through while attenuating high frequencies; a high-pass filter does the reverse. A bandpass filter passes only frequencies in a band while rejecting components with frequencies above and below the band; a band-reject (or notch) filter does the reverse.

Although Eq. (113) is simple, it is not trivial. Both the input spectrum $X(\omega)$ and the output spectrum $Y(\omega)$ are complex functions with real and imaginary parts. The transfer function is also complex. Therefore, the product $H(\omega)X(\omega)$ is actually the sum of four terms, two of them real and two of them imaginary. A more convenient way to think about filtering is to represent the transfer function in polar form:

$$H(\omega) = |H(\omega)|e^{i\phi(\omega)}. \tag{117}$$

This form separates two aspects of the filtering function: the amplitudes of the signals passed through the filter are multiplied by the real number $|H(\omega)|$, and the phases of the signals are shifted by the filter phase shift $\phi(\omega)$. The functional dependence of $|H|$ and ϕ on frequency represents the effect of the filter on an input component with the corresponding frequency. We illustrate that with a few examples.

1. Examples of Pure Tone Response and Complex Tone Response

The response of a filter to a pure tone with frequency of 1000 Hz might be

$$H(2\pi1000) = 1/\sqrt{2} \tag{118}$$

$$\phi(2\pi1000) = -45°. \tag{119}$$

This means that if we pass a 1000 Hz sine tone through the filter its amplitude will be multiplied by the factor $1/\sqrt{2}$ and its phase will be shifted by $-45°$.

The response at a different frequency will be different. At 2000 Hz the filter transfer function might be

$$H(2\pi 2000) = 1/\sqrt{5} \tag{120}$$

$$\phi(2\pi 2000) = -63°, \tag{121}$$

which means that, if the frequency of the input is changed to 2000 Hz, there will be more attenuation (smaller factor $|H|$) and the phase will be shifted by more; namely, by $-63°$.

Suppose now that the input is a complex signal with a fundamental frequency of 1000 Hz and a second harmonic at 2000 Hz. We let both components have amplitude 1 and cosine phase:

$$x(t) = \cos(2\pi 1000t) + \cos(2\pi 2000t) \tag{122}$$

The spectrum of this tone is given in Figure 12(a). The spectrum of the output is given in Figure 12(b).

The output of the filter as a function of time is a complex tone described by

$$x(t) = \frac{1}{\sqrt{2}} \cos\left(2\pi 1000t - \frac{\pi}{4}\right) + \frac{1}{\sqrt{5}} \cos\left(2\pi 2000t - \frac{2\pi}{360}\,63\right). \tag{123}$$

Input and output signals, as functions of time, are shown in Figures 13(a) and 13(b), respectively.

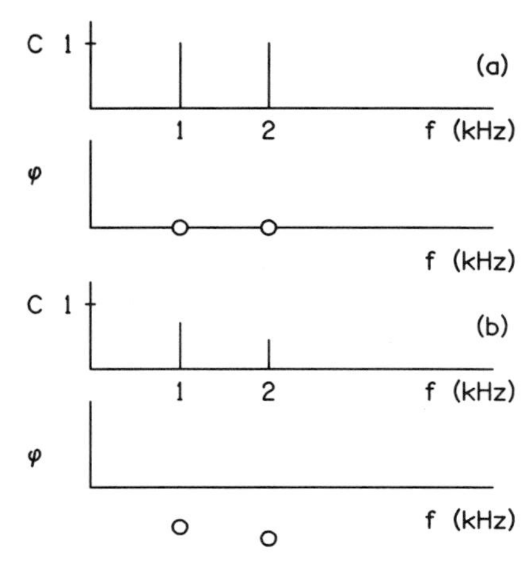

FIGURE 12 (a) Input spectrum, magnitude, and phase for the sum of two cosines. (b) Spectrum of the filtered output.

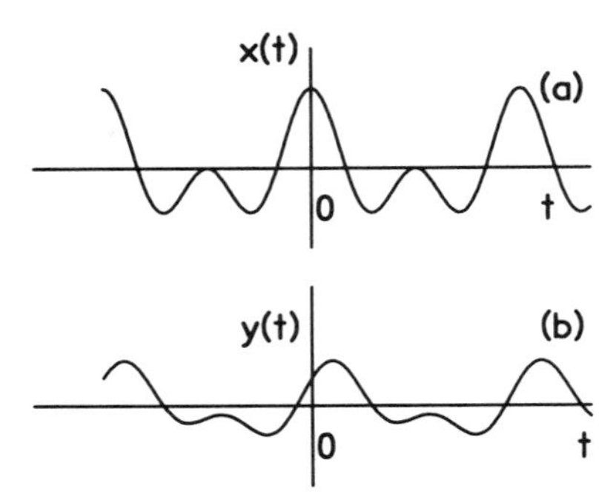

FIGURE 13 (a) Time waveform for the wave of Figure 12(a). (b) Time waveform for the filtered wave of Figure 12(b).

The solution to the complex tone problem has combined the solutions for the pure tones. The fact that the pure tone response can be entirely taken over in calculating the response to a complex tone is a direct result of the fact that a filter is linear. No such simple treatment is possible if the device is nonlinear.

This example illustrates several important general points about filters.

1. Whatever frequencies go into a filter also come out. The filter cannot change frequencies nor can it introduce components with frequencies not present in the input.
2. The act of filtering changes the shape of a complex time-dependent signal so that the output looks different from the input. There are two reasons for this. First, the relative amplitudes of the components are changed. Second, the phases are changed.

B. The Impulse Response

Filters were introduced previously in a frequency-domain representation; namely, by the transfer function. However, any operation described in the frequency domain has a corresponding description in the time domain. Because the operation in the frequency domain is multiplication, the operation in the time domain is convolution. Therefore,

$$y(t) = \int_{-\infty}^{\infty} dt'\ h(t - t')x(t') \tag{124}$$

or, equivalently,

$$y(t) = \int_{-\infty}^{\infty} dt'\ h(t')x(t - t'). \tag{125}$$

Function $h(t)$ is called the *impulse response*. It is the inverse Fourier transform of $H(\omega)$:

$$h(t) = \frac{1}{2\pi}\int_{-\infty}^{\infty} d\omega\; e^{i\omega t}H(\omega). \tag{126}$$

Equations (124) and (125) give the output as a function of time in terms of a convolution between the input and the impulse response. Either of these equations can be derived from the other by a change in the definition of the dummy variable t'.

The form in Eq. (124) particularly leads to insight into the way in which the convolution integral works. The integral there expresses the output at time t in terms of the input, $x(t')$, at all other times, represented by the moving variable t'. The degree to which the input at time t' affects the output at time t is given by the response function $h(t - t')$. This kind of reasoning leads to intuition about the response function itself: It seems likely that what is most important in determining the output right now should be the input right now. Therefore, we expect $h(t - t')$ to be large when $t = t'$. Just as "time heals all wounds" we expect that input events in the distant past should not affect the present value of the output. Therefore, $h(t - t')$ should become small when $t - t'$ becomes large.

Common sense tells us that the output of a physical system cannot depend upon the input at future times. Therefore, in Eq. (124), whenever t' is greater than t, the contribution to the integral must be 0. The requirement that the output cannot depend upon the input of the future is known as *causality,* and it leads to a restriction on the form of the impulse response:

$$h(t) = 0 \qquad t < 0. \tag{127}$$

As a result, it is possible to rewrite the integrals with new limits

$$y(t) = \int_{-\infty}^{t} dt'\; h(t - t')x(t'), \tag{128}$$

or, equivalently,

$$y(t) = \int_{0}^{\infty} dt'\; h(t')x(t - t'). \tag{129}$$

This restriction on $h(t)$ leads to restrictions on $H(\omega)$, which limits the kind of filters that are physically realizable.

The function $h(t)$ is called the *impulse response* because this function is exactly what the output of the filter becomes if input to the filter is an ideal impulse. If the input is a delta function, $x(t) = \delta(t)$, then

$$y(t) = \int_{-\infty}^{\infty} dt'\ h(t - t')\delta(t'). \tag{130}$$

But by the sampling property of the delta function, this gives

$$y(t) = h(t). \tag{131}$$

Therefore, the response, $y(t)$, of the system to an impulse is just function $h(t)$ itself.

C. The One-Pole Low-Pass Filter

The one-pole low-pass filter is important as a concept in psychoacoustics because it is often used to represent a perceptual operation that averages an input over time. It is sometimes called a *leaky integrator*. To describe this system we return to the mechanical oscillators introduced in Section II.

We suppose that there is an oscillatory force, F_{ext}, with a particular frequency ω. We know that the system will respond at this particular frequency, and we use the concept of the filter transfer function to tell us the amplitude and phase of the response. The particular response of interest is the displacement of the mass, $X(\omega)$. Figure 14(a) shows the low-pass filter. It consists of a spring and a viscous damping dashpot. There is no mass. An external force from the left drives the system.

A problem with the mechanical system defined here is that we have an output displacement but an input force. To put the two on an equivalent basis so that we can apply the concept of a transfer function, we define the force by an equivalent displacement, using the mechanism of Figure 14(b). It is the same as Figure 14(a), but there is no damping. The effective input displacement X_{ext} is the displacement that the external force would achieve with the spring alone as load; namely, $X_{ext} = F_{ext}/k$.

To return to the low-pass filter of Figure 14(a), the equation of motion for displacement x is Eq. (16) with the spring constant set equal to 0. Fourier transforming and solving, we find the output in terms of the input:

$$X(\omega) = X_{ext}(\omega)H(\omega) \tag{132}$$

where

$$H(\omega) = \frac{1}{1 + i\omega\tau} \tag{133}$$

and $\tau = m/r$. This filter is called a *one-pole filter* because ω appears only to the first power in the denominator. Its magnitude and phase responses are isolated by the polar form

$$H(\omega) = \sqrt{\frac{1}{1 + \omega^2\tau^2}}\ e^{i\phi}, \tag{134}$$

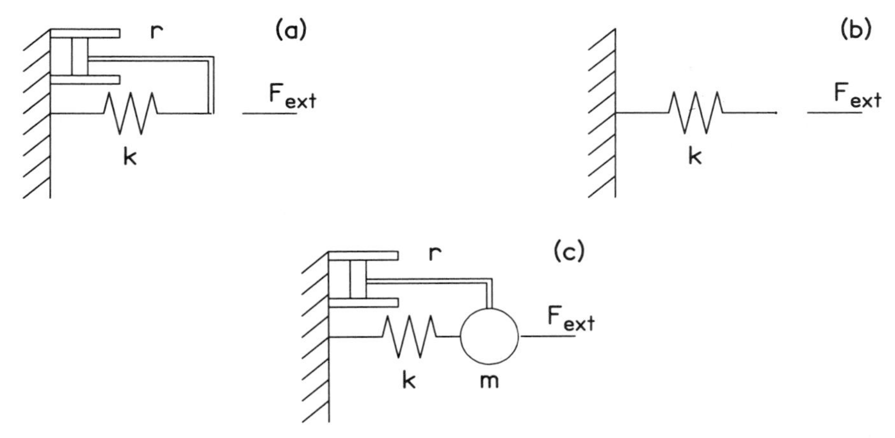

FIGURE 14 (a) The spring and dashpot make a one-pole low-pass filter for displacement. (b) The effective input displacement is determined by driving the spring alone. (c) The spring, mass, and dashpot make a two-pole filter for displacement.

where

$$\phi(\omega) = -\tan^{-1}(\omega\tau). \tag{135}$$

It is actually this filter, with time constant $\tau = 1$ ms, that appears in the previous example (Eqs. 118 and 120) and in Figures 12 and 13.

Filters can be described by their asymptotic magnitude response. At high frequency the response of the one-pole low-pass filter is proportional to $1/\omega$, or -6 dB per octave.

D. The Two-Pole Low-Pass Filter

The two-pole filter is the simplest system that can exhibit sharp tuning. We use it here to illustrate concepts of resonance, bandwidth, and the selectivity measure Q.

The physical system is shown in Figure 14(c). The equation of motion for the displacement is given by Eq. (16). Fourier transforming and solving, we find the output in terms of the input:

$$X(\omega) = X_{ext}(\omega)H(\omega), \tag{136}$$

where X_{ext} is the effective displacement, as earlier, and the transfer function is

$$H(\omega) = \frac{\omega_0^2}{\omega_0^2 - \omega^2 + i\omega\omega_0/Q}. \tag{137}$$

The parameters of the equation are determined by the physical quantities: The resonance frequency ω_0 is the natural frequency of vibration of the spring and mass system, given by $\omega_0 = \sqrt{k/m}$. The sharpness parameter Q is inversely proportional to the damping, $Q = \omega_0 m/r$.

At resonance, where the frequency of the external force equals the natural frequency of the system, the response is limited only by the damping. If r becomes small so that Q becomes large, then the response H grows indefinitely. The larger is the Q, the sharper is the tuning.

The ratio of output power to input power is given by

$$|H(\omega)|^2 = \frac{\omega_0^4}{(\omega_0^2 - \omega^2)^2 + (\omega\omega_0/Q)^2}. \tag{138}$$

The power has its maximum value near resonance. The bandwidth of the resonance region is normally defined as the difference between those frequency values, above and below the resonance frequency, at which the power response has fallen to one half of the response obtained at the resonance frequency. From the power response function in Eq. (138), it is easy to show that the Q parameter is simply the ratio of the center frequency to the bandwidth. This bandwidth is sometimes called the *full bandwidth at half power* or *half-power bandwidth*.

It is also possible to relate this bandwidth to an "equivalent rectangular bandwidth," or ERB. We imagine a rectangular power response function whose maximum height is equal to the peak of the power response function of a two-pole bandpass filter (for larger values of Q it does not matter much whether the filter is bandpass or low pass). We then choose a width for this rectangle so that its area is equal to the area under the two-pole power response function. This width is the ERB. The result of this calculation is that the ERB is equal to $\pi/2$ (= 1.57) times the full bandwidth at half power. The equivalent rectangular bandwidth is a convenient construction that will be used in chapters of this book that follow.

Reference

Ronken, D. A. (1971). Some effects of bandwidth–duration constraints on frequency discrimination. *Journal of the Acoustical Society of America, 49,* 1232–1242.

CHAPTER 2

Cochlear Structure and Function

Graeme K. Yates

I. FUNCTION AND STRUCTURE OF THE COCHLEA

A. Transduction of Acoustic Stimuli

The cochlea is required to transduce minute, rapid fluctuations in the atmospheric baseline pressure into a neural code on the auditory nerve. In doing so it must make available to the brain as much as possible of the information contained in those fluctuations. Sound is a mode of energy transfer by longitudinal motion (i.e., in the direction of propagation) of air molecules, and typical fluctuations occur on a time scale from tens of milliseconds down to microseconds. The amplitudes are extremely small fractions of the baseline pressure; the conventional physical reference level of 2×10^{-5} Pa is equivalent to the pressure change caused by driving a standard 10 ml syringe into its barrel by a mere 10^{-8} mm while a sound pressure level (SPL) of 120 dB is equivalent to pushing such a barrel in by 0.1 mm. The lower level is close to the noise level expected as a consequence of simple shot noise across a receiving window the size of the tympanic membrane. The larger value, although still extremely small, is some 10^6 times larger in pressure and highlights the extreme range that is apparently of interest to hearing animals. This combination of rapid change, the very small magni-

Hearing

tude, and the wide range of pressures places special demands on the cochlea that have been solved in an elegant and efficient way.

B. Hair Cells and Mechanical-to-Electrical Transduction

The basic mechanical sensory unit is the hair cell (Russell 1981). Hair cells take a wide variety of forms in the acoustico–lateralis system but share a common configuration in having two types of ciliary processes protruding from the apical ends: a kinocilium and several stereocilia. The former is a true cilium with a characteristic structure, while the latter are actually microvilli with rootlets projecting into their supporting base. Many hair cells lack the kinocilium, showing only the vestigial basal body of the kinocilium, and it is not required for mechanical transduction. Apparently, the stereocilia are responsible for the mechanical sensitivity in the cochlear hair cells. Mechanically sensitive ion channels exist near the tips of the stereocilia (Jaramillo & Hudspeth, 1991) and deflection of the stereocilia toward or away from the basal body modulates the standing current through them, resulting in a receptor current (Corey & Hudspeth, 1979). This current then develops a receptor potential across the basal membrane of the hair cell, which in turn modulates transmitter release from the afferent synapse. The basal end of the cell receives afferent innervation with typically many afferent terminals to each hair cell.

1. Speed Limitations

If mechanically induced opening and closing of the ion channels of the stereocilia is to modulate the transmembrane potential by changing the resting current through the hair cells, then the channels of any one hair cell must collectively have an electrical impedance approximately equal to that of the base of the cell. This expectation is confirmed by the measurements of Sellick and Russell (1978), who showed that the resistance of guinea pig inner hair cells (IHCs) was reduced by at most 50% when driven at very high SPLs by low-frequency stimuli. Thus, the receptor *current* is determined by the state of the ion channels and by the basal properties of the cell, and the receptor *potential* is determined by the electrical impedance of the cell membrane. Typically, cell membranes have large shunt electrical capacitances and so the receptor potentials are low-pass filtered representations of the receptor current, restricted to rise times on the order of a millisecond or so. Because the transmembrane potential determines the afferent synaptic response, this means that a simple hair cell is incapable of encoding sounds that vary on a time scale significantly faster than a millisecond.

Similarly, the afferent nerve cannot drive action potentials at a rate faster than, at most, a thousand per second and in particular cannot rapidly modulate its rate of action potential production. This is because it too is limited by the electrical properties of its membranes.

2. Dynamic Range Limitations

A second problem for hearing imposed by the limitations of hair cells is dynamic range, the range of sound intensities that is sufficient to stimulate the receptor and yet not overload it. Typically, an afferent synapse of the acoustico–lateralis system requires as much as 1 mV of receptor potential to stimulate an increase in transmitter release above the spontaneous rate (Sand, Ozawa, & Hagiwara, 1975), yet the receptor potential saturates at a level of a few tens of millivolts, implying a dynamic range of at most 30:1, or about 30 dB. The lower limit is determined by the threshold properties of the synapse while the upper limit is determined in part by saturation of the mechano–electrical transduction process itself, partly by the reduced electro–chemical potential across the transduction channels caused by the change in the internal potential of the cell and partly by saturation of the transmitter-releasing mechanism of the synapse.

The afferent nerve similarly has a restriction on its dynamic range. The maximum sustainable rate of action potentials is on the order of a few hundred to a thousand per second while a reasonable minimum would be on the order of ten or so per second. Rates slower then this could not pass information quickly enough. Hence, the dynamic range of an afferent nerve fiber would be around 10–100:1, or between 20 and 40 dB.

Both of these limitations, speed and dynamic range, would be a major problem for hearing. Without some mechanism to overcome them a large amount of acoustic information, valuable for survival, would be lost; and so most animals have evolved specialized preprocessing mechanisms to defeat these limitations.

C. Acoustic Preprocessing

1. Time-Domain Filtering of the Stimulus

The problem of coding a wideband acoustic signal when only very low-pass channels (the hair-cell/afferent nerve channels) are available may be solved by breaking the wideband signal up into many narrowband signals and transmitting each signal separately on an independent narrowband channel. Thus, if a suitable preprocessor could analyze the wideband acoustic signal through a series of parallel, overlapping, narrowband filters it could then pass on all the information in the original signal by transmitting the amplitude and phase of each of its constituent filters. Because filters can change their amplitudes and phases only slowly, at a rate inversely proportional to their bandwidths, the information rate on each channel would now be quite low, easily handled by the narrowband channels. In effect, the preprocessor would convert a single, wideband signal into a number of narrowband signals. All information contained in the original signal could be preserved to be reconstructed at the receiving end of the channels.

This is precisely what the cochlea does. The acoustical signal received by the middle ear is passed through to a partial Fourier analyzer, a parallel series of narrowband filters. This converts the information contained in the rapid temporal variations of the stimulus into a parallel set of information channels with a slow temporal variation. Apparently the cochlea preserves only the amplitude information, at least for the higher frequencies, and discards the phase of the stimulus.

2. Dynamic Range Compression

The range of sound intensities that the cochlea can handle could be increased by some form of mechanical amplitude–compression before the stimulus is applied to the stereocilia of the IHCs. That is, if appropriate mechanical preprocessing could reduce the change in vibration of the basilar membrane (BM) produced by a given change in the sound pressure, then the dynamic range of the IHCs would be increased.

Again, this is precisely what is accomplished by the cochlea. The nonlinear transduction properties of the outer hair cells (OHCs), presumably similar to those of the IHCs, are used as a template to compress the BM motion over the useful range of hair cell transduction.

D. Structure of the Cochlea

Functionally, the mammalian cochlea consists of a transduction organ, the *organ of Corti,* stimulated by a hydrodynamic surface wave propagating on the BM. In most mammals it exists as a cavity in the petrous temporal bone of the skull, a section through which is represented diagrammatically in Fig. 1A. It is a long, tapered tubular structure, divided into three chambers. The uppermost chamber of Fig. 1A is the scala vestibuli, which is in direct mechanical communication with the stimulus-induced displacements of the middle ear. It is filled with perilymph, similar to other extracellular fluids and high in sodium, low in potassium. The middle chamber, scala media, is mechanically probably a part of scala vestibuli but is chemically and electrically quite distinct. It is filled with endolymph, similar to intracellular fluids, being high in potassium and low in sodium. The membrane separating the two upper chambers, *Reissner's membrane,* appears to be very compliant mechanically but provides chemical isolation between the compartments and some electrical isolation for low frequencies. The lower chamber, scala tympani, is terminated at the basal end by the *round window,* which functions as the mechanical pressure release for the cochlea. Scala tympani is physi-

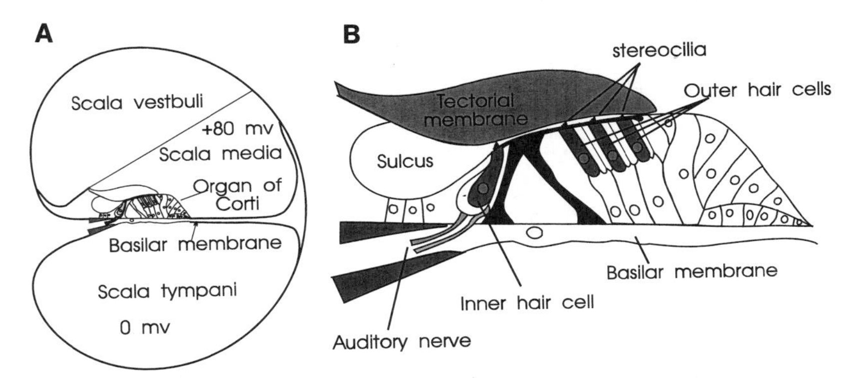

FIGURE 1 (A) Section through the cochlea; (B) section through the organ of Corti.

cally contiguous with scala vestibuli, the two being joined at the apical end by a small hole known as the helicotrema. Scalas media and tympani are separated by the BM, which carries the organ of Corti.

Acoustical stimulation results in small fluctuations in the volume of scala vestibuli, caused by movement of the stapes footplate into and out of this chamber. This volume change causes both vertical displacement of the BM and longitudinal displacement of perilymph, and interactions between these two result in a surface wave disturbance propagating along the basilar membrane from the basal (stimulating) end toward the apical end. Because of the mechanical properties of the BM, the velocity of propagation is strongly dispersive, i.e., dependent on both frequency and place along the cochlea, so that different frequency components of the propagating wave are separated as they travel.

The vertical vibration of the BM results in vibration of the organ of Corti (Fig. 1B). This structure supports two types of receptor cells, the IHCs and the OHCs. The IHCs are the true afferent receptor cells, responding to mechanical displacements by modulating their standing current in sympathy with the displacement. The modulation is not symmetrical, however, resulting in a pronounced rectification of the current waveform, which is then low-pass filtered by the cell basal membrane. The high-frequency fluctuations are attenuated, leaving only an ac voltage representation of the slower movements and a dc representation of the faster movements. The OHCs contribute to the cochlear amplifier, discussed later.

E. Active Processes within the Cochlea

Much of the processing within the cochlea is dominated by the so-called *cochlear amplifier,* an hypothesized, nonlinear mechanism that acts simultaneously to assist the filtering process and to compress the vibration of the BM over most of the input amplitude range. The idea is summarized in Fig. 2.

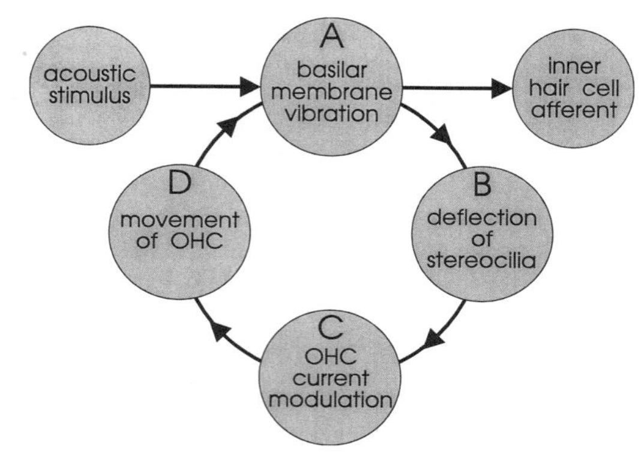

FIGURE 2 Diagrammatic representation of the positive feedback loop within the cochlea, the basis of the cochlear amplifier. If the loop is broken at any point the BM is driven by only the input stimulus. The IHCs play no part on the amplification, being passive motion detectors.

Acoustic power entering the cochlea induces a pressure difference across the BM and a traveling wave motion that propagates from the basal end towards the apex (A), Displacement of the BM causes deflection of the stereocilia of the OHCs (B), which in turn modulates the current through the OHCs (C). The next stage (D) is less well-understood, but some sort of mechanical motion is induced in the OHCs and this produces a direct effect on the BM in such a way as to assist the original displacement. This loop, A–B–C–D–A, is the cochlear amplifier: stage C is known as the *forward transduction,* or mechanical-to-electrical transduction, stage; and stage D is known as the *reverse transduction,* or electrical-to-mechanical transduction, stage. The ratio of the output of stage A to its input with the loop closed is called the *closed-loop gain,* whereas the same ratio with the loop opened or disconnected is called the *open-loop gain.*

II. MACROMECHANICS

The term *macromechanics* usually refers to the gross motion of the cochlear partition, as measured by direct observation of the BM. It includes the hydrodynamics of the wave motion but generally excludes details of the motion of various components of the cochlear partition.

A. Experimental Observations

1. Von Békésy

Georg von Békésy described the first direct observations of the traveling wave in his book *Experiments in Hearing* (1960). Using direct visual observations of the BM in explanted guinea pig cochleas, he observed the motion to be in the form of waves traveling from the basal end toward the apical end. Waves of different stimulus frequency peaked in amplitude at a different locations, characteristic of the stimulus frequency, and then attenuated to zero amplitude. Lower frequencies traveled further to locations nearer the apex than did higher frequencies.

Von Békésy also established the graded elastic properties of the BM by probing it with fine, calibrated hairs, showing that it behaved as an elastic membrane with very little coupling in the longitudinal direction (i.e., along the length of the cochlea) and that it was more compliant at the apical end than at the basal end. He also demonstrated that the traveling wave invariably traveled from the region of low compliance (the most rigid end) to the region of high compliance, explaining why the traveling wave in mechanical models always appeared to travel from base to apex regardless of the site of the stapes.

Thus, his pioneering observations established concepts of the traveling wave, the CF for different places along the cochlea, and the elastic gradient along the BM.

2. Early Measurements

The traveling wave concept was confirmed by the early measurements of Johnstone and his group using the Mössbauer technique (Johnstone & Boyle, 1967; Johnstone *et al.*, 1970; Johnstone & Yates, 1974), a technique that uses a small radiactive source of gamma rays placed on the BM. The radiactive counts registered by an appropriately placed counter tube vary nonlinearly with instantaneous velocity of the BM; and by counting the activity into histogram bins phase-locked to the stimulus, it is possible to estimate the velocity waveform (Yates & Johnstone, 1979). Johnstone and his coworkers confirmed the continuous phase accumulation with frequency at the site of measurement on the membrane and assumed, reasonably, that this represented continuous accumulation of phase along the membrane for a fixed frequency. Wilson and Johnstone (1975) used the capacitive probe, a device that records displacement by detecting the change in electrical capacitance between the membrane and the tip of a probe, to further confirm these results. Failure by both groups to find the sharply tuned response that had been observed in auditory nerve frequency-threshold curves (FTCs, Evans, 1972) was taken to imply the presence of a "second filter" between the BM and the afferent nerve (Evans & Wilson, 1975). Rhode's (1971) Mössbauer measurements on squirrel monkeys, however, gave the first hints that a physiologically vulnerable sharp tuning might be present in the mechanics, and more recent results have confirmed this. It is now clear that the sharp tuning observed in neural FTCs is present, to some degree at least, in the BM mechanics.

It is now generally believed that these early measurements were done on physiologically compromised cochleas. With the development of the compound action potential (CAP) technique it became possible to monitor the physiological condition of the cochlea from a gross electrode placed outside the cochlea (Johnstone *et al.*, 1979). This guided improvements in technique, and damage to the cochlea, previously undetected, is now easily recognized and avoided.

These early measurements all showed a tonotopically organized frequency response that was basically low pass, similar to, but somewhat sharper than those of von Békésy. Comparison of the results from different authors is somewhat complicated by the presentation of the results: some authors plotted the BM frequency responses at a fixed SPL at the tympanic membrane whereas others plotted it against stapes displacement. The SPL-referenced results are more easily compared with neural FTCs but the stapes-referenced data are more easily compared with theoretical results. Typically, however, the tuning curves showed high-frequency slopes of the order of 70–100 dB/octave and low-frequency slopes, for the stapes-corrected data, of approximately 9 dB/octave.

3. Recent Measurements

More recent results all show tuning at least comparable with neural FTCs (LePage & Johnstone, 1980; Sellick, Patuzzi, & Johnstone, 1982; Khanna & Leonard, 1982; Robles, Ruggero, & Rich, 1986; Cooper & Rhode, 1992a, b) although whether it is sufficient to explain fully the neural response is still a subject of debate (Allen, 1980). Few comparisons exist in the literature between the parameters of the measured tuning curves and those of neural FTCs.

Sellick *et al.* (1982) compared five of their guinea pig BM curves with two guinea pig FTCs, and although the data are somewhat sparse for the BM, the comparisons suggest that the neural tuning might be somewhat sharper. In their Fig. 11, for example, two of the four sets of data show significantly broader BM tuning while the other two are comparable. In their Fig. 10 they compare another FTC with their best BM tuning curve plotted as either iso-displacement or iso-velocity SPL curves. The FTC matches the iso-displacement curve best in shape and falls roughly between the iso-displacement and iso-velocity curves in depth of the tuning curve tip. Bandwidths are comparable. Comparison with their Fig. 3, which plots CAP thresholds for each animal represented in Figs. 10 and 11, shows that the match between the frequency threshold and the BM curves is better for those animals that had the best CAP thresholds. Therefore, the conclusion might be drawn that the FTCs and the BM displacement tuning would be identical if BM responses could be measured in a guinea pig in perfect condition.

Khanna and Leonard's (1982) laser-measured iso-displacement curves did not have tip-to-tail ratios equivalent to typical cat neural tuning curves, but the overall shape of the displacement curve matched the equivalent FTCs very well. Again this suggests that their preparations were somewhat compromised and cochleas in better condition might have yielded mechanical displacement curves that matched the neural curve, a point Khanna and Leonard make themselves. In particular, the slope of the low-frequency tail is very similar to the neural curve and different from an iso-velocity curve.

Similarly, Robles *et al.* (1986), using the Mössbauer technique, compared BM responses with neural FTCs and came to the conclusion that iso-displacement curves gave a better match with neural data than did iso-velocity curves.

The cat, however, may be a little different from other species in that the neural FTCs sometimes appear to have higher thresholds than BM tuning curves in the middle-frequency range, between one half and two octaves below the tuning peak. This is well illustrated by Cooper and Rhode (1992a) Fig. 21, where the cat neural tuning is compared directly with their BM data. Cooper and Rhode warn of the possibility of sound pressure calibration errors, however, and conclude that it is quite possible the mismatch is not real. A similar, though less pronounced, mismatch is evident in Khanna and Leonard's data, but again the differences are small (on the order of 10 dB) and may be due to differences in preparations used to measure the two types of curves; for example, in the bulla opening necessary to gain access to the BM.

B. The Traveling Wave

We now turn to a discussion of the mechanics of cochlear wave motion and of the possible role of the cochlear amplifier in maintaining sharp tuning. We attempt to understand the way in which the BM interacts with the fluids (essentially water) in the cochlea and how its properties determine much of the behavior of the traveling wave. We will then consider the various ways in which the active process behind the cochlear amplifier might operate and how it might couple energy into the traveling wave.

1. Hydrodynamics of Surface-Wave Motion

The cochlear traveling wave is a hydrodynamic surface wave, similar to waves propagating on the ocean and completely different from the motion that characterizes sound waves. Any disturbance of a liquid surface subject to the pull of gravity will cause that surface to return to its equilibrium position. If the return takes place at any reasonable speed the mass (inertia) of the fluid will result in overshoot, displacing fluid and elevating the surface nearby. This self-regenerating process will continue until damped out by viscous forces. The essential properties of the system that permit wave propagation are the restoring force (gravity), which acts always to restore the surface to its resting position, and the inertia of the water itself.

The velocity with which such a surface wave propagates depends on the depth of the channel, relative to the wavelength (Elmore & Heald, 1969). (1) If it is very deep, the wave will propagate at a velocity independent of the depth but that decreases with decreasing wavelength. Thus, the wave velocity depends on the frequency of the waves; i.e., the motion is dispersive. (2) For very shallow channels, however, the velocity of wave propagation becomes independent of the wavelength but inversely proportional to the depth.

For sufficiently low frequencies the BM traveling wave presents a similar picture, with the role of gravity being taken by the elasticity of the cochlear partition, but at higher frequencies the mass of the partition becomes significant. The elasticity of the BM is known to taper from base to apex, with the basal end being the stiffer, but the mass is probably approximately constant.

The surface mass provides the major difference between waves that travel on the ocean surface and along the cochlear partition. The velocity of ocean waves decreases with frequency, but no faster than $1/f$ so that there is no upper limit to the frequency with which such waves might propagate (Fig. 3a). For a mass-loaded surface wave, however, there is a critical frequency, the frequency at which the mass and elasticity resonate, above which no wave propagation is possible (Fig. 3b). For frequencies above this critical value, oscillatory motion of the surface simply decays exponentially with distance (Yates, 1986). Hence, resonance in the cochlear partition determines the highest frequency at which wave motion is possible at any place along the cochlea. Above this frequency the phase rotation must cease, to be replaced by an in-phase vibration.

This behavior is seen directly in measurement of BM frequency response, in the plateau region at the foot of the high-frequency cutoff. Here the high-frequency slope is seen to break and the rapid accumulation of phase ceases (Rhode, 1971; Johnstone & Yates, 1974; Wilson & Johnstone, 1975), indicating that the traveling wave has been replaced by an exponentially decaying, in-phase motion. This frequency is determined uniquely by the elastic and mass properties of the cochlear partition and is of particular theoretical significance. Other parameters of cochlear tuning, such as the frequency of maximum amplitude, are determined by a variety of factors such as fluid viscosity, the presence of a cochlear amplifier and whether perilymph is present on both sides of the partition, but the plateau frequency is quite invariant with these factors.

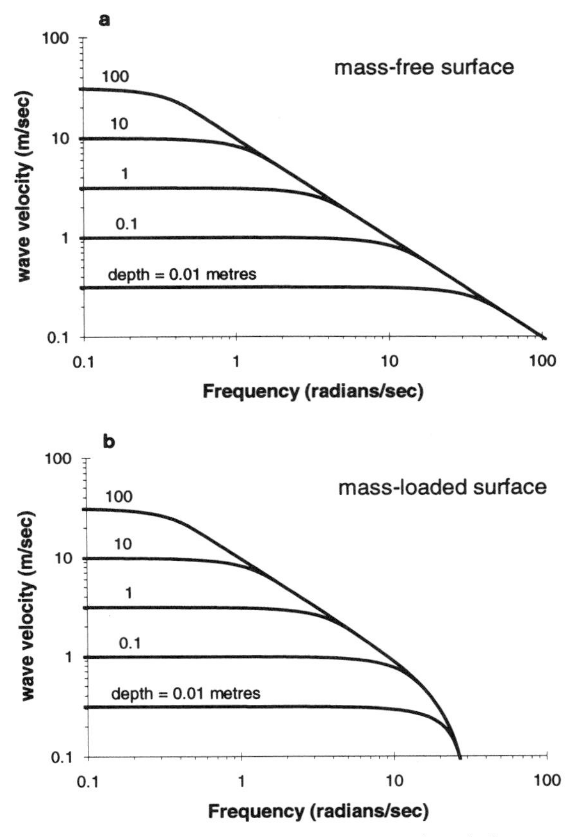

FIGURE 3 (a) Dispersion in a surface water wave; wavelength decreases with frequency but has a real value for all frequencies. (b) Dispersion in a mass-loaded surface wave; the wavelength falls to zero at the frequency at which the mass and compliance of the surface resonate.

2. Hydrodynamics of the Passive Traveling Wave

As emphasized earlier, the main mode of wave motion in the cochlea is a surface wave, compressional sound (longitudinal) waves being unimportant. Pressure differences exist within the cochlea because of bulk fluid flow (which implies acceleration) rather than fluid compression. When the stapes is pushed into scala vestibuli, it *displaces* and does not compress the cochlear fluid. To displace the fluid, however, it must either push fluid along scala vestibuli or, alternatively, displace the BM. In fact, both occur. Accelerating fluid along scala vestibuli implies a pressure gradient in the same direction; a pressure gradient implies a pressure difference across the BM; and a pressure difference across the BM implies a displacement of the BM.

For sinusoidal displacements a steady state traveling wave develops with small vibration amplitudes near to the stapes and larger amplitudes toward the apex.

a. Wave amplitude and velocity

The power in a surface wave passing a given point on a surface is given by the product of the energy density and the group velocity (Elmore & Heald, 1969). As the wave travels along the cochlea it encounters progressively greater and greater membrane compliance and the wave velocity falls. The amplitude of the wave grows to maintain power flux. As the wave approaches the point of membrane resonance the membrane impedance begins to fall rapidly and the velocity falls even faster. In the complete absence of damping the wave velocity would approach zero and the amplitude would tend to infinity, but in fact resistance becomes important near to resonance and damps the wave. The amplitude falls rapidly. The result is a poorly tuned peak some one or two millimeters basal to the point of membrane resonance, the plateau region. Some reflection may take place at the point of resonance, resulting in small notches at the plateau region of the tuning curve.

Pressure fluctuations at frequencies higher than the local resonance frequency will not couple well into a traveling wave (Kirk and Yates, 1994).

3. Hydrodynamics of the Active Traveling Wave

The hydrodynamics of the traveling wave at low frequencies appears to be unaffected by the presence of the cochlear amplifier. Measurements of BM motion reveal little or no difference at low frequencies between cochleas with and without a functioning cochlear amplifier, and most models of cochlear macromechanics permit the active process to act only locally, close to the characteristic place for a given frequency. Since the active process is supposed only to overcome resistive damping, the magnitude of the forces involved will be small relative to the compliance of the partition except close to CF, so the effect of the active process might be expected to be minimal. This suggests that little of the energy injected near the characteristic place propagates basally.

Close to the characteristic place, however, the presence of the amplifier has a profound effect. The amplitude rises dramatically, increasing the maximum in the vibration envelope and accompanied by moderate phase changes. There is also a shift, toward the apex, of the place at which maximum amplitude is achieved (Rhode, 1971; Sellick *et al.*, 1982). The amplitude increase does not appear to be accompanied by very large phase changes.

The actual increase in amplitude at any one place may be as much as 60 dB or more, but because of the shift in CF place, this does not necessarily imply that the active process boosts the overall power of the traveling wave by 60 dB. The velocity of the traveling wave is slowing close to CF, implying an increase in amplitude *for the same power flux,* so for the same power flux we would expect the maximum vibration amplitude to increase simply because of the shift in CF. That is, the amplitude at any one point may increase by 60 dB without necessarily implying the same power increase relative to a more basal location; it may simply be that the same power is reaching a region where the wave velocity is slower and the amplitude is consequently larger, i.e., the impedance is lower.

III. MICROMECHANICS

The term *micromechanics* has been coined to describe the mechanical motions that take place within the organ of Corti in response to the BM displacement and includes deflection of the stereocilia of the OHCs, the motion of the tectorial membrane and the action of the hypothesized cochlear amplifier. Direct observation of the micromechanics is presently impossible *in vivo,* but attempts have been made to study at least the gross behavior of the organ of Corti in various explant preparations. In this section we look at the properties required of the reverse transduction process and the properties observed for the putative motor behind it. We also consider the way in which this process could couple its energy into the macromechanics.

A. OHC Mechanical Activity

Implicit in the proposal for an active cochlea is the existence of a "motor" that may convert metabolic energy into vibratory mechanical motion of sufficient power to influence the vibration of the cochlear partition. This imposes certain physical restrictions on the process, which we now consider.

1. What Is Required of the Active Process?

a. It must be fast

To influence the vibration pattern of the cochlear partition, the active process must act on a cycle-by-cycle basis. That is, it must generate a pressure difference across the BM on each cycle of the vibration. It might be permitted to skip some cycles between operations but must at least act on the same phase of the vibration. In this latter case, however, it must generate a force with a rise-time less than the cycle time of the oscillation. Tonic pressure or motion cannot inject power into the vibration of the partition, in the same way as a steady push on a child's swing will not increase its swinging amplitude.

It is not sufficient, however, simply to demonstrate that a candidate active process is capable of generating *some* power at high audio frequencies: it is further necessary that the force generated in response to stimulation of the stereocilia must be substantially independent of frequency over the audio range. For example, consider a candidate process that, when stimulated by a given displacement of the BM, is shown to produce a force sufficient to affect the membrane motion at low frequencies. If this force falls off at higher frequencies, then at those higher frequencies the force may not be great enough to be effective. In this case, some additional mechanism must be imposed between the BM vibration and the active force-generating mechanism to bring the high-frequency force back to an effective level again. Thus, any hypothesized force-generating mechanism must include mechanisms, if necessary, to maintain at least a flat frequency response. In particular, any reverse transduction process that is driven by the voltage developed across the basal membrane of the OHCs must somehow compensate for the frequency dependence of the voltage.

b. It must oppose friction

In fact, the frequency response requirement is even stronger than stated previously: the force generated by the active process must *increase* at a rate of 6 dB/octave if it is to cancel the cochlear forces due to friction that, for sinusoidal motion of fixed amplitude, increase at this rate. And even this may be insufficient, in fact, because it would appear that the gain of the cochlear amplifier is greater at the base of the cochlea, i.e., at higher frequencies, so the force generated must rise even faster than 6 dB/octave.

Furthermore, energy may be supplied to a system only if the point of application of the force moves, because work done is equal to force times distance moved. If the cochlear amplifier is to amplify by injecting energy into the traveling wave, it must therefore generate a force that has at least some component in phase with the velocity of the vibrations.

2. Possible Modes of Movement

The motor process has been associated with the OHCs for several reasons. First, the OHCs have sparse afferent innervation but significant efferent contacts, suggesting some sort of effector role for them. Second, damage to and loss of the OHCs results in elevation of thresholds without complete loss of hearing (Dallos and Harris, 1978), suggesting that they are not necessary for afferent activation but are directly responsible for maintenance of thresholds. Third, reduction of the endocochlear potential, the driving force for current through the hair cells and consequently the stimulus for the reverse transduction, results in loss of cochlear sensitivity and reduction of BM amplitude (Ruggero and Rich, 1991). Fourth, it has been shown that when the transduction gates at the top of the stereocilia of the OHCs are damaged by any one of a range of insults, there is a direct correlation between the loss of cochlear microphonic and the elevation of threshold (Patuzzi, Yates, & Johnstone, 1989a, b). Although circumstantial, taken together these make a powerful argument that the source of power for the active process resides within the OHCs.

Within the OHCs, two candidates have emerged for the role of the motor process; only one has much direct supporting evidence, but neither is entirely satisfactory. Basically, the two split between a speculative motor process that might cause the stereocilia of the hair cells to move, or "twitch," and an electrically driven contractile process in the base of the hair cell.

a. OHC body contraction

Initial suggestions that the active process might be localized to the OHCs were supported by Brownell *et al.*'s (1986) demonstration that electrical stimulation of isolated OHCs could cause them to expand or contract. Since then considerable evidence has accumulated that the OHC contains a motor associated with its outer wall that, when the cell is depolarized, causes it to shorten in length while simultaneously increasing its diameter. Electrical potential changes across the cell membrane are sensed by an electrical dipole that shifts and causes an accompanying contraction of a submembranous network (Zenner, 1986; Kachar *et al.*, 1986; Ashmore, 1987; Santos-Sacchi & Dilger, 1988; Santos-Sacchi, 1989; Arima *et al.*, 1991; Forge, 1991; Dallos, Hallworth, & Evans, 1993; Pollice & Brownell, 1993). There is now little doubt that a highly specialized system exists within the OHC, capable of producing a fast contraction of the cell when it is stimulated by displacement of its stereocilia.

Maximum contraction amplitudes of this mechanism are at least approximately compatible with the vibration amplitude of the BM, being on the order of 1–2 μm when the cell is fully depolarized, whereas sensitivities of around 4–20 μm/V are typical (Ashmore, 1987; Santos-Sacchi, 1989; Gitter, Rudert, & Zenner, 1993). There is, however, more doubt about the speed of the contraction. When driven by imposed sinusoidal voltages, Ashmore (1987) found the amplitude of the contraction remained independent of frequency only up to approximately 35 Hz, but fell off at a rate of 6 dB/octave above approximately 1 kHz. Santos-Sacchi (1992) essentially confirmed Ashmore's result, although he found no evidence of the small

35 Hz corner. Both authors were confident of their voltage-clamping circuits and yet neither could push the corner frequency out further. It would appear, therefore, that the motor is limited to corner frequencies of around 1 kHz, at least for isotonic contractions (both recorded from hair cells that were physically constrained at only one end).

It might be argued that if viscosity, either internal or external, limits the amplitude of contraction at higher frequencies, then an isometric contraction might be faster. That is, if the cell were to be coupled to the membrane at a point of high mechanical impedance then a large force with little displacement would provide energy for the partition and viscosity might not be a problem. The advantage gained in such a way would be minimal, however, because in the limiting case of infinite impedance, full force might be developed over the full frequency range but no work could be done by the cell (because it did not move) and therefore no energy could be coupled into the partition. Working into lower impedances could increase the amount of work coupled from the cell into the membrane but would reduce the frequency range over which the contraction operates.

Aside from the question of whether the active motor located in the base of the OHCs is mechanically fast enough to influence the BM motion at high frequencies, there remains the problem of electrical stimulation. It is now clear that the force develops in response to voltage changes across the basal membrane of the cell (Ashmore, 1987) and the source of such voltage changes can be only the modulated current generated by the stereocilia. It is clear, however, that such currents will be low-pass filtered by the capacitance of the membrane and so the force generated will similarly fall with frequency; indeed such a fall-off has been directly observed (Russell, Cody, & Richardson, 1986). The cochlear amplifier theory, however, demands that the force should *increase* with frequency (see Fig. 4).

Several proposals have been made in an attempt to resolve this dilemma, all involving some form of high-frequency current boosting. Neely and Kim (1983, 1986) and Mammano and Nobili (1993) are typical. Both invoke a resonance between the stiffness of the stereocilia and the mass of the tectorial membrane to modify the displacement of the stereocilia under stimulation. Such a resonance necessarily predicts that the displacement of the stereocilia, and consequently the current through the OHCs, will rise at a rate of +12 dB/octave at least up to a frequency somewhat below the local CF, and such a current would result in a transmembrane voltage that rises with frequency, as is required to overcome resistance. This would resolve the problem but for the facts that (1) no such rise in current is observed in the frequency response of the gross cochlear microphonic, which is believed to reflect the current flowing through the OHCs, nor is there seen a 180° phase shift, which would accompany such a rise; (2) as mentioned earlier, the transmembrane voltage is directly observed to decrease with frequency

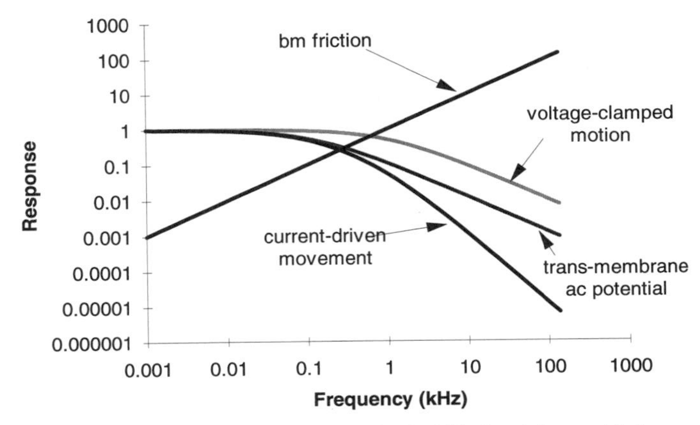

FIGURE 4 Diagram illustrating the increase in BM frictional force with frequency at any one place. Compare this with the typical low-pass filtering of receptor voltages by the basal membrane of the OHCs and with the likely additional behavior of any active force-generating mechanism within the cell.

(Russell *et al.*, 1986); and (3) because the OHCs would be driven harder for the same BM displacement, current saturation would severely limit the dynamic range of the active gain at high frequencies. It is therefore unlikely that such a mechanism is operating in the mammalian cochlea. Neely (1993) took a different approach in his later model and simply increased the magnitude of the forces at all frequencies. Thus, although the magnitude of the active force fell at high frequencies, it did so from such a large low-frequency value that it was still effective at CF. Such an approach works because the point impedance of the BM at low frequencies is much greater than it is close to CF and the increased low-frequency feedback is ineffective. This model, however, also has some inconsistencies with observation: it predicts the BM displacement, OHC current, and neural FTCs to be different from one another, in contradiction with direct measurements (Sellick, *et al.*, 1982; Robles *et al.*, 1986).

b. Stereociliar movement

Another possibility for the active process is direct movement or tilting of the stereocilia. To date little direct evidence for this process has been accumulated and no mechanism has been demonstrated. There are, however, some intriguing hints.

Lizards lack a well-developed BM, possessing only a short basilar papilla to carry the sensory cells. Direct measurement has confirmed that little or no frequency selectivity is present in the motion of the papilla, and yet neural FTCs are quite similar to those of mammals, if a little less sensitive (Manley, Yates, & Köppl, 1988). The morphology suggests little possibility that contraction of the hair cells (there are not two distinct types in these species) would influence the vibration of the papilla, and yet the Australian bobtail lizard (*Tiliqua rugosa*) has all the signs of an active process found in mammals: easily damaged sensitivity and tuning, reversible apoxia sensitivity, and spontaneous emissions (see later for a description of emissions). Furthermore, their tuning mechanism appears to involve a resonance between the stereocilia and their sallets, small tectorial structures orthogonal to rows of hair cells and individual to a few hair cells, and for such a mechanism cellular contraction would seem to be ineffective as a stimulus. A much more direct mechanism would be motion of the stereocilia, which would inject energy directly into the sallet–stereocilia resonant unit.

Movement of stereocilia in response to mechanical stimulation has been directly observed in the turtle hair cell (Crawford & Fettiplace, 1985). When a step force was applied to the stereocilia bundle of isolated turtle hair cells, the bundle of many cells was seen to ring at a frequency between 31–171 Hz, close to the frequency of electrical resonance in the cells. The oscillations could be abolished by large depolarizing currents that reduced the receptor currents, so it is unlikely that the ringing was purely mechanical. Fettiplace and Crawford concluded that the movement of the stereocilia was driven by the electrical response and that the turtle cochlear hair cells therefore contained an active force generating process.

There are at least two possible mechanisms by which direct motion of the stereocilia might be produced by the hair cell.

i. Tilting of cuticular plate If the motor that exists in the cell body were somehow coupled to the reticular plate or to the top surface of the hair cell and its forces were asymmetric with respect to the cell axis or the top of the hair cell was more rigid on one side than the other, then contraction of the cell would be accompanied by a tilting of the top of the hair cell and a concomitant tilting of the stereocilia. Such a mechanism would, however, be subject to most of the criticisms of the hypothetical cell contraction motor itself, such as poor frequency response and reduction of electrical drive at high frequencies. It might, however, overcome the problem of viscous drag reducing the contraction at high frequencies because the cell itself would not have to contract. Viscous effects would be limited to a simple tilting of the top of the cell. A tilting mechanism such as this has been described (Zenner, Zimmermann, & Gitter, 1988) and an effect consistent with it has been observed in explanted cochleas (Reuter & Zenner, 1990; Reuter *et al.*, 1992).

ii. A motor in the stereocilia An alternative motor mechanism has been proposed by Hudspeth and Gillespie (1994). The transduction channels at the top of the stereocilia are thought to be connected to adjacent stereocilia by "strings" that, when the stereocilia are displaced toward the basal pole, pull on the channel gates to bias them open (Pickles, Comis, & Osborne,

1984). Some mechanism must be provided to adjust the tension in the string to maintain the operating point against changes due to environmental factors. Such a mechanism has been described and is thought to involve a myosin motor sliding over an actin filament within the stereocilium (Howard & Hudspeth, 1987a, 1987b; Eatock, Corey, & Hudspeth, 1987; Crawford, Evans, & Fettiplace, 1989; Assad *et al.*, 1989, 1991; Shepherd *et al.*, 1990). When the stereocilia bundle is displaced and the gating channels are biased open, calcium is thought to enter the stereocilia where it functions as an internal messenger to control the motor (Eatock *et al.*, 1987; Assad *et al.*, 1989). The gating channels are proposed to be attached to insertional plaques that themselves form part of a myosin motor unit riding along one of the actin filaments within the stereocilium. The myosin motor is assumed to be working continuously to ascend the stereocilium while calcium entry stimulates slipping of the motor. Thus a negative feedback is achieved between myosin-driven tension in the string, which opens the channels, and internal calcium concentrations, which increase when the gating channels open, promoting slipping of the motor and reduction of tension. Hudspeth and Gillespie propose that this adaptation motor could, under appropriate circumstances, constitute a motor for reverse transduction.

Such a motor based on actin–myosin interaction would have many interesting characteristics and solve some of the problems with other candidate motors. Of particular interest is the fact that such a mechanism would eliminate the problem of the phase relationship between BM displacement and the active force necessary to do net work on the membrane. If the force generated by the myosin molecules when slipping is less than that generated when locked, the rest would be automatic. Consider the situation of a stereocilium containing such a putative motor being rocked back and forth. At the start of the forward stroke, toward the basal body of the kinocilium, when the BM is displaced towards scala vestibuli, the myosin would be locked to the actin filaments. As the tension in the tip links increases, it will reach a point where the myosin starts to slip, the tension drops dramatically (Fig. 5), and the myosin motor slips. Under these conditions the tension generated by the motor would be relatively small. At the end of the forward stroke the myosin will attach to the actin again and, on the return stroke, the force will rise again and the motor will move to recover the distance lost to slippage on the forward stroke.

Such a process has an analogy in the way a violin bow supplies energy to a violin string. As the bow is pushed forward the string sticks to the bow hair and is carried with it. At some point, however, the string starts to slip and, because the slipping friction is much less than the sticking friction, it slips easily back to its starting position. When it comes to rest it again sticks to the bow and is carried forward again. In this way, the linear motion of the bow is converted to oscillatory movement of the string. The timing of the

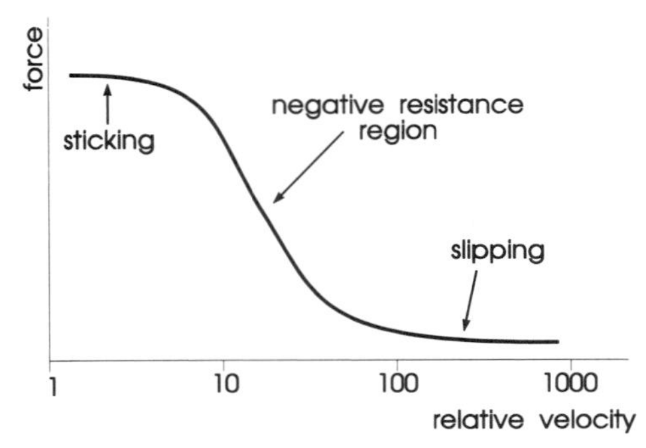

FIGURE 5 Hypothetical "frictional" force between myosin adaptation motor and velocity of slip between myosin motor and actin filament. At low relative velocities the force is high (sticking) but when slippage become greater, at higher relative velocities, the force drops dramatically (slipping).

stick–slip cycle is, however, controlled not by the rising frictional force but rather by the resonance characteristics of the string. The force that ultimately causes the string to break free of the bow is due more to the wave motion reflected from the nut or bridge of the violin than to the linear displacement of the string. The stick–slip cycle cannot take place without both components, the bow and the string, and it must have an oscillatory motion imposed externally.

If the stick–slip mechanism were to be applicable to the actin–myosin motor of the adaptation plaque then this might explain why oscillatory mechanical activity has not been observed in isolated hair cell stereocilia: because it would not work for an isolated hair cell unless it were appropriately loaded with its tectorial membrane or an equivalent, or unless the stereocilia were driven sinusoidally and the reaction measured. This situation is seen in insect asynchronous flight muscle. When isolated muscles are tested isometrically, they demonstrate a length–tension relationship typical of normal muscle fibres and show no sign of oscillatory motion. When a small mass is substituted for the isometric transducer, however, rapid oscillations begin immediately, at a frequency determined by the mass of the attachment and the stiffness of the muscle (Pringle, 1967). Such oscillations have been recorded at frequencies up to 1 kHz.

B. Coupling Energy to the BM

Any candidate reverse transduction process must be mechanically configured in such a way that it can couple its power source into the displacement of the BM. Whatever the power source, if it cannot displace the BM it cannot function as the reverse transduction stage of the active process. Thus, it must, when operating, produce an alternating torque about the inner foot of the arch of Corti or else a pressure difference in the fluids either side of the BM. Furthermore, at least some component of the force produced by the reverse transduction must phase-lead the BM displacement by 90° to assist the vibrations.

1. OHC Contraction

The OHCs are supported at their upper poles by the reticular lamina and at their bottom ends by the supporting cells of the organ of Corti. Leaving aside for the moment the question of whether there is a strong bond between the hair cells and their supporting cells, it is clear that contraction of the basal part of the OHCs will produce a force pulling the BM and the reticular lamina closer together. This is clearly an inefficient mechanism for injecting mechanical energy into the motion of the BM because it will produce an upward force on the membrane itself but a downward force on the reticular lamina and, if the stereocilia are tightly inserted into the tectorial membrane, the tectorial membrane, too (Fig. 6, middle panel). It is not clear what might be the net result on a hydraulically loaded cochlear partition. The inertia of the upper end of the organ of Corti, i.e., the reticular lamina and the tectorial membrane, might be large enough to provide some momentum to the fluid that must be displaced if the BM is pulled upward, but the center of mass of the organ itself might well be pulled in the opposite direction.

2. OHC Stereociliar Twitching

If the reverse transduction mechanism is in fact a displacement of the stereocilia, then the coupling problem is greatly reduced. If the displacement of the BM can result, through the geometry of the organ of Corti, in a shearing of the stereocilia then it follows that a torque tilting the stereocilia could displace the BM. In fact, because of the mechanical displacement advantage between displacement of the BM and shearing of the stereocilia (Rhode & Geisler, 1967), there would be a mechanical force advantage in the reverse mode. Typically, this will amplify the force between the stereocilia and the BM by a small factor, at the cost of reducing the displacement of the BM by a similar factor.

Such a motion has not been demonstrated directly and with present techniques is unlikely to be. Even if independent motion of the stereocilia could be demonstrated, it would be difficult to demonstrate directly that this was responsible for the reverse transduction.

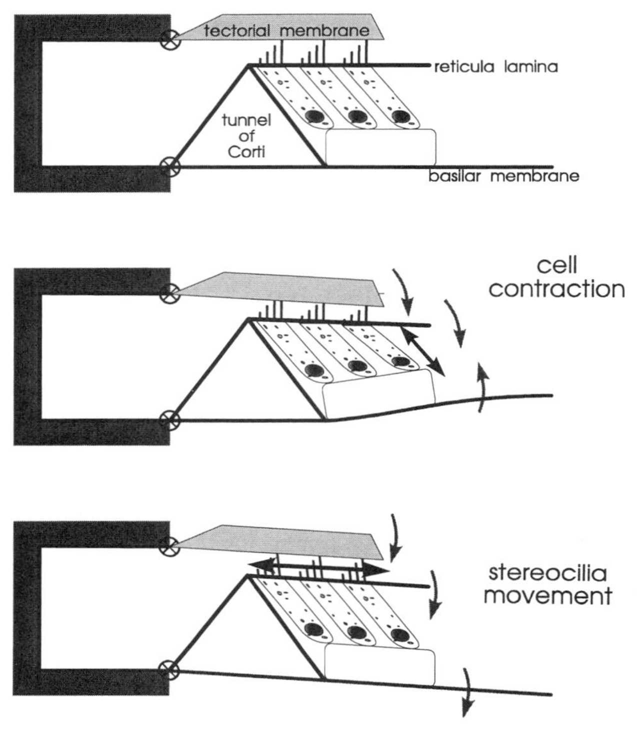

FIGURE 6 Two possible modes by which OHC activity might be coupled into the mechanics. The top panel is a diagram of the partition at rest. The other two panels show the probable motion of the partition (middle panel) when the OHCs contract and (bottom panel) when the stereocilia move.

C. Localization of the Cochlear Amplifier along the BM

1. Is the Active Process Localized?

Present models and theory of the active process call for a localization of the active process to a region just basal of the characteristic place on the BM. It has not yet been determined experimentally, however, just what the distribution of the active process is, but several experiments have provided hints. Robertson and Johnstone (1981) recorded sharply tuned frequency-threshold tuning curves from basal-turn ganglion cells of the guinea pig in which damage had been caused by acoustic trauma. Cells from the edge of the affected region had normal FTCs and normal low-frequency suppression (see later, and also the chapter by Palmer in this volume), but high-frequency suppression was greatly reduced. In a similar preparation Cody (1992) showed that near-normal tuning curves could be recorded within

0.5 mm of a lesioned area of the organ of Corti, suggesting that the active process was important only out to 0.5 mm from the characteristic place. Finally, Allen and Fahey (1992), using an ingenious distortion-tone experiment, concluded that the active process, if it existed at all, was confined to within 1 mm of the characteristic place.

Such experiments, however, do not necessarily demonstrate that the active process is limited to acting within a prescribed region, because the impedance of the BM might well prevent any mechanism that behaves as a resistance, positive or negative, from having a significant effect far from the characteristic place. That is, since the active process is thought to act as a negative resistance and to be on the same order of magnitude as the positive resistance caused by viscosity, it might well be that it has minimal effect anywhere except close to the resonance place of the BM. Hence, the experimental results do not necessarily preclude the possibility that the active process acts along the entire cochlea basal to the characteristic place. It may simply be ineffective except close to the characteristic place. This is not a trivial point because any satisfactory explanation of the active process must explain how it is spatially limited, and such spatial limiting must be frequency dependent because the region to which it is limited depends upon the stimulus frequency.

The concept that the active process is localized appears to have started with Kim *et al.* (1980). They incorporated a localized active process simply because it was necessary to maintain numerical stability. All models and theories since then appear to have adopted the idea, but to date no convincing evidence that it is necessary appears to have emerged.

2. Mechanisms for Localization

The mechanisms postulated to localize the active process have relied mostly on some form of resonance in the stereocilia or between the stereocilia and tectorial membrane. Zwislocki and Kletsky (1979) proposed such a mechanism as a "second filter" and Strelioff, Flock and Minser (1985) suggested that the same mechanism could contribute to the frequency selectivity of the cochlea. Kim *et al.* (1980) did not specify a mechanism for localization but simply prescribed the active process to be limited in its extent. Later Neely and Kim (1983) incorporated a resonance of the stereocilia of the OHCs and later still (1986) adapted that to resonance between the stereocilia and the tectorial membrane. Mammano and Nobili (1993) made similar assumptions but placed the resonance frequency well above the local CF.

Experimental support for these models appears in the lizards. Resonance of free-standing stereocilia appears to be the basis for the frequency selectivity in the alligator lizard (Frishkopf & De Rosier, 1983), which otherwise has no substrate for a filtering mechanism. The bobtail lizard has a cochlea closer to the mammalian cochlea in that it has tectorial structures supported on the stereocilia and the difference between the frequency selectivity of its auditory nerve and its basilar papilla is suggestively similar to a simple high-pass filter with resonance. Therefore, the case for micromechanical resonance appears strong in the lizards, but could this be an example of species difference?

All suggestions of resonance between the stereocilia and the tectorial membrane pose two fundamental problems, both of which are a consequence of the additional filtering imposed on the hair cells by such a system. First, such a resonance would require the tails of the auditory nerve tuning curves to be 12 dB/octave steeper than those of the BM mechanics; and for the guinea pig and chinchilla at least, this does not seem to be the case. Some argument has taken place over whether the neural tuning is closer to displacement or velocity of the BM but there appears to be no possibility that the difference could amount to a high-pass resonance system. Second, such a resonance would require a 12 dB/octave increase in the cochlear microphonic relative to BM, accompanied by a 180° phase lead. Various studies have shown this definitely not to be the case (Russell & Sellick, 1983): the phase in particular is easily shown to be as expected, with displacements toward scala vestibuli producing an increase in current through the OHCs. Thus, attractive as such models might be, tectorial resonance can be rejected on solid experimental grounds.

IV. COCHLEAR NONLINEARITY

The final section of this review will examine the mechanical nonlinearities within the cochlea. It has become widely accepted that an active process implies nonlinearity and vice versa, but this is not so. The active process *happens* to be nonlinear, but nothing inherent in the concept of a positive feedback mechanism requires it to be so and nothing in the observed nonlinearities implies an active process. Indeed, the most well-known of the nonlinearities, the cubic distortion product, may be independent of the cochlear amplifier.

A. Sources of Nonlinearity

We consider now the various potential sources of distortion within the cochlea and how that nonlinearity may manifest itself. It can be difficult to identify distinct sources of nonlinearity within a closed feedback loop but studies of isolated cells and cochleas wherein the feedback loop has been opened by various techniques have given us a fair understanding of the roles of each of the stages in generating nonlinear effects within the cochlea.

The generally accepted concept of the cochlear amplifier has significant implications for nonlinearity. Although negative feedback is well understood to reduce nonlinearities and stabilize a system against small parameter changes, it is less well understood that positive feedback has the opposite implication. With positive feedback the closed-loop gain dependence on loop parameters is actually enhanced so that, if the loop gain changes by even a small amount, the closed loop gain may be strongly affected. Small nonlinearities within the loop are magnified.

1. Forward Transduction

The first stage in transduction, the mechanical to electrical transduction, is mediated by stretch-induced gating of ion channels at the tops of the stereocilia. These ion channels appear to be gated on and off by mechanical tension in the thin links seen connecting the extreme tips of some stereocilia to the sides of adjacent, taller, stereocilia (Pickles *et al.*, 1984). Because the energy required to open a transduction gate is comparable with the thermal energy thermodynamically associated with physiological temperatures, the channels are not in a simple open or closed state but rather are rapidly fluctuating between open and closed states with the mean value determined by the tension. Furthermore, because the channels cannot be synchronized, channels in one stereocilium might be in the open state at the instant that those of an adjacent stereocilium might be closed. Hence, it is appropriate to talk to the mean state of the gates or the mean proportion of time that the gate spends in the open state or, less accurately, of gates being partly opened. The function describing the probability of being in the open state versus the instantaneous displacement of the BM is, for OHCs at least, very close to a two-state Boltzmann function, which is, in turn, the type of function to be expected of a simple two-state system in thermodynamic equilibrium (Corey & Hudspeth, 1983). At one extreme the gates will be always closed and at the other always open, and there exists a smooth transition between the two.

Inefficacy of the forward transduction channels is probably the most commonly observed cause of threshold elevation in both experimental animals and in humans. Anything that alters the probability of a channel being in one state or another—be that mechanical damage to the gating protein itself, blockage of the channel by pharmacological agents, reduction in the driving voltage across the hair cell, or interference from another stimulus of high intensity—will reduce the gain of that stage of the feedback loop (C in Fig. 2) and hence reduce the BM amplitude at the CF. Experimentally, the ion channels appear to be susceptible to a wide range of insults from acoustic and mechanical trauma through to ototoxic agents. Reduction of the endocochlear potential also reduces the current passing through the channels and elevates thresholds.

2. Reverse Transduction

The relationship between transmembrane voltage and OHC contraction has also been shown to approximate a Boltzmann function (Santos-Sacchi, 1992; Dallos *et al.*, 1993). Again this is to be expected on the grounds that the motor appears to be driven by a charge movement for which the energy difference is comparable with the thermodynamic energy at physiological temperatures (Dallos *et al.*, 1993). The characteristic length for the Boltzmann function, the point at which saturation has reduced the force significantly, is around 1 μm.

Because the forward, mechanical-to-electrical transduction is in series with the reverse, electrical-to-mechanical, transduction, either would be expected to contribute to the closed-loop nonlinearity, but the forward transduction saturation is evident at much smaller BM displacements and hence is expected to be the dominant nonlinearity in the cochlea over much of its dynamic range (Patuzzi *et al.*, 1989a, b; Santos-Sacchi, 1993).

3. Gating Forces

The Boltzmann transfer function associated with the mechanotransduction gates in the stereocilia contributes to nonlinearity in the cochlea because of its effect of reducing the active gain as it saturates. It has been shown however, to manifest itself in a second way through a direct modulation of the compliance of the stereocilia during deflection (Fig. 7). In effect, as the stereocilia are displaced in the excitatory direction (toward the basal body, away from the shorter stereocilia) the tip-links initially are stretched while the gates remain closed; the only elastic component is the tip-link itself (A of Fig. 7, left panel). Further displacement results in a gradual (in a statistical sense) opening of the gates, and during this stage of incomplete opening there are two elastic components in series: the tip-links and the gates themselves (A and B of Fig. 7, middle panel). Still further displacement results in the gates being fully opened; they become rigid again and the tip-links again become the only contributor to the elasticity (Fig. 7, right panel). In the transition range, when the gates are only partially opened, the overall stiffness is reduced.

This nonlinear compliance of the stereocilia has been demonstrated in hair cells from the sacculus (Howard & Hudspeth, 1988) but its contribution to BM nonlinearity has not yet been demonstrated explicitly. Its importance in the overall cochlear mechanics will depend on the (as yet unknown) extent to which stereociliar stiffness contributes to the overall BM stiffness. If stereociliar stiffness has a significant influence on BM stiffness, then this gating stiffness will manifest itself as a nonlinearity in cochlear mechanics. It will, however, be independent of the cochlear amplifier in that its effect will be the same regardless of the state of the cochlear amplifier and regardless of

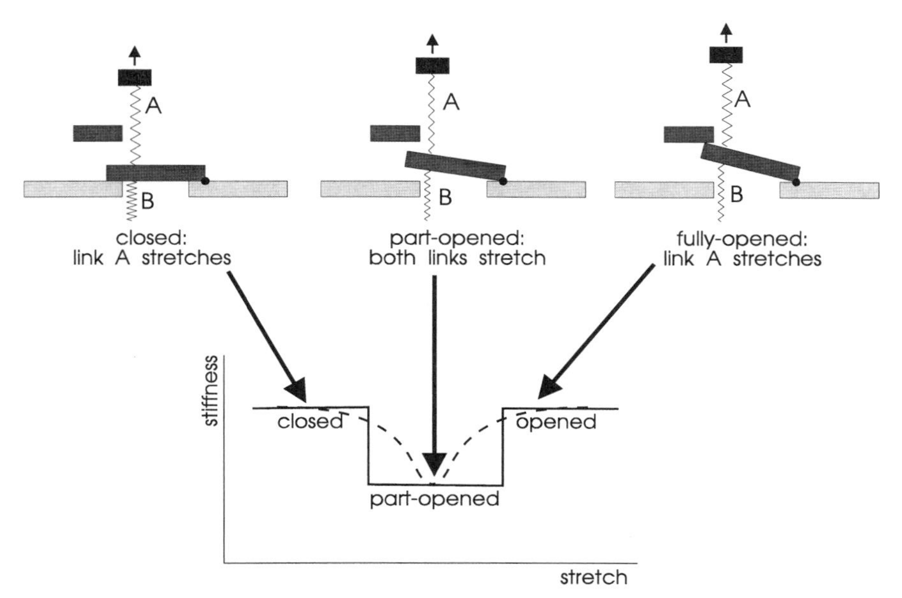

FIGURE 7 Simplified model for gating stiffness. A and B are two elastic links: A is the external tip-link of the stereocilia, B is the internal spring biasing the ion-channel gate closed. Tension is applied to link A. In this representation the compliance is a discontinuous function of stretch but in the stereocilia gates it is a smooth transition, because of the Boltzmann statistics of opening. The dashed line shows a more realistic shape for the gating stiffness.

frequency, *for the same BM displacement*. That is, its contribution to the overall nonlinearity will depend upon the absolute BM displacement and be independent of any amplification mechanisms that contribute to that displacement.

B. Input–Output Functions

Saturation of the mechanical-to-electrical transduction stage in the positive feedback loop also leads to compression of the dynamic range. It has been shown (Yates, 1990) that a saturating nonlinearity will lead to highly compressive input–output curves for the following reason. As stimulus intensity is increased, the loop gain gradually reduces, mostly due to saturation of the forward transduction (stage C of Fig. 2) so the overall gain falls dramatically. That is, the overall gain becomes a decreasing function of input level and the BM amplitude grows more slowly than the stimulus. Quantitatively, if the overall closed-loop gain is α and large, then it will start to decrease when the total gain around the loop is reduced by a factor of $1/\alpha$ by saturation of any stage.

Such saturation effects are seen in the experimental data when BM input–output functions are plotted (Sellick *et al.*, 1982; Robles *et al.*, 1986), and in the rate-vs-intensity functions of high-threshold nerve fibers (Sachs & Abbas, 1974; Winter, Robertson, & Yates, 1990; Yates *et al.*, 1990).

C. Distortion in the Cochlea

Harmonic distortion might be expected of a system such as the cochlea, because phase-related fluctuations in the feedback force must occur at intervals directly related to multiples of the period of the stimulus. However, whenever such harmonic distortion is generated at the characteristic place, it will be above the local BM cutoff frequency and would be expected to have no influence on the mechanics. Hence, harmonic distortion is not a feature of the cochlea. Other types of distortion, particularly intermodulation distortion, would be expected, however, and this is discussed later.

D. Otoacoustic Emissions

When an acoustic stimulus is presented to a normally functioning ear a variety of extra products may be demonstrated in the sound field of the ear canal. Of these additional components, some are essentially nonlinear, others are not, but the only way in which most can be identified is by taking advantage of their nonlinear properties.

1. Cochlear Echoes

Kemp (1978) first demonstrated cochlear echoes when he presented short transient stimuli to guinea pig ears while recording the sound field during and after the stimulus. After the initial stimulus had subsided he was able to record another wideband signal, delayed from the stimulus by approximately 10 ms, which grew in amplitude more slowly than the stimulus. This particular form of emission, the transiently evoked otoacoustic emission (TEOAE), is probably the least well understood of all the emissions, and no good theory of its generation exists.

One explanation of the TEOAE is that it reflects a weighted, summed response of all active generators along the cochlea and hence much of the response for any one frequency component must be generated at the characteristic place for that frequency. It is suggested that some of the energy stimulated by the cochlear amplifier travels in the reverse direction back along the cochlea to the stapes, where it reemerges into the ear canal sound field. If so, the presence or absence of the particular frequency component in the response would be expected to provide some information about the state of the cochlear amplifier at that particular characteristic place.

Even so, the behavior of the transient-evoked emission is not clearly interpretable. There is evidence (Avan *et al.*, 1991) that the low-frequency components of the TEOAE are strongly influenced by the physiological state of the basal region of the cochlea when it might be expected to be influenced only by the more apical regions, near to its characteristic place. There is also evidence that changes in the TEOAE may be evident before changes in threshold are observed at corresponding frequencies. If this is so, some revision of our ideas about the function of the cochlear amplifier would be necessary because the present concept permits no room for such an effect.

2. Single-Tone Emissions

If a single-frequency tone burst is presented to the ear, it is possible to demonstrate the presence of a component at the same frequency generated within the cochlea. Notwithstanding statements made in the introduction to Section IV, this single-tone emission (STE) is *demonstrable* only by virtue of its nonlinearity and, in particular, its suppression by a second tone. Brass and Kemp (1991) recorded the sound field in front of the tympanic membrane while they stimulated alternately with a probe tone alone and with probe tone and suppressor. The suppressor suppressed the nonlinear component from within the cochlea and with some arithmetic manipulation they were able to reveal the nonlinear component by itself. It had a latency of about 5–10 ms, appropriate for a response traveling from the stimulus frequency characteristic place, and they interpreted this as leaked energy from the cochlear amplifier.

If this interpretation is correct, then the STE should be analogous to the TEOAE, with the stimulus traveling to the characteristic place for each frequency involved and some energy from the cochlear amplifier traveling from each characteristic place to the stapes. However, in the case of the STE the possible interactions between frequency components are much simpler to interpret because there are at most two of them. More study of the relationship between the STE and the TEOAE is required, however, before we can precisely interpret them.

3. Intermodulation Products

If two stimulus frequencies are simultaneously presented to an ear, it is possible to measure in the external ear canal a third component at the cubic intermodulation frequency, $2f_1-f_2$. Again the general interpretation of this emission product is that it is generated at the characteristic place of one of the primary components (evidence suggests the f_2 characteristic place) due to nonlinear modulation of its BM response at the f_1 frequency and in this case the interpretation is fairly straightforward.

It has been suggested that there are two independent generators of the $2f_1-f_2$ cubic distortion tone (CDT), one active at low intensities and another active at high. This interpretation is based on differences between the CDT responses when driven by low- and high-intensity stimuli, but there is an alternative explanation that also appears consistent with the observations: that there is one generator only and that the amount of CDT depends, to first order at least, only on the displacement amplitude of the BM and not on how that amplitude is achieved.

For example, the low-level CDT grows approximately linearly with stimulus intensity whereas the high-intensity product grows at a greater rate, with a slope approaching 3. It has also been observed that the growth is steep when the cochlea is damaged. One interpretation is that there are two generators with different slopes, but another explanation is that there is one generator but two stages of growth of the BM input–output function. When any nonlinear transfer function is expanded as a polynomial, the cubic term chiefly gives rise to the $2f_1-f_2$ distortion, which is the reason for the name *cubic distortion product*. Further analysis indicates that the CDT generated by a simple cubic polynomial transfer function should grow approximately as the cube of the input level so that we might expect the cochlear CDT to grow as the cube of the stimulus intensity. But, experimentally, this is found to be the case only for the high-level CDT or the CDT from damaged cochleas; the low-level CDT grows more slowly. At low intensities, however, the BM amplitude grows at a rate of between 0.2 and 0.3 dB/dB (Sellick *et al.*, 1982; Robles *et al.*, 1986; Sachs & Abbas, 1978; Yates *et al.*, 1990; Cooper & Yates, 1994), and if the CDT depends upon BM displacement, then the CDT might be expected to grow as (0.2 to 0.3) × 3.0 or around 0.6 to 1, as is in fact the case.

4. Spontaneous Emissions

While not strictly a sign of nonlinearity, the presence of sounds spontaneously generated within the cochlea has been interpreted as evidence of an active process within the cochlea, limited from growing indefinitely in amplitude by an amplitude-limiting nonlinearity.

E. Two-Tone Suppression

Two-tone suppression, the reduction in response to one tone caused by the simultaneous presentation of a second tone of suitable frequency and intensity, is also to be expected from a positive feedback loop. The second tone, if of appropriate frequency and amplitude, will drive the nonlinearity into even greater saturation, thereby reducing the loop gain for the first (Geisler *et al.*, 1990), and hence it will reduce the amplitude of the first tone.

V. SUMMARY

The cochlea uses filtering to transform an incoming wideband signal into a set of parallel narrowband channels to match the channel capacity to the neural system. Filtering is accomplished by mechanical means assisted by mechanical positive feedback from the OHCs, with the IHCs taking the role of passive displacement transducers. The active process within the OHCs is still under debate with two significant candidates: (1) contraction of the cell body under the control of the transmembrane receptor potential and (2) motion of the stereocilia. Although the details of the active process are unknown, much is known about how the cochlear amplifier operates at a macroscopic level. Saturation of the forward transduction stage is probably the most significant nonlinearity and may explain the dynamic range of the cochlea as well as various forms of distortion detectable by psychoacoustic and physical means.

References

Allen, J. B. (1980). Cochlear micromechanics—a physical model of transduction. *Journal of the Acoustical Society of America, 68,* 1660–1670.

Allen, J. B., & Fahey, P. F. (1992). Using acoustic distortion products to measure the cochlear amplifier gain on the BM. *Journal of the Acoustical Society of America, 92,* 178–188.

Arima, T., Kuraoka, A., Toriya, R., Shibata, Y., & Uemura, T. (1991). Quick-freeze, deep-etch visualization of the 'cytoskeletal spring' of cochlear outer hair cells. *Cell Tissue Research, 263,* 91–97.

Ashmore, J. F. (1987). A fast motile response in guinea pig outer hair cell: The cellular basis of the cochlear amplifier. *Journal of Physiology, 388,* 323–347.

Assad, J. A., Hacohen, N., & Corey, D. P. (1989). Voltage dependence of adaptation and active bundle movement in bullfrog saccular hair cells. *Proceedings of the National Academy of Sciences USA, 86,* 2918–2922.

Assad, J. A., Shepherd, G. M. G., & Corey, D. P. (1991). Tip-link integrity and mechanical transduction in vertebrate hair cells. *Neuron, 7,* 985–994.

Avan, P., Bonfils, P., Loth, D., Narcy, P., & Trotoux, J. (1991). Quantitative assessment of human cochlear function by evoked otoacoustic emissions. *Hearing Research, 52,* 99–112.

Brass, D., & Kemp, D. T. (1991). Time-domain observation of otoacoustic emissions during constant tone stimulation. *Journal of the Acoustical Society of America, 90,* 2415–2427.

Brownell, W. E., Bader, C. R., Bertrand, D., & Ribaupierre, Y. (1986). Evoked mechanical responses of isolated cochlear outer hair cells. *Science, 227,* 194–196.

Cody, A. R. (1992). Acoustic lesions in the mammalian cochlea: Implications for the spatial distribution of the "active process." *Hearing Research, 62,* 166–172.

Cooper, N. P., & Rhode, W. S. (1992a). BM mechanics in the hook region of cat and guinea-pig cochleae—Sharp tuning and nonlinearity in the absence of baseline position shifts. *Hearing Research, 63,* 163–190.

Cooper, N. P., & Rhode, W. S. (1992b). BM tonotopicity in the hook region of the cat cochlea. *Hearing Research, 63,* 191–196.

Cooper, N. G., & Yates, G. K. (1994). Nonlinear input–output functions derived from the responses of guinea-pig cochlear nerve fibres: Variations with CF. *Hearing Research, 78,* 221–234.

Corey, D. P., & Hudspeth, A. J. (1979). Response latency of vertebrate hair cells. *Biophysics Journal, 26,* 499–506.
Corey, D. P., & Hudspeth, A. J. (1983). Kinetics of the receptor current in bullfrog saccular hair cells. *Journal of Neuroscience, 3,* 962–976.
Crawford, A. C., Evans, M. C., & Fettiplace, R. (1989). Activation and adaptation of transducer currents in turtle hair cells. *Journal of Physiology, 419,* 405–434.
Crawford, A. C., & Fettiplace, A. C. (1985). The mechanical properties of ciliary bundles of turtle cochlear hair cells. *Journal of Physiology, 364,* 359–379.
Dallos, P., Hallworth, R., & Evans, B. N. (1993). Theory of electrically driven shape changes of cochlear OHCs. *Journal of Neurophysiology, 70,* 299–323.
Dallos, P., & Harris, D. (1978). Properties of auditory nerve responses in absence of outer hair cells. *Journal of Neurophysiology, 41,* 365–383.
Eatock, R. A., Corey, D. P., & Hudspeth, A. J. (1987). Adaptation of mechanoelectrical transduction in hair cells of the bullfrog's sacculus. *Journal of Neuroscience, 7,* 2821–2836.
Elmore, W. C., & Heald, M. A. (1969). *Physics of waves.* New York: McGraw-Hill.
Evans, E. F. (1972). The frequency response and other properties of single fibres in the guinea pig cochlear nerve. *Journal of Physiology, 226,* 263–287.
Evans, E. F., & Wilson, J. P. (1975). Cochlear tuning properties: Concurrent BM and single nerve fibre measurements. *Science, 190,* 1218–1221.
Forge, A. (1991). Structural features of the lateral walls in mammalian cochlear outer hair cells. *Cell Tissue Research, 265,* 473–483.
Frishkopf, L. S., & DeRosier, D. J. (1983). Mechanical tuning of free-standing stereociliary bundles and frequency analysis in the alligator lizard cochlea. *Hearing Research, 12,* 393–404.
Geisler, C. D., Yates, G. K., Patuzzi, R. B., & Johnstone, B. M. (1990). Saturation of outer hair cell receptor current causes two-tone suppression. *Hearing Research, 44,* 241–256.
Gitter, A. H., Rudert, M., & Zenner, H. P. (1993). Forces involved in length changes of cochlear OHCs. *Pflugers Archiv–European Journal of Physiology, 424,* 914.
Howard, J., & Hudspeth, A. J. (1987a). Mechanical relaxation of the hair bundles mediates adaptation to mechanoelectrical transduction by the bullfrog's saccular hair cell. *Proceedings of the National Academy of Sciences USA, 84,* 3064–3068.
Howard, J., & Hudspeth, A. J. (1987b). Adaptation of mechanoelectrical transduction in hair cells. In A. J. Hudspeth, P. R. MacLeish, F. L. Margolis, & T. N. Wiesel (Eds.), *Sensory transduction,* (pp. 138–145) Geneva: Fondation pour l'Etude du Systèmе Nerveaux Central et Périphèrique.
Howard, J., & Hudspeth, J. A. (1988). Compliance of the hair bundle associated with gating of the mechanoelectrical transduction channels in the bullfrog's saccular hair cell. *Neuron, 1,* 189–199.
Hudspeth, A. J., & Gillespie, P. G. (1994). Pulling strings to tune transduction: Adaptation by hair cells. *Neuron, 12,* 1–9.
Jaramillo, F., & Hudspeth, A. J. (1991). Localization of the hair cell's transduction channels at the hair bundle's top by iontophoretic application of a channel blocker. *Neuron, 7,* 409–420.
Johnstone, B. M., & Boyle, A. J. F. (1967). BM vibration examined with the Mössbauer technique. *Science, 158,* 389–390.
Johnstone, B. M., & Yates, G. K. (1974). BM tuning curves in the guinea pig. *Journal of the Acoustical Society of America, 55,* 584–587.
Johnstone, B. M., Taylor, K. J., & Boyle, A. J. F. (1970). Mechanics of the guinea pig cochlea. *Journal of the Acoustical Society of America, 47,* 504–509.
Johnstone, J. R., Alder, V. A., Johnstone, B. M., & Yates, G. K. (1979). Cochlear action potential threshold and single unit thresholds. *Journal of the Acoustical Society of America, 65,* 254–257.

Kachar, B., Brownell, W. E., Altschuler, R., & Fex, J. (1986). Electrokinetic shape changes of cochlear outer hair cells. *Nature, 322,* 365–367.

Kemp, D. T. (1978). Stimulated acoustic emissions from within the human auditory system. *Journal of the Acoustical Society of America, 64,* 1386–1391.

Khanna, S. M., & Leonard, D. G. B. (1982). BM tuning in the cat cochlea. *Science, 215,* 305–306.

Kim, D. O., Neely, S. T., Molnar, C. E., Matthews, J. W. (1980). An active cochlear model with negative damping in the partition: Comparison with Rhode's ante- and post-mortem observations. In G. Van den Brink & F. A. Bilsen (Eds.), *Psychophysical, physiological and behavioral studies in hearing* (pp. 7–13). Delft: Delft University Press.

Kirk, D. L., & Yates, G. K. (1994). Evidence for electrically evoked travelling waves in the guinea pig cochlea. *Hearing Research, 74,* 38–50.

Lepage, E. L., & Johnstone, B. M. (1980). Nonlinear mechanical behavior of the BM in the basal turn of the guinea pig cochlea. *Hearing Research, 2,* 183–189.

Mammano, F., & Nobili, R. (1993). Biophysics of the cochlea: Linear approximation. *Journal of the Acoustical Society of America, 93,* 3320–3332.

Manley, G. A., Yates, G. K., & Köppl, C. (1988). Auditory peripheral resonance: Evidence for a simple resonance phenomenon in the lizard Tiliqua. *Hearing Research, 33,* 181–190.

Neely, S. T. (1993). A model of cochlear mechanics with OHC motility. *Journal of the Acoustical Society of America, 94,* 137–146.

Neely, S. T., & Kim, D. O. (1983). An active cochlear model showing sharp tuning and high sensitivity. *Hearing Research, 9,* 123–130.

Neely, S. T., & Kim, D. O. (1986). A model for active elements in cochlear biomechanics. *Journal of the Acoustical Society of America, 79,* 1472–1480.

Patuzzi, R. B., Yates, G. K., & Johnstone, B. M. (1989a). Changes in cochlear microphonic and neural sensitivity produced by acoustic trauma. *Hearing Research, 39,* 189–202.

Patuzzi, R. B., Yates, G. K., & Johnstone, B. M. (1989b). OHC receptor current and sensorineural hearing loss. *Hearing Research, 42,* 47–72.

Pickles, J. O., Comis, S. D., & Osborne, M. P. (1984). Cross-links between stereocilia in the guinea pig organ of Corti, and their possible relation to sensory transduction. *Hearing Research, 15,* 103–112.

Pollice, P. A., & Brownell, W. E. (1993). Characterization of the outer hair cell's lateral wall membranes. *Hearing Research, 70,* 187–196.

Pringle, J. W. S. (1967). The contractile mechanism of insect fibrilar muscle. *Progress in Biophysics and Molecular Biology, 17,* 1–60.

Reuter, G., & Zenner, H. P. (1990). Active radial and transverse motile responses of OHCs in the organ of Corti. *Hearing Research, 43,* 219–230.

Reuter, G., Gitter, A. H., Thurm, U., & Zenner, H. P. (1992). High frequency radial movements of the reticular lamina induced by OHC motility. *Hearing Research, 60,* 236–246.

Rhode, W. S. (1971). Observations of the vibration of the BM using the Mössbauer technique. *Journal of the Acoustical Society of America, 49,* 1218–1231.

Rhode, W. S., & Geisler, C. D. (1967). Model of the displacement between opposing points on the tectorial membrane and reticular lamina. *Journal of the Acoustical Society of America, 42,* 185–190.

Robertson, D., & Johnstone, B. M. (1981). Primary auditory neurons: Nonlinear responses altered without changes in sharp tuning. *Journal of the Acoustical Society of America, 69,* 1096–1098.

Robles, L., Ruggero, M. A., & Rich, N. C. (1986). BM mechanics at the base of the chinchilla cochlea. I. Input-output functions, tuning curves, and phase responses. *Journal of the Acoustical Society of America, 80,* 1364–1374.

Ruggero, M. A., & Rich, N. C. (1991). Furosemide alters organ of Corti mechanics: Evidence for feedback of OHCs upon the BM. *Journal of Neuroscience, 11,* 1057–1067.

Russell, I. J. (1981). The responses of vertebrate hair cells to mechanical stimulation. In A. Roberts & B. M. H. Bush (Eds.), *Neurons without impulses* (pp. 117–145). Cambridge: Cambridge University Press.

Russell, I. J., Cody, A. R., & Richardson, G. P. (1986). The responses of inner and OHCs in the basal turn of the guinea-pig cochlea and in the mouse cochlea grown in vitro. *Hearing Research, 22,* 199–216.

Russell, I. J., & Sellick, P. M. (1983). Low-frequency characteristics of intracellularly recorded receptor potentials in guinea-pig cochlear hair cells. *Journal of Physiology, 338,* 179–206.

Sachs, M. B., & Abbas, P. J. (1974). Rate versus level functions for auditory-nerve fibers in cats: Tone-burst stimuli. *Journal of the Acoustical Society of America, 56,* 1835–1847.

Sand, O., Ozawa, S., & Hagiwara, S. (1975). Electrical and mechanical stimulation of hair cells in the mudpuppy. *Journal of Comparative Physiology A, 102,* 13–26.

Santos-Sacchi, J. (1989). Asymmetry in voltage-dependent movements of isolated outer hair cells from the organ of Corti. *Journal of Neuroscience, 9,* 2954–2962.

Santos-Sacchi, J. (1992). On the frequency limit and phase of OHC motility: Effects of the membrane filter. *Journal of Neuroscience, 12,* 1906–1916.

Santos-Sacchi, J. (1993). Harmonics of outer hair cell motility. *Biophysics Journal, 65,* 2217–2227.

Santos-Sacchi, J., & Dilger, J. P. (1988). Whole cell currents and mechanical responses of isolated outer hair cells. *Hearing Research, 35,* 143–150.

Sellick, P. M., Patuzzi, R., & Johnstone, B. M. (1982). Measurements of BM motion in the guinea pig using the Mössbauer technique. *Journal of the Acoustical Society of America, 72,* 131–141.

Sellick, P. M., & Russell, I. J. (1978). Intracellular studies of cochlear hair cells: Filling the gap between BM mechanics and neural excitation. In R. F. Naunton & C. Fernandez (Eds.), *Evoked electrical activity in the auditory nervous system* (pp. 113–139). New York: Academic Press.

Sellick, P. M., Yates, G. K., & Patuzzi, R. (1983). The influence of Mössbauer source size and position on phase and amplitude measurements of the guinea pig BM. *Hearing Research, 10,* 101–108.

Shepherd, G. M. G., Corey, D. P., & Block, S. M. (1990). Actin cores of hair-cell stereocilia support myosin motility. *Proceedings of the National Academy of Sciences USA, 87,* 8627–8631.

Strelioff, D., Flock, A., & Minser, K. E. (1985). Role of inner and OHCs in mechanical frequency selectivity of the cochlea. *Hearing Research, 18,* 169–176.

von Békésy, G. (1960). *Experiments in hearing.* New York: McGraw-Hill.

Winter, I. J., Robertson, D., & Yates, G. K. (1990). Diversity of CF rate-intensity functions in guinea pig auditory nerve fibres. *Hearing Research, 45,* 191–202.

Wilson, J. P., & Johnstone, J. R. (1975). BM and middle ear vibration in guinea pig measured by capacitive probe. *Journal of the Acoustical Society of America, 57,* 705–723.

Yates, G. K. (1986). Frequency selectivity in the auditory periphery. In B. C. J. Moore (Ed.), *Frequency selectivity in hearing* (pp. 1–50). London: Academic Press.

Yates, G. K. (1990). BM nonlinearity and its influence on auditory nerve rate-intensity functions. *Hearing Research, 50,* 145–16.

Yates, G. K., & Johnstone, B. M. (1979). Measurement of BM movement. In H. A. Beagley (Ed.), *Auditory Investigation: The scientific and technological basis* (pp. 418–430). Oxford: Clarendon Press.

Yates, G. K., Winter, I. M., & Robertson, D. (1990). Basilar membrane nonlinearity determines auditory nerve rate-intensity functions and cochlear dynamic range. *Hearing Research, 45,* 203–219.

Zenner, H.-P. (1986). Motile responses in outer hair cells. *Hearing Research, 22,* 83–90.

Zenner, H.-P., Zimmermann, R., & Gitter, A. H. (1988). Active movements of the cuticular plate induce sensory hair motion in mammalian OHCs. *Hearing Research, 34,* 233–240.

Zwislocki, J. J., & Kletsky, E. J. (1979). Tectorial membrane: A possible effect on frequency analysis in the cochlea. *Science, 204,* 639–641.

Neural Signal Processing

Alan R. Palmer

Following detection of a sound the auditory nervous system must decipher "what is it?" and "where is it?" The first stage of processing to answer these questions involves breaking down complex sounds into their components by the cochlea (see Chapter 2). Thus, the processing of a complex sound can be thought of as the processing of the frequencies and intensities of its components. However, real sounds are characterized by spectro–temporal variations, and the processing of individual frequency components is neither independent nor linear. We cannot, therefore, predict the responses to complex sounds by simple summation of the responses to their frequency components. This has prompted the use of stimuli with increasing complexity such as two-tone complexes, frequency and amplitude modulated stimuli, and vocalizations. "Where is it?" additionally involves computations based on differences in the timing and level of the signals at the two ears. The account given here is highly selective and is limited to non-specialized mammals.

I. SOUND FREQUENCY

A. Frequency Selectivity

The cochlea operates as a short-term Fourier analyzer separating complex acoustic signals into their frequency components. The very sharply tuned

Hearing

mechanical vibrations of the cochlear partition are transduced by the hair cells, which exhibit similarly sharply tuned receptor potentials. These receptor potentials provide the driving force for the release of neurotransmitter at the base of the hair cells, which in turn generates action potentials in the fibers of the auditory nerve (see Chapter 2). Although the axons of *outer* hair cells project to the cochlear nucleus, most, and probably all, responses from auditory nerve fibers must have been recorded from the axons that innervate the *inner* hair cells. The only proven recording from an outer hair cell afferent showed no spontaneous activity and no activation by sounds at a moderate level (Robertson, 1984).

The only response of auditory nerve fibers in mammals to a single tone is excitation; that is, an increase in the rate of generation of action potentials above the resting or spontaneous rate (Kiang, Watanabe, Thomas, & Clark, 1965). The range of frequencies capable of exciting a single fiber is restricted in exactly the same way as the vibration pattern of the basilar membrane and the receptor potentials of the hair cells (see Chapter 2). Thus one can measure a tuning curve of a fiber that represents the sound level of tones that evoke the same just-detectable increase in the firing rate of the fiber; this is often referred to as a *frequency-threshold curve* (FTC). The frequency at the minimum of the FTC is termed the best or *characteristic frequency* (CF), and it indicates the position along the cochlear partition of the hair cell that the fiber innervates. The FTCs of the fibres innervating hair cells along the length of the cochlea form an overlapping series of band-pass filters that encompass the hearing range of the animal, as can be seen in Figure 1(a), which shows FTCs of 127 fibres recorded from the auditory nerve of a single cat.

Examination of the ten fibers in Figure 1(b) (selected from those in Figure 1(a)) reveals a variation in the shape of the FTC with CF. These variations have been quantified using several parameters, such as the slope of the FTC above and below the CF, the bandwidth of the FTC at 3 or 10 dB above the minimum threshold, and the length of the sharply tuned tip region (see Evans, 1975; Evans, Pratt, Spenner & Cooper, 1992). For the fibers shown in Figure 1(a) all these variables covary with CF. The relative sharpness of the fibers' FTC is shown as the variation in $Q_{10\ dB}$ with CF in Figure 1(c) ($Q_{10\ dB}$ = CF/bandwidth at 10 dB above the threshold at CF). The sharper is the filter, the larger is the $Q_{10\ dB}$. Thus, for auditory nerve fibers, although the absolute bandwidth increases, the relative sharpness increases also with CF up to about 15–20 kHz. It has been demonstrated in several studies that the tuning present in the cochlea, as manifested in the tuning of auditory nerve fibers, accounts well for psychophysical frequency selectivity, as shown in Figure 2. In this figure, the bandwidth of auditory nerve fibers is compared with two behavioral measures of frequency selectivity from the same animal (the guinea pig, Evans et al., 1992). See Chapter

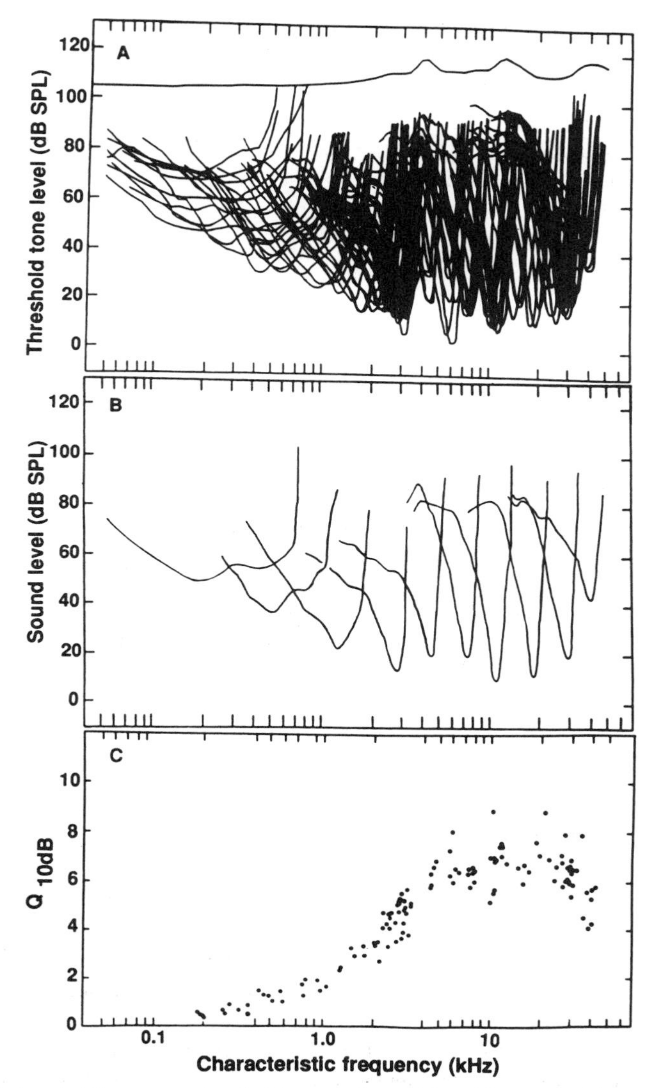

FIGURE 1 (A) Frequency threshold curves from 127 cochlear nerve fibers in a single cat obtained by an automated threshold tracking procedure. The continuous line at the top shows the maximum output levels of the sound system. (B) Ten of the curves extracted from the top figure to illustrate the progressive changes in shape with characteristic frequency. (C) $Q_{10\ dB}$ measures of tuning for this population (the characteristic frequency divided by the bandwidth at 10 dB above the threshold). (From Palmer & Evans, unpublished data.)

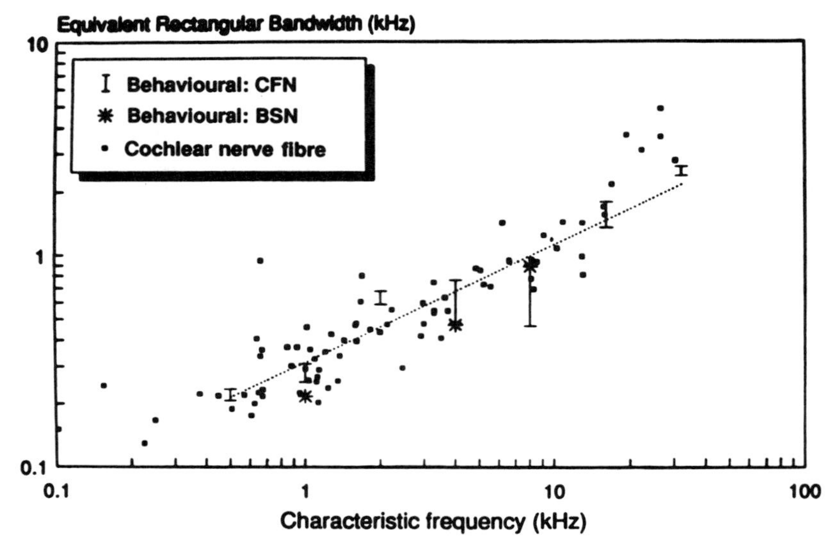

FIGURE 2 Comparison of the equivalent rectangular bandwidths (ERBs) of physiological (cochlear nerve fiber) tuning curves with behavioral filter functions obtained from the same species. Each square represents the data from a single cochlear nerve fiber. The asterisks show ERBs obtained behaviorally from the masked thresholds for detecting pure tones in the presence of a comb-filtered noise masker and the bracket symbols (±1 S.E.) show ERBs obtained from masked thresholds in a bandstop noise masker. The dotted line is a regression line fitted through the comb-filtered noise data. (Reprinted from Evans et al., 1992, p. 162, with permission from Pergamon Press, Ltd, Headington Hill Hall, Oxford OX3 OBW, UK.)

5 for further discussion of this point and a description of the behavioral measures.

Although the most sensitive auditory nerve fibers have minimum thresholds that match the behavioral audiogram of the animal (e.g., Liberman, 1978), the thresholds at the CF of the auditory nerve fibers innervating a restricted region of the basilar membrane vary with the fibers' spontaneous discharge rate. When large numbers of fibers are recorded from a single nerve, it is possible to distinguish three populations of auditory nerve fibers that differ according to their spontaneous rate (16% of fibers have rates below 0.5 spikes/s, 23% have rates between 0.5 and 18 spikes/s, and 61% have rates exceeding 18 spikes/s; Liberman, 1978). The most sensitive fibers are those with high spontaneous rates, followed by the medium (about 10 dB higher), with the highest thresholds found in the fibers with low spontaneous rates (about 30 dB higher). These values are typical of studies using only acoustic search stimuli; a narrower threshold distribution is found if signal detection theory is used to establish the threshold and a wider distribution of thresholds at the CF (over an 80 dB range, see Figure 3) has been

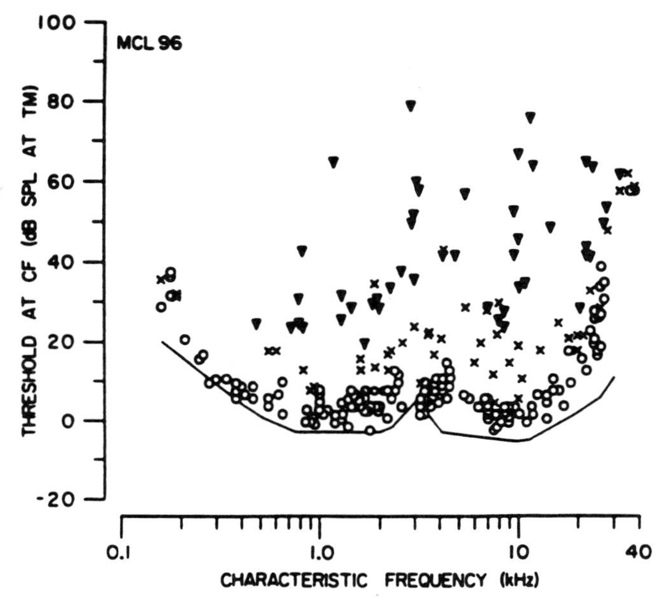

FIGURE 3 Thresholds of fibers from a single cat cochlear nerve as a function of characteristic frequency. Fibers had high (circles), medium (crosses), or low (triangles) rates of spontaneous discharge. The continuous line is the best threshold curve, which represents the lowest CF thresholds seen in a large sample of fibers from 43 animals. (From Liberman & Kiang, 1978, with permission.)

described when an electrical search stimulus was used (Liberman, 1978; Liberman & Kiang, 1978). The significance of the minority group of very high threshold auditory nerve fibers remains obscure.

B. Population Responses to Single Tones

It is clear from Figure 1 that there is increasing and substantial overlap in the FTCs of auditory nerve fibers innervating disparate regions of the basilar membrane as the sound level is increased. This implies that, although at threshold a single tone will activate only a small group of fibers with CFs at the tone frequency, at higher sound levels fibers with CFs away from the tone frequency will be activated. That this is the case may be seen in Figure 4. At the highest sound levels, nearly the whole population is activated by a 1 kHz tone, whereas the very steep high-frequency slopes of the FTCs ensure that an 8 kHz tone largely activates fibers with CFs above the tone frequency. Such a spread in activity with level is a major factor causing increased masking of high frequencies by lower frequencies at high sound level (the upward spread of masking).

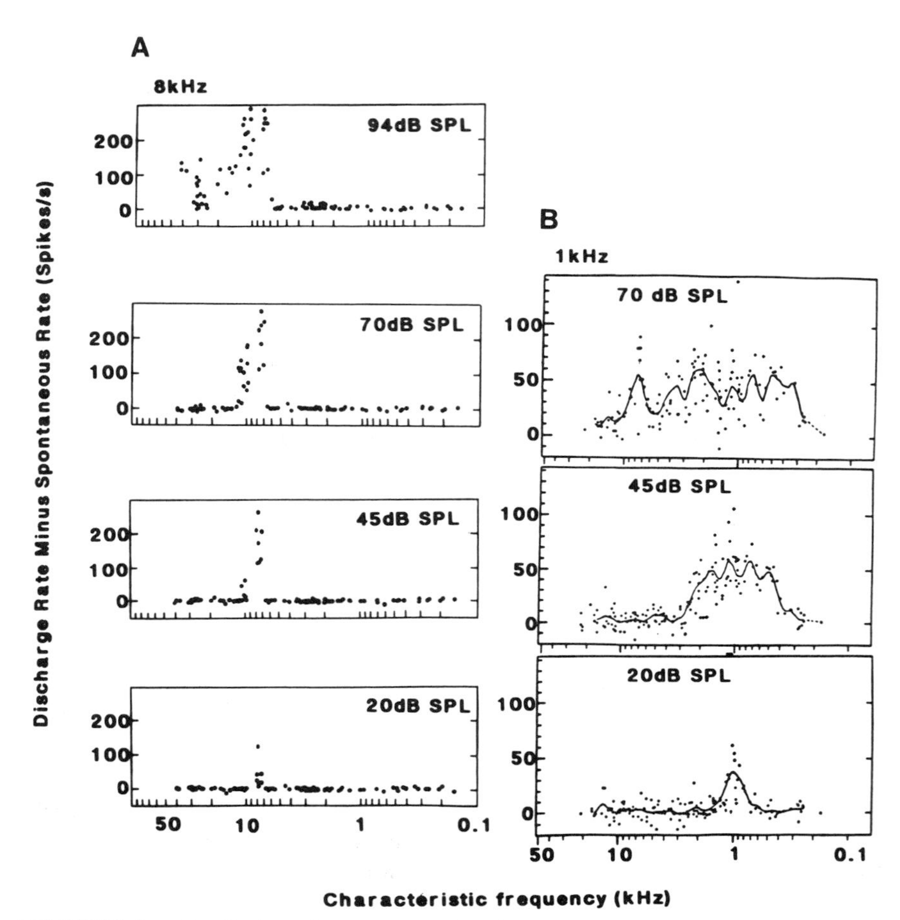

FIGURE 4 Distribution of activity across the fiber array evoked by an 8 kHz tone (A) and a 1 kHz tone (B) as a function of sound level (as indicated). Each data point represents the increase in discharge rate above spontaneous activity for a single fiber, plotted as its characteristic frequency; all fibers for each frequency were recorded from a single cat. Curves in B are moving window averages of the activity of fibers with high (>15 spikes/s) spontaneous rates. Note the reversed frequency axis. (Adapted from Irvine, 1986, with permission. Original data were from Palmer & Evans, unpublished data, and Kim & Molnar, 1979, with permission from Springer-Verlag.)

C. Cochleotopic Organization

A microelectrode passing through the cochlear nucleus (the first auditory relay in the brainstem), in a dorso–ventral direction, first encounters neurons with high CFs then progressively lower CFs (Rose, Galambos, &

Hughes, 1959). Such orderly frequency mapping is termed *tonotopic* or, more correctly, *cochleotopic* organization; every major nucleus between the cochlea and the cortex has been found to be cochleotopically organized, as illustrated diagrammatically in Figure 5. In the central nervous system large areas of tissue may be most sensitive to the same frequency thus forming iso-frequency laminae in the brainstem, midbrain, and thalamus and iso-frequency bands in the cortex.

D. Frequency–Intensity Response Areas

Although a best frequency can generally be attributed to most auditory neurons, this does not imply that their tuning resembles that of auditory nerve fibers. The homogeneous excitatory response of auditory nerve fibers to simple tones, which results in V-shaped tuning curves, no longer applies for the majority of higher order neurons. Even at the first relay stage of the cochlear nucleus there is evidence of convergence to produce wider and more complicated tuning curves, and single tones often evoke inhibition of the neural activity (Rose et al., 1959; Evans & Nelson, 1973). A more useful representation for describing the responsiveness of central neurons is the frequency–intensity response area or response map. This form of analysis demonstrates the frequencies and intensities of single tones that produce excitation or inhibition of the neuron's output. In the cochlear nucleus such response areas have been used to classify neuron responses according to the strength and prevalence of inhibition evoked by single tones (Evans & Nelson, 1973; see Young, 1984; Young, Shofner, White, Robert, & Voigt, 1988) as can be seen in Figure 6.

Sensitivity to a wider range of frequencies, and admixtures of excitation and inhibition across frequency, are common throughout the central auditory pathway, although simple, narrowly tuned, excitatory response areas may be found in all nuclei including the cortex (see Calford, Webster, & Semple, 1983). Above the level of the cochlear nucleus the picture is further complicated by the sensitivity of the neurons to signals at either ear (see Section V.). Careful mapping studies at the inferior colliculus and auditory cortex within iso-frequency sheets have revealed that in addition to the cochleotopic organization there is a topographic organization according to sharpness of neural tuning and the symmetry of the lateral inhibitory side-bands (e.g., Schreiner & Mendelson, 1990). Schreiner and Mendelson hypothesize that a given spectral component will be simultaneously processed by analyzers with a variety of bandwidths centered on that component. The result would be equivalent to a "multiple-bandwidth spectral analyzer," which could be of particular value in the discrimination of natural sounds that differ in spectral shape, tilt, or contrast.

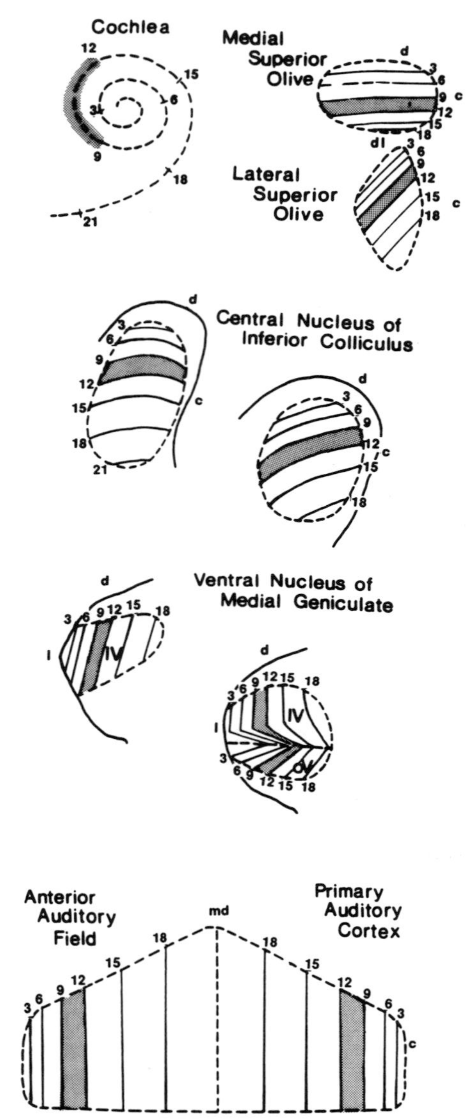

FIGURE 5 Schematic representation of the cochleotopic organization within each of the major nuclei of the auditory nervous system. The bands indicate 3 mm sections of the cochlear sensory epithelium with the 9–12 mm section shaded. Abbreviations: *r*, rostral; *c*, caudal; *d*, dorsal; *v*, ventral; *l*, lateral. (From Merzenich, Roth, Anderson, Knight & Colwell, 1977, with permission.)

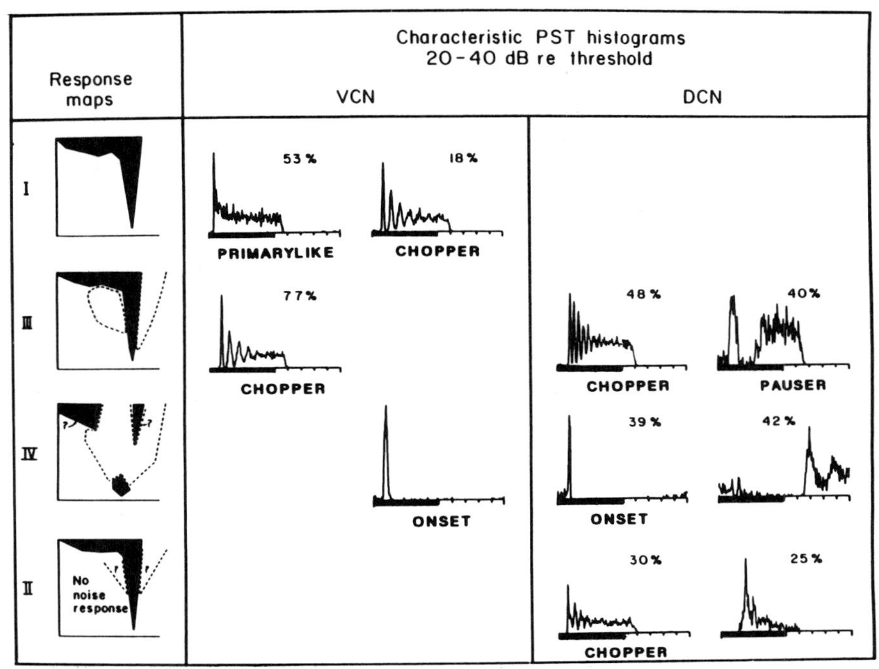

FIGURE 6 Relationship between the response area type, post-stimulus time histogram type, and unit location within the cochlear nucleus complex. The percentages indicate the proportions of the units sampled within the ventral and dorsal cochlear nucleus with particular combinations of response area and post-stimulus time histogram. Areas of excitation in the response maps are shown hatched and inhibitory areas are delimited by dashed lines. (From Young et al., 1988, with permission.)

E. Time Course of Activation by Single Tones

1. Adaptation and Shapes of Post-Stimulus Time Histograms

When a single excitatory tone is presented, the discharge rate of an auditory nerve fiber is maximal at the stimulus onset, gradually decreases, and reaches a steady state after some tens of milliseconds (e.g., Kiang et al., 1965; see histogram labeled *Primarylike* in Figure 6). Such adaptation is typical of most sensory neurons and is observed in all fibers of the auditory nerve. It is thought to originate in the synapse at the base of the hair cell. Adaptation does not, however, seem to be simply due to the depletion of neurotransmitter, as small step increases in intensity generate about the same increase in discharge irrespective of the state of adaptation (depletion of the transmitter should cause a decrease in the gain and therefore a multiplicative effect on step intensity changes; Smith & Zwislocki, 1975). The time course of the decline from the high rate of discharge at the onset is not a simple exponential, but is characterized by more than one time constant: one very rapid (<10 ms), one of some tens of milliseconds, and possibly others even longer (Kiang et al., 1965; Westerman & Smith, 1984; Yates, Robertson, & Johnstone, 1985). Following the cessation of an excitatory tone there is a depression of the excitability, which has been suggested to be the cause of various forward masking phenomena (see Smith, 1979). Complete recovery after a stimulus takes of tens of milliseconds for high spontaneous rate fibers, but may extend to seconds for low spontaneous rate fibers (Relkin & Doucet, 1991).

In the cochlear nucleus the time course of the response to a single tone is different for the principal cell types. Figure 6 illustrates the time courses of the responses of neurons in the dorsal (DCN) and ventral (VCN) subdivisions of the cochlear nucleus and the way in which these responses relate to the response areas of the neurons. The post-stimulus time histogram (PSTH) classification scheme is not definitive, because the responses of some units fall between categories, and others change from one pattern to another with changes in stimulus conditions. Nevertheless, it does provide a convenient segregation of the responses, which, taken together with the response area, does in some cases correlate with the underlying cell morphology. The different PSTHs are referred to by discriptive names as shown in Figure 6. The separation of the peaks of chopper responses to single tones is not related to the stimulus frequency, but reflects intrinsic membrane properties coupled with a sustained input excitation derived from multiple synaptic contacts (Oertel, Wu, & Hirsh, 1988). Primarylike responses in the rostral pole of the VCN are the consequence of secure synaptic activation of the cells by large synapses. Primarylike-with-a-notch PSTHs (not shown in the Figure 6) are the likely consequence of multiple, secure inputs from the auditory nerve that generate an onset spike with high probability, thus revealing a refractory gap of <2 ms. The pauser and build-up responses in the dorsal cochlear nucleus (DCN) result from the temporal overlap of excitatory inputs from the auditory nerve and inhibitory inputs from within the cochlear nucleus (see Young, 1984). The likely functions of these response types are discussed later.

A similar range of PSTH types is found in the superior olive to monaural stimulation, but with some additional transformations of the responses (Guinan, Norris, & Guinan, 1972; Tsuchitani & Johnson, 1991). At higher levels in the pathway a distinction is often drawn between neurons that respond only at the stimulus onset and those that show sustained discharge throughout the tone burst (primarylike, chopper, and pauser types). Both types of response have been reported in all higher nuclei with the proportion responding only at the onset becoming greater in more central nuclei (see Irvine, 1986; Phillips, Reale, & Brugge, 1991).

2. Fine Time Structure of the Responses to Single Tones (Phase Locking)

Cochleotopy implies a "place" coding of frequency, but the frequency of a stimulus or of the components of a complex sound may also be signaled in the timing of the impulses. Impulses are initiated in auditory nerve fibers when the hair cell stereocilia are bent toward the longest stereocilium (see Chapter 2). Thus, in response to low-frequency sounds the impulses in auditory nerve fibers occur preferentially at a particular phase of the stimulus waveform. This phenomenon, termed *phase locking* (Rose, Brugge, Anderson, & Hind, 1967), has been demonstrated to occur in all vertebrate classes (see Palmer & Russell, 1986, for a review). In the guinea pig, phase locking begins to decline at 600 Hz and is no longer detectable at 3.5 kHz, whereas in the cat phase locking persists at frequencies about an octave higher: the decline begins at about 1 kHz and phase locking is not detected above 5 kHz as shown in Figure 7 (Rose et al., 1967; Palmer & Russell, 1986; Kiang et al., 1965; Johnson, 1980). Several reasons have been suggested for the limit in the ability of auditory nerve fibers to phase lock to high-frequency tones. One major contributory factor is the capacitance and resis-

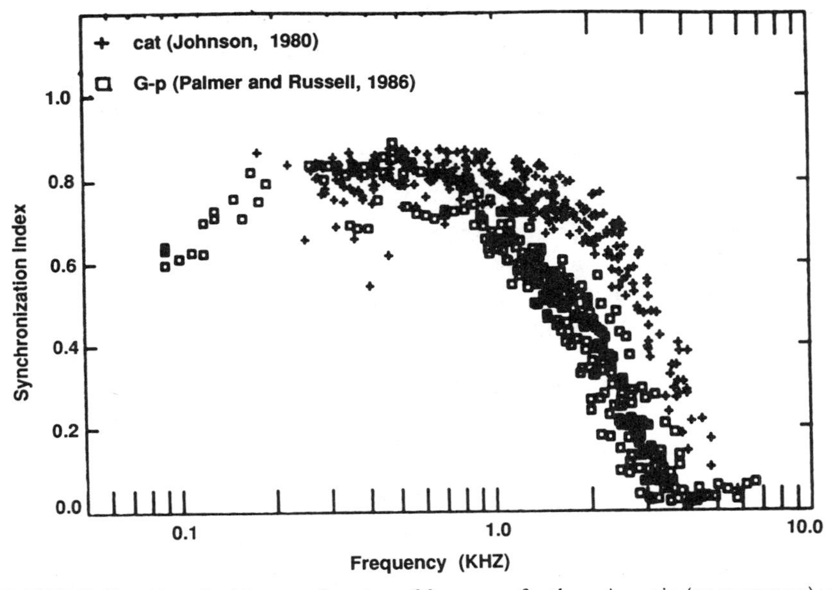

FIGURE 7 Phase locking as a function of frequency for the guinea pig (open squares) and cat (crosses). The synchronization index is calculated by normalizing the vectorial sum of the bins of a period histogram locked to the tone waveform, each bin being assigned a vector angle based on its position within the cycle and an amplitude equal to the number of spikes in the bin. (From Palmer & Russell, 1986. Reprinted with permission from Elsevier Science Publishers.)

tance of the hair cell membrane; these act as a low-pass filter to attenuate the sinusoidal components of the receptor potential that periodically activates the nerve fiber synapse (Palmer & Russell, 1986). Phase locking appears as an entrainment of spontaneous activity up to 20 dB below the threshold for an increase in discharge rate, but this threshold difference disappears when signal detection theory is used to determine both the mean rate and phase locking thresholds. Phase locking persists with no indication of clipping at levels above those producing saturation of the fiber discharge rate (Rose et al., 1967; Johnson, 1980; Palmer & Russell, 1986).

Both the proportion of neurons exhibiting phase locking and the highest frequency at which phase locking can be detected generally decline with ascent toward the cortex. Several factors contribute to this decline in phase locking (see Rouiller, De Ribaupierre, & De Ribaupierre, 1979) and even in the cochlear nucleus, different cell groups vary in their phase locking capability. Thus, spherical bushy cells and some onset responding multipolar cells phase lock as well as do auditory nerve fibers (Bourk, 1976; Winter & Palmer, 1990; Blackburn & Sachs, 1989). Other multipolar cells (which have chopper PSTHs) have a lower cut-off frequency for phase locking than auditory nerve fibers; the decline starts at a few hundred Hertz and no phase locking is detectable above 2 kHz (Bourk, 1976; Winter & Palmer, 1990). Phase locking in the DCN occurs only to very low frequencies (e.g., Goldberg & Brownell, 1973). At the medial superior olive, a nucleus that contains a majority of neurons with low CFs, the neurons do show phase locking but probably not as well as the spherical bushy cells in the VCN from which they derive their input (see Tsuchitani & Johnson, 1991). Only 18% of inferior colliculus cells show phase locking, and it is seldom seen to frequencies above 600 Hz (Kuwada, Yin, Syka, Buunen, & Wickesberg, 1984). At the level of the medial geniculate body only 2% of neurons show phase locking (Rouiller et al., 1979), and phase locking has not been reported to occur in the primary auditory cortex to frequencies above about 100 Hz (Phillips et al., 1991). Clearly, any signal processing dependent on the fine timing of the impulses must be accomplished early in the auditory pathway.

F. Two-Tone Rate and Synchrony Suppression

The discharge evoked in an auditory nerve fiber by a tone may be reduced or eliminated by the simultaneous presentation of a second tone situated within prescribed areas of frequency and intensity either side of the FTC (Kiang et al., 1965, Arthur, Pfeiffer, & Suga, 1971). The spectral dependence of this *rate suppression* is illustrated in Figure 8(a) and its time course by the PSTH in Figure 8(b). The suppressive areas defined by reduction of discharge rate extend only to the edges of the FTC because, once within the

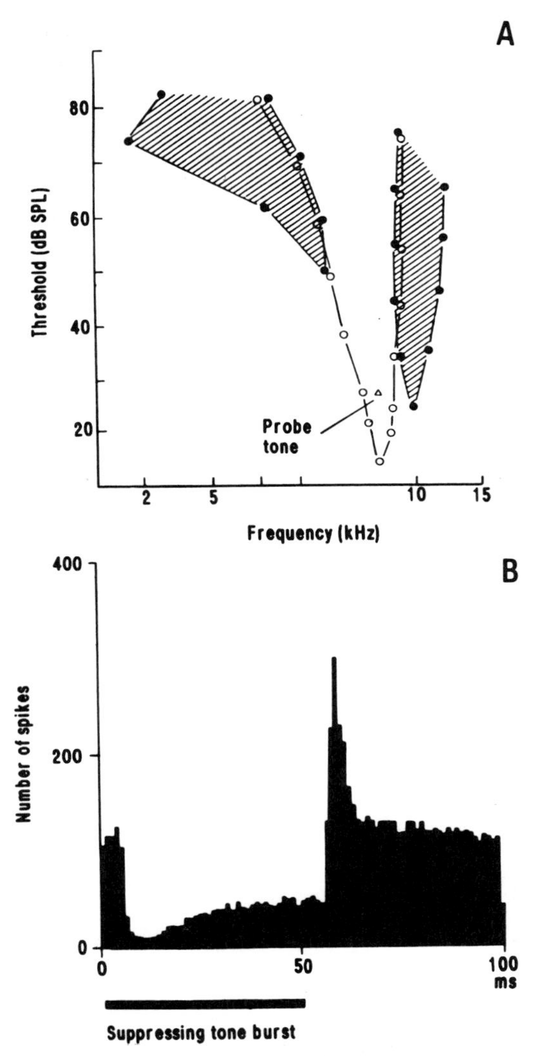

FIGURE 8 (A) Two-tone suppression areas of a cochlear nerve fiber in the cat. The shaded areas show the frequencies and intensities of a second tone that will reduce the mean firing rate to a CF tone (at the level shown by the open triangle) by 20% or more. The excitatory response area for single tones is bordered by the open circles (From Arthur et al., 1971, with permission.) (B) The time course of two-tone suppression in a cochlear nerve fiber of the cat shown as a peristimulus time histogram in which activity evoked by a continuous excitatory tone is suppressed by a burst of a second tone. (From Kiang et al., 1965, with permission.)

FTC, the suppressor also excites the fiber: the high-frequency suppressive area may extend down to very close to the fiber's CF threshold, but the low-frequency suppressive area is generally 15–40 dB higher. The time course of the suppression shows a maximum at the suppressor tone onset with a gradual recovery of the firing rate, followed by a large overshoot in the rate when the suppressor is turned off (cf. the adaptation described previously). The similarity of the latencies for the onset of the excitation and suppression suggests that the involvement of an inhibitory synapse is unlikely, and since the suppression survives sectioning of the olivocochlear bundle it is not an effect of the descending system (Kiang et al., 1965; Arthur et al., 1971). Recent evidence suggests that the source of this form of suppression is in the interaction of the mechanical responses on the basilar membrane (see Chapter 2). At low frequencies, it is possible to investigate the suppressive interactions during complex sound stimulation by measuring the phase locking to the constituents of the complex. In the case of two tone stimulation, the number of discharges phase locked to one tone is reduced when the second tone is also presented. Such "synchrony suppression" is not limited to the regions causing rate suppression, but rather extends throughout the fiber frequency–intensity response area with maximum synchrony suppression occurring at or near CF (see Javel, 1981).

At levels above the auditory nerve, suppressive sidebands are a common finding. However, in the presence of spontaneous activity these sidebands are seen with single tones and reflect neurally mediated inhibition. When stimulation consists of spectrally rich sounds, the sidebands must reflect not only the suppressive effects taking place in the cochlea but also the neurally mediated lateral inhibition, which is often the greater effect.

II. SOUND LEVEL

A. Rate versus Level Functions to CF Tones

Increasing the level of a single tone above threshold causes a monotonic increase in the rate at which auditory nerve fibers discharge action potentials. For fibers with high rates of spontaneous activity and low thresholds, the discharge rate increases with level in a sigmoidal fashion, reaching a maximum or saturated discharge rate (Kiang et al., 1965; Sachs & Abbas, 1974; Palmer & Evans, 1980). The dynamic range (i.e., the range of levels from the threshold to the point of saturation) of the mean discharge rate (measured over tens of milliseconds) for the majority of nerve fibers is limited to about 40 dB (for fibers with spontaneous rates in excess of 15/s; Evans & Palmer, 1980). However, many fibers with low rates of spontaneous discharge do not fully saturate over the ranges of level used in most laboratories. Instead their discharge continues to increase with level at the highest levels, although the slope of the rate-level function is greatly reduced (Sachs & Abbas, 1974; Palmer & Evans, 1980). Examples of both these types of rate versus level function from the cat are shown in Figure 9. In the guinea pig there is a third type of function, for fibers with zero spontaneous rates, which shows no evidence of saturation (Winter, Robertson, & Yates, 1990). The different shapes of rate–level functions are well predicted by a sigmoid shaped saturating nonlinearity (the exact cause is unknown, but may be a limitation in the neurotransmission at the hair–cell synapse) following the nonlinear basilar membrane input–output function (Sachs & Abbas, 1974; Yates, Winter, & Robertson, 1990). One conclusion from these studies is that the parameter determining the shape of the rate–level function is not the spontaneous rate per se, but the threshold of the fiber relative to the basilar membrane input–output function.

B. The Dynamic Range Problem

The apparent disparity between the dynamic range measured physiologically for the majority of auditory nerve fibers and that measured psycho-

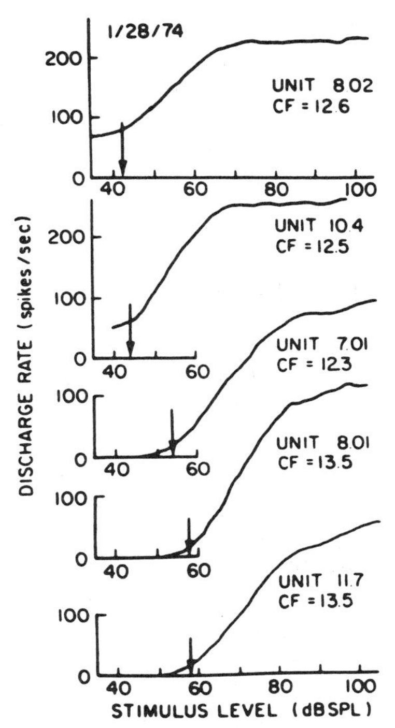

FIGURE 9 Rate-level functions (at CF) for five cochlear nerve fibers from a single cat. The fibers have CFs of 12.3–13.5 kHz. The arrows indicate the mean rate thresholds. (From Sachs and Abbas, 1974, with permission.)

physically has been extensively reviewed by a several authors (Evans, 1981; R. L. Smith, 1988; Irvine, 1986). More recently, there have been several theoretical studies of the ability of auditory nerve fibers to provide an adequate basis for psychophysically measured intensity difference limens. In these studies, optimal combination of the rate–level functions, from small groups of fibers of similar CF and with plausible distributions of threshold, has proven sufficient to account for human psychophysical performance, when the statistics of the discharge are taken into account (Viemeister, 1988; Delgutte, 1987; Winslow & Sachs, 1988). This issue is discussed more fully in Chapter 4.

C. Wider Dynamic Range of the Onset Response

Smith and his colleagues have explored the time course over which the saturating synaptic function limits the discharge rate. Noting that the receptor potentials of inner hair cells had wider dynamic ranges than the majority of auditory nerve fibers, they suggested that the transfer function of the hair cell synapse could be responsible for the extra compression (see Smith, 1988). If the rate versus intensity function is plotted only for spikes occurring over the first few millisecond of stimulation, the dynamic range of the function is considerably wider than that of the steady-state rate (measured over tens of milliseconds) and more closely resembles the input–output function of the hair cell receptor potentials (R. L. Smith & Brachman, 1980; Westerman & Smith, 1984). This wider dynamic range to transient stimuli is evident when using amplitude-modulated stimuli or speech sounds (see Sections III.A and IV.).

D. Rate versus Level Functions in the Central Auditory System

The rate–level function to a single tone is a vertical slice through the response area. At the cochlear nucleus the response areas become complicated with sideband or center band inhibition or both (type III and IV units, Figure 6), and the rate–level function therefore reflects the level dependence of this excitation and inhibition at any single frequency. Thus, although many units in the cochlear nucleus show monotonic sigmoidally shaped rate–level functions (primiarylike type I units and many type IIIs to CF tones), others show more or less severe reductions in discharge rate as increases in level evoke stronger inhibitory inputs (type IV, type II and type V, see Figure 6; Young, 1984; Young et al., 1988). Onset responding units in the VCN are characterized by wide response areas and by rate–level functions that show (in many cases) little indication of saturation and hence exhibit wide dynamic ranges. It has been suggested that both the broad tuning and the wide dynamic range of onset units result from integration of the inputs from auditory nerve fibers with different CFs (Godfrey, Kiang, & Norris, 1975; Bourk, 1976; Rhode & Smith, 1986a).

Most nuclei above the cochlear nucleus receive input from both ears, and the shape and dynamic range of rate–level functions of central neurons often depends on the relative level and spectral content of the signals at the two ears (see Section V.A.). It is nevertheless the case that, in nuclei throughout the auditory pathway, for both monaural and binaural stimulation, rate–level functions of all the types described previously have been reported: monotonic sigmoid, strongly nonmonotonic, and nonsaturating wide dynamic ranges. At the inferior colliculus and cortex, the peak firing rate in nonmonotonic rate–level functions occurs over a wide range of sound levels in different units. Thus the "best" sound level for a cortical neuron may be as low as 15 dB SPL or as high as 106 dB SPL (Ehret & Merzenich, 1988; Brugge & Merzenich, 1973; Pfingst & O'Connor, 1981). A range of such nonmonotonic functions from the cortex of an awake monkey performing a behavioral task is shown in Figure 10. To date, with the exception of the bat

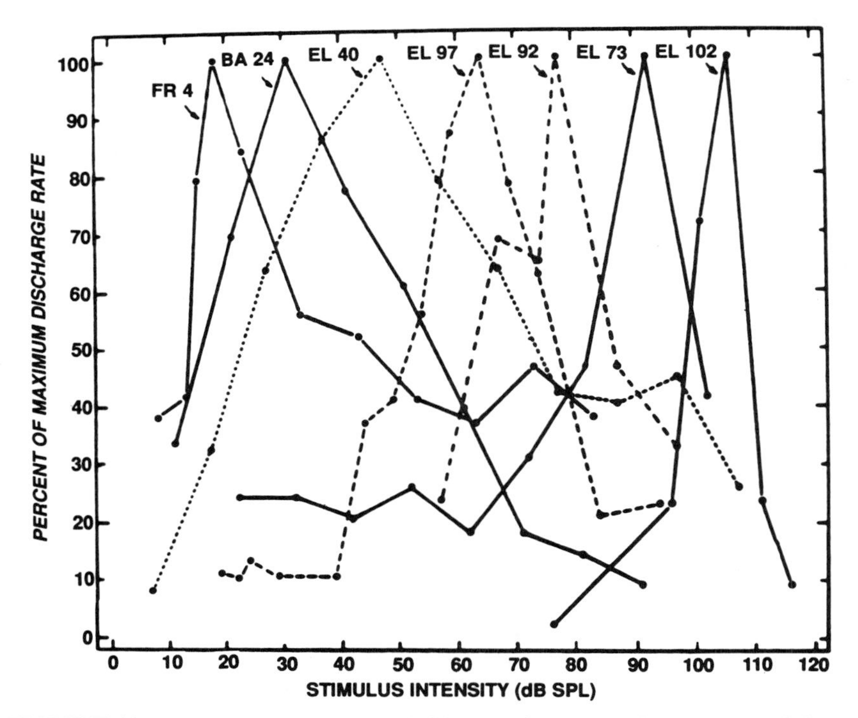

FIGURE 10 Nonmonotonic rate–level functions from the auditory cortex of the unanaesthetized monkey performing an auditory task. (From Pfingst and O'Connor, 1981, with permission.)

(Suga, 1988), there has been no indication of an orderly topographical distribution of best intensities in the cortex.

E. Effect of Background Noise on Rate–Level Functions

When rate–level functions of auditory nerve fibers to CF tones are measured in the presence of broadband noise, they are found to be shifted to higher sound levels (e.g., Costalupes, Young, & Gibson, 1984). As the noise background is progressively increased in level, the baseline discharge rate increases, the saturated discharge decreases, and the operating range shifts to higher level. All of these effects may be seen in Figure 11. The shift in the threshold appears to be a result of competition between the noise and the tone for capture of the fiber's activity (Young & Barta, 1986; Rhode, Geisler, & Kennedy, 1978). The shift in the level at which saturation occurs is

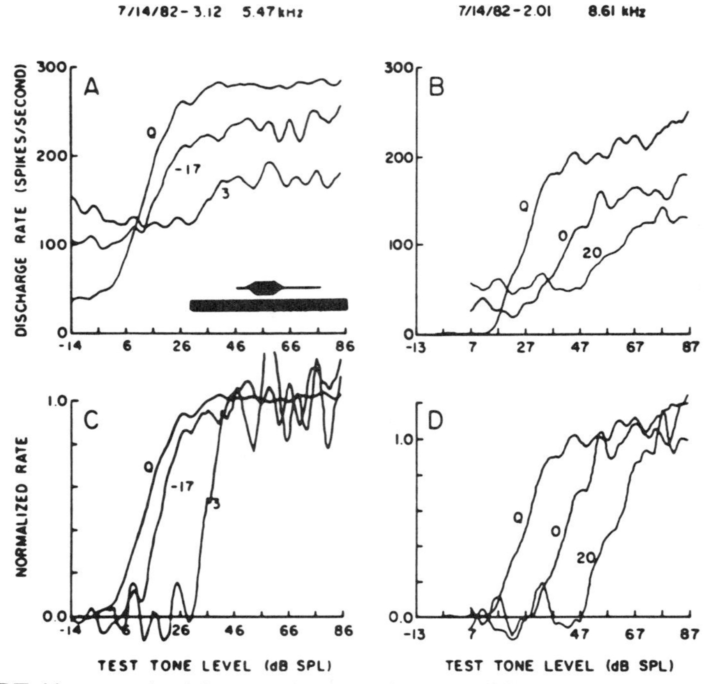

FIGURE 11 Rate–level functions for two auditory nerve fibers in response to CF tones in quiet (Q) and in the presence of various levels of continuous background noise. A and B show the measured functions and C and D show the same functions normalized, by subtracting the baseline rate and dividing by the saturation rate, to emphasize the shift along the ordinate produced by the noise (at the spectrum levels indicated in dB re 20 μPa). (From Costalupes et al., 1984, with permission.)

due to suppressive effects in the cochlea (see Costalupes et al., 1984; Costalupes, Rich, & Ruggero, 1987). The raised baseline is due to the activation of the fiber by the noise signal, and the reduced saturation rate is due to the adaptation caused by a continuous noise (Costalupes et al., 1984). The net effect of all of these is a shift to higher levels of the rate–level function by 0.6 dB for each 1 dB increase in noise level and a compression of the rate–level function. Because of the compression, at high sound levels increments in level will be signaled by relatively small increases in discharge rate despite the shift. Winslow and Sachs (1988) have shown that the compression of the rate–level function in background noise can be reduced by stimulating the olivocochlear bundle. Thus the reduced ability of auditory nerve fibers to signal intensity increments in the presence of noise, resulting from the rate–level function compression, may not be as profound in the awake behaving animal (with active descending systems) as in the anaesthetized preparation. Consistent with this suggestion are the recent data of May, Aleszczyk, and Sachs (1991), who showed less reduction of the saturated discharge in the ventral cochlear nucleus of the awake animal.

A shift to higher sound levels of rate–level functions measured in noise has also been demonstrated in the cochlear nucleus (Gibson, Young, & Costalupes, 1985). For most units in the VCN and DCN the shift was similar to that found in the cochlear nerve. However, for some DCN units (type IV) the shift was close to 1 dB for a 1 dB increase in noise level, indicating an additional contribution from inhibitory sidebands. At both inferior colliculus and cortex, although the rate intensity functions also shift by 1 dB for each 1 dB increase in noise level, the compression of the rate–level function is less extreme than in the nerve: the noise does not itself drive the units so there is no increase in the baseline, and the reduction of the saturated firing rate is smaller (Rees & Palmer, 1988; Phillips & Cynader, 1985). The shifts were found to be the same for monotonic and nonmonotonic rate–level functions at the cochlear nucleus and in the higher nuclei.

III. MODULATION

A. Amplitude Modulation

Naturally occurring sounds are characterized by more-or-less rapid changes in their amplitude and spectral content. Speech sounds, for example, exhibit a range of fluctuation rates in the amplitude of their envelope, corresponding to the different speech segments such as syllables, words, and sentences. Steady-state harmonic sounds (such as voiced vowels) can also produce amplitude modulation as a result of interaction between the harmonics passing through a single cochlear nerve fiber filter (see Section IV). To investigate sensitivity to dynamically varying sounds, investigators have more often employed simpler signals such as sinusoidally amplitude modulated tones (whose spectrum consists of a carrier frequency and two sidebands separated from the carrier by a frequency equal to the modulation rate; see Chapter 1) and have described the ability of the neural discharge to signal the modulation as a function of stimulus parameters such as modulation frequency, depth of modulation, mean sound level, and carrier frequency. In general, auditory neurons throughout the auditory pathway are able to signal amplitude modulations as a modulation of their discharge. However, the range of modulation frequencies over which they are able to do so decreases from the periphery to the cortex (see Rees & Moller, 1983; Schreiner & Langner, 1988a). A useful summary of the neural sensitivity is given by the modulation transfer function (MTF), which plots either the degree of response modulation (assessed from period histograms) or the total response as a function of the modulation rate. The degree of neural modulation is quantified as the ratio of the depth of stimulus modulation and the depth of response modulation, and it is expressed as a gain in decibels (0 dB indicates equal modulation depth in stimulus and response while ±6 dB indicates response modulations double or half that in the stimulus, etc.).

1. Amplitude Modulation Sensitivity in Auditory Nerve Fibers

MTFs of auditory nerve fibers are low-pass functions (see for example Figure 12(a)) irrespective of modulation depth. Slopes in the passband rarely exceed 1 dB/octave, while the cutoff has an initial slope of about −12 dB/octave between the 3 and 10 dB points, beyond which the slope increases considerably (Palmer, 1982; Frisina, Smith, & Chamberlain, 1990; Kim, Sirianni, & Chang, 1990; Joris & Yin, 1992). Variation in the rate of modulation has little effect on the mean discharge rate. The modulation of the discharge is maximal at about 10 dB above the rate threshold and declines as the fiber is driven into saturation, as can be seen in Figure 12(a) (Frisina et al., 1990). The sound level at which the maximum modulation occurs (for low modulation depths) is better predicted by the rate–level function measured over a few milliseconds at tone onset (see Section II.C) than by that for steady-state tones (R. L. Smith & Brachman, 1980). The cutoff frequency of the MTF is dependent upon the fiber CF, which probably reflects the attenuation of the signal sidebands by cochlear filtering, but increases in fiber bandwidth beyond 4 kHz (in fibers with CFs above 10 kHz) are not accompanied by increases in MTF cutoff frequency, thus implying some additional limitation on response modulation in these fibers (Palmer, 1982; Joris & Yin, 1992). The maximum frequency at which the MTFs are 3 dB down for auditory nerve fibers is on the order of 1500 Hz (Palmer, 1982; Kim et al., 1990; Joris & Yin, 1992).

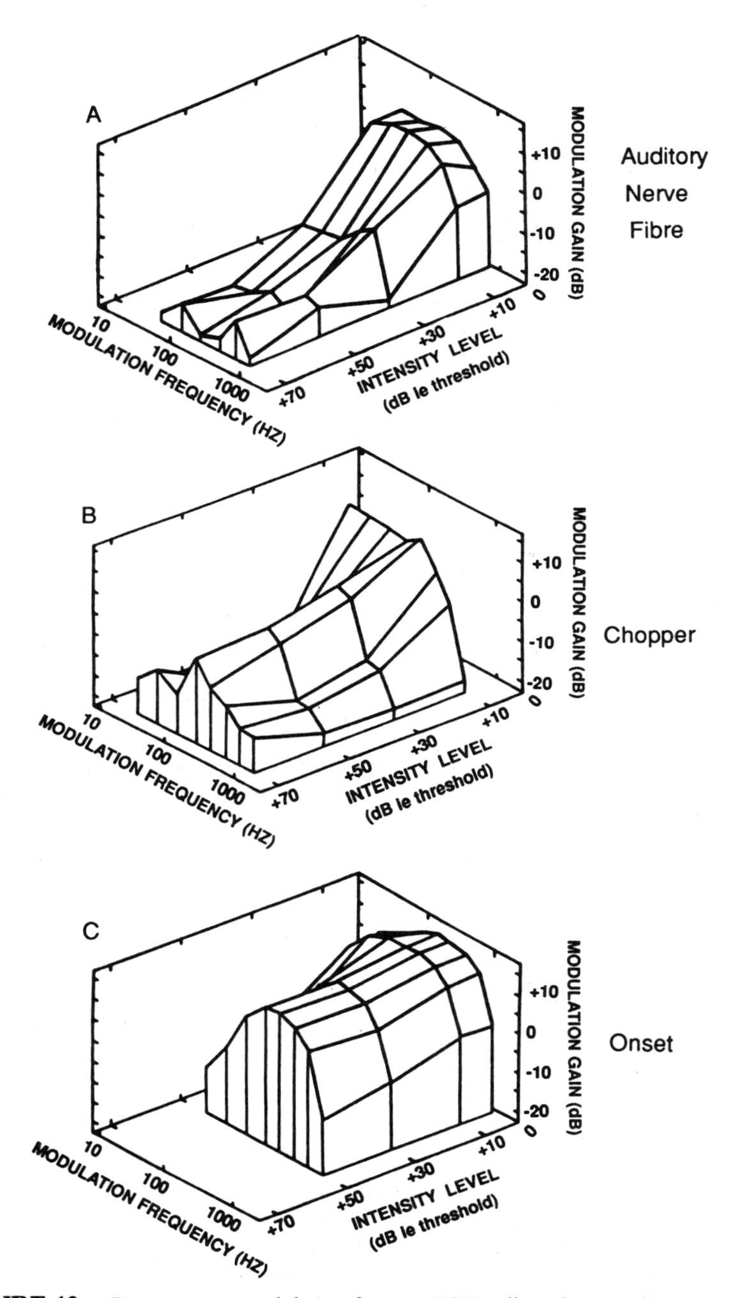

FIGURE 12 Responses to modulation for two VCN cells and one auditory nerve fiber. The carrier frequency in all cases was the CF: (A) auditory nerve fiber, CF = 6.5 kHz; (B) chopper unit, CF = 9.9 kHz; (C) onset-L unit, CF = 12 kHz. (Data from Frisina et al., 1990; figure from Smith, 1988, with permission.)

2. Amplitude Modulation Sensitivity in the Cochlear Nucleus

The pioneering studies of Møller (e.g., 1972, 1977) established that the MTFs of cochlear nucleus neurones in the rat are often bandpass functions showing considerable amplification near the peak in the response modulation relative to the signal modulation. These early studies did not specify the exact location or response type of the units, but the findings have been repeatedly confirmed in more recent studies. Frisina et al. (1990), for example, have demonstrated that in the VCN the degree of enhancement of the discharge modulation is different for units of different response classifications. Such differences are illustrated in Figures 12(b) and 12(c) for chopper and onset units, respectively. Figure 12(c) shows the response to amplitude modulation of tones at the CF of an onset unit as a function of both modulation rate and sound level. Near threshold the MTF is low-pass in shape with a passband gain of more than 10 dB (i.e., the discharge is about three times more modulated than the stimulus). As the sound level is increased, the MTF becomes bandpass in shape with little or no decrease in the gain at the peak for levels up to 90 dB above threshold. For the chopper unit shown in Figure 12(b), the MTF at low sound level is again low pass, becoming bandpass at higher levels, but the modulation gain is severely reduced at the higher levels, even at the peak of the MTF. The auditory nerve fiber responses shown in Figure 12(a) indicate a low-pass MTF at low stimulus levels, with a passband gain that is low and further decreases as level increases. Frisina et al. suggested that the ability to encode amplitude modulation (measured by the amount of gain in the MTF) is best in onset units followed by choppers, primarylike-with-a-notch, and finally primarylike and auditory nerve fibers.

The peak of the MTF function is often referred to as the best modulation frequency (BMF). In the study of Frisina et al. in VCN, the BMFs varied over different ranges for the various unit types: 180–240 Hz for onset units, 120–380 Hz for primiarylike-with-a-notch, 80–520 Hz for choppers, and 80–700 Hz for primarylike units. Changes in the MTF shape, from low pass at low sound levels to bandpass at high levels (with BMFs ranging from 50–500 Hz), have also been reported for units in the DCN of unanaesthetized, decerebrate cats (Kim et al., 1990).

3. Amplitude Modulation Sensitivity in the Inferior Colliculus

The MTFs of units in the inferior colliculus are also low-pass at low sound levels, becoming bandpass at high sound levels, with BMFs generally lower than those in the cochlear nucleus. In both rat and guinea pig the BMFs are less than 200 Hz, but in the cat, although BMFs of the majority of units are below 100 Hz, BMFs of 300–1000 Hz are also found (Rees & Møller, 1983; Langner & Shcreiner, 1988; Rees & Palmer, 1989). At the cochlear nucleus the MTF computed from the synchronized responses shows tuning to modulation rate without any corresponding variation in the mean discharge rate. In the inferior colliculus, the mean discharge rate of many units also varies with the modulation rate. Thus MTFs determined from the synchronized activity *or* the mean discharge rate are similar and thus a significant recoding of the modulation information has occurred (Langner & Schreiner, 1988; Rees & Palmer, 1989).

The most striking difference between the modulation sensitivity in the inferior colliculus and that at reported at lower levels is in the topographical distribution of the units with different BMFs (Schreiner & Langner, 1988b). This topographical distribution is illustrated in Figure 13, in which is shown the distribution of BMFs across the 3 and 12 kHz iso-frequency laminae. The 3-D plots in Figure 13 show the BMF as a function of position in an iso-frequency lamina and the 2-D plots are the contour representations derived from these data. Such topographical distributions of best modulation frequencies have only been found in the cat inferior colliculus, but it would be surprising if such an organization were not ubiquitous across higher animals.

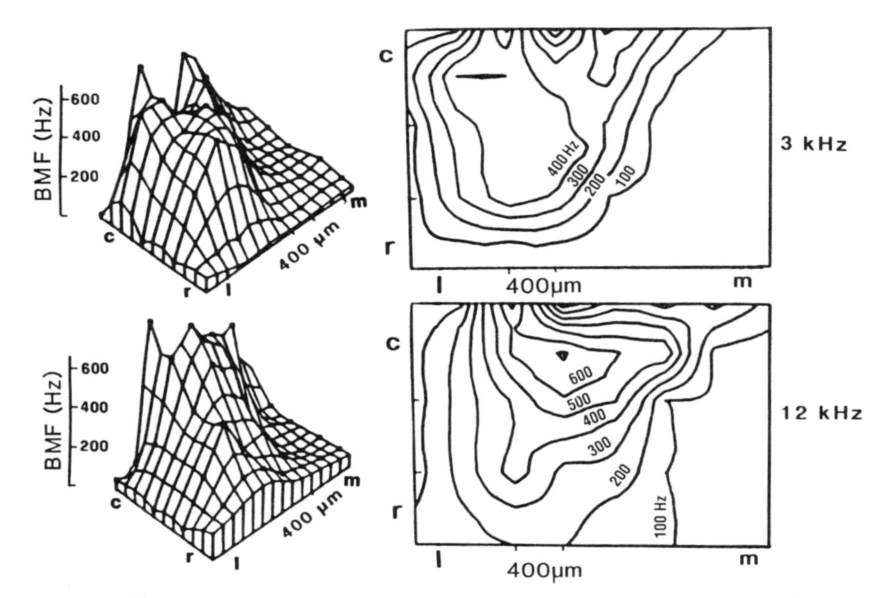

FIGURE 13 Representation of the best modulation frequency within two iso-frequency laminae of the central nucleus of the inferior colliculus: *c, r, l, m* indicate the caudal, rostral, lateral, and medial directions. (top) 3 ± 0.5 kHz lamina, (bottom) 12 ± 1 kHz. Increment for iso-best-modulation frequency contours is 100 Hz. (From Schreiner & Langner, 1988b, with permission.)

4. Amplitude Modulation Sensitivity in the Auditory Cortex

The majority of neurons in the auditory cortex are unable to signal envelope modulation at modulation rates of much more than 20 Hz and, although there is no topographic organization with respect to the BMF, the several divisions of the auditory cortex have different distributuons of BMFs (Whitfield & Evans, 1965; for a review, see Schreiner & Langner, 1988a).

B. Frequency Modulation

Auditory nerve fibers respond to frequency-modulated tones (swept tones) in ways that are generally predictable from their responses to stationary tones (Britt & Starr, 1975; Sinex & Geisler, 1981), that is, in a manner determined by their frequency selectivity, with modification by saturation and adaptation effects. The fibers respond to the short-term frequency values, which fall within their response areas, not to the long-term spectral characteristics. The direction of frequency change has little effect on the responses other than a shift in the frequency evoking the maximum firing rate in a manner consistent with adaptation by earlier components of the sweep.

In many cases, the responses of cochlear nucleus neurons to frequency-modulated tones are also consistent with their responses to stationary tones. However, gross asymmetries have been reported in the responses of some cochlear nucleus neurones to upward and downward frequency sweeps, which were often related to asymmetry in the inhibitory regions of the response area (Evans, 1975; Britt & Starr, 1975; Rhode & Smith, 1986a, 1986b). The responses in cochlear nucleus to frequency sweeps show a tuning for the rate of frequency change, producing maximum responses to frequency sweeps changing at 10–30 Hz/s (Moller, 1977). When a carrier at the unit CF is sinusoidally frequency modulated by a *small* amount, analyses can be applied to produce an MTF for frequency-modulated signals, which in many cases appears qualitatively and quantitatively similar to that produced by amplitude modulation of a CF carrier (i.e., having BMFs in the cochlear nucleus in the range 50–300 Hz; Moller, 1972). The tuning to the rate of frequency modulation shows changes similar to the tuning for amplitude modulation rate in the different nuclei of the auditory pathway. Thus MTFs in the inferior colliculus are bandpass functions, but their BMFs are lower than at the cochlear nucleus (below 80 Hz; Rees & Moller, 1983), and cells in the primary auditory cortex are sensitive to still lower rates of frequency modulation (below 15 Hz; Whitfield & Evans, 1965). The degree and variety of asymmetries in the response to upward and downward frequency transitions increases from inferior colliculus to cortex (Nelson, Erulkar, & Bryan, 1966; Whitfield & Evans, 1965). Nevertheless, even at the cortex there are neurons whose responses to frequency-modulated signals are largely predictable from their responses to stationary tones. Some cortical units show responses to frequency-modulated tones even though they do not respond to steady tones, while others respond to frequency sweeps that are entirely outside the unit's response area determined with steady tones. For many cells, only one direction of frequency sweep was effective irrespective of the relationship of the sweep to the cells' CF (Whitfield & Evans, 1965). Phillips, Mendelson, Cynader, and Douglas (1985) also found sensitivity to sweep direction in primary auditory cortex, but the preferred sweep direction (for relatively narrow sweep excursions; 2 kHz) was toward the CF, and some part of the sweep had to be within the pure tone response area. They also reported profound directional sensitivity to frequency sweeps covering a wide range and concluded that the mechanisms responsible for the sensitivity to sweep direction were different for wide and narrow sweeps.

IV. SPEECH AND VOCALIZATION

A. Representation of Speech Signals in the Auditory Nerve

Naturally occurring communication signals such as speech are characterized by a high degree of complexity in terms of their spectral richness and their variation as a function of time. As we have seen in Section I.C, the auditory nervous system is organized cochleotopically, and it is appropriate with spectrally rich sounds to consider the activity across homogeneous cochleotopic continua rather than dwell on the details of individual neuron responses.

I begin here with the activity evoked in the fibers of the auditory nerve by one of the simpler speech sounds, a steady-state voiced vowel. Vowels are often relatively stable and periodic even in natural utterances and can be made completely so in synthetic tokens, which have been widely used as stimuli. Their spectrum consists of harmonics of the voice fundamental frequency (F0: in the range of 80–400 Hz) some of which have greater amplitude, producing peaks (formants) that correspond to the resonant frequencies of the vocal tract (an example is shown for /ɛ/ in Figure 14(a)). The perceptual identification of such vowel sounds depends on the frequencies of the formants. A major issue here is whether the pattern of gross neural activity evoked at "places" within cochleotopically organized populations of neurons is sufficient to signal the vowel identity or whether the fine timing (phase locking) of the discharges is also important. This issue has been dealt with in detail elsewhere and is therefore recounted here only briefly (see Sachs, 1984).

Presentation of a voiced vowel sound at moderate levels (i.e., levels insuf-

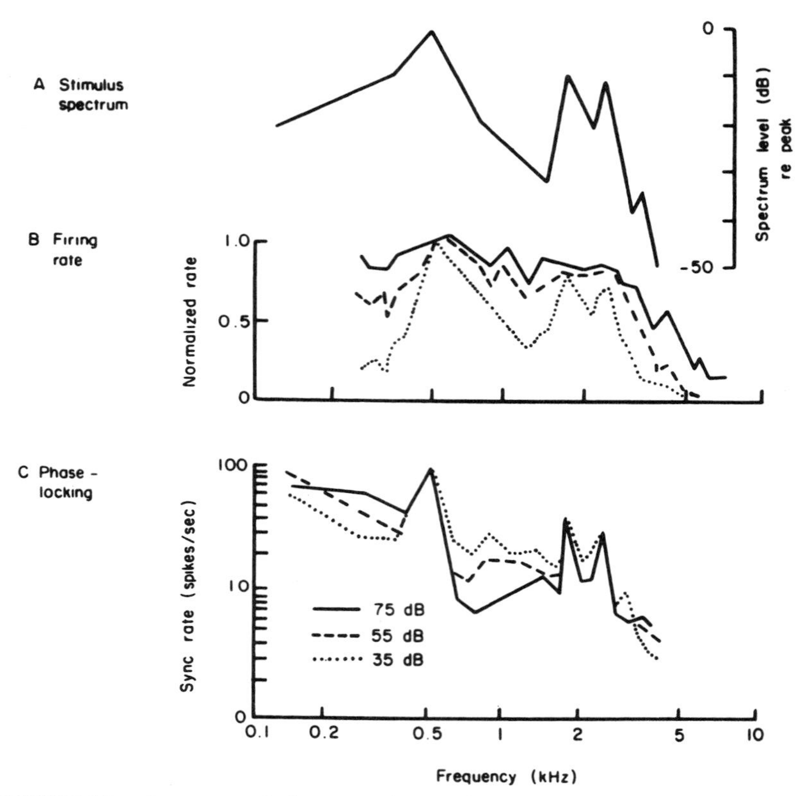

FIGURE 14 Responses of a large population of auditory-nerve fibers to the vowel /ɛ/. (A) The spectral envelope of the vowel. (B) The distribution of mean discharge rates of the high spontaneous rate fibers. The lines are a moving-window average of the mean rates of a population of individual fibers in response to the vowel at the levels indicated. (C) The ALSR (see text) function for the same population of fibers in response to the same vowel at the three sound levels. (From Pickles, 1981, adapted from Sachs, Young, Schalk, & Bernardin, 1980, with permission.)

ficient to cause saturation of the fiber discharge by any but the strongest components, see Section II.A) evokes more discharges in fibers with CFs near the formants than in those with CFs away from the formants, as can be seen in Figure 14(b), which shows moving-window averages of the mean discharge rate evoked by the vowel /ɛ/ in a large population of fibers in a single auditory nerve (the vast majority of these fibers are low-threshold, high spontaneous rate fibers). At the lowest stimulus level the frequency positions of the first two or three formants are clearly signaled by regions of increased discharge. However, at the higher presentation levels the fibers with CFs between the formants increase their discharge, while the fibers at the formant frequencies reach saturation, causing the formant-related peaks to become obscured (an additional factor in this loss of definition is rate suppression of fibers with CFs above the first formant by the energy at the first formant; Sachs & Young, 1979). Since human vowel identification is unchanged at the highest sound levels used for Figure 14(b), it is tempting to conclude that the distribution of mean discharge rates is inadequate as an internal representation of the vowel. This is too simplistic for a number of important reasons. First, these plots of mean discharge rate include only the fibers with high rates of spontaneous discharge (and hence low thresholds and narrow dynamic ranges, see Sections I.A and II.A). If a similar plot is made for fibers with low spontaneous discharge (fewer in number but having higher thresholds and wider dynamic ranges) formant-related peaks are still discernible at the highest levels used (see Young & Sachs, 1979). Second, the mean rates shown here are for steady-state vowels; the wider dynamic range at onset (see Section II.C) provides some extension to the range over which the mean rates signal the formant frequencies (Sachs, Young, & Miller, 1982). Third, the data have been collected in anaesthetized animals and it is possible that the action of various feedback pathways (the middle ear muscles and the efferents to the cochlea) may affect the fibers' activity (see Section II.E), preserving their ability to signal the formant peaks in their discharge rate at high sound levels. Finally, even at the highest levels, at which the formant structure is no longer evident, the mean rate distribution is, nevertheless, different for different vowels, and discrimination could be made on the basis of the gross mean rate profile (Winslow, 1985). However, when mean discharge rates to vowels are measured in the presence of background noise (at levels that do not prevent detection of changes in the second formant frequency), neither the onset rates nor the low-spontaneous-rate fibers seem capable of sustaining an adequate mean-rate representation of the formant structure (Sachs, Voigt, & Young, 1983). This result would seem to present a severe problem for any simple place coding scheme.

The distribution of mean discharge rates takes no account of the temporal patterning of the impulses. Since the spectra of voiced vowels are largely restricted to the frequency range below 5 kHz, responses of individual nerve fibers are phase locked to components of the vowel within their response area (see Section I.E). It is a general finding that fibers with CFs near a formant are phase locked to the harmonic nearest to the formant peak. The periodicity of this harmonic dominates the temporal response of the fibers, excluding phase locked responses to other weaker components (Young & Sachs, 1979; Sinex & Geisler, 1983; Delgutte & Kiang, 1984a; Palmer, Winter, & Darwin, 1986). Fibers with CFs remote from formant frequencies at low (below the first formant) and middle (between the first and second formants) frequencies are dominated either by the harmonic closest to the fiber CF or by modulation at the voice pitch, indicating a beating of the harmonics of the voice pitch that fall within their response area. At CFs above the second formant, the discharge is dominated either by an intense harmonic near the second formant or again by the modulations at the voice pitch caused by interactions of several harmonics.

A useful summary of the pattern of phase locking within the population of auditory nerve fibers has been developed, which allows more direct comparison with the stimulus spectrum and the mean rate distributions (Young & Sachs, 1979; Delgutte, 1984). The first stage in these analyses is the construction of histograms of the fiber responses to the vowel sounds, which are Fourier transformed to provide measures of the phase locking to individual harmonic components. These analyses revealed that phase locking to individual harmonics occurred at the appropriate "place," that is, in fibers with CFs close to the harmonic frequency, and that, as level increased, phase locking to the intense harmonics near the formant peaks spreads from their place to dominate fibers of other CFs and suppress the responses to the weaker harmonics in those fibers. By forming an average of the phase locking to each harmonic in turn, in fibers at the appropriate place for the harmonic, Young and Sachs (1979) were able to compare the amount of temporal response to the various harmonics of the signal. The "average localized synchronized rate" (ALSR) function so derived for the vowel /ɛ/ is shown in Figure 14(c) for a series of sound levels. The similarity of the functions in Figure 14(c) to the spectrum of this vowel is evident, as is the fact that this form of representation (which combines phase locking, cochlear place, and discharge rate) is robust and retains well-defined peaks at the formant-related frequencies at high stimulus levels. This internal representation is unaffected by background noise (Sachs et al., 1983; Delgutte & Kiang, 1984d), can also be computed for unvoiced vowels (Voigt, Sachs, & Young, 1982), and preserves the details of the spectra of two simultaneously presented vowels with different F0s (Palmer, 1990). It is salutary to remember, however, that, as yet, no evidence suggests that mechanisms in the central nervous system can use (or transform) the information about the vowel spectrum contained in the variation in phase locking across the population of auditory nerve fibers.

The other major group of speech sounds is the consonants (which include fricatives, e.g., *s, sh;* stops, e.g., *p, d;* and nasals, e.g., *m, n*), which are diverse in spectro–temporal terms and often time varying. In general, the major spectral components of nasal and voiced stop consonants are well represented in the temporal patterning (phase locking) of populations of auditory nerve fibers (Miller & Sachs, 1983; Sinex & Geisler, 1983; Delgutte & Kiang, 1984c; Deng & Geisler, 1987). Additionally, the mean discharge rates are able to signal the formant positions during transitions (Miller & Sachs, 1983; Delgutte & Kiang, 1984c), at sound levels where the mean rate distributions to vowels do not have formant-related peaks. Since, first, transitions in consonants are relatively brief and occur at the start of a syllable, and second, they entail changes in frequency and therefore excite a succession of different CF fibers, the place representation of consonants over a wide dynamic range is presumably a result of the wider onset dynamic range (see Section II.C). The voiceless fricative consonants are generally distinguished by the frequency position of a single broad band of energy, which results in a distinctive distribution of mean discharge rate across the cochleotopically ordered array of the most sensitive auditory nerve fibers. The frequency range in which the mean rates are highest corresponds with the regions of maximal stimulus energy (Delgutte & Kiang, 1984b). One reason why this scheme is successful, in this instance, is that the levels of fricatives in running speech are low compared to those of vowels. Processing schemes based on distribution of temporal patterns were less successful for fricatives, because the energy in most fricatives is above the limit of phase locking (Delgutte & Kiang, 1984b). Delgutte and Kiang (1984c) also investigated the effect of the speech context in which the consonant–vowel syllable /da/ is placed and found context dependent changes in both the temporal and mean rate measures, with the major effects limited to those frequency regions in which the context had considerable energy. While the average rate profile was radically altered by the context, the major components of the synchronized response were little affected.

A potential cue for the voice pitch is the modulation of the fiber discharge at the F0, in fibers not dominated by a single, strong stimulus component. However, this cue is not robust and is reduced or eliminated for the majority of fibers by the presence of background noise (Miller & Sachs, 1984; Delgutte & Kiang, 1984d; Palmer & Winter, 1992). A second cue arises from the fact that the phase locking of auditory nerve fibers occurs only at the frequencies of harmonics of the vowel F0. Thus, if the temporal response of the population of auditory nerve fibers is computed at high resolution, peaks are found at each of the harmonic frequencies, which are resistant to noise and which can provide the basis for a spectral computation of F0 (see Chapter 8). This type of analysis can track the change in F0 during formant transitions and distinguish two simultaneously present F0s (Miller & Sachs, 1983; Palmer, 1990). Profiles of mean discharge rate do not appear to contain information related to the voice pitch (Miller & Sachs, 1983).

B. Representation of Speech Signals in the Central Nervous System

Whether the mean discharge rate or the timing of the impulses constitute the means by which auditory nerve fibers signal important speech elements, the neurons of the cochlear nucleus must either faithfully transmit the information or perform some kind of transformation. Recent studies have measured the responses to speech sounds for the different unit response types in the cochlear nucleus (Palmer et al., 1986; Blackburn & Sachs, 1990; Palmer, Winter, & Stabler, 1993; Kim, Rhode, & Greenberg, 1986; Kim & Leonard, 1988; Palmer & Winter, 1992). Only the spherical bushy cells in the VCN faithfully transmit the temporal activity and show population temporal responses (quantified by ALSR functions) similar to those in the nerve (Blackburn & Sachs, 1990). A most intriguing finding is that the distribution of mean discharge rates across a population of chopper units exhibits peaks at the positions of the formants even for sound levels at which such peaks are no longer visible in the responses of the high spontaneous rate auditory-nerve fibers (Blackburn & Sachs, 1990). For vowels at low stimulus levels the mean-rate profiles of choppers resemble the near-threshold profiles of high spontaneous auditory nerve fibers, and at high sound levels they resemble the profiles of low spontaneous rate fibers. This led Sachs and his colleagues to suggest that the choppers respond selectively to high spontaneous rate auditory nerve fibers at low sound levels and to low spontaneous rate fibers at high sound levels.

While the discharge of all of the unit types in the cochlear nucleus in response to speech sounds is modulated at the F0 (Kim & Leonard, 1988), the response of onset units is so precisely locked to the F0 that Kim and his colleagues (1986, 1988) have described it as "pitch-period following." All evidence points to the conclusion that this precise locking to the F0 is achieved by a coincidence detection mechanism following a very wide convergence across frequency. The output of the onset units appears to be consistent with the perceived pitch for a wide range of signals (Palmer & Winter, 1992). Units in the dorsal cochlear nucleus do not phase lock well to single sinusoids, and thus no temporal representation of the spectrum of speech sounds is expected here. However, the disposition of strong inhibitory sidebands and the asymmetry of responses to frequency sweeps (Sections I.D and III.B) could provide some basis for differential responses to consonants in which the formant transitions sweep across the response areas of the units. The only detailed study of the responses of identified dorsal cochlear nucleus neurones to consonant–vowel syllables (/ba/, /da/, and /ga/) failed to detect any specific sensitivity to the particular formant transitions used, over and above a linear summation of the excitation and inhibition evoked by each of the formants separately (Palmer et al., 1993).

In nuclei more central than the cochlear nucleus, the use of speech stimuli has been limited to date. At the level of the inferior colliculus the responses to speech sounds are context dependent and consist of discharges locked to the F0 irrespective of the CF or response type (Watanabe & Sakai, 1978). Multiunit activity in the cortex in response to speech signals suggests that both low- and high-frequency units respond at the F0, but also that the distribution of activity reflects the energy at the formant frequencies with nonlinear combination of the activity evoked by each formant separately (Steinschneider, Arezzo, & Vaughan, 1990). Detailed studies of the responses of monkey cortical cells to conspecific vocalizations suggest that, rather than responding to the spectra of the sounds, the cells follow the time structure of individual stimulus components in a very context dependent manner. The specificity of some cells for particular vocalizations may result from overlap of the spectra of transient parts of the stimulus with the neuron's response area (for a review, see Phillips et al., 1991).

V. CUES FOR LOCALIZATION

The cues used for localizing a sound source derive from the fact that we have two ears that possess pinnae and are separated over a significant distance by an acoustically opaque medium. The result is that the sounds arriving at the two ears are characterized by an interaural delay caused by the longer sound path to one ear and a difference in level caused by the shadowing effect of the head combined with spectral alterations within the pinnae. The pinna effects and the head shadowing are minimal for low-frequency sounds for which the wavelength is longer than the head (or pinna) width. For such low-frequency tones the time difference is manifested as an interaural phase difference (IPD). For tones of wavelength shorter than the head width, the IPD presents an ambiguous cue, but interaural level differences (ILDs) may be as much as 20 dB. For high-frequency tones, the interaural time delay can be extracted from the time of arrival of the first wavefront or from the ongoing delay of the envelope of complex sounds. The psychophysical investigation of these localization cues is covered in detail in Chapter 9.

Anatomical and behavioral evidence suggests that the processing of interaural time and level differences involves separate pathways (see Yin & Chan, 1988; and Irvine, 1986 for detailed reviews). These pathways begin with the projections of different cell groups in the ventral cochlear nucleus to the superior olivary complex, which is the first major site of convergence of activity from the two ears. At levels above the superior olive, responses to binaural stimulation reflect mainly these first binaural interactions (Yin & Chan, 1990), but some further elaboration of responses and binaural convergence may also occur (e.g., Semple & Aitkin, 1979). There is space here only for a cursory treatment, but copious detailed reviews of the neural coding of interaural cues for localization are available (Phillips & Brugge, 1985; Yin & Chan, 1988; Irvine, 1986; Caspary & Finlayson, 1991; Tsuchitani & Johnson, 1991; Phillips et al., 1991).

A. Interaural Level Differences

The lateral superior olive (LSO), which is innervated mainly by high-frequency neurons, is the brainstem nucleus in which most of the initial processing of interaural level differences takes place. The principal cells of the LSO receive excitatory inputs from the ipsilateral VCN and inhibitory inputs from the medial nucleus of the trapezoid body of the same side, which in turn receives excitatory input from the contralateral VCN (see Cant, 1991). The pathway from the contralateral VCN is characterized by large synaptic endings that result in very secure short latency responses and therefore near coincident arrival of the excitation from the ipsilateral VCN and the indirect inhibition from the contralateral VCN. The majority of LSO cells with CFs above 1 kHz receive IE binaural input (I = contralateral inhibition, E = ipsilateral excitation). IE cells in the LSO are sensitive to the balance of intensity at the ears (and hence the ILD) as can be seen from Figure 15(a), in which the excitation resulting from an ipsilateral CF tone is reduced by increasing levels of a contralateral CF tone (Boudreau & Tsuchitani, 1968, 1970; Guinan et al., 1972). The form of this curve as a function of ILD is a sigmoid, varying from the excitatory response to the ipsilateral tone alone to complete inhibition by the contralateral tone.

The high-frequency part of the LSO projects to the contralateral inferior colliculus (IC) and the sensitivity of cells in the IC (and above) to ILDs is therefore the mirror image of that in LSO; that is, they generally have EI type responses (Rose, Gross, Geisler, & Hind, 1966; Semple & Aitkin, 1979; Caird & Klinke, 1987; Yin, Kuwada, & Sajaku, 1984). The slope of the ILD function and the ILD at which the inhibition takes effect vary across the different cells in the IC even for stimulation with best frequency tones, as can be seen in Figure 15(b). Similar sensitivities of EI cells to ILDs are found in the primary auditory cortex (for a review, see Phillips et al., 1991).

At the level of the IC and cortex, sensitivity to ILDs of cells excited by both ears has been reported (EE cells) and even of cells only responsive to binaural stimuli. The form of this sensitivity is a sharply peaked non-monotonic curve with a maximum at zero interaural level difference (see Benevento, Coleman, & Loe, 1970; Semple & Aitkin, 1979).

B. Interaural Phase Differences

The initial processing of IPDs takes place in the principal cells of the medial superior olive (MSO, a predominantly low-frequency sensitive nucleus), which receive excitatory input from the large spherical bushy cells in the rostral pole of the ventral cochlear nucleus of each side (see Cant, 1991). Unfortunately, this nucleus is notoriously difficult to study electrophysiologically, and our knowledge of its action is therefore based on only a few studies, each of which included only relatively small samples of MSO neurons, and some of which did not equivocally identify the recording sites within the MSO. Nevertheless, all the evidence we have to date suggests that the MSO cells are performing a coincidence detection between the

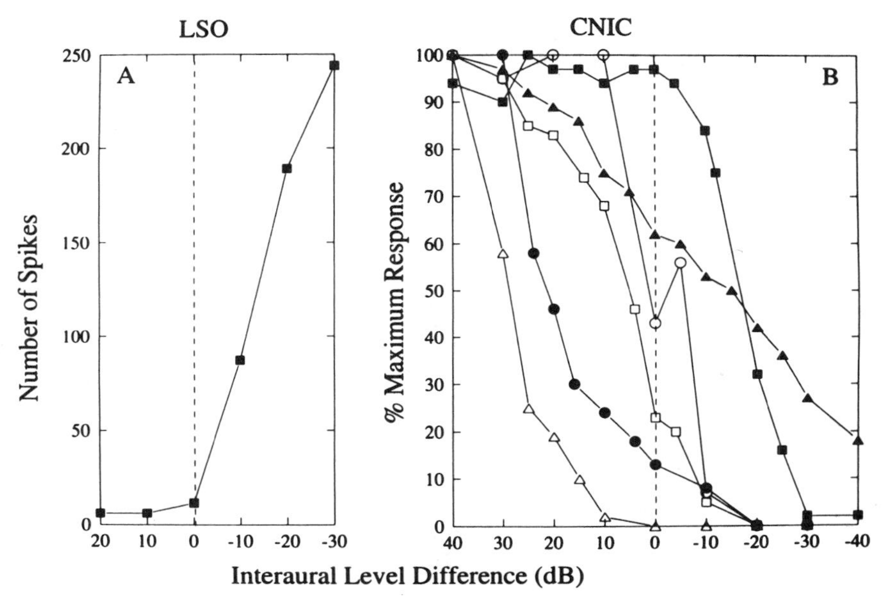

FIGURE 15 (A) The response of a principal cell in the lateral superior olive to variations in the level of a contralateral CF tone. As the level of the contralateral tone is increased, the response is progressively inhibited often becoming totally inhibited when the tones to the two ears are of equal level. (From Caspary & Finlayson, 1991, with permission.) (B) Responses of six neurons in the central nucleus of the inferior colliculus (CNIC) to variations in the level of a contralateral CF tone. Notice the variation in the value of interaural level difference at which the response switched from excitation to inhibition. (From Irvine, 1986, with permission.)

excitatory inputs from each ear (as originally proposed by Jeffress, 1948; see Goldberg & Brown, 1969; Yin & Chan, 1988, 1990, and Chapter 10). The principal cells of the MSO are *insensitive* to the interaural level difference or the onset time delay. The responses of a cell in the MSO as a function of the IPD of a CF tone are illustrated in Figure 16(a). (This is a more recent demonstration of the classical results of Goldberg & Brown, 1969.) Notice first that the discharge rate for monaural stimulation from either ear, indicated by the arrows marked *C* for contralateral and *I* for ipsilateral in Figure 16(a), is only on the order of about 50 spikes/s. As the delay between the tones at each ear is varied, the discharge rate is greatly facilitated (350 spikes/s: much more than the sum of the monaural responses) at some delays and is inhibited at others (near zero spikes/s). The response cycles at the frequency of the stimulus, indicating that it is sensitive to the relative phase between the two ears. The period histograms shown in Figure 16(b) indicate that the cell phase locks to 1000 Hz tones to either the ipsilateral (top panel) or contralateral (bottom panel) ear alone. However, the phase of

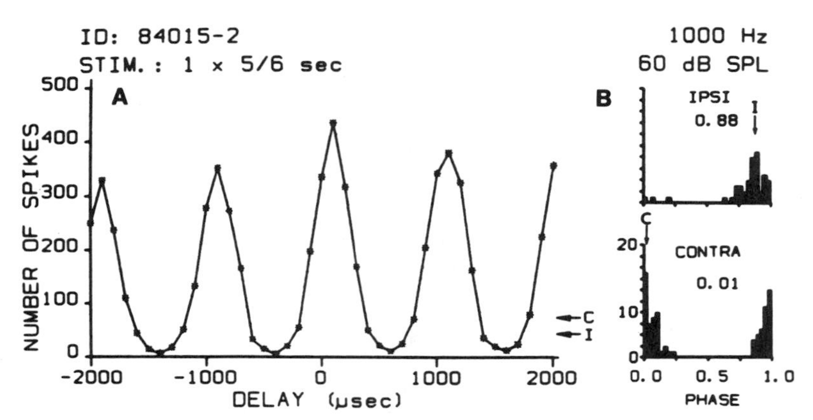

FIGURE 16 (A) The response of a neuron in the medial superior olive as a function of the interaural time difference between two CF tones, one at each ear. Positive delays represent delays of the ipsilateral stimulus. The stimulus consisted of a single presentation of a 5 s tone at each of the interaural delays and at the frequency and level indicated. The arrows marked *I* and *C* show the monaural response levels to ipsilateral and contralateral tones. (B) Period histograms in response to monaural stimulation. The arrows and numbers indicate the interaural phase giving the highest response for each histogram. (From Yin & Chan, 1988, with permission.)

the stimulus that evokes the maximum number of discharges is different for each ear (the arrows indicate the phase angle at the maximum; 0.88 cycles for the ipsilateral response and 0.01 cycles for the contralateral response). The coincidence detector hypothesis predicts that the maximum output of the MSO cell should occur when a delay is introduced, such that the phase locked inputs from each ear reach this neuron simultaneously (i.e., at a delay of the ipsilateral stimulus of $1.01 - 0.88 = 0.13$ cycles). Good agreement between phase delays at the maximum of the IPD curve and those predicted from monaural phase locking have been found (Goldberg & Brown, 1969; Yin & Chan, 1990). For the unit in Figure 16, for example, the measured phase delay for maximum facilitation (0.09 cycles) compares well with the predicted value (0.13 cycles).

Yin and Chan (1990) have reported the most complete (though still limited in terms of the number of units) and detailed study to date of MSO neurons. They were able to demonstrate that the form of binaural interaction was indeed a coincidence detection between phase locked inputs from the two ears. However, because of the difficulties in recording from MSO and the fact that the MSO projects directly to the ipsilateral IC, where similar responses as a function of IPD are found, much of our detailed knowledge of the processing of interaural time differences is derived from extensive studies at the level of the IC.

In a pioneering study, Rose et al. (1966) recorded the responses of IC neurons as a function of the IPD, using different frequencies of pure-tone stimuli. They noted that cells responded equally to the different frequencies only at a particular value of interaural delay, which reflected the fixed physiological delay of the input from one ear with respect to the other and which they termed the *characteristic delay*. The original findings of Rose et al. have been confirmed and extended, and it has been shown that, in the anaesthetized cat, the characteristic delay can occur on the slopes of the function relating firing rate to interaural delay rather than only at a peak or trough. For the majority of units the characteristic delay is within the normally encountered range of interaural delays (Kuwada & Yin, 1983; Yin & Kuwada, 1983, 1984). Using a wide range of frequencies to test single cells in the central nucleus of the IC, Yin, Chan, and Irvine (1986) were able to demonstrate that linear summation of the delay functions to tones gave a "composite curve" very similar to the delay function obtained in response to wideband noise, particularly with respect to the position of the central peak, which usually occurred at delays within the animals' physiological range. The positions of the peaks of both the noise delay function and the composite curve were relatively invariant with sound level, and Yin and Kuwada have proposed that the peak of the delay curve for a wideband stimulus is more likely to be a functionally relevant parameter than the characteristic delay. Whatever is the most functionally relevant parameter of the IPD function, it seems clear that different IPDs will result in different patterns of activation across the population of IPD sensitive neurons. Further studies by Yin and his colleagues (both at MSO and IC), using noise signals at the two ears with different degrees of correlation, have extended the concept of the characteristic delay and are consistent with the suggestion that the sensitivity of low-frequency cells to interaural time delays involves a process of cross-correlation following peripheral filtering (Chan, Yin, & Musicant, 1987; Yin, Chan, & Carney, 1987; Yin & Chan, 1988, 1990).

Interaural time delay sensitivity at the level of the medial geniculate body and primary auditory cortex seems to reflect processing at lower levels without degradation (Aitkin & Webster, 1972; Rerale & Brugge, 1990).

The characteristic delays and the peaks of the wideband noise delay functions for the cat fall within the animals' physiological delay range, but similar values for these parameters have been obtained for animal's with smaller heads, for which the delays exceed their physiological ranges (e.g., guinea pig, Palmer, Rees, & Caird, 1990). However, the greatest acuity for localizing a sound source is on the midline, well away from the peak of the delay functions of most mammals. Peak sensitivity, to small changes in delay, occurs at the steepest part of the delay function, which in nearly all animals is well within the physiological range and generally passes through the midline (see Phillips & Brugge, 1985, for detailed discussion).

Masking of binaural signals by broadband noise at low frequencies depends strongly on the interaural phase relationship of the signal and masker (i.e., the binaural masking level difference; see Chapters 9 and 10). Such characteristics implicate IPD sensitive cells as part of the physiological mechanisms responsible for this masking, and recent physiological data provide some direct evidence that this is the case (Caird, Palmer, & Rees, 1991).

C. Onset and Ongoing Time Differences

At the levels of the MSO, LSO, IC, and auditory cortex, it has been demonstrated that high-frequency cells that are sensitive to ILDs are also sensitive to onset time differences and the delays of the envelope of complex sounds (Caird & Klinke, 1987; Yin et al., 1984; Benevento et al., 1970; Yin & Chan, 1990). Presumably, these cells mediate our abilities to localize high-frequency sounds on the basis of the time delay of their envelopes.

D. Pinna Spectral Effects

In recent years there have been detailed psychophysical and acoustical investigations of the role of the pinna in altering the spectrum of sounds reaching the ears (see Chapters 9 and 10). The transfer function of the pinna introduces sharp high-frequency notches into the spectra of wideband sounds and the frequency of these notches depends on the position of the sound in space (Rice, May, Spirou, & Young, 1992). Recently, Young and his colleagues (Spirou & Young, 1991; Young, Spirou, Rice, and Voigt, 1992) have hypothesized a role for the type IV units of the DCN in detecting these spectral notches. The response areas of type IV units are characterized by strong inhibitory inputs (from type II units) at frequencies just below their CF (see Spirou & Young, 1991). This response area organization results in large changes in the output of the type IV units with small changes in the frequency position of sharp notches in wideband noise. It is further suggested that the input to the DCN via the parallel fibers of the superficial layer may allow somatosensory information about the orientation of the pinna to be integrated with the localization cues within the DCN (Young et al., 1992). Support for this role of the DCN is provided by a recent study in which deficits in the ability to localize elevated sound sources were shown by cats after section of the output pathway of the DCN (Sutherland, 1991).

E. Topographical Distribution of Interaural Sensitivities and Spatial Hearing

At the brainstem level there is a fairly clear separation of the sensitivities to ILDs and ITDs between the LSO and MSO, respectively. Furthermore, the Jeffress (1948) model suggested that an orderly arrangement of the neural delay lines, of the sort implied by the ITD sensitivities, should result in a spatial mapping of delay across the MSO (see Chapter 10). There is physiological and anatomical evidence at the MSO that this is indeed the case (Yin & Chan, 1990; Smith, Joris, & Yin, 1993). Neurons with composite-curve peaks near zero delay lie in rostral MSO locations and those with progressively longer ipsilateral delays in more caudal positions. Additional evidence for this proposition comes from anatomical and evoked potential studies in the nucleus laminaris, which is the avian homologue of the MSO (Konishi, Takahasi, Wagner, Sullivan, & Carr, 1988). Details of the topographical distribution of IPD sensitivity across the IC are very limited, but there have been reports of systematic changes along some electrode penetrations within iso-frequency laminae (Yin, Chan, & Kuwada, 1983). There is certainly segregation within the IC of responsiveness to different binaural cues that mainly, but not entirely, reflects the cochleotopic organization (Semple & Aitkin, 1979). At the level of the primary auditory cortex, several studies have revealed bands alternating across the cortex, approximately orthogonal to the cochleotopic axis, of cells of the EI and EE type. Since these are within the high-frequency regions these bands represent the alternative forms of ILD sensitivity (for a review see Phillips et al., 1991).

From the level of the brainstem right up to the cortex, the cells respond to signals favoring the contralateral ear: each half of the brain appears to deal only with localization cues for the contralateral hemifield. Furthermore, Jenkins and Merzenich (1984) were able to demonstrate by a combination of behavioral, physiological, and lesioning techniques that the representation of a band of frequencies in the primary auditory cortex is necessary and sufficient for the correct localization of sounds within that frequency band.

In 1978 Knudsen and Konishi demonstrated the existence in the barn owl of a topographic representation or "map" of auditory space in which neurons responded only when the sound was within a relatively small area of three-dimensional space. This demonstration renewed interest in the use of free-field stimuli to investigate the representation of sounds in space and recent free-field studies in the cat have measured the spatial response areas (i.e., the spatial limits over which a specified response criterion is exceeded) of neurons in IC and auditory cortex to low level tones (Semple, Aitkin, Calford, Pettigrew, & Phillips, 1983; Moore, Semple, Addison, & Aitkin, 1984; Middlebrooks & Pettigrew, 1981). These studies revealed three classes of response: omnidirectional, hemifield, and units with well circumscribed spatial response areas. The circumscribed response areas were all recorded from high-frequency neurons and fell on the axis of the contralateral pinna. The omnidirectional response areas corresponded to low-frequency neurons and the hemifield areas to units with intermediate best frequencies. At these low sound levels, the spatial response areas reflected almost entirely the monaural effects of the pinna, which in the cat produces considerable amplification of high-frequency sounds on the pinna axis, but has little effect on low-frequencies. An alternative to the measurement of spatial response areas is to determine the response magnitude of the cell as a function of spatial position. This analysis revealed that about half of the low-frequency neurons were selective for azimuth and many were sharply tuned for azimuthal positions from 0–80° (Aitkin, Gates, & Phillips, 1984; Aitkin, Pettigrew, Calford, Phillips, & Wise, 1985). These responses were generally those expected from interaural time-sensitive neurons, given the limitations on the values of interaural time imposed by the use of free-field stimuli. There is evidence of a topographical distribution of azimuthal sensitivity with cells responsive to peripheral azimuths located rostrally and medially in the IC and cells responsive to midline azimuths located caudally and laterally (Aitkin et al., 1985).

To date, there has been no demonstration in the central nucleus of the IC or in the auditory cortex of a map of auditory space of the kind found in the barn owl mesencephalicus lateralis dorsalis or optic tectum. However, there have been demonstrations of such a topographic "auditory space map" in the external nucleus of the IC of the guinea pig (Binns, Grant, Withington, & Keating, 1991) and in the deep layers of the superior colliculus of the guinea pig, cat, and ferret (see Middlebrooks, 1988). The basis for the map in the deep layers of the superior colliculus is a topographical distribution of sensitivity of the cells to binaural stimuli (see Irvine, 1986).

VI. SUMMARY AND CONCLUDING REMARKS

At the level of the cochlear nerve, clear differences exist in the responses of nerve fibers distinguishable by their relative threshold and spontaneous rate. Nevertheless, the activity of the ensemble of auditory nerve fibers is relatively homogeneous and explicable in terms of the vibration patterns of the basilar membrane, the nature of the receptor potentials of inner hair cells, and the function of the synapse at the base of the hair cells. In contrast, the responses of the different principal cell types of the cochlear nucleus are characterized by their diversity. It seems reasonable to suppose that the different principal projection neurones of the cochlear nucleus constitute parallel output pathways for analysis of different aspects of the auditory signal. We can propose several hypotheses for the functions of these output pathways: (1) It seems reasonably clear that the bushy cell system in the ventral cochlear nucleus directs activity to the superior olivary complex where interaural time and level differences are analyzed in largely separate but overlapping pathways. If, however, the spectra of complex sounds are coded in terms of the distribution of phase locked activity, the spherical bushy cell pathway is the only one capable of faithfully transmitting this information to higher levels. (2) The stellate (chopper) cells of the ventral cochlear nucleus may encode the spectra of complex sounds, including speech, as a place code by selectively responding to high and low threshold auditory nerve fibers depending on the sound level. (3) Onset responding cells in the cochlear nucleus often respond to complex sounds in a way that is consistent with the pitchs heard. It seems premature at this time, however, to conclude that these cells really are involved in the analysis of the pitch of complex sounds. They may have more of an alerting or arousal function or may have other, as yet undefined, functions within wider networks of cells. (4) Finally, the principal cells of the dorsal cochlear nucleus may process the spectral cues for localization generated by the pinnae.

In the more central nuclei of the auditory system, recent studies have revealed topographical organizations of signal parameters such as frequency, binaural sensitivity, interaural time delay, width of tuning, sensitivity to frequency sweeps, and sensitivity to modulation. We do not as yet know the functional significance of this organization. Why, for example, is there a map of modulation sensitivity in the inferior colliculus and not at higher levels? If the lack of such organization is a real finding, then the need for this organization must be met at the midbrain level.

Present evidence in most mammals would not seem to favor a radical remapping of auditory information onto functional dimensions akin to that demonstrated in animals such as the bat. However, Suga (1988) has cogently argued that understanding the functional basis of the organization of central nuclei will depend on the use of ethologically appropriate sounds. Thus, although the use of speech sounds may be justifiable in studies of the *auditory periphery* of animals, their use would reveal only the central organization of animals for whom speech was important. Although this may prove to be the case, the use of conspecific stimuli has as yet not radically changed our view of auditory processing (Symmes, 1981). It may equally be that, in animals that are not specialized for tasks such as biosonar, a more general purpose organization has been retained, in which signal attributes are represented by the spatio–temporal patterns of activity across populations of neurons still organized on a basically cochleotopic axis.

Acknowledgments

I am grateful to P. Moorjani for assistance in the preparation of this chapter. I would also like to thank Ian Winter, Adrian Rees, and Quentin Summerfield, whose critical comments assisted me in greatly reducing the size of my earlier attempts.

References

Aitkin, L. M., Gates, G. R., & Phillips, S. C. (1984). Responses of neurons in the inferior colliculus to variations in sound-source azimuth. *Journal of Neurophysiology, 52,* 1–17.

Aitkin, L. M., Pettigrew, J. D., Calford, M. B., Phillips, S. C., & Wise, L. Z. (1985).

Representation of stimulus azimuth by low-frequency neurons in the inferior colliculus of the cat. *Journal of Neurophysiology, 53,* 43–59.

Aitkin, L. M., & Webster, W. R. (1972). Medial geniculate body: Unit responses in the awake cat. *Journal of Neurophysiology, 36,* 275–283.

Arthur, R. M., Pfeiffer, R. R., & Suga, N. (1971). Properties of "two-tone inhibition" in primary auditory neurones. *Journal of Physiology (London), 212,* 593–609.

Benevento, L. A., Coleman, P. D., & Loe, P. R. (1970). Responses of single cells in cat inferior colliculus to binaural click stimuli: Combinations of intensity levels, time differences and intensity differences. *Brain Research, 17,* 387–405.

Binns, K. E., Grant, S., Withington, D. J., & Keating, M. J. (1991). A topographic representation of auditory space in the external nucleus of the inferior colliculus of the guinea pig. *Brain Research, 589,* 231–242.

Blackburn, C. C., & Sachs, M. B. (1989). Classification of unit types in the anteroventral cochlear nucleus: PST histograms and regularity analysis. *Journal of Neurophysiology, 62,* 1301–1329.

Blackburn, C. C., & Sachs, M. B. (1990). The representation of the steady-state vowel /ϵ/ in the discharge pattern of cat anteroventral cochlear nucleus neurons. *Journal of Neurophysiology, 63,* 1191–1212.

Boudreau, J. C., & Tsuchitani, C. (1968). Binaural interaction in the cat superior olive S segment. *Journal of Neurophysiology, 31,* 442–454.

Boudreau, J. C., & Tsuchitani, C. (1970). Cat superior olive S-segment cell discharge to tonal stimulation. In W. D. Neff (Ed.), *Contributions to sensory physiology* (Vol. 4, pp. 143–213). New York: Academic Press.

Bourk, T. R. (1976). Electrical responses of neural units in the anteroventral cochlear nucleus of the cat. Ph.D. thesis, MIT, Cambridge, MA.

Britt, R., & Starr, A. (1975). Synaptic events and discharge patterns of cochlear nucleus cells. II. Frequency-modulated tones. *Journal of Neurophysiology, 39,* 179–194.

Brugge, J. F., & Merzenich, M. M. (1973). Response of neurons in auditory cortex of the macaque monkey to monaural and binaural stimulation. *Journal of Neurophysiology, 36,* 1138–1158.

Caird, D., & Klinke, R. (1987). Processing of interaural time and intensity differences in the cat inferior colliculus. *Experimental Brain Research, 68,* 379–392.

Caird, D. M., Palmer, A. R., & Rees, A. (1991). Binaural masking level difference effects in single units of the guinea pig inferior colliculus. *Hearing Research, 57,* 91–106.

Calford, M. B., Webster, W. R., & Semple, M. N. (1983). Measurement of frequency selectivity of single neurones in the central auditory pathway. *Hearing Research, 11,* 395–401.

Cant, N. B. (1991). Projections to the lateral and medial superior olivary nuclei from spherical and globular bushy cells of the anteroventral cochlear nucleus. In R. A. Altschuler, R. P. Bobbin, B. M. Clopton, & D. W. Hoffman (Eds.), *Neurobiology of hearing: The central auditory system* (pp. 99–120). New York: Raven Press.

Caspary, D. M., & Finlayson, P. G. (1991). Superior olivary complex: Functional neuropharmacology of the principal cell types. In R. A. Altschuler, R. P. Bobbin, B. M. Clopton, & D. W. Hoffman (Eds.), *Neurobiology of hearing: The central auditory system* (pp. 141–162). New York: Raven Press.

Chan, J. C. K., Yin, T. C. T., & Musicant, A. D. (1987). Effects of interaural time delays of noise stimuli on low-frequency cells in the cat's inferior colliculus. II. Responses to bandpass filtered noises. *Journal of Neurophysiology, 58,* 543–561.

Costalupes, J. A., Rich, N. C., & Ruggero, M. A. (1987). Effects of two-tone rate suppression in auditory nerve fibers. *Hearing Research, 26,* 155–164.

Costalupes, J. A., Young, E. D., & Gibson, D. J. (1984). Effects of continuous noise backgrounds on rate response of auditory nerve fibers in cat. *Journal of Neurophysiology, 51,* 1326–1344.

Delgutte, B. (1984). Speech coding in the auditory nerve: II. Processing schemes for vowel-like sounds. *Journal of the Acoustical Society of America, 75,* 879–886.

Delgutte, B. (1987). Peripheral processing of speech information: implications from a physiological study of intensity discrimination. In M. E. H. Schouten (Ed.), *Psychophysics and speech perception* (pp. 333–353). Dortrecht: Nijhoff.

Delgutte, B., & Kiang, N. Y. S. (1984a). Speech coding in the auditory nerve. I. Vowel-like sounds. *Journal of the Acoustical Society of America, 75,* 866–878.

Delgutte, B., & Kiang, N. Y. S. (1984b). Speech coding in the auditory nerve. III. Voiceless fricative consonants. *Journal of the Acoustical Society of America, 75,* 887–896.

Delgutte, B., & Kiang, N. Y. S. (1984c). Speech coding in the auditory nerve. IV. Sounds with consonant-like dynamic characteristics. *Journal of the Acoustical Society of America, 75,* 897–907.

Delgutte, B., & Kiang, N. Y. S. (1984d). Speech coding in the auditory nerve. V. Vowels in background noise. *Journal of the Acoustical Society of America, 75,* 908–918.

Deng, L., & Geisler, C. D. (1987). Responses of auditory-nerve fibers to nasal consonant-vowel syllables. *Journal of Acoustical Society of America, 82,* 1977–1988.

Ehret, G., & Merzenich, M. M. (1988). Neuronal discharge rate is unsuitable for encoding sound intensity at the inferior-colliculus level. *Hearing Research, 35,* 1–8.

Evans, E. F. (1975). Cochlear nerve and cochlear nucleus. In W. D. Keidel & W. D. Neff (Eds.), *Handbook of sensory physiology* (pp. 1–108). New York: Springer-Verlag.

Evans, E. F. (1981). The dynamic range problem: place and time coding at the level of the cochlear nerve and nucleus. In J. Syka & L. Aitkin (Eds.), *Neuronal mechanisms of hearing* (pp. 69–85). New York: Plenum.

Evans, E. F., & Nelson, P. G. (1973). The responses of single neurons in the cochlear nucleus of the cat as a function of their location and anesthetic state. *Experimental Brain Research, 17,* 402–427.

Evans, E. F., & Palmer, A. R. (1980). Relationship between the dynamic range of cochlear nerve fibres and their spontaneous activity. *Experimental Brain Research, 40,* 115–118.

Evans, E. F., Pratt, S. R., Spenner, H., & Cooper, N. P. (1992). Comparisons of physiological and behavioural properties: auditory frequency selectivity. In *Advances in the biosciences, 83,* 159–169.

Frisina, R. D., Smith, R. L., & Chamberlain, S. C. (1990). Encoding of amplitude modulation in the gerbil cochlear nucleus. I. A hierarchy of enhancement. *Hearing Research, 44,* 99–122.

Gibson, D. J., Young, E. D., & Costalupes, J. A. (1985). Similarity of dynamic range adjustment in auditory nerve and cochlear nuclei. *Journal of Neurophysiology, 53,* 940–958.

Godfrey, D. A., Kiang, N. Y. S., & Norris, B. E. (1975). Single unit activity in the posteroventral cochlear nucleus of the cat. *Journal of Comparitive Neurology, 162,* 247–268.

Goldberg, J. M., & Brown, P. B. (1969). Response of binaural neurons of dog superior olivary complex to dichotic tonal stimuli: Some physiological mechanisms of sound localization. *Journal of Neurophysiology, 32,* 613–636.

Goldberg, J. M., & Brownell, W. E. (1973). Discharge characteristics of neurons in anteroventral and dorsal cochlear nuclei of the cat. *Brain Research, 64,* 35–64.

Guinan, J. J., Norris, B. E., & Guinan, S. S. (1972). Single auditory units in the superior olivary complex. I. Response to sounds and classifications based on physiological properties. *Journal of Neuroscience, 4,* 101–120.

Irvine, D. R. F. (1986). *The auditory brainstem. Progress in sensory physiology* 7 (Ed. D. Ottoson). Berlin: Springer-Verlag.

Javel, E. (1981). Suppression of auditory nerve responses. I. Temporal analysis, intensity effects and suppression contours. *Journal of the Acoustical Society of America, 69,* 1735–1745.

Jeffress, L. A. (1948). A place theory of sound localization. *Journal of Comparitive Physiology and Psychology, 41,* 35–39.

Jenkins, W. M., & Merzenich, M. M. (1984). Role of cat primary auditory cortex for sound localization behaviour. *Journal of Neurophysiology, 47,* 987–1016.

Johnson, D. H. (1980). The relationship between spike rate and synchrony in responses of auditory-nerve fibers to single tones. *Journal of the Acoustical Society of America, 68,* 1115–1122.

Joris, P. X., & Yin, T. C. T. (1992). Responses to amplitude-modulated tones in the auditory nerve of the cat. *Journal of the Acoustical Society of America, 91,* 215–232.

Kiang, N. Y. S., Watanabe, T., Thomas, E. C., & Clark, L. F. (1965). *Discharge Patterns of Fibers in the Cat's Auditory Nerve.* Cambridge, MA: MIT Press.

Kim, D. O., & Leonard, G. (1988). Pitch-period following response of cat cochlear nucleus neurons to speech sounds. In H. Duifhuis, J. W. Horst, & H. P. Wit (Eds.), *Basic issues in hearing* (pp. 252–260). London: Academic Press.

Kim, D. O., & Molnar, C. E. (1979). A population study of cochlear nerve fibers: Comparison of spatial distributions of average-rate and phase-locking measures of responses to single tones. *Journal of Neurophysiology, 42,* 16–30.

Kim, D. O., Rhode, W. S., & Greenberg, S. R. (1986). Responses of cochlear nucleus neurons to speech signals: Neural coding of pitch, intensity and other parameters. In B. C. J. Moore & R. D. Patterson (Eds.), *Auditory frequency selectivity* (pp. 281–288). New York: Plenum.

Kim, D. O., Sirianni, J. G., & Chang, S. O. (19990). Responses of DCN-PVCN neurons and auditory nerve fibers in unanesthetized decerebrate cats to AM and pure tones: Analysis with autocorrelation/power-spectrum. *Hearing Research, 45,* 95–113.

Knudsen, E. I., & Konishi, M. (1978). A neural map of auditory space in the owl. *Science, 200,* 795–797.

Konishi, M., Takahasi, T. T., Wagner, H., Sullivan, W. E., & Carr, C. E. (1988). Neurophysiological and anatomical substrates of sound localization in the owl. In G. M. Edelman, W. E. Gall, & W. M. Cowan (Eds.), *Auditory function* (pp. 721–746) New York: Wiley.

Kuwada, S., & Yin, T. C. T. (1983). Binaural interaction in low-frequency neurons in inferior colliculus of the cat: Effects of long interaural delays, intensity, and repetition rate on interaural delay function. *Journal of Neurophysiology, 50,* 981–999.

Kuwada, S., Yin, T. C. T., Syka, J., Buunen, T. J. F., & Wickesberg, R. E. (1984). Binaural interaction in low-frequency neurons in the inferior colliculus of the cat. IV. Comparison of monaural and binaural response properties. *Journal of Neurophysiology, 51,* 1306–1325.

Langner, G., & Schreiner, C. E. (1988). Periodicity coding in the inferior colliculus of the cat. I. neuronal mechanisms. *Journal of Neurophysiology, 60,* 1799–1822.

Liberman, M. C. (1978). Auditory-nerve response from cats raised in a low-noise chamber. *Journal of the Acoustical Society of America, 63,* 442–455.

Liberman, M. C., & Kiang, N. Y. S. (1978). Acoustic trauma in cats. *Acta Otolaryngology* (Suppl. 358), 1–63.

May, B. J., Aleszczyk, C. M., & Sachs, M. B. (1991). Single-unit recording in the ventral cochlear nucleus of behaving cats. *Journal of Neuroscience Methods, 40,* 155–169.

Merzenich, M. M., Roth, G. L., Anderson, R. A., Knight, P. L., & Colwell, S. A. (1977). Some basic features of the organization of the central nervous system. In E. F. Evans & P. J. Wilson (Eds.), *Psychophysics and physiology of hearing* (pp. 485–497). London: Academic Press.

Middlebrooks, J. C. (1988). Auditory mechanisms underlying a neural code for space in the cat's superior colliculus. In G. M. Edelman, W. E. Gall, & W. M. Cowan (Eds.), *Auditory function* (pp. 431–455). New York: Wiley.

Middlebrooks, J. C., & Pettigrew, J. D. (1981). Functional classes of neurons in primary auditory cortex of the cat distinguished by sensitivity to sound location. *Journal of Neuroscience, 1,* 107–120.

Miller, M. I., & Sachs, M. B. (1983). Representation of stop constonants in the discharge patterns of auditory-nerve fibers. *Journal of the Acoustical Society of America, 74,* 502–517.

Miller, M. I., & Sachs, M. B. (1984). Representation of voice pitch in discharge patterns of auditory nerve fibers. *Hearing Research, 14,* 257–259.

Møller, A. R. (1972). Coding of amplitude and frequency modulated sounds in the cochlear nucleus of the rat. *Acta Physiological Scandanavica, 86,* 223–238.

Møller, A. R. (1977). Coding of time varying sounds in the cochlear nucleus. *Audiology, 17,* 446–468.

Moore, D. R., Semple, M. N., Addison, P. D., & Aitkin, L. M. (1984). Properties of spatial receptive fields in the cat inferior colliculus. I. Responses to tones of low intensity. *Hearing Research, 13,* 159–174.

Nelson, P. G., Erulkar, S. D., & Bryan, J. S. (1966). Responses of units of the inferior colliculus to time-varying acoustic stimuli. *Journal of Neurophysiology, 29,* 834–860.

Oertel, D., Wu, S. H., & Hirsch, J. A. (1988). Electrical characteristics of cells and neuronal circuity in the cochlear nuclei studied with intracellular recordings from brain slices. In G. M. Edelman, W. E. Gall, & W. M. Cowan (Eds.), *Auditory function* (pp. 313–336). New York: Wiley.

Palmer, A. R. (1982). Encoding of rapid amplitude fluctuations by cochlear nerve fibres in the guinea pig. *Archives of Otorhinolaryngology, 236,* 197–202.

Palmer, A. R. (1990). The representation of the spectra and fundamental frequencies of steady-state single- and double-vowel sounds in the temporal discharge patterns of guinea pig cochlear-nerve fibers. *Journal of the Acoustical Society of America, 88,* 1412–1426.

Palmer, A. R., & Evans, E. F. (1980). Cochlear fiber rate-intensity functions: no evidence for basilar membrane nonlinearities. *Hearing Research, 2,* 319–326.

Palmer, A. R., Rees, A., & Caird, D. (1990). Interaural delay sensitivity to tones and broad band signals in the guinea pig inferior colliculus. *Hearing Research, 50,* 71–86.

Palmer, A. R., & Russell, I. J. (1986). Phase-locking in the cochlear nerve of the guinea pig and its relation to the receptor potential of inner hair-cells. *Hearing Research, 24,* 1–15.

Palmer, A. R., & Winter, I. M. (1992). Coding of the fundamental frequency of voiced speech and harmonic complexes in the cochlear nerve and cochlear nucleus. In M. A. Merchan, J. M. Juiz, D. A. Godfrey, & E. Mugnaini (Eds.), *The mammalian cochlear nuclei: Organization and function.* New York: Plenum.

Palmer, A. R., Winter, I. M., & Darwin, C. J. (1986). The representation of steady-state vowel sounds in the temporal discharge patterns of the guinea pig cochlear nerve and primarylike cochlear nucleus neurons. *Journal of the Acoustical Society of America, 79,* 100–113.

Palmer, A. R., Winter, I. M., & Stabler, S. E. (1995). Responses to simple and complex sounds in the cochlear nucleus of the guinea pig. In W. A. Ainsworth (Ed.), *Advances in speech, hearing and language processing.* London: JAI Press.

Pfingst, B. E., & O'Connor, T. A. (1981). Characteristics of neurons in auditory cortex of monkeys performing a simple auditory task. *Journal of Neurophysiology, 45,* 16–34.

Phillips, D. P., & Brugge, J. F. (1985). Progress in neurophysiology of sound localization. *Annual Review of Psychology, 36,* 245–274.

Phillips, D. P., & Cynader, M. S. (1985). Some neural mechanisms in the cat's auditory cortex underlying sensitivity to combined tone and wide-spectrum noise stimuli. *Hearing Research, 18,* 87–102.

Phillips, D. P., Mendelson, J. R., Cynader, M. S., & Douglas, R. M. (1985). Responses of single neurones in the cat auditory cortex to time-varying stimuli: frequency-modulated tones of narrow excursion. *Experimental Brain Research, 58,* 443–454.

Phillips, D. P., Reale, R. A., & Brugge, J. F. (1991). Stimulus processing in the auditory cortex. In R. A. Altschuler, R. P. Bobbin, B. M. Clopton, & D. W. Hoffman (Eds.), *Neurobiology of hearing: The central auditory system* (pp. 335–366). New York: Raven Press.

Pickles, J. O. (1981). *An introduction to the physiology of hearing*. London: Academic Press.

Reale, R. A., & Brugge, J. F. (1990). Auditory cortical neurons are sensitive to static and continuously changing interaural phase cues. *Journal of Neurophysiology, 64,* 1247–1260.

Rees, A., & Møller, A. R. (1983). Responses of neurons in the inferior colliculus of the rat to AM and FM tones. *Hearing Research, 10,* 301–330.

Rees, A., & Palmer, A. R. (1988). Rate-intensity functions and their modification by broadband noise for neurons in the guinea pig inferior colliculus. *Journal of the Acoustical Society of America, 83,* 1488–1498.

Rees, A., & Palmer, A. R. (1989). Neuronal responses to amplitude-modulated and pure-tone stimuli in the guinea pig inferior colliculus, and their modification by broadband noise. *Journal of the Acoustical Society of America, 85,* 1978–1994.

Relkin, E. M., & Doucet, J. R. (1991). Recovery from prior stimulation. 1. Relationship to spontaneous firing rates of primary auditory neurons. *Hearing Research, 55,* 215–222.

Rhode, W. S., Geisler, C. D., & Kennedy, D. T. (1978). Auditory nerve fiber responses to wideband noise and tone combinations. *Journal of Neurophysiology, 41,* 692–704.

Rhode, W. S., & Smith, P. H. (1986a). Encoding timing and intensity in the ventral cochlear nucleus of the cat. *Journal of Neurophysiology, 56,* 261–286.

Rhode, W. S., & Smith, P. H. (1986b). Physiological studies of neurons in the dorsal cochlear nucleus of the cat. *Journal of Neurophysiology, 56,* 287–307.

Rice, J. J., May, B. J., Spirou, G. A., & Young, E. D. (1992). Pinna-based spectral cues for sound localization in cat. *Hearing Research, 58,* 132–152.

Robertson, D. (1984). Horseradish peroxidase injection of physiologically characterized afferent and efferent neurones in the guinea pig spiral ganglion. *Hearing Research, 15,* 113–121.

Rose, J. E., Brugge, J. F., Anderson, D. J., & Hind, J. E. (1967). Phase locked response to low-frequency tones in single auditory nerve fibers of the squirrel monkey. *Journal of Neurophysiology, 30,* 769–793.

Rose, J. E., Galambos, R., & Hughes, J. R. (1959). Microelectrode studies of the cochlear nuclei of the cat. *Bulletin of the Johns Hopkins Hospital, 104,* 211–251.

Rose, J. E., Gross, N. B., Geisler, C. D., & Hind, J. E. (1966). Some neural mechanisms in the inferior colliculus of the cat which may be relevant to localization of a sound source. *Journal of Neurophysiology, 29,* 288–314.

Rouiller, E., De Ribaupierre, Y., & De Ribaupierre, F. (1979). Phase-locked responses to low frequency tones in the medial geniculate body. *Hearing Research, 1,* 213–226.

Sachs, M. B. (1984). Speech encoding in the auditory nerve. In C. Berlin (Ed.), *Hearing science,* (pp. 263–308). London: Taylor and Francis.

Sachs, M. B., & Abbas, P. J. (1974). Rate versus level functions for auditory nerve fibers in cats: Tone burst stimuli. *Journal of the Acoustical Society of America, 56,* 1835–1847.

Sachs, M. B., Voigt, H. F., & Young, E. D. (1983). Auditory nerve representation of vowels in background noise. *Journal of Neurophysiology, 50,* 27–45.

Sachs, M. B., & Young, E. D. (1979). Encoding of steady-state vowels in the auditory nerve: Representation in terms of discharge rate. *Journal of the Acoustical Society of America, 66,* 470–479.

Sachs, M. B., Young, E. D., & Miller, M. I. (1982). Speech encoding in the auditory nerve:

Implications for cochlear implants. *Annals of the New York Academy of Science, 405,* 94–113.

Sachs, M. B., Young, E. D., Schalk, T. B., & Bernardin, C. P. (1980). Suppression effects in the responses of auditory nerve fibers to broadband stimuli. In G. van den Brink & F. A. Bilsen (Eds.), *Psychophysical, physiological and behavioural studies in hearing* (pp. 284–291). Delft: Delft University Press.

Schreiner, C. E., & Langner, G. (1988a). Coding of temporal patterns in the central auditory nervous system. In G. M. Edelman, W. E. Gall, & W. M. Cowan (Eds.), *Auditory function* (pp. 337–361). New York: Wiley.

Schreiner, C. E., & Langner, G. (1988b). Periodicity coding in the inferior colliculus of the cat. II. Topographical organization. *Journal of Neurophysiology, 60,* 1823–1840.

Schreiner, C. E., & Mendelson, J. R. (1990). Functional topography of cat primary auditory cortex: Distribution of integrated excitation. *Journal of Neurophysiology, 64,* 1442–1459.

Semple, M. N., & Aitkin, L. M. (1979). Representation of sound frequency and laterality by units in the central nucleus of cat inferior colliculus. *Journal of Neurophysiology, 42,* 1626–1639.

Semple, M. N., Aitkin, L. M., Calford, M. B., Pettigrew, J. D., & Phillips, D. P. (1983). Spatial receptive fields in the cat inferior colliculus. *Hearing Research, 10,* 203–215.

Sinex, C. D., & Geisler, C. D. (1981). Auditory-nerve fiber responses to frequency-modulated tones. *Hearing Research, 4,* 127–148.

Sinex, C. D., & Geisler, C. D. (1983). Responses of auditory nerve fibers to consonant-vowel syllables. *Journal of the Acoustical Society of America, 73,* 602–615.

Smith, P. H., Joris, P. X., & Yin, T. C. T. (1993). Projections of physiologically characterized spherical bushy cell axons from the cochlear nucleus of the cat: evidence for delay lines to the medial superior olive. *Journal of Comparitive Neurology, 331,* 245–260.

Smith, R. L. (1979). Adaptation, saturation, and physiological masking in single auditory-nerve fibers. *Journal of the Acoustical Society of America, 65,* 166–178.

Smith, R. L. (1988). Neural intensity coding. In G. M. Edelman, W. E. Gall, & W. M. Cowan (Eds.), *Auditory function* (pp. 213–274). New York: Wiley.

Smith, R. L., & Brachman, M. L. (1980). Operating range and maximum response of single auditory nerve fibers. *Brain Research, 184,* 499–505.

Smith, R. L., & Zwislocki, J. J. (1975). Short-term adaptation and incremental responses of single auditory-nerve fibers. *Biological Cybernetics, 17,* 169–182.

Spirou, G. A., & Young, E. D. (1991). Organization of dorsal cochlear nucleus type IV response maps and their relationship to activation by bandlimited noise. *Journal of Neurophysiology, 66,* 1750–1768.

Steinschneider, M., Arezzo, J. C., & Vaughan, H. G. (1990). Tonotopic features of speech-evoked activity in primate auditory cortex. *Brain Research, 519,* 158–168.

Suga, N. (1988). Auditory neuroethology and speech processing; complex sound processing by combination-sensitive neurons. In G. M. Edelman, W. E. Gall, & W. M. Cowan (Eds.), *Auditory function* (pp. 679–720). New York: Wiley.

Sutherland, D. P. (1991). A role of the dorsal cochlear nucleus in the localization of elevated sound sources. *Abstracts of the Association for Research in Otolaryngology, 14,* 33.

Symmes, D. (1981). On the use of natural stimuli in neurophysiological studies of audition. *Hearing Research, 4,* 203–214.

Tsuchitani, C., & Johnson, D. H. (1991). Binaural cues and signal processing in the superior olivary complex. In R. A. Altschuler, R. P. Bobbin, B. M. Clopton, & D. W. Hoffman (Eds.), *Neurobiology of hearing, the central auditory system* (pp. 163–194). New York: Raven Press.

Viemeister, N. F. (1988). Psychophysical aspects of auditory intensity coding. In G. M.

Edelman, W. E. Gall, & W. M. Cowan (Eds.), *Auditory function* (pp. 213–241). New York: Wiley.
Voigt, H. F., Sachs, M. B., & Young, E. D. (1982). Representation of whispered vowels in discharge patterns of auditory nerve fibers. *Hearing Research, 8,* 49–58.
Watanabe, T., & Sakai, H. (1978). Responses of the cat's collicular auditory neuron to human speech. *Journal of the Acoustical Society of America, 64,* 333–337.
Westerman, L. A., & Smith, R. L. (1984). Rapid and short-term adaptation in auditory nerve responses. *Hearing Research, 15,* 249–260.
Whitfield, I. C., & Evans, E. F. (1965). Responses of auditory cortical neurons to stimuli of changing frequency. *Journal of Neurophysiology, 28,* 655–672.
Winslow, R. L. (1985). A quantitative analysis of rate coding in the auditory nerve. Doctoral dissertation, Johns Hopkins University, Baltimore.
Winslow, R. L., & Sachs, M. B. (1988). Single tone intensity discrimination based on auditory-nerve rate responses in backgrounds of quiet, noise and stimulation of the olivocochlear bundle. *Hearing Research, 35,* 165–190.
Winter, I. M., & Palmer, A. R. (1990). Responses of single units in the anteroventral cochlear nucleus of the guinea pig. *Hearing Research, 44,* 161–178.
Winter, I. M., Robertson, D., & Yates, G. K. (1990). Diversity of characteristic frequency rate-intensity functions in guinea pig auditory-nerve fibers. *Hearing Research, 45,* 191–202.
Yates, G. K., Robertson, D., & Johnstone, B. M. (1985). Very rapid adaptation in the guinea pig auditory nerve. *Hearing Research, 17,* 1–12.
Yates, G. K., Winter, I. M., & Robertson, D. (1990). Basilar membrane nonlinearity determines auditory-nerve rate-intensity functions and cochlear dynamic range. *Hearing Research, 45,* 203–220.
Yin, T. C. T., & Chan, J. C. K. (1988). Neural mechanisms underlying interaural time sensitivity to tones and noise. In G. M. Edelman, W. E. Gall, & W. M. Cowan (Eds.), *Auditory function* (pp. 385–430). New York: Wiley.
Yin, T. C. T., & Chan, J. C. K. (1990). Interaural time sensitivity in medial superior olive of cat. *Journal of Neurophysiology, 64,* 465–488.
Yin, T. C. T., Chan, J. C. K., & Carney, L. H. (1987). Effects of interaural delays of noise stimuli on low-frequency cells in the cat's inferior colliculus. III. Evidence for cross-correlation. *Journal of Neurophysiology, 58,* 562–583.
Yin, T. C. T., Chan, J. C. K., & Irvine, D. R. F. (1986). Effects of interaural time delays of noise stimuli on low-frequency cells in the cat's inferior colliculus. I. Responses to wide-band noise. *Journal of Neurophysiology, 55,* 280–300.
Yin, T. C. T., Chan, J. C. K., & Kuwada, S. (1983). Characteristic delays and their topographical distribution in the inferior colliculus of the cat. In W. R. Webster & L. M. Aitkin (Eds.), *Mechanisms of hearing* (pp. 94–99). Clayton, Victoria, Australia: Monash University Press.
Yin, T. C. T., & Kuwada, S. (1983). Binaural interaction in low-frequency neurons in inferior colliculus of the cat. III. Effects of changing frequency. *Journal of Neurophysiology, 50,* 1020–1042.
Yin, T. C. T., & Kuwada, S. (1984). Neuronal mechanisms of binaural interaction. In G. M. Edelman, W. E. Gall, & W. M. Cowan (Eds.), *Dynamic aspects of neocortical function* (pp. 263–313). New York: Wiley.
Yin, T. C. T., Kuwada, S., & Sujaku, Y. (1984). Interaural time sensitivity of high-frequency neurons in the inferior colliculus. *Journal of the Acoustical Society of America, 76,* 1401–1410.
Young, E. D. (1984). Response characteristics of neurons in the cochlear nuclei. In C. Berlin (Ed.), *Hearing science* (pp. 423–460). San Diego: College-Hill Press.
Young, E. D., & Barta, P. E. (1986). Rate responses of auditory nerve fibers to tones in noise near masked threshold. *Journal of the Acoustical Society of America, 79,* 426–442.

Young, E. D., & Sachs, M. B. (1979). Representation of steady-state vowels in the temporal aspects of the discharge patterns of populations of auditory-nerve fibers. *Journal of the Acoustical Society of America, 66,* 1381–1403.

Young, E. D., Shofner, W. P., White, J. A., Robert, J.-M., & Voigt, H. F. (1988). Response properties of cochlear nucleus neurons in relationship to physiological mechanisms. In G. M. Edelman, W. E. Gall, & W. M. Cowan (Eds.), *Auditory Function* (pp. 277–312). New York: Wiley.

Young, E. D., Spirou, G. A., Rice, J. J., & Voigt, H. F. (1992). Neural organization and responses to complex stimuli in the dorsal cochlear nucleus. *Philosophical Transactions of the Royal Society of London, B, 336,* 407–413.

CHAPTER 4

Loudness Perception and Intensity Coding

Christopher J. Plack
Robert P. Carlyon

I. INTRODUCTION

It is almost a truism to say that all perceptually relevant sounds are characterized by differences in intensity across time or frequency. Intensity perception can be regarded, therefore, as one of the basic concerns of hearing research, and one that has generated a great deal of interest. Despite many decades of work in this area, the field is still progressing rapidly, and the last ten years have seen significant developments in our understanding of a very fundamental problem, how sound intensity is represented, or coded, in the auditory system.

In this chapter we will consider first the factors that determine the subjective intensity or the loudness of sounds. We will discuss the techniques used to measure loudness, and how loudness varies with intensity, frequency, and spectral extent. We will then describe listeners' ability to detect differences in intensity, and what discrimination experiments have told us about the coding of intensity in the auditory system. These results will be discussed in terms of the various models that have been proposed to account for intensity perception.

Hearing

II. THE PERCEPTION OF LOUDNESS

A. Definition of Loudness

Loudness may be defined as that attribute of auditory sensation that corresponds most closely to the physical measure of sound intensity, although, as we shall see, this definition is not accurate in all circumstances. A looser definition is that loudness is a psychological description of the magnitude of an auditory sensation (Fletcher & Munson, 1933). Loudness is often regarded as a global attribute of a sound, so that we usually talk about the overall loudness of a sound rather than describe separately the loudness in individual frequency regions. An exception to this arises from the models of loudness described in Section II.D that calculate the "specific loudness" of a sound in each frequency channel it excites to obtain an overall loudness measure by summation.

Loudness is a subjective quantity, and any measurement technique used is based on assumptions and open to interpretation. Our discussion of loudness will be centered on the techniques that have been used to obtain a reliable measure of auditory sensation.

B. Loudness Matching

Probably the least controversial measurement technique is that of *loudness matching,* in which the listener is required to vary the intensity of one stimulus so that it sounds as loud as a standard stimulus with a fixed intensity. This procedure can reveal how the physical parameters of a sound, such as its frequency and bandwidth, affect its loudness. It can also shed light on how loudness is affected by factors intrinsic to the listener, such as the existence of a sensory hearing loss in one ear. The main strength of this technique is that it makes no major assumptions, beyond the reasonable one that the listener can equate the loudness of sounds that differ on some other dimension (e.g., frequency or bandwidth).

In the following subsections we will describe some of the experimental results that have been obtained using the loudness matching procedure.

1. Equal Loudness Contours

Equal loudness contours are descriptions of the frequency dependence of the loudness of pure tones (Fletcher & Munson, 1933). They can be measured fairly easily by requiring listeners to match the intensity of a comparison tone of variable frequency to the intensity of a standard tone at 1 kHz. An equivalent technique often used is to present a series of levels of the comparison tone and ask listeners to judge for each level whether the comparison is "louder" or "softer" than the standard. The transition level from "louder" judgments to "softer" judgments can be taken as the point of subjective equality.

The loudness level (in *phons*) of a tone at any frequency is taken as the level (in dB SPL) of the 1 kHz tone to which it sounds equal in loudness. This means that, for example, any tone that has the same loudness as a 40 dB, 1 kHz tone has, by definition, a loudness of 40 phons. An equal loudness contour, then, is a line joining the levels of tones of different frequencies that have the same loudness in phons.

Figure 1 shows equal loudness contours as measured by Robinson and Dadson (1956). It can be seen that, although the equal loudness contours tend to follow the absolute threshold curve at low loudness levels, at high loudness levels the contours flatten somewhat; this is the result, principally, of a steeper function relating loudness to intensity at low frequencies than at medium frequencies. A familiar consequence of this is that recorded music has, subjectively, a greater relative amount of bass at high intensities than at low intensities. The finding that the loudness function is steeper at low than at medium frequencies is explained by loudness models mainly in terms of the increase in absolute threshold at low frequencies (see Section II.D).

It should be noted that there is currently some argument over whether the equal loudness contours derived by Robinson and Dadson accurately reflect subjective equality (Suzuki & Sone, 1994). It appears that the measurements are not free from bias and can be affected, for example, by the range of comparison levels used (Gabriel, Kollmeier, & Mellert, 1994).

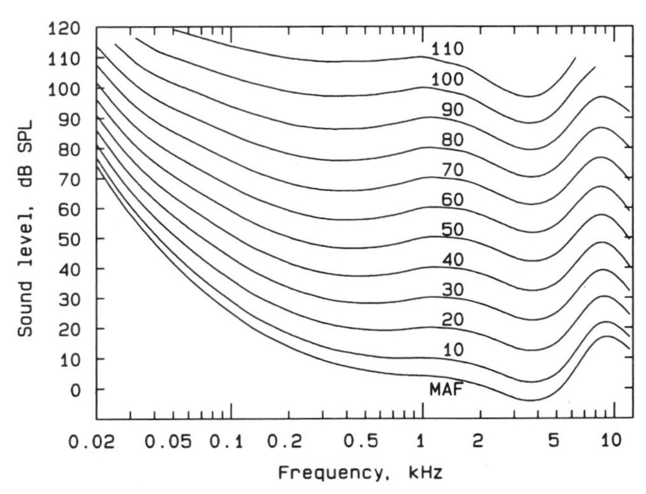

FIGURE 1 Equal loudness contours illustrating the variation in loudness with frequency. Each curve represents one loudness level. (Redrawn from Robinson and Dadson, 1956.)

2. Spectral Factors

Zwicker, Flottorp, and Stevens (1957) asked listeners to match the loudness of a band of noise with that of a standard. The total intensity of the noise was held constant. As the bandwidth of the noise was increased, its loudness stayed roughly constant until the bandwidth was greater than a certain value, termed the *critical bandwidth for loudness summation,* similar in magnitude to the critical bandwidth measure of frequency selectivity (Fletcher, 1940; see Chapter 5). Beyond this point the loudness increased with increasing bandwidth. Cacace and Margolis (1985) have argued that, if the results are plotted on an octave bandwidth scale, rather than the logarithmic frequency bandwidth scale employed by Zwicker, then no sharp transition corresponding to the critical bandwidth is apparent and loudness appears to grow steadily with increasing bandwidth. These results can be accounted for, qualitatively, simply in terms of the total firing rate of fibers in the auditory nerve (Pickles, 1983). As the bandwidth is increased more fibers are excited, and this more than compensates for the reduction in excitation in each critical band. In Zwicker's terms, if the loudness in each critical band follows Stevens's power law (a *compressive* function, see Section II.C), then distributing the intensity of a sound across n critical bands, as opposed to concentrating the intensity in one critical band, will result in an increase in the total loudness, simply because $kI^{0.3} < nk(I/n)^{0.3}$, when $n > 1$.

3. Duration Effects

Measurements of the variation of absolute threshold with duration indicate that, over a certain range of durations, the threshold corresponds to a constant energy rather than a constant power (Garner and Miller, 1947). In other words, over this range of durations, the ear behaves *as if* it were a perfect energy integrator, although there is some debate as to whether the actual neural mechanisms involved include a "true" long time-constant integration device. This issue is discussed in detail in Chapter 6. Not surprisingly, loudness is also related to stimulus duration, although, as one might expect, there is considerable variability in the results. Using a procedure in which listeners were required to match the loudness of tone bursts of variable duration to that of a continuous reference tone, Boone (1973) showed that loudness is also proportional to the total energy of the tone, so that as the duration of a tone of constant power is increased, its loudness also increases. Stephens (1974) replicated these results but showed in addition that this relationship is highly susceptible to the experimental procedure and the instructions given to the listener. In particular, at long durations it is hard to make a judgment of the total loudness of the sound, rather than the loudness at a particular instant or over a short time period. This inevitably leads to a departure from the energy integration rule at long durations.

4. Effects of Sensory Hearing Loss on Loudness: Recruitment

Sensorineural hearing impairment is characterized by elevated thresholds for the detection of sounds in quiet. Despite this loss in sensitivity, a sound at a high intensity might sound equally loud to a hearing impaired listener as it does to a normally hearing listener. In other words, there is an abnormally steep growth of loudness with intensity in the impaired ear. This phenomenon, called *recruitment,* is illustrated in Figure 2. If the listener has an impairment in one ear only, then recruitment can be measured simply by asking the listener to match the loudness of a pure tone with variable intensity presented to the impaired ear, with a pure tone of the same frequency and a fixed intensity presented to the normal ear. Alternatively, the variable tone can be presented to the normal ear and the fixed tone to the impaired ear. To reduce bias effects, both these procedures can be used and the results averaged. The two tones should be presented alternately to minimize the effects of the abnormally fast loudness adaptation that sometimes occurs in impaired ears (Hood, 1950).

Evans (1975) has suggested that recruitment may be a result of the reduced frequency selectivity usually associated with hearing loss (Tyler, 1986). As the intensity of a pure tone is increased, excitation spreads across the basilar membrane so that the number of nerve fibers excited also increases. The broad auditory filters in impaired ears will give rise to a greater spread of excitation with increasing intensity than occurs in normal ears. We

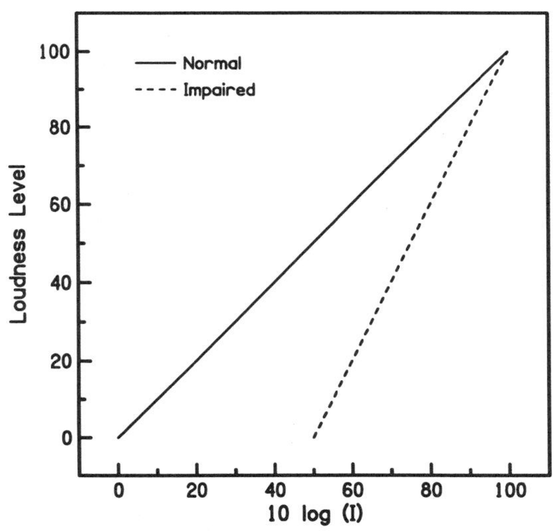

FIGURE 2 Schematic of loudness growth in normal and impaired ears, illustrating recruitment.

have seen in Subsection II.B.2 how loudness increases with spectral extent. Recruitment, therefore, may simply be a consequence of the broader excitation patterns at high intensities in impaired ears: indeed, the phenomenon was so named because it was assumed to be due to the impaired ear "recruiting" an abnormally large number of off-frequency fibers as intensity was raised. Moore, Glasberg, Hess, and Birchall (1985) tested this hypothesis by measuring the loudness of pure tones presented between two flanking bands of noise that were designed to mask the spread of excitation. Loudness matches between the two ears of unilaterally impaired listeners showed that the noise had very little effect on the loudness function in the impaired ears, suggesting that recruitment is not due to abnormal spread of excitation. It seems more probable that recruitment is directly related to damage to the outer hair cells (OHCs) and consequently the loss of the active mechanism in the cochlea (Moore, 1989). Animal models of hearing impairment using ototoxic drugs suggest that damage to the OHCs is the main cause of cochlear hearing loss (Liberman & Dodds, 1984). The OHCs seem to be responsible for enhancing sensitivity to low intensity sounds (Kiang, Moxon, & Levine, 1970), but their response saturates at high intensites (Russell, Cody, & Richardson, 1986), where the effect of the OHCs is probably minimal. We can imagine, therefore, that the effect of OHC damage will be to reduce the loudness of low intensity sounds while leaving the loudness of high intensity sounds unaffected, leading to recruitment.

C. Loudness Scales

Although loudness matching techniques have proven successful in determining the factors that affect loudness, they provide no direct measure of sensation. For example, although equal loudness contours (Figure 1) reveal that loudness grows more steeply with intensity at low than at high frequencies, they cannot provide an absolute measure of the slope of this function. This is a much thornier issue, because perceptual parameters, unlike physical parameters such as frequency or bandwidth, are hidden away inside the listener's head.

Perhaps the most straightforward approach to this problem is that of *magnitude estimation,* in which the listener is simply required to assign numbers to sounds of different intensities. An alternative to magnitude estimation is the technique of *magnitude production,* in which, conversely, the listener is given a number and asked to adjust the intensity of a sound so that its loudness matches that number. In a variant of this method, the listener is presented with a standard sound and asked to adjust the intensity of a second sound so that the ratio between the loudness of the two sounds corresponds to a number given by the experimenter. Stevens (1957) used these techniques to develop *loudness scales,* which describe the relationship between loudness and intensity. He introduced the *sone* as the unit of loudness, where 1 sone was defined, arbitrarily, as the loudness of a 1 kHz pure tone with a level of 40 dB SPL. He then used techniques such as magnitude estimation to determine the loudness of other sounds. For example, a sound judged by listeners to be twice as loud as the 1 kHz 40 dB tone would have a loudness of 2 sones, a sound judged to be half as loud would have a loudness of 0.5 sones, and so on. Using this technique, Stevens argued that the loudness of pure tones is scaled, approximately, as a power function of intensity with exponent 0.3. Specifically,

$$S = kI^{0.3}, \tag{1}$$

where S is loudness, I is intensity, and k is a constant dependent on the listener and the units employed. Under this law, a 10 dB increase in intensity corresponds to roughly a twofold increase in loudness. This relationship holds above about 40 dB SL; for levels between absolute threshold (which has a loudness of 0 sones) and 40 dB the function is steeper.

Magnitude production or magnitude estimation experiments have an advantage over matching experiments in that they might provide a direct link between perception and a physical parameter (intensity), but they both make the important assumption that listeners can use numbers in a consistent and linear way. There are several reasons why this assumption may not always be correct, and listeners can display a number of biases that affect loudness estimations. One example is the "centering" bias, in which listeners center their range of numerical responses on the intensity range of the stimuli. Another is the "contraction" bias, in which large differences between stimuli are underestimated and small differences are overestimated (Poulton, 1979). Responses can also be affected by the instructions to the listener: if the experimenter suggests a large range of numbers for responses ("call it four if it's four times as loud" rather than "call it two if it's two times as loud") this will tend to produce a large range of responses from the listener, artificially expanding the loudness scale. An additional problem has been highlighted by Krueger (1989), who argued that numerical estimates do not scale linearly with sensation, but that in fact $S = N^e$, where S is the "real" sensation, N is the numerical estimate produced by listeners, and e is an exponent with a value less than 1.

Despite these problems, magnitude estimation is still widely used to measure loudness. One reason for this is that many of the problems, such as the contraction bias and those associated with the instructions given to the listener, can be minimized by the use of quite straightforward precautions. These include the use of a variant of magnitude estimation in which no explicit standard is presented and in which listeners are given free rein in their generation of numbers (Stevens, 1971) and the rather more tedious technique of requesting only a single judgment from each listener (Warren,

1970). A second reason, which partially overcomes Krueger's objection, is that measures of magnitude estimation obey the principle of *transitivity*. Hellmann and Meiselman (1990) asked normal and hearing-impaired listeners to estimate the loudnesses of a set of tones of different intensities and estimate the perceived lengths of a set of lines of different actual lengths. They measured the slopes of the functions relating "assigned number" to magnitude on each of these two dimensions. Using the techique of *cross-modality matching,* they also presented listeners with a tone and asked them to adjust the length of a line until its perceived length was "equal," subjectively, to the loudness of the tone. Despite its seemingly bizarre nature, this task can be performed quite consistently, and more important, Hellmann and Meiselman could accurately predict the slopes of the functions relating assigned number to line length from the combination of the slopes obtained in magnitude estimation for loudness and in cross-modality matching. This is illustrated in Figure 3. What their results show is that, although there may

FIGURE 3 Schematic representation of the principle of transitivity as it applies to magnitude estimation and cross-modality matching. The two upper plots show numerical magnitude estimations as a function of sound intensity and line length, respectively. The lower plot shows line length adjusted to match sound intensity. Only if physical magnitude is related to sensation in the same way for the two modalities will the measures combine linearly, as illustrated.

be some nonlinearity relating assigned number to sensation, the nonlinearity must be similar for the two modalities involved. In other words, even though there may be a disparity between the data obtained with magnitude estimation and "true" sensation, we can at least say that magnitude estimates obtained in different modalities and with different groups of listeners are consistent with each other and with other estimates of loudness growth such as cross-modality matching.

D. Models of Loudness

Zwicker (1958; Zwicker & Scharf, 1965) developed a model of loudness based on the excitation pattern. The model consists of a number of stages. First, the input stimulus is passed through a fixed filter representing the transfer characteristics of the outer and middle ear. Above 2 kHz the form of the filter is given by the inverted absolute threshold curve. Below 2 kHz, Zwicker assumed that the transfer function is flat. The rise in absolute threshold with decreasing frequency is assumed to be caused by an internal low-frequency noise. In the second stage, an excitation pattern for the stimulus is calculated (see Chapter 5). Zwicker based his calculation of the excitation pattern on masking patterns for narrowband noises. Excitation is plotted as a function of frequency on a Bark scale (see Chapter 5). Finally, excitation is converted into specific loudness (or loudness per critical band), N'. Following Stevens, N' is assumed to be related to excitation intensity, E, by a power law. Basically,

$$N' = CE^{\alpha}, \tag{2}$$

where C and α are constants and $\alpha < 1$. Zwicker and Fastl (1990) estimated α to be 0.23. This relationship works for excitation levels well above absolute threshold. To account for the steep growth of loudness near absolute threshold the equation was modified as follows:

$$N' = C(E_{THRQ})^{\alpha}[(0.5E_{SIG}/E_{THRQ} + 0.5)^{\alpha} - 1], \tag{3}$$

where E_{SIG} is the excitation produced by the stimulus and E_{THRQ} is the excitation at absolute threshold. The overall loudness of the sound is defined as the *area* under the specific loudness pattern. In other words, the total loudness is the sum of the loudness across each critical band.

This model has been modified by Moore and Glasberg (1986, 1994). They assumed that, below 1 kHz, the form of the initial filter is given by the inverted equal loudness contour at 100 phon (see Figure 1). Above 1 kHz the filter shape is given by the inverted absolute threshold curve. Excitation patterns are calculated from auditory filter shapes they derived in earlier work (Moore & Glasberg, 1983; see Chapter 5 for details). Excitation is converted into specific loudness according to the following relationship:

$$N' = C[(E_{SIG})^{\alpha} - (E_{THRQ})^{\alpha}], \quad (4)$$

the symbols being the same as in Eqs. (2) and (3). This is a simplified version of Eq. (3). Notice that when $E_{SIG} = E_{THRQ}$ the specific loudness is 0. Hence, near absolute threshold, a small change in excitation produces a large *proportional* change in specific loudness. Equation (4) can account, therefore, for the steep (proportional) growth of loudness with level near absolute threshold. When E_{SIG} is much greater than E_{THRQ}, specific loudness (expressed as a logarithm) is almost unaffected by E_{THRQ}. In this way, Eq. (4) can also account for the steep loudness function at low frequencies, as seen in Figure 1, where the specific loudness has to increase from a value of 0 at the (high) absolute threshold to a value at high levels similar to that for higher frequencies.

In the model of Moore and Glasberg, the overall loudness of the sound is calculated by integrating *positive* specific loudness values across the specific loudness pattern, as before. In this case, however, the specific loudness pattern is plotted on an "equivalent rectangular bandwidth" (ERB) frequency scale rather than on the Bark scale (see Chapter 5).

The modified model is quite successful at predicting the variation in loudness with intensity, frequency, and bandwidth (Moore & Glasberg, 1994), supporting the view that loudness is intimately related to the frequency selectivity of the peripheral auditory system, and not just to the physical intensity of the sound per se.

E. Other Factors That Affect Loudness

1. Loudness Adaptation

It is a general property of sensory systems that the neural response to long duration stimulation decays rapidly after stimulus onset to reach a steady "equilibrium" firing rate some time after. It is surprising, therefore, that the loudness of moderate- to high-intensity sounds does not appear to decay over time. Using the technique of successive loudness estimation, it has been shown that only sounds within 39 dB of the absolute threshold show loudness adaptation (Scharf, 1983). High-frequency tones adapt more than low-frequency tones and steady tones adapt more than modulated tones. However, there is a considerable between-listener variability in the results, with some listeners showing no adaptation at all. It is not clear that the loudness estimation technique gives a reliable indication of the actual percept in this instance, although techniques that do indicate considerable adaptation, such as simultaneous dichotic loudness balance, are more seriously flawed (see Moore, 1989).

2. Partial Masking

The experiments described so far have required listeners to judge the loudness of a single "target" sound presented in isolation. However, this does not always provide a good estimate of the loudness of targets presented in the context of other sounds. If, for example, a sinusoidal target is presented in noise, the noise generally reduces the perceived loudness of the target. This phenomenon is termed *partial masking*. In the limiting case, of course, where the noise renders the target undetectable, the loudness of the signal is 0. It has been observed that, just as loudness grows steeply when the intensity of the target is increased just above the absolute threshold, loudness also grows steeply when the intensity of the target is increased just above the masked threshold (Zwicker, 1963; Stevens & Guirao, 1967). The modification of Zwicker's model by Moore and Glasberg (1994; see Section II.D) can account for this effect by assuming that specific loudness is additive:

$$N_{TOT}' = N_{SIG}' + N_{NOISE}', \qquad (5)$$

where N_{SIG}' is the specific loudness of the target, N_{NOISE}' is the specific loudness of the noise and N_{TOT}' is the specific loudness of the noise and the target combined. Moore and Glasberg showed that, by making a few simple assumptions, it is possible to derive N_{TOT}', N_{NOISE}', and hence, by subtraction, N_{SIG}'. When the noise level is high relative to the signal level, N_{NOISE}' is almost identical to N_{TOT}'; hence, N_{SIG}' is low, the effect observed in partial masking experiments.

3. Loudness Enhancement

In contrast to the reduction in loudness produced by sounds presented simultaneously with the target, sounds presented before the target can sometimes produce an increase in loudness. This is the phenomenon of *loudness enhancement,* which was first reported in the early 1970s by two groups of researchers (Irwin & Zwislocki, 1971; Galambos, Bauer, Picton, Squires, & Squires, 1972; Zwislocki, Ketkar, Cannon, & Nodar, 1974; Zwislocki & Sokolich, 1974; Elmasian & Galambos, 1975; Elmasian, Galambos, & Bernheim, 1980). Typically, the listener is presented with two brief tone bursts separated by an interval Δt, followed by a third, "comparison" burst, presented after a much longer silent interval. The task is to adjust the intensity of the comparison burst so that its loudness is equal to that of the second burst. When the first burst is, say, 20 dB more intense than the second, the results show that the loudness of the second burst is enhanced by as much as 15 dB, although the size of this effect decreases with increases in Δt, and the effect disappears after a few hundred milliseconds. Interestingly, two lines of evidence suggest that the phenomenon is not simply due to the loudnesses of the two tones being judged separately and then combined. First, the maximum effect occurs when the enhancer and target tones have the same frequency (Zwislocki & Sokolich, 1974), and, second, the effect is attenuated (but not eliminated) when the two tones are presented to different ears (Galambos et al., 1972; Elmasian & Galambos, 1975). Thus it seems that the sounds must be perceptually similar in some way for the enhancement effect to occur.

III. PARAMETRIC STUDIES OF INTENSITY DISCRIMINATION

A. Measurement Techniques

Intensity discrimination refers to the ability of the auditory system to detect differences in the intensity of sounds, and a large number of experiments have attempted to measure the limits of this important aspect of auditory function. Most of these experiments have used one of two techniques, termed *modulation detection* and *increment detection*. In modulation detection, listeners are required to detect the presence of slow amplitude modulation (AM), threshold being taken as the smallest detectable depth of AM. In increment detection, listeners are required to detect a change in the intensity of a standard stimulus (the *pedestal*). The pedestal can be presented either continuously or gated with the increment. In the latter case, the task is usually to disciminate a stimulus containing the increment from one with the pedestal alone (e.g., which of two stimuli sounds "louder"). Thresholds measured using both the modulation detection and increment detection techniques have been interpreted as reflecting the accuracy with which intensity is encoded in the auditory nerve. In later sections, we will question whether this interpretation is in fact valid, but we will first describe the basic findings on which this discussion is based.

Several different definitions of the "just noticeable difference" (jnd) for intensity have been employed in the past. At present the most common are ΔL, equal to $10 \log(1 + \Delta I/I)$, and the Weber fraction, defined either as $\Delta I/I$, or as $10 \log(\Delta I/I)$, where I is the intensity of the pedestal and ΔI is the intensity of the smallest detectable increment. Green and colleagues have advocated the use of the pressure ratio expressed in dB [$20 \log(\Delta p/p)$] for their studies of "profile analysis" (see Green, 1988, for a discussion of this and other measures). These units produce *relative* measures of the jnd, so that, in an increment detection task, if the intensity of the increment is a constant proportion of the intensity of the pedestal, then ΔL, the Weber fraction, and the pressure ratio will also be constant.

B. Weber's Law and the Near Miss

For wideband noise, the smallest detectable change in intensity, ΔI, is approximately proportional to the intensity of the stimulus, I. That is, $\Delta I/I$, the Weber fraction, is a constant. This is an example of *Weber's Law.* This relationship holds for intensities from about 20 dB above the absolute threshold to about 100 dB above the absolute threshold (Miller, 1947). Within this range a plot of 10 log(ΔI) against 10 log(I) (i.e., both expressed in dB) will give a straight line with a slope of 1. In contrast to the results for wideband noise, for pure tones the Weber fraction decreases slightly at high levels, so that a plot of 10 log(ΔI) against 10 log(I) gives a slope of approximately 0.9. This is referred to as the *near miss* to Weber's Law (McGill & Goldberg, 1968) and is probably due to the spread of excitation along the basilar membrane associated with an increase in the intensity of a pure tone (see Subsection IV.B.1). Weber's Law and the near miss are illustrated schematically in Figure 4. Near absolute threshold the Weber fraction increases dramatically, particularly at frequencies below 200 Hz (Ward & Davidson, 1993).

C. Frequency Effects

An exception to the near miss to Weber's Law for pure tones occurs at high frequencies. Florentine, Buus, and Mason (1987) have provided a comprehensive description of the variation in intensity discrimination with frequency. Their results show that the Weber fraction for pure tone pedestals is generally independent of frequency for frequencies from 250 Hz to 4 kHz (measured at equal sensation level), but that, at higher frequencies, there is a small maximum at medium pedestal intensities, as reported originally by Carlyon and Moore (1984).

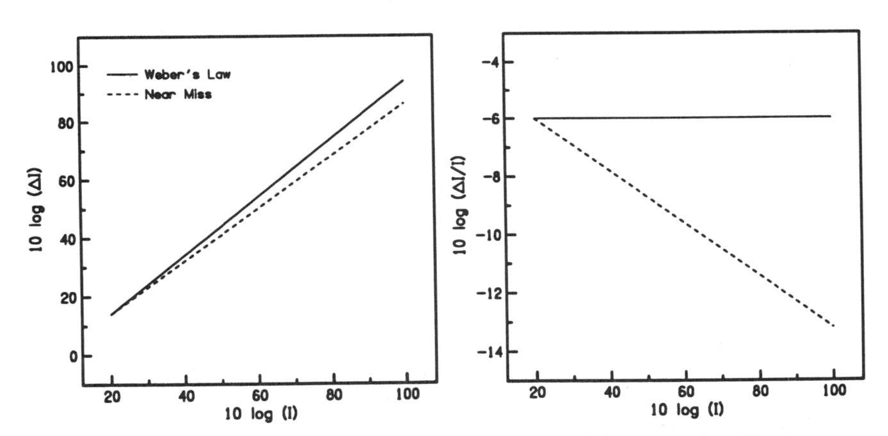

FIGURE 4 A schematic illustration of Weber's Law and the near miss. The curves are plotted in terms of the smallest detectable change in intensity (ΔI, left panel), and the Weber fraction ($\Delta I/I$, right panel).

D. Duration Effects

Henning (1970) reported that, up to a certain duration, the *energy* of the smallest detectable increment (ΔE) on gated pure tone pedestals was constant. Hence, the Weber fraction decreased by 3 dB with every doubling of duration within this limit. Beyond the critical duration, the Weber fraction was constant. The value of the critical duration decreased with increasing pedestal frequency, from 100 ms at 250 Hz to 10 ms at 4 kHz. Florentine (1986), on the other hand, reported critical durations ranging from 500 ms at 250 Hz to over 2 s at 8 kHz. She suggested that the reason for the discrepancy was the much smaller range of stimulus durations used by Henning.

The rate of decrease of the Weber fraction with duration seems to be inversely related to the bandwidth of the pedestal, so that intensity discrimination measured with wideband noise pedestals shows only a slight improvement with duration (Raab & Goldberg, 1975). Another factor affecting the relationship between Weber fractions and duration may be the intensity of the pedestal. Carlyon and Moore (1984) reported that the mid-level deterioration in intensity discrimination at high frequencies was more marked for short than for long stimuli, and so we might expect that, at high frequencies, the variation in the Weber fraction with duration would be most marked at intermediate intensities.

IV. MODELS OF PERIPHERAL INTENSITY CODING

The study of intensity coding is concerned with determining how the ear tells the brain how intense a particular sound is, or, more specifically, how the physical intensity of a sound is represented in terms of the activity, or pattern of activity, of nerve fibers in the auditory system. Intensity coding is inherently related to intensity discrimination. The ability to "hear" two sounds of 120 dB and 130 dB, for example, does not imply that these two intensities are represented differently in the auditory system; they may produce identical percepts. If a listener can detect a difference between the two sounds, however, that difference must be represented at all stages in the auditory pathway between the cochlea and the decision process. The fidelity of the coding mechanism will determine the smallest difference that can be detected. Intensity discrimination experiments are therefore the primary psychophysical tools for testing models of intensity coding.

A successful model of intensity coding has to take account of the frequency selectivity of the cochlea, so that intensity is encoded independently for different frequency channels, even though this information may be combined at some later stage. The firing rates of fibers in the auditory nerve are generally monotonically related to physical intensity. A simplistic hypothesis, therefore, is that the intensity in any given frequency region is coded purely by the firing rate of fibers tuned to that frequency region; the higher the intensity, the higher the firing rate. Although this account may turn out to be accurate in some circumstances, there are several complications that will be discussed in the following sections.

A. The Dynamic Range Problem

As we have seen, the human auditory system can detect differences in the intensity of sounds over a very wide dynamic range; as much as 120 dB in normal hearing listeners (Viemeister & Bacon, 1988). This performance is even more remarkable, however, when we consider the information present in the auditory nerve, as measured in other mammals. The majority of auditory nerve fibers, those with a relatively high spontaneous rate (SR), have low thresholds but relatively small dynamic ranges, with most showing a saturation in their firing rate above an intensity of around 60 dB SPL when stimulated by a tone at their characteristic frequency (CF) (Palmer & Evans, 1979; see also Chapter 3). That is to say, increases in stimulus intensity beyond this point will not result in a change in the firing rate of the majority of auditory nerve fibers. The fact that Weber's Law continues to hold even at high stimulus intensities has prompted a number of researchers to examine ways in which intensity may be coded other than by the firing rate of nerve fibers tuned to the pedestal frequency.

B. Coding by Spread of Excitation

One possible explanation for the apparent paradox is that, at least for narrowband pedestals, information regarding the intensity of a sound is available from nerve fibers tuned to frequencies above and below the pedestal frequency (Siebert, 1965). Although most fibers tuned to the pedestal frequency will be saturated by an intense pedestal, fibers with CFs remote from the pedestal frequency will receive less excitation and may not be saturated. It has been suggested that these "off-frequency" fibers are responsible for coding intensity at high levels (Zwicker, 1956).

1. Masking Spread of Excitation

This hypothesis has been tested in experiments that have used masking to limit the information available from off-frequency fibers. High-pass noise, low-pass noise, and notched noise centered on the frequency of the (pure tone) pedestal, have been added to mask spread of excitation and to force listeners to use nerve fibers with CFs close to the pedestal frequency (Viemeister, 1972; Moore & Raab, 1974). These maskers produced a slight increase in the Weber fraction at high intensities, removing the near miss, but performance overall was relatively unimpaired by limiting spread of excitation. Both low-pass noise and high-pass noise were effective in increasing the Weber fraction at high intensities, and notched noise was more effective still (Moore & Raab, 1974), suggesting that spread of excitation on both sides of the excitation pattern is involved in the near miss.

This conclusion has been supported by attempts to model intensity discrimination using excitation patterns. Zwicker (1956, 1970) described a single-band model of intensity discrimination in which performance is assumed to be determined by the output of the auditory filter in which the change in excitation (in dB) is greatest. Because of the steeper growth of excitation with increasing intensity on the high-frequency side of the excitation pattern (see Chapter 5), for pure tone pedestals the optimum filter will generally have a CF above the frequency of the pedestal. Florentine and Buus (1981) proposed a multiband version of this model in which information is optimally combined from all regions of the excitation pattern. Although both models predict qualitatively the near miss and its removal by masking with notched noise, the multiband model gives predictions that are in closer quantitative agreement with the data (Florentine & Buus, 1981).

The results from masking experiments have been taken as evidence that, although spread of excitation may aid intensity discrimination at high intensities, producing the near miss, the auditory system can code intensity over a large dynamic range on the basis of the information from a small range of CFs. This is perhaps not surprising, as most complex sounds in the environment have relatively broad spectra and occur not in isolation but in the presence of other sounds. Both of these factors will limit the usefulness of information from spread of excitation in everyday situations, where listeners would need to code intensity over a wide dynamic range. As we shall see in the next subsection, however, the argument for a large dynamic range at a single CF is not entirely watertight.

2. Role of Suppression and Adaptation

One way for the auditory system to maintain a large dynamic range despite the limited dynamic range of the majority of auditory nerve fibers would be to adjust the operating ranges of individual fibers according to the input level, so that the intensity of the incoming stimulus always fell on the steep part of their input–output functions. We are all familiar with this principle of *automatic gain control* (AGC) in vision, where the phenomenon of adaptation allows us to see in bright sunlight or in the darkest cinema. We are also familiar with the fact that the recovery from this adaptation is fairly slow, as anyone who has entered a cinema from bright sunlight will testify. However, in hearing we have to process sounds that vary dramatically in intensity over a very short time scale: we need the entire dynamic range of hearing "on tap." (For example, it may be advantageous to identify an intense speech signal, "The psychopath is coming down the corridor!" and then immediately listen for those quiet footsteps.) One way out of this dilemma would be to have an AGC system with a short time constant, and one process that could fulfill this role is suppression.

Suppression is a nonlinear process whereby intense excitation at one region of the basilar membrane reduces the excitation in neighboring regions. It has a very short time constant (Arthur, Pfeiffer, & Suga, 1971), so that suppression occurs only when the suppressor and the suppressee are presented simultaneously. Auditory nerve recordings have shown that a notched noise has a strong suppressive effect on the response of nerve fibers to a pure tone presented in the center of the notch, effectively shifting the rate–intensity functions to higher intensities (Palmer & Evans, 1982; Costalupes, Young, & Gibson, 1984). In effect, the excitation produced by the pedestal is reduced by the notched noise, and this may increase the apparent dynamic range of the system, accounting, qualitatively, for the results of the experiments described in the previous subsection that used noise to mask spread of excitation. In this way it is possible that the compressive process on the basilar membrane that produces suppression acts as a form of automatic gain control for wideband stimuli.

Making use of the fact that suppression only occurs between simultaneously presented stimuli, Plack and Viemeister (1993) attempted to mask the spread of excitation while avoiding suppressive effects by using nonsimultaneous masking. They measured intensity discrimination for a brief pure tone pedestal presented in the silent interval between two bursts of an intense masking complex, consisting of a notched noise and two pure tones with frequencies either side of the pedestal frequency. This complex masked spread of excitation without suppressing the pedestal. Figure 5 shows that intensity discrimination was largely unimpaired at high intensities under these conditions, suggesting that suppression is not necessary for the maintenance of a large dynamic range in a limited frequency region. This conclusion seems justified, even though the maskers used by Plack and Viemeister were very intense and would have caused considerable adaptation in the auditory nerve. Although adaptation is, theoretically, a possible additional mechanism that could extend dynamic range, physiological evidence suggests that it does not shift the rate–intensity functions of high-SR fibers to higher intensities (see Chapter 3).

C. Coding by Neural Synchrony

Carlyon and Moore (1984) suggested that, in some circumstances, intensity might be coded by the pattern of phase locking in auditory nerve fibers.

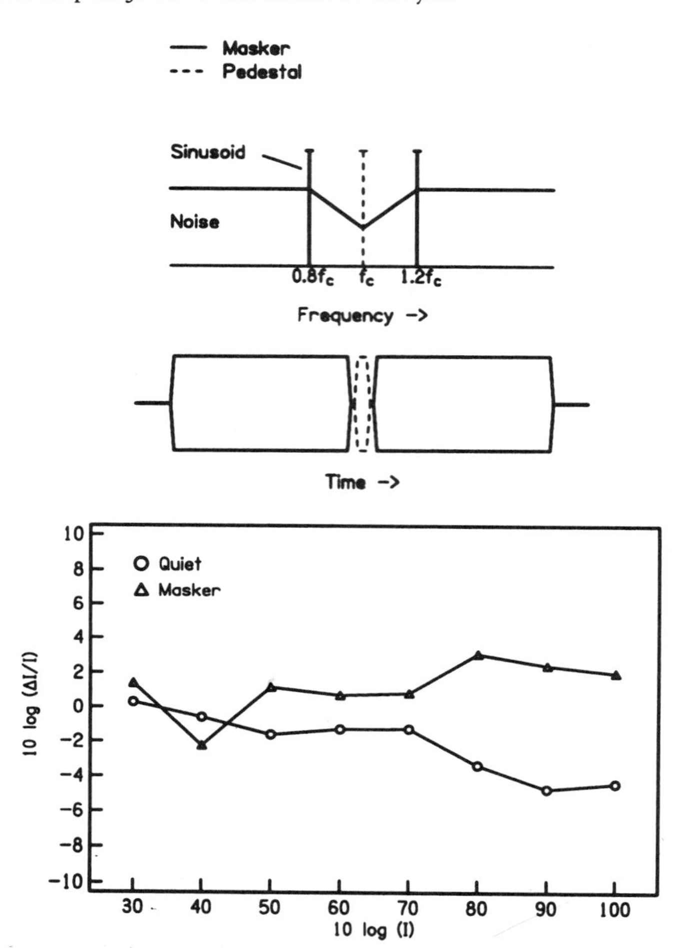

FIGURE 5 Weber fractions for a 30 ms tone burst in quiet (circles) and when presented in the silent interval between two bursts of a masking complex (triangles). The spectral and temporal characteristics of the stimuli are illustrated schematically above the graph. (Data are from Plack and Viemeister, 1993.)

Increasing the intensity of a pure tone in the presence of a noise can produce an increase in the synchronization to the fine structure of the pure tone, away from the fine structure of the noise, even though the overall firing rate of the fiber does not change; in other words, the fiber is saturated (Javel, 1981). In particular, in the notched noise experiment of Moore and Raab (1974), intensity differences at high intensities may have been detected by virtue of an increase in the degree of synchrony in the firing of a neuron tuned to the pedestal frequency. Carlyon and Moore (1984) tested this hypothesis by measuring intensity discrimination for 30 ms pure tone pedestals presented in notched noise for pedestal frequencies of 0.5, 4, and 6.5 kHz. At 6.5 kHz phase locking to the fine structure of the pedestal would be completely absent. Carlyon and Moore demonstrated that there was a large increase in the Weber fraction at high frequencies, but only at medium intensities; at both high and low intensities performance was still relatively good. These results are illustrated in Figure 6. Thus, although neural synchrony may play a role in intensity coding, there is sufficient intensity information from other sources at low and high intensities. In other words, neural synchrony is not *solely* responsible for the large dynamic range observed in these experiments.

D. Models Based on Rate–Intensity Functions

The discussion in Subsections IV.B.1 and IV.B.2 has indicated that the dynamic range in a single frequency channel is probably much greater than that observed in the high-SR fibers. The dynamic range of hearing must depend, therefore, on the minority of auditory nerve fibers with low SRs. These fibers have higher thresholds than the high-SR fibers and have larger dynamic ranges, many of which extend up to very high intensities (Sachs & Abbas, 1974). A plausible hypothesis is that the high-SR fibers are responsible for conveying intensity information at low stimulus intensities (below the thresholds of the low-SR fibers), and the low-SR fibers are responsible for coding intensity at high stimulus intensities (above the saturation intensities of the high-SR fibers). We will refer to this as the *dual population model* of intensity coding.

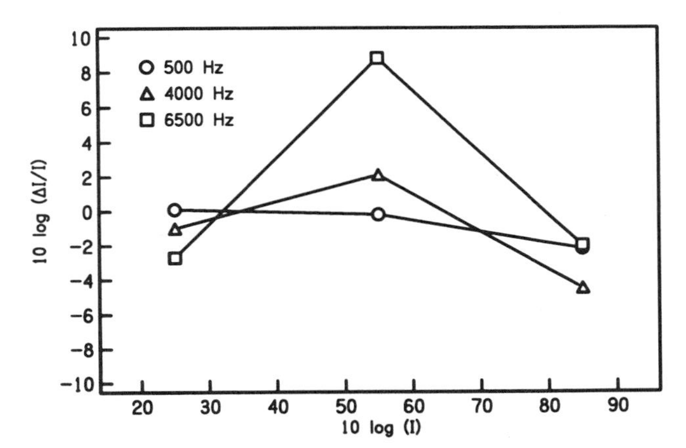

FIGURE 6 The Weber fraction as a function of intensity for 30 ms tone bursts presented in notched noise at two frequencies. (Data are from Carlyon and Moore, 1984.)

The main drawback of this model is that it has difficulty explaining why the Weber fraction is roughly constant for a wide range of stimulus intensities. On the basis of the dual population model it might be expected that intensity resolution would be much more acute at low intensities than at high intensities. This is both because of the far greater numbers of high-SR fibers than low-SR fibers and the fact that the rate–intensity functions (plotted as firing rate against intensity in dB) of low-SR fibers are shallower than those for high-SR fibers. This means that a change in intensity at high intensities will result in a smaller change in the firing rate of the low-SR fibers than the same proportional change in intensity at low intensities produces in the firing rate of the high-SR fibers. To illustrate this point we will consider next a model of intensity coding based on the properties of auditory nerve fibers.

Numerous attempts have been made to model intensity discrimination using the rate–intensity functions of auditory nerve fibers (e.g., Winslow & Sachs, 1988; Young & Barta, 1986; Delgutte, 1987; Viemeister, 1988). As an example, we will consider Viemeister's model here. The first stage of this model calculates the sensitivity measure, d', for a given change in intensity based on the rate–intensity function and the variability in firing rate of a single auditory nerve fiber. The shallower the rate–intensity function and the larger the variability, the smaller the value of d', and the poorer the sensitivity. The performance of a group of nerve fibers is then calculated from an optimal combination of the information from the individual fibers. The bottom panel of Figure 7 shows the predicted Weber fractions based on the performance of 10 or 50 fibers with rate–intensity functions distributed according to the curves in the top panel. These functions are based on the physiological data of Liberman (1978) and Evans and Palmer (1980). It can be seen that the predictions of the model do not obey Weber's Law. Performance at low-to-medium intensities is predicted to be far superior to that at high intensities. Also plotted are psychophysical Weber fractions for a 6–14 kHz noiseband presented in a notched noise to mask spread of excitation (Viemeister, 1983). The neural data lead to very poor predictions of human performance. On the other hand, Figure 7 demonstrates that the information from only ten nerve fibers is enough to achieve performance superior to that of the human listener over a fairly wide dynamic range. With 50 fibers (still a very small percentage of the total number that would be activated in most situations), human performance is matched or bettered over a range of almost 100 dB. These results suggest that sufficient information *is* available in the firing rates of auditory-nerve fibers to encode intensity over a wide range of intensities. Perhaps we should not be asking why human performance is so good at high intensities, but rather, why it is not better at low intensities. We will return to this issue in Section V.

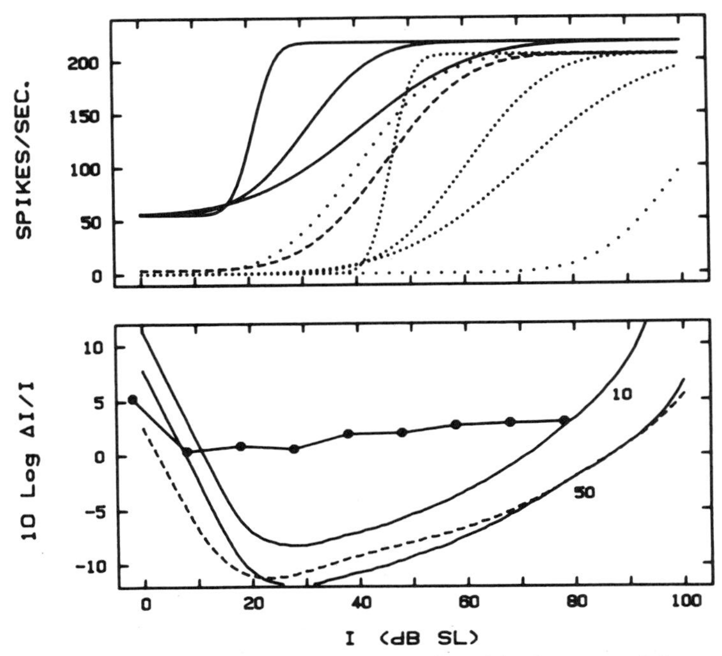

FIGURE 7 The predictions of Viemeister's neural model. The top panel shows the rate–intensity functions for the representative auditory nerve fibers used in the simulation. The lower panel shows psychophysical data (filled circles) and the Weber fractions predicted by the model. The solid lines show the predictions for CF tones using populations of 10 and 50 fibers. The dashed line shows the predictions for broadband noise with a population of 50 fibers.

E. Intensity Discrimination under Nonsimultaneous Masking

A recent discovery by Zeng and colleagues (Zeng, Turner, & Relkin, 1991; Zeng & Turner, 1992) has aroused considerable interest in the effects of nonsimultaneous masking on intensity discrimination. They measured intensity discrimination for 30 ms pure-tone pedestals presented 100 ms after a 90 dB SPL narrowband noise. The Weber fraction was unaffected compared to the value in quiet for low and high pedestal intensities, but was increased by 5–10 dB at pedestal intensities between about 40 and 70 dB. The continuous lines in the left panel of Figure 8 illustrate this phenomenon. Zeng et al. argued that the "mid-level elevation" might be related to the physiological finding that low-SR fibers take several hundred milliseconds to recover their sensitivity after intense stimulation (Relkin & Doucet, 1991). During this period, the thresholds of the low-SR fibers will be elevated. On the basis of the dual population model, Zeng et al. suggested that intense

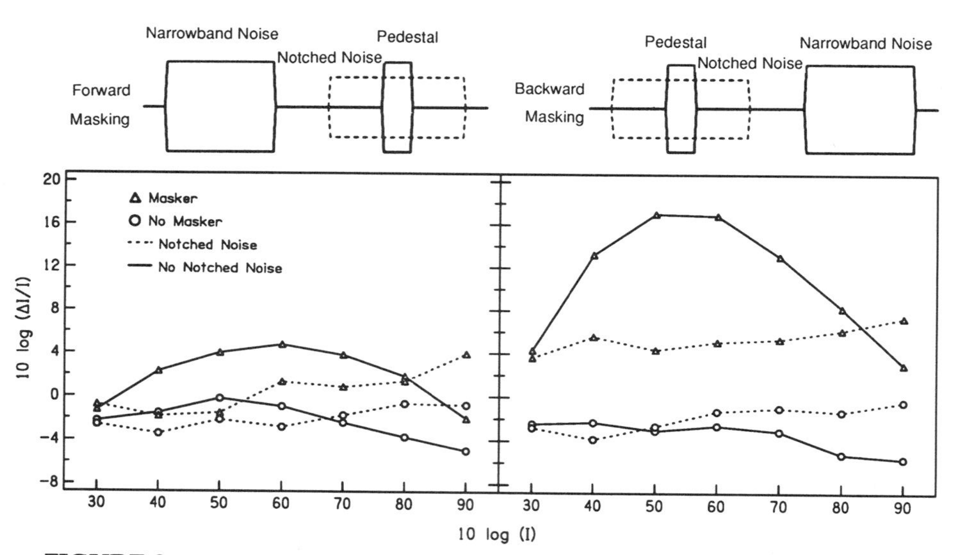

FIGURE 8 The Weber fraction for a 30 ms tone burst in quiet (circles) and under forward and backward masking (triangles). Also shown are the effects of adding notched noise to the pedestal for each of these two conditions (dashed lines). Schematic illustrations of the temporal characteristics of the stimuli are shown above each panel. (Data are from Plack and Viemeister, 1992b.)

stimulation creates a discontinuity in the coding of intensity between the saturation level of the high-SR fibers and the elevated thresholds of the low-SR fibers. A change in intensity between these two levels will not result in a change in the firing rate of either group of fibers, and hence intensity discrimination will be poor. This might account for the mid-level elevation under forward masking.

The hypothesis of Zeng et al. appeared very appealing and of significant theoretical interest because, if true, then their experiment would provide the first direct psychophysical evidence for the dual population model. Unfortunately, subsequent experiments have cast doubt on these claims. Plack and Viemeister (1992b) demonstrated that an even larger mid-level elevation was observed in backward masking conditions, where the masker cannot have affected the representation of the pedestal at the level of the auditory nerve. Furthermore, they showed that the elevation observed in both forward and backward masking could be reduced, or removed entirely, by presenting notched noise with the pedestal. Both these findings are illustrated in Figure 8. Plack and Viemeister (1992a) argued that there is no known physiological mechanism at the level of the auditory nerve that can account for the effect of the notched noise and hence that the processes responsible for the elevation, and its reduction by the notched noise, are probably located more centrally. Two theories being considered are that (1) forward and backward maskers disrupt the memory trace for the pedestal (Mori & Ward, 1992; Plack & Viemeister, 1992b; Carlyon & Beveridge, 1993) or that (2) the effect is related to variability in the loudness enhancement produced by the masker (see Subsection II.E.3) (Carlyon & Beveridge, 1993). In either case the notched noise may improve performance by providing a within-interval reference for the intensity of the pedestal (Plack & Viemeister, 1992b), a form of *context coding*. This will be discussed further in Subsection V.B.1.

V. WHAT LIMITS INTENSITY DISCRIMINATION?

We have described the ways in which the intensity of sounds may be represented in the auditory periphery. We will now consider more generally the factors that may limit our ability to detect changes in the intensity of sounds, including processes central to the auditory nerve. This account will incorporate a description of how intensity may be represented in memory and how memory limitations affect performance on discrimination tasks. Much of the discussion in this section is still speculative, although the hypotheses proposed are open to experimental investigation and it is to be hoped that the account will become clearer and more specific over the next few years.

A. Peripheral and Central Limitations

It is clear from the neural model described in Section IV.D that the information available in the auditory nerve is more than sufficient to account for human psychophysical performance over a wide range of intensities. In fact, when spread of excitation is taken into account, a model based on an optimum combination of the activity of auditory nerve fibers produces far superior performance to that of humans, even at high intensities (Delgutte, 1987). It seems likely from the results of masking experiments, however, that spread of excitation is not necessary for the maintenance of a large dynamic range. Indeed, these experiments suggest that intensity discrimination at single CF is characterised by Weber's Law (except, perhaps, at high frequencies), whereas the physiological models predict much smaller Weber fractions at low and medium intensities than at high intensities.

The richness of the representation in the auditory nerve, and the failure of models of intensity coding based on the properties of auditory nerve fibers to account for Weber's Law, implies that some process central to the auditory nerve must not make optimal use of the neural information (Carlyon & Moore, 1984). Presumably, this *central limitation* determines discrimination performance in most circumstances and prevents human performance from being better at low and medium intensities than at high intensities. The central limitation could be modeled as a constant internal "noise," or variability, added to the decision processes. However, a single central noise could not limit performance based on a single channel (e.g., for intensity discrimination of a tone in notched noise) and, at the same time, allow for any improvement from spread of excitation cues (the near miss to Weber's Law). A more plausible explanation is that there is an independent noise for each frequency channel (Carlyon, 1984; Carlyon & Moore, 1984). This is reasonable if the central noise arises from synaptic transmission throughout the auditory pathway. If the noise is channel specific, then spread of excitation would improve performance by increasing the number of independent "looks" at the intensity change.

While it seems naive to assume that there is a single unitary central limitation, it is also naive to assume that there is just one for each frequency channel. For example, each synapse at every stage in the auditory pathway will probably make some contribution to the total variability of the representation of the stimulus. We should stress, therefore, that we use the term *central limitation* to distinguish central processes from limitations at the level of the auditory nerve, rather than to imply the existence of a single process. In addition, we cannot exclude the possibility that a further source of limitation occurs after information from the different channels has been combined (Carlyon & Moore, 1984).

1. Effects of Restricting Peripheral Information

Experiments that have reduced or degraded the information in the physical stimulus available to the auditory system have provided valuable clues as to the nature of limitations in intensity coding. Carlyon and Moore (1984) went to extreme lengths to remove potential cues for intensity discrimination. They used short tone-burst pedestals presented in notched noise (to mask spread of excitation) at high frequencies (to remove phase locking information) and with the onset and offset of the tone burst masked with bursts of noise to prevent listeners using "transient" cues, such as the physiological onset response. The result of these manipulations was not an overall degradation in performance at all pedestal intensities, as might be anticipated, but an increase in the Weber fraction at *medium intensities only*. There would seem to be two broad explanations for this finding.

1. In the absence of information from spread of excitation and neural synchrony, the coding in the auditory nerve is less accurate at medium intensities. Under normal circumstances this is not evident because the central limitation determines performance. When the information in the stimulus is degraded sufficiently, however, the information in the auditory nerve becomes the limiting factor and hence the coding deficiency at medi-

um intensities is evident in the psychophysical data. A possible reason why there might be such a deficiency is that the basilar membrane, while being relatively linear at low and at high intensities, is compressive in the range from 40 to 80 dB (see Chapter 2). In this region, therefore, the same proportional change in intensity will not produce as large a change in basilar membrane displacement as at low or high intensities (Klitzing & Kohlrausch, 1994). The main problem with this hypothesis is that none of the physiological models demonstrates a coding deficiency at medium intensities although all of the single channel versions show a deficiency at high intensities. It might be expected, therefore, that degrading peripheral information would degrade performance more at high intensities than at medium intensities.

2. Some aspect of the central limitation itself is intensity dependent. Reducing the peripheral information might have the effect of exposing this dependency. A possible mechanism for this will be considered in Subsection V.B.1.

B. Memory for Intensity

Most intensity discrimination experiments employ a two-alternative task in which the listener is required to choose which of two *observation intervals*, separated by an *interstimulus interval* (ISI), contains the more intense stimulus. These three intervals constitute a *trial*. This task could be performed in two different ways. First, the listener could directly compare the intensities of the stimuli in the two observation intervals. This requires that the listener store a representation of the intensity of the first stimulus in short-term memory. Second, if the pedestal, or standard, has the same intensity across a number of trials, then the listener may form a long-term representation of this intensity that can be used to perform the task on a *within-interval* basis, avoiding a direct comparison of the stimuli in the two observation intervals. For example, the listener may select the stimulus that sounds more intense than the representation of the standard in long-term memory.

The absence of a substantial effect of ISI in intensity discrimination tasks employing a fixed standard supports the idea that listeners use long-term memory. A short-term store would be expected to decay over time, producing a large effect of ISI. As an extreme example, Pollack (1955) used an ISI of *24 hours* and found only a small deterioration in intensity discrimination performance. The long-term memory cue can be removed by randomly varying, or *roving*, the intensity of the standard between trials, so that listeners are forced to make a comparison between the two stimuli within each trial. When this is done, performance consistently worsens with increasing ISI (Berliner & Durlach, 1973; Green, Kidd, & Picardi, 1983).

For a 100 ms, 1 kHz sinusoid, the Weber fraction increases from about −2 dB to 5 dB as the ISI is increased from 250 ms to 8 s. Presumably, this reflects the decay of a short-term memory store. If this is the case, then this memory limitation may be an important component of the central limitation in some circumstances.

1. Trace Coding and Context Coding

Durlach and Braida (1969) described a model of intensity coding that includes two different modes of memory operation: the trace mode and the context-coding mode. In the trace mode the direct sensations produced by the stimuli are stored. These sensations have the tendency to decay over time, leading to an increase in "memory noise" and accounting for the effects of ISI. In the context-coding mode, on the other hand, intensity is coded *relative* to a reference intensity or *relative* to internal "perceptual anchors" (Braida, Lim, Berliner, Durlach, Rabinowitz, & Purks, 1984); for example, the absolute threshold or the discomfort threshold. The accuracy of the coding is supposedly dependent on the "distance" on the sensation axis between the sensation of the target stimulus and the sensation of the anchor.

This hypothesis can account, qualitatively at least, for several of the phenomena we have described earlier. First, we have seen that there are several cases in which reducing or masking intensity information causes an elevation in the Weber fraction at *medium intensities only;* for example, short tone bursts at high frequencies in notched noise (Carlyon & Moore, 1984), or tone bursts in nonsimultaneous masking conditions (Zeng et al., 1991; Plack & Viemeister, 1992a; 1992b). If the effect of these manipulations is to degrade the memory trace for the pedestal in some way, then discrimination at low and high intensities may be affected less because the intensity of the pedestal can be context coded with respect to the absolute or discomfort thresholds. This creates a more robust memory trace that is less susceptible to degradation. The context coding hypothesis may provide, therefore, a mechanism whereby the central limitation may be intensity dependent, as postulated in hypothesis 2 in Subsection V.A.1. In normal circumstances the memory trace is rich enough to be relatively immune to the effects of degradation. If the information in the memory trace is reduced, however, then the system relies more and more on context coding, which is not effective at medium intensities because of the distance from the internal anchors at the extremes of sensation.

Second, Carlyon and Moore (1986a) showed that the mid-level elevation in the Weber fraction for short pure tone pedestals at 6 kHz was eliminated if the pedestal was continuous rather than gated with the increment. In the former case, the increment can be context coded with respect to the pedestal intensity before and after the increment. Similarly, notched noise may reduce the mid-level elevation in nonsimultaneous masking by providing a proximal context (Plack & Viemeister, 1992b); for example, the intensity of the pedestal may be coded across frequency with respect to the intensity of the notched noise. Recently, Plack, Carlyon, and Viemeister (1995) showed that presenting a stable "proximal" tone burst shortly before or after the pedestal in nonsimultaneous masking could also reduce the mid-level deterioration in intensity discrimination.

It is fairly obvious that the auditory system needs to code intensity in a relative way at some stage, simply because auditory objects (for example, syllables) are defined by the *relative* intensity of features either simultaneously present (as in the pattern of spectral peaks and dips of a vowel sound) or proximal in time (for example, the dip in intensity that characterizes a stop consonant). Consistent with the relationship of relative intensity to object identification, the context code is often regarded as a "categorical" type of memory trace, so that, for example, in the discrimination of spectral shape (Green et al., 1983; see Chapter 7), the listener may categorize each stimulus as "bumped" or "flat" and use these distinctions as the basis for discrimination. In a sense, the classic intensity discrimination task is an extremely artificial one: Absolute intensity does not affect the identity of auditory objects in most circumstances. It seems plausible, therefore, that the auditory system should be good at "short-range" comparisons of intensity rather than comparisons over several hundred milliseconds, even when the greater short-term memory load in the latter case is taken into account.

VI. INTENSITY DISCRIMINATION AND LOUDNESS

A. The Relationship between Intensity Discrimination and Loudness

Loudness is the subjective correlate of intensity, so we might expect to find some relationship between the loudness of sounds and listeners' ability to detect differences in intensity between them. In particular, it might be expected that the steeper the function relating loudness to intensity, the smaller the intensity jnd. *Fechner's Law* goes one step further to suggest that the Weber fraction, $\Delta I/I$, is associated with a constant increment in loudness, ΔS, so that

$$\Delta S = k\Delta I/I, \tag{6}$$

where k is a constant. In fact, this relationship has been shown to be incorrect and there appears to be no correlation between the Weber fraction and the steepness of the loudness function (Zwislocki & Jordan, 1986). This seemingly paradoxical finding can be understood by considering the factors influencing discriminability. Whereas the difference between the loudness of two sounds may be determined by the difference in the mean level of activity (e.g., firing rate) in nerve fibers that they each produce, their discriminability will also be affected by the variability of that activity. Zwislocki and Jordan suggested that this variability depends not only on the magnitude of loudness, but also on the steepness of the loudness function. In other words, a steeper loudness function may be associated with a greater variability so that there is no reduction in the Weber fraction. Zwislocki and Jordan suggested further that the Weber fraction is equal for stimuli that produce equal loudness. While this relationship seems to hold in many situations (Zwislocki & Jordan, 1986; Schlauch & Wier, 1987), Schlauch (1994) has shown that a high-pass noise has a greater effect on the Weber fraction for a 1 kHz tone than on the loudness function for the tone. This suggests that loudness is less dependent on excitation spread than the Weber fraction is.

B. Intensity Discrimination in Impaired Ears

It is not clear a priori what the consequences of sensorineural hearing impairment should be with regard to intensity discrimination. The steep growth of loudness with intensity (see Subsection II.B.4) suggests that impaired ears might have superior intensity resolution compared to normal ears, since a given change in intensity presumably produces a larger change in loudness in impaired ears. On the other hand, although it is quite possible that impaired ears show an abnormally large difference in, say, the mean auditory-nerve firing rates elicited by two sounds, if the standard deviations of those firing rates are also abnormally large, then the resulting Weber fractions may not be less than those for normally hearing listeners.

Early studies using pure tone pedestals showed little difference between impaired and normal ears (Hirsh, Palva, & Goodman, 1954). Recent results, however, have indicated that listeners with increasing hearing loss above the pedestal frequency have impaired intensity discrimination, particularly at high intensities, whereas listeners with increasing hearing loss below the pedestal frequency show relatively normal intensity discrimination functions (Florentine, Reed, Rabinowitz, Braida, Durlach, & Buus, 1993). These results suggest that the restricted frequency range associated with hearing loss is of far more consequence with regard to intensity discrimination that the shape of the loudness function. They also support the hypothesis that information on the high frequency side of the excitation pattern is important in intensity coding at high intensities. These conclusions are consistent with the idea that intensity discrimination is limited by processes central to the auditory nerve (see Section V.A).

VII. DETECTION OF TONES IN NOISE

A task that is at least superficially similar to that used in many intensity discrimination experiments is that of detecting a tone added to a band of noise. In each case, the listener has to distinguish between two sounds, one of which has a tone added to it: The only difference is that in intensity discrimination the baseline sound is another sinusoid, whereas for the detection of tones in noise it is, of course, a noise. The psychophysical data reveals at least two similarities between the two tasks. First, thresholds for the detection of long-duration (>about 100 ms) tones in noise usually follow Weber's Law, a finding similar to that obtained for tone-on-tone intensity discrimination when the spread of excitation is masked by notched noise. Second, for brief tones added to synchronous bursts of noise, there is a mid-level deterioration in performance at high frequencies, similar to that seen for intensity discrimination (Carlyon & Moore, 1986b).

A. The Overshoot Effect

In contrast to the case where the masker is turned on and off synchronously with the signal, Weber fractions for brief tones presented in continuous noise are uniformly low, even at high frequencies and intermediate intensities (compare the two columns of Figure 9(a)). Under these conditions, then, thresholds are much lower with continuous than with synchronous maskers, a finding that can be thought of as an extreme example of a phenomenon first reported by Zwicker in 1965, termed the *overshoot effect*. The term refers to the fact that the threshold for a brief tone presented shortly after the onset of a burst of noise can be more than 10 dB higher than when it is presented, say, 200 ms after the beginning of the noise (Figure 9(b)). Bacon and Smith (1991) have recently shown that the overshoot effect is greatest at intermediate intensities.

As Figure 9 shows, differences in timing critically affect Weber functions at medium intensities both for intensity discrimination and the detection of tones in noise. One possible reason for this comes from an influential explanation for the overshoot effect, derived by Smith and Zwislocki (1975) from their measurements of adaptation in the auditory nerve. They showed that, when a signal was presented Δt ms after the onset of a masker, the increase in firing rate produced at the beginning of the signal was independent of Δt. As the response to the masker decreased over time, this meant that, at longer values of Δt, the same increment in firing rate was being detected against a lower background rate: in effect, the "neural signal-to-masker ratio" for a given signal intensity was higher at longer values of Δt, and this could have reduced the intensity of the signal necessary for detection. Al-

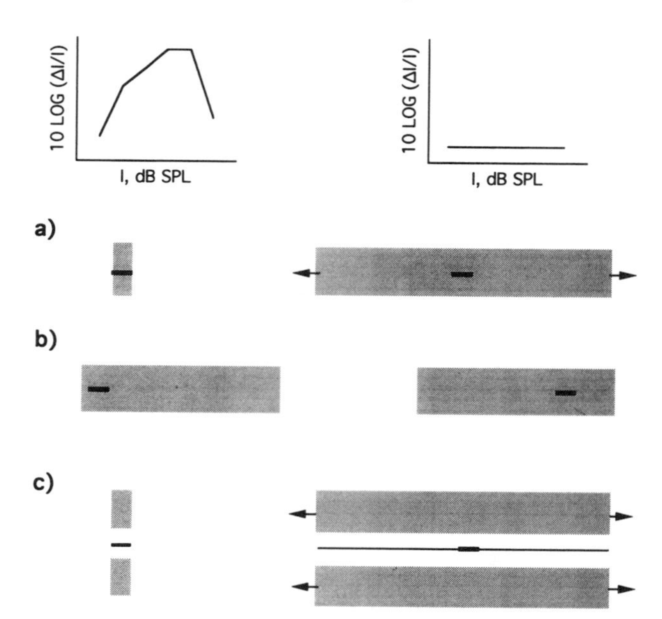

FIGURE 9 Schematic spectrograms of conditions producing a mid-level deterioration (left column) or Weber's Law (right column): (a) synchronous vs. continuous masker, (b) the overshoot effect, (c) intensity discrimination with pedestal and notched noise synchronous with increment, vs. continuous pedestal and noise.

though the explanation of Smith and Zwislocki was formulated before the intensity and frequency dependence of the overshoot effect was fully determined, it is not too difficult to imagine how such a peripheral explanation could be extended to account for the psychophysical results. For example, if peripheral information were degraded enough to limit performance only at intermediate intensities and at high frequencies, then the beneficial effects of adaptation would affect thresholds only under those conditions.

Unfortunately, the hypothesis of Smith and Zwislocki cannot provide a complete account of the overshoot effect. Some more recent experiments performed by McFadden (1989) make the important point that, for a large overshoot effect to occur, the masker must contain energy at frequencies *remote* from the signal frequency and that this energy must be turned on shortly before the signal (Figure 10) (Much of this information was actually presented by Zwicker, 1965, but was largely neglected until McFadden's study.) As the adaptation in fibers maximally responsive to the signal is produced mainly by masker energy close to, rather than remote from, the signal frequency, the effect of off-frequency energy must be mediated by a different, probably more central, mechanism. There are a number of ways in which this could occur. For example, Bacon and Moore (1987) have suggested that the occurrence of a large number of onset responses in fibers tuned away from the signal frequency might somehow prevent a central detector from efficiently processing information in a single channel. However, it is hard to see why this "transient masking" should be greatest at high frequencies and medium intensities.

The two explanations for the overshoot effect that we have discussed can, individually, account for only part of the phenomenon. On the one hand, adaptation could possibly account for the dependence of the effect on frequency and intensity, but not for the role of off-frequency energy. On the other, transient masking can explain the importance of off-frequency energy, but not the dependence of overshoot on masker level or signal frequen-

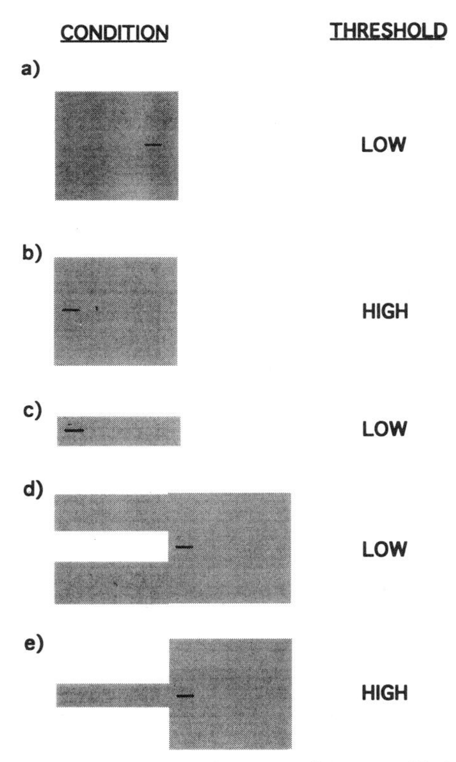

FIGURE 10 Schematic spectrograms of some conditions used in (or relevant to) the study by McFadden (1989). Parts (a) and (b) show the basic overshoot effect, characterized by a high threshold at the shorter Δt. Part (c) shows that this high threshold does not occur with a narrowband masker. Part (d) shows that it does not occur when there is a delay between the onset of the off-frequency band and the signal. Part (e) shows the high threshold *does* occur when only the energy immediately surrounding the signal frequency is turned on before the signal.

cy. One way out of this dilemma has recently been proposed by Carlyon and White (1992), who showed that, although having a wideband masker was essential for a large overshoot in most conditions, this was not the case at high frequencies and intermediate intensities. They suggested that there are two components to the overshoot effect, one of which is mediated by masker energy close to the signal frequency, might be due to adaptation, and has a large effect only at high frequencies and intermediate intensities. According to them, the other component raises thresholds at short values of Δt for all combinations of signal frequency and masker intensity and is mediated by masker energy remote from the signal frequency.

Finally, it is worth noting two important pieces of evidence indicating that the overshoot effect is affected by the way sounds are encoded in the auditory periphery: Bacon and Takahashi (1992) have shown that the mid-level maximum in the overshoot is reduced in listeners with cochlear hearing loss, and McFadden and Champlin (1990) have shown a reduction in overshoot produced by the ingestion of aspirin, which is believed to affect the action of an active mechanism in the cochlea.

B. Intensity-Independent Cues

In a tone-on-tone intensity discrimination task, the only difference between the signal and standard tones is that the former contains more energy. This cue is, of course, also available in a tone-in-noise task, but, despite the similarities in the form of the Weber function for the two tasks, there is now evidence that the detection of a tone in noise is not always strongly dependent on there being a intensity difference between the noise alone and the noise plus tone.

Richards (1992) asked listeners to detect a 200 ms tone added to a 40 Hz wide band of noise in an experiment where the overall level of each stimulus (noise or noise plus tone) was either varied randomly (roved) between the two intervals to be compared or was not roved. The 30 dB rove meant that, if listeners relied on intensity differences between the masker and the masker plus signal, thresholds should have been much higher than in the condition where the level was not roved: The signal would have had to be much more intense for the masker plus signal interval to have consistently more energy than the "masker alone" interval. Instead, thresholds increased by only about 4 dB. Richards suggested that listeners could detect the signal by virtue of it decreasing the "average envelope slope" of the masker; in effect, the signal filled in the dips in the masker envelope, providing a cue independent of overall masker intensity.

What do Richards's data tell us about the mechanisms underlying the different forms of the Weber functions described in previous sections? Richards and Nekrich (1993) have suggested that envelope cues are usable only for long-duration signals: When the signal duration is short relative to the average period of the fluctuations in the masker's envelope, then the cues related to a change in this envelope are not available. In other words, a brief signal might be perceived simply as a random fluctuation in the masker's envelope. As overshoot experiments necessarily use brief signals, it is likely that the results from overshoot experiments are unaffected by envelope cues and that the detection process is essentially the same as for intensity discrimination experiments.

VIII. SUMMARY

The perception of sound intensity underlies all aspects of auditory function, and many of the concepts we have considered here are relevant to the topics discussed in subsequent chapters of this volume. In this chapter we have examined several interrelated aspects of intensity perception. In the early sections we described how the perceived magnitude, or loudness, of a sound is related to its physical magnitude. There are severe methodological problems with trying to measure objectively an inherently subjective quantity, although many of these can be overcome with careful choice of stimuli and procedure. These procedures have been used to provide fairly reliable estimates of the variation in loudness with intensity, frequency, and spectral extent. Models that summate loudness over the entire excitation pattern give an accurate account of the psychophysical data.

The way in which the large dynamic range of hearing is coded or represented in the auditory system is still a matter of speculation. It seems that, although phase locking and spread of excitation may play a role in intensity discrimination, there is sufficient information in the firing rates of auditory nerve fibers tuned to the signal frequency to account for human performance. Indeed, at moderate sound intensities the intensity information in the auditory nerve is much richer than is needed to account for the psychophysical data. The fact that there is more information at moderate intensities than is available above about 70 dB means that neural models are poor predictors of Weber's Law. This has lead to the idea that there is a central limitation to intensity coding that determines Weber's Law. The central limitation may take the form of a channel-specific internal noise. Part of the central limitation may result from our volatile memory for intensity that can limit performance when comparisons are made over long time intervals or perhaps when the memory trace is degraded by nonsimultaneous masking. It is possible that the auditory system can form a more robust memory trace by coding intensity relative to internal or external references. These relative representations may be used in a stage in the identification of auditory objects where the *relative* intensities of features within the sound are more important than the *absolute* intensity of the sound. It is important to remind ourselves that sound intensity is not of great interest to the auditory system in itself, but is vital for the higher function of extracting meaning from sounds in the environment.

References

Arthur, R. M., Pfeiffer, R. R., & Suga, N. (1971). Properties of "two tone inhibition" in primary auditory neurones. *Journal of Physiology, 212,* 593–609.

Bacon, S. P., & Moore, B. C. J. (1987). Transient masking and the temporal course of simultaneous tone-on-tone masking. *Journal of the Acoustical Society of America, 81,* 257–266.

Bacon, S. P., & Smith, M. A. (1991). Spectral, intensive, and temporal factors influencing overshoot. *Quarterly Journal of Experimental Psychology, 43A,* 373–399.

Bacon, S. P., & Takahashi, G. A. (1992). Overshoot in normal-hearing and hearing-impaired subjects. *Journal of the Acoustical Society of America, 91,* 2865–2871.

Berliner, J. E., & Durlach, N. I. (1973). Intensity perception. IV. Resolution in roving-level discrimination. *Journal of the Acoustical Society of America, 53,* 1270–1287.

Boone, M. M. (1973). Loudness measurements on pure tone and broad band impulsive sounds. *Acustica, 29,* 198–204.

Braida, L. D., Lim, J. S., Berliner, J. E., Durlach, N. I., Rabinowitz, W. M., & Purks, S. R. (1984). Intensity perception. XIII. Perceptual anchor model of context coding. *Journal of the Acoustical Society of America, 76,* 722–731.

Cacace, A. T., & Margolis, R. H. (1985). On the loudness of complex stimuli and its relationship to cochlear excitation. *Journal of the Acoustical Society of America, 78,* 1568–1573.

Carlyon, R. P. (1984). Intensity discrimination in hearing. Ph.D. dissertation, University of Cambridge, England.

Carlyon, R. P., & Beveridge, H. A. (1993). Effects of forward masking on intensity discrimination, frequency discrimination, and the detection of tones in noise. *Journal of the Acoustical Society of America, 93,* 2886–2895.

Carlyon, R. P., & Moore, B. C. J. (1984). Intensity discrimination: A severe departure from Weber's law. *Journal of the Acoustical Society of America, 76,* 1369–1376.

Carlyon, R. P., & Moore, B. C. J. (1986a). Continuous versus gated pedestals and the severe departure from Weber's law. *Journal of the Acoustical Society of America, 79,* 453–460.

Carlyon, R. P., & Moore, B. C. J. (1986b). Detection of tones in noise and the severe departure from Weber's Law. *Journal of the Acoustical Society of America, 79,* 461–464.

Carlyon, R. P., & White, L. J. (1992). Some experiments relating to the overshoot effect. In Y. Cazals, L. Demany, & K. Horner (Eds.), *Auditory physiology and perception* (pp. 271–278). Oxford: Pergamon.

Costalupes, J. A., Young, E. D., & Gibson, D. J. (1984). Effects of continuous noise backgrounds on rate response of auditory nerve fibers in cat. *Journal of Neurophysiology, 51,* 1326–1344.

Delgutte, B. (1987). Peripheral auditory processing of speech information: Implications from a physiological study of intensity discrimination. In M. E. H. Schouten (Ed.), *The psychophysics of speech perception* (pp. 333–353). Dordrecht: Nijhof.

Durlach, N. I., & Braida, L. D. (1969). Intensity perception. I. Preliminary theory of intensity resolution. *Journal of the Acoustical Society of America, 46,* 372–383.

Elmasian, R., & Galambos, R. (1975). Loudness enhancement: monaural, binaural, and dichotic. *Journal of the Acoustical Society of America, 58,* 229–234.

Elmasian, R., Galambos, R., & Bernheim, A. (1980). Loudness enhancement and decrement in four paradigms. *Journal of the Acoustical Society of America, 67,* 601–607.

Evans, E. F. (1975). The sharpening of cochlear frequency selectivity in the normal and abnormal cochlea. *Audiology, 14,* 419–442.

Evans, E. F., & Palmer, A. R. (1980). Relationship between the dynamic range of cochlear nerve fibres and their spontaneous activity. *Experimental Brain Research, 40,* 115–118.

Fletcher, H. (1940). Auditory patterns. *Reviews of Modern Physics, 12,* 47–65.

Fletcher, H., & Munson, W. A. (1933). Loudness, its definition, measurement and calculation. *Journal of the Acoustical Society of America, 5,* 82–108.

Florentine, M. (1986). Level discrimination of tones as a function of duration. *Journal of the Acoustical Society of America, 79,* 792–798.

Florentine, M., & Buus, S. (1981). An excitation-pattern model for intensity discrimination. *Journal of the Acoustical Society of America, 70,* 1646–1654.

Florentine, M., Buus, S., & Mason, C. R. (1987). Level discrimination as a function of level for tones from 0.25 to 16 kHz. *Journal of the Acoustical Society of America, 81,* 1528–1541.

Florentine, M., Reed, C. M., Rabinowitz, W. M., Braida, L. D., Durlach, N. I., & Buus, S. (1993). Intensity perception. XIV. Intensity discrimination in listeners with sensorineural hearing loss. *Journal of the Acoustical Society of America, 94,* 2575–2586.

Gabriel, B., Kollmeier, B., & Mellert, V. (1994). Einfluss verschiedener Messmethoden auf Kurven gleicher Pegellautstärke, in *Fortschritte der Akustik, DAGA'94.* Bad Honnef: DPG-GmbH.

Galambos, R., Bauer, J., Picton, T., Squires, K., & Squires, N. (1972). Loudness enhancement following contralateral stimulation. *Journal of the Acoustical Society of America, 52,* 1127–1130.

Garner, W. R., & Miller, G. A. (1947). The masked threshold of pure tones as a function of duration. *Journal of Experimental Psychology, 37,* 293–303.

Green, D. M. (1988). *Profile Analysis.* New York: Oxford University Press.

Green, D. M., Kidd, G., & Picardi, M. C. (1983). Successive versus simultaneous comparison in auditory intensity discrimination. *Journal of the Acoustical Society of America, 73,* 639–643.

Hellman, R. H., & Meiselman, C. H. (1990). Loudness relations for individuals and groups in normal and impaired hearing. *Journal of the Acoustical Society of America, 88,* 2596–2606.

Henning, G. B. (1970). Comparison of the effects of signal duration on frequency and amplitude discrimination. In R. Plomp & G. F. Smoorenberg (Eds.), *Frequency analysis and periodicity detection in hearing.* Leiden: A. W. Sijthoff.

Hirsh, I., Palva, T., & Goodman, A. (1954). Difference limen and recruitment. *Archives of Otolaryngology, 60,* 525–540.

Hood, J. D. (1950). Studies in auditory fatigue and adaptation. *Acta Otolaryngology,* Suppl. *92,* 1–57.

Irwin, R. J., & Zwislocki, J. J. (1971). Loudness effects in pairs of tone bursts. *Perception and Psychophysics, 10,* 189–192.

Javel, E. (1981). Suppression of auditory nerve responses. I. Temporal analysis, intensity effects, and suppression contours. *Journal of the Acoustical Society of America, 69,* 1735–1745.

Kiang, N. Y.-S., Moxon, E. C., & Levine, R. A. (1970). Auditory-nerve activity in cats with normal and abnormal cochleas. In G. E. W. Wolstenhome & J. Knight (Eds.), *Sensorineural Hearing Loss* (pp. 241–268). London: Churchill.

Klitzing, R. V., & Kohlrausch, A. (1994). Effect of masker level on overshoot in running- and frozen-noise maskers. *Journal of the Acoustical Society of America, 95,* 2192–2201.

Krueger, L. E. (1989). Reconciling Fechner and Stevens: Toward a unified psychophysical law. *Behavioral and Brain Sciences, 12,* 251–320.

Liberman, M. C. (1978). Auditory-nerve response from cats raised in a low-noise chamber. *Journal of the Acoustical Society of America, 63,* 442–455.

Liberman, M. C., & Dodds, L. W. (1984). Single neuron labeling and chronic cochlear pathology. II. Stereocilia damage and alterations of threshold tuning curves. *Hearing Research, 16,* 55–74.

McFadden, D. (1989). Spectral differences in the ability of temporal gaps to reset the mechanisms underlying overshoot. *Journal of the Acoustical Society of America, 85,* 254–261.

McFadden, D., & Champlin, C. A. (1990). Reductions in overshoot during aspirin use. *Journal of the Acoustical Society of America, 87,* 2634–2642.

McGill, W. J., & Goldberg, J. P. (1968). A study of the near-miss involving Weber's Law and pure-tone intensity discrimination. *Perception and Psychophysics, 4,* 105–109.

Miller, G. A. (1947). Sensitivity to changes in the intensity of white noise and its relation to masking and loudness. *Journal of the Acoustical Society of America, 19,* 609–619.

Moore, B. C. J. (1989). *An introduction to the psychology of hearing.* New York: Academic Press.

Moore, B. C. J., & Glasberg, B. R. (1983). Suggested formulae for calculating auditory-filter bandwidths and excitation patterns. *Journal of the Acoustical Society of America, 74,* 750–753.

Moore, B. C. J., & Glasberg, B. R. (1986). The role of frequency selectivity in the perception of loudness, pitch and time. In B. C. J. Moore (Ed.), *Frequency Selectivity in Hearing* (pp. 251–308). London: Academic Press.

Moore, B. C. J., & Glasberg, B. R. (in press). A revision of Zwicker's loudness model. *Acustica.*

Moore, B. C. J., Glasberg, B. R., Hess, R. F., & Birchall, J. P. (1985). Effects of flanking noise bands on the rate of growth of loudness of tones in normal and recruiting ears. *Journal of the Acoustical Society of America, 77,* 1505–1513.

Moore, B. C. J., & Raab, D. H. (1974). Pure-tone intensity discrimination: Some experiments relating to the "near miss" to Weber's Law. *Journal of the Acoustical Society of America, 55,* 1049–1054.

Mori, S., & Ward, L. M. (1992). Intensity and frequency resolution: Masking of absolute identification and fixed and roving discrimination. *Journal of the Acoustical Society of America, 91,* 246–255.

Palmer, A. R., & Evans, E. F. (1979). On the peripheral coding of the level of individual frequency components of complex sounds at high sound levels. In O. Creutzfeldt, H. Scheich, & C. Schreiner (Eds.), *Hearing mechanisms and speech* (pp. 19–26). Berlin: Springer-Verlag.

Palmer, A. R., & Evans, E. F. (1982). Intensity coding in the auditory periphery of the cat: Responses of cochlear nerve and cochlear nucleus neurons to signals in the presence of background noise. *Hearing Research, 7,* 305–323.

Pickles, J. O. (1983). Auditory-nerve correlates of loudness summation with stimulus bandwidth, in normal and pathological cochleae. *Hearing Research, 12,* 239–250.

Plack, C. J., Carlyon, R. P., & Viemeister, N. F. (1995). Intensity discrimination under forward and backward masking: Role of referential coding. *Journal of the Acoustical Society of America, 97,* 1141–1149.

Plack, C. J., & Viemeister, N. F. (1992a). The effects of notched noise on intensity discrimination under forward masking. *Journal of the Acoustical Society of America, 92,* 1902–1910.

Plack, C. J., & Viemeister, N. F. (1992b). Intensity discrimination under backward masking. *Journal of the Acoustical Society of America, 92,* 3087–3101.

Plack, C. J., & Viemeister, N. F. (1993). Suppression and the dynamic range of hearing. *Journal of the Acoustical Society of America, 93,* 976–982.

Pollack, I. (1955). "Long-time" differential intensity sensitivity. *Journal of the Acoustical Society of America, 27,* 380–381.

Poulton, E. C. (1979). Models for the biases in judging sensory magnitude. *Psychological Bulletin, 86,* 777–803.

Raab, D. H., & Goldberg, I. A. (1975). Auditory intensity discrimination with bursts of reproducible noise. *Journal of the Acoustical Society of America, 57,* 437–447.

Relkin, E. M., & Doucet, J. R. (1991). Recovery from prior stimulation. I. Relationship to spontaneous firing rates of primary auditory neurons. *Hearing Research, 55,* 215–222.

Richards, V. M. (1992). The effects of level uncertainty on the detection of a tone added to narrow bands of noise. In Y. Cazals, L. Demany, & K. Horner (Eds.), *Auditory physiology and perception* (pp. 337–344). Oxford: Pergamon.

Richards, V. M., & Nekrich, R. D. (1993). The incorporation of level and level-invariant cues for the detection of a tone added to noise. *Journal of the Acoustical Society of America, 94,* 2560–2574.

Robinson, D. W., & Dadson, R. S. (1956). A redetermination of the equal-loudness relations for pure tones. *British Journal of Applied Physics, 7,* 166–181.

Russell, I. J., Cody, A. R., & Richardson, G. P. (1986). The responses of inner and outer hair cells in the basal turn of the guinea-pig cochlea grown *in vitro*. *Hearing Research, 22,* 199–216.

Sachs, M. B., & Abbas, P. J. (1974). Rate versus level functions for auditory-nerve fibres in cats: Tone burst stimuli. *Journal of the Acoustical Society of America, 56,* 1835–1847.

Scharf, B. (1983). Loudness adaptation. In J. V. Tobias & E. D. Schubert (Eds.), *Hearing research and theory* (Vol. 2, pp. 1–56). New York: Academic Press.

Schlauch, R. S. (1994). Intensity resolution and loudness in high-pass noise. *Journal of the Acoustical Society of America, 95,* 2171–2179.

Schlauch, R. S., & Wier, C. C. (1987). A method for relating loudness-matching and intensity-discrimination data. *Journal of Speech and Hearing Research, 30,* 13–20.

Siebert, W. M. (1965). Some implications of the stochastic behavior of primary auditory neurons. *Kybernetik, 2,* 205–215.

Smith, R. L., & Zwislocki, J. J. (1975). Short-term adaptation and incremental response of single auditory-nerve fibers. *Biological Cybernetics, 17,* 169–182.

Stephens, S. D. G. (1974). Methodological factors influencing loudness of short duration sounds. *Journal of Sound and Vibration, 37,* 235–246.

Stevens, S. S. (1957). On the psychophysical law. *Psychological Review, 64,* 153–181.

Stevens, S. S. (1971). Issues in psychophysical measurement. *Psychological Bulletin, 78,* 426–450.

Stevens, S. S., & Guirao, M. (1967). Loudness functions under inhibition. *Perception and Psychophysics, 2,* 459–465.

Suzuki, Y., & Sone, T. (1994). Frequency characteristics of loudness perception: Principles and applications. In A. Schick (Ed.), *Contributions to psychological acoustics* (pp. 193–221). Oldenburg: Bibliotheks und Informationssystem der Universitat Oldenburg.

Tyler, R. S. (1986). Frequency resolution in hearing-impaired listeners. In B. C. J. Moore (Ed.), *Frequency selectivity in hearing* (pp. 309–371). London: Academic Press.

Viemeister, N. F. (1972). Intensity discrimination of pulsed sinusoids: the effects of filtered noise. *Journal of the Acoustical Society of America, 51,* 1256–1269.

Viemeister, N. F. (1983). Auditory intensity discrimination at high frequencies in the presence of noise. *Science, 221,* 1206–1208.

Viemeister, N. F. (1988). Intensity coding and the dynamic range problem. *Hearing Research, 34,* 267–274.

Viemeister, N. F., & Bacon, S. P. (1988). Intensity discrimination, increment detection, and magnitude estimation for 1-kHz tones. *Journal of the Acoustical Society of America, 84,* 172–178.

Ward, L. M., & Davidson, K. P. (1993). Where the action is: Weber fractions as a function of sound pressure at low frequencies. *Journal of the Acoustical Society of America, 94,* 2587–2594.

Warren, R. M. (1970). Elimination of biases in loudness judgements for tones. *Journal of the Acoustical Society of America, 48,* 1397–1403.

Winslow, R. L., & Sachs, M. B. (1988). Single tone intensity discrimination based on auditory-nerve rate responses in backgrounds of quiet, noise and stimulation of the olivocochlear bundle. *Hearing Research, 35,* 165–190.

Young, E. D., & Barta, P. E. (1986). Rate responses of auditory-nerve fibers to tones in noise near masked threshold. *Journal of the Acoustical Society of America, 79,* 426–442.

Zeng, F.-G., & Turner, C. W. (1992). Intensity discrimination in forward masking. *Journal of the Acoustical Society of America, 92,* 782–787.

Zeng, F.-G., Turner, C. W., & Relkin, E. M. (1991). Recovery from prior stimulation. II. Effects upon intensity discrimination. *Hearing Research, 55,* 223–230.

Zwicker, E. (1956). Die Elementaren Grundlagen zur Bestimmung der Informationskapazität des Gehörs. *Acustica, 6,* 365–381.

Zwicker, E. (1958). Über psychologische und methodische Grundlagen der Lautheit. *Acustica, 8,* 237–258.

Zwicker, E. (1963). Über die Lautheit von ungedrosselten und degrosselten Schallen. *Acustica, 13,* 194–211.

Zwicker, E. (1965). Temporal effects in simultaneous masking by white-noise bursts. *Journal of the Acoustical Society of America, 37,* 653–663.

Zwicker, E. (1970). Masking and psychological excitation as consequences of the ear's frequency analysis. In R. Plomp & G. F. Smoorenburg (Eds.), *Frequency analysis and periodicity detection in hearing* (pp. 376–396). Leiden: A. W. Sijthoff.

Zwicker, E., & Fastl, H. (1990). *Psychoacoustics—Facts and models.* Berlin: Springer-Verlag.

Zwicker, E., Flottorp, G., & Stevens, S. S. (1957). Critical bandwidth in loudness summation. *Journal of the Acoustical Society of America, 29,* 548–557.

Zwislocki, J. J., & Jordan, H. N. (1986). On the relations of intensity jnd's to loudness and neural noise. *Journal of the Acoustical Society of America, 79,* 772–780.

Zwislocki, J. J., Ketkar, I., Cannon, M. W., & Nodar, R. H. (1974). Loudness enhancement and summation in pairs or short sound bursts. *Perception and Psychophysics, 16,* 91–95.

Zwicker, E., & Scharf, B. (1965). A model of loudness summation. *Psychological Review, 72,* 3–26.

Zwislocki, J. J., & Sokolich, W. G. (1974). On loudness enhancement of a tone burst by a preceding tone burst. *Perception and Psychophysics, 16,* 87–90.

CHAPTER 5

Frequency Analysis and Masking

Brian C. J. Moore

I. INTRODUCTION

Frequency analysis refers to the ability of the auditory system to separate or resolve (to a certain extent) the components in a complex sound. For example, if two tuning forks, each tuned to a different frequency, are struck simultaneously, two different tones can usually be heard, one corresponding to each frequency. This ability is also known as *frequency selectivity* and *frequency resolution;* these terms will be used interchangeably in this chapter.

It seems likely that frequency analysis depends to a large extent on the filtering that takes place in the cochlea (see Chapters 2 and 3, this volume). Thus, any complex sound, such as a note produced by a musical instrument or a vowel sound produced by the human voice, undergoes such an analysis at an early stage of auditory processing; the sinusoidal components of the sound are separated, and coded independently in the auditory nerve, provided that their frequency separation is sufficiently large. Furthermore, this stage of analysis cannot be bypassed; all sounds are subject to frequency analysis within the cochlea. Hence, the percept of such sounds as a coherent whole depends upon the representations of the individual components being "reassembled" at some later stage in the auditory system (see Chapters 8, 11 and 12).

Hearing

Frequency analysis is most often demonstrated and quantified by studying masking. Masking may be regarded as reflecting the limits of frequency analysis. If a sound of a given frequency is masked by another sound with a different frequency, then the auditory system has failed to resolve the two sounds. Hence, by measuring when one sound is just masked by another, it is possible to characterize the frequency analysis capabilities of the auditory system.

II. THE POWER SPECTRUM MODEL AND THE CONCEPT OF THE CRITICAL BAND

Fletcher (1940) measured the threshold for detecting a sinusoidal signal as a function of the bandwidth of a bandpass noise masker. The noise was always centered at the signal frequency, and the noise power density was held constant. Thus, the total noise power increased as the bandwidth increased. This experiment has been repeated several times since then, with similar results (Hamilton, 1957; Greenwood, 1961a; Spiegel, 1981; Schooneveldt & Moore, 1989; Bernstein & Raab, 1990; Moore, Shailer, Hall, & Schooneveldt, 1993). An example of the results, taken from Moore et al. (1993), is given in Figure 1. The threshold of the signal increases at first as the noise bandwidth increases, but then flattens off; further increases in noise bandwidth do not change the signal threshold significantly.

To account for this pattern of results, Fletcher (1940) suggested that the peripheral auditory system behaves as if it contained a bank of bandpass filters, with overlapping passbands. These filters are now called the *auditory filters*. Fletcher suggested that the signal was detected by attending to the output of the auditory filter centered on the signal frequency. Increases in noise bandwidth result in more noise passing through that filter, as long as the noise bandwidth is less than the filter bandwidth. However, once the noise bandwidth exceeds the filter bandwidth, further increases in noise bandwidth will not increase the noise passing through the filter. Fletcher called the bandwidth at which the signal threshold ceased to increase the *critical bandwidth* (CB). It is usually assumed that this bandwidth is closely related to the bandwidth of the auditory filter at the same center frequency.

Traditionally, the value of the CB has been estimated by fitting the data with two straight lines, a horizontal line for large bandwidths where thresholds are roughly constant, and a sloping line for smaller bandwidths. However, this approach has two problems. First, the data often show no distinct "break point" at which the slope abruptly decreases to 0. Rather, the slope gradually decreases as the bandwidth increases. Second, small errors of measurement can lead to rather large errors in the estimated CB. Thus, Fletcher's band-widening experiment does not provide a precise way of estimating the bandwidth of the auditory filter (Patterson & Moore, 1986).

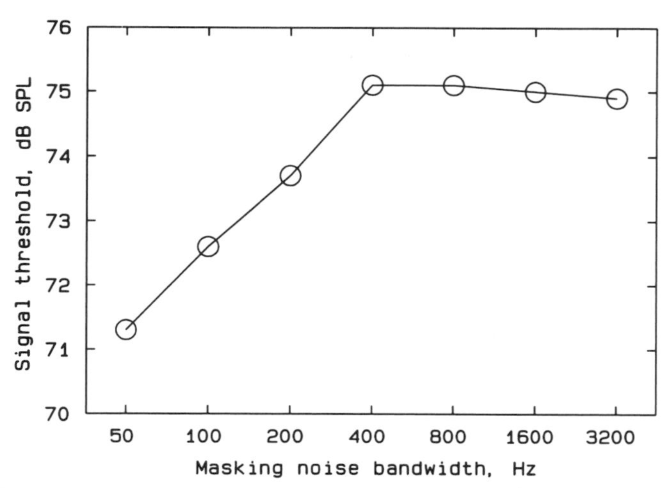

FIGURE 1 The threshold of a 2 kHz sinusoidal signal plotted as a function of the bandwidth of a noise masker centered at 2 kHz. (From Moore et al., 1993.)

Nevertheless, the experiment is important for the concepts to which it gave rise.

Fletcher's experiment led to a model of masking known as the *power-spectrum model,* which is based on the following assumptions:

1. The peripheral auditory system contains an array of linear overlapping bandpass filters.
2. When trying to detect a signal in a noise background, the listener is assumed to use just one filter with a center frequency close to that of the signal. Usually, it is assumed that the filter used is the one that has the highest signal-to-masker ratio at its output.
3. Only the components in the noise that pass through the filter have any effect in masking the signal.
4. The threshold for detecting the signal is determined by the amount of noise passing through the auditory filter; specifically, the threshold is assumed to correspond to a certain signal-to-noise ratio, K, at the output of the filter. The stimuli are represented by their long-term power spectra, that is, the relative phases of the components and the short-term fluctuations in the masker are ignored.

We now know that none of these assumptions is strictly correct: The filters are not linear, but are level dependent (Moore & Glasberg, 1987b); listeners can combine information from more than one filter to enhance signal detection (Spiegel, 1981; Buus, Schorer, Florentine, & Zwicker, 1986; see also

Chapter 4); noise falling outside the passband of the auditory filter centered at the signal frequency can affect the detection of that signal (Hall, Haggard, & Fernandes, 1984; see also Chapter 7); and fluctuations in the masker can play a strong role (Patterson & Henning, 1977; Kohlrausch, 1988; Moore, 1988).

These failures of the model do not mean that the basic concept of the auditory filter is wrong. Indeed, the concept is widely accepted and has proven to be very useful. Although the assumptions of the model do sometimes fail, it works well in many situations. Nevertheless, it should be remembered that simplifying assumptions are often made in attempts to characterize and model the auditory filter.

In analyzing the results of his experiment, Fletcher made a simplifying assumption. He assumed that the shape of the auditory filter could be approximated as a simple rectangle, with a flat top and vertical edges. For such a filter all components within the passband of the filter are passed equally, and all components outside the passband are removed totally. The width of the passband of this hypothetical filter would be equal to the CB described previously. However, it should be emphasized that the auditory filter is *not* rectangular and that the data rarely show a distinct break point corresponding to the CB. It is surprising how, even today, many researchers talk about the critical band as if the underlying filter were rectangular.

Fletcher pointed out that the value of the CB could be estimated indirectly, by measuring the power of a sinusoidal signal (P_s) required for the signal to be detected in broadband white noise, given the assumptions of the power-spectrum model. For a white noise with power density N_0, the total noise power falling within the CB is $N_0 \times$ CB. According to assumption 4,

$$P_s/(\text{CB} \times N_0) = K \tag{1}$$

and

$$\text{CB} = P_s/(K \times N_0). \tag{2}$$

By measuring P_s and N_0, and by estimating K, the value of the CB can be evaluated.

Fletcher estimated that K was equal to 1, indicating that the value of the CB should be equal to P_s/N_0. The ratio P_s/N_0 is now usually known as the *critical ratio*. Unfortunately, Fletcher's estimate of K has turned out not to be accurate. More recent experiments show that K is typically about 0.4 (Scharf, 1970). Thus, at most frequencies the critical ratio is about 0.4 times the value of the CB estimated by more direct methods, such as the band-widening experiment. Also, K varies with center frequency, increasing markedly at low frequencies, so the critical ratio does not give a correct indication of how the CB varies with center frequency (Patterson & Moore, 1986; Moore, Peters, & Glasberg, 1990).

One other aspect of the data in Figure 1 should be noted. If the assumptions of the power spectrum model were correct and if the auditory filter were rectangular, then for subcritical bandwidths the signal threshold should increase by 3 dB per doubling of bandwidth; each doubling of bandwidth should lead to a doubling of the noise power passing through the filter, which corresponds to a 3 dB increase in level. In fact, the rate of change is markedly less than this. The exact slope of the function varies from study to study, but it has often been found to be less than the theoretical 3 dB per doubling of bandwidth (Bernstein & Raab, 1990). The deviation from the theoretical value can probably be explained by two factors: the filter is not actually rectangular, but has a rounded top and sloping edges; and for narrow noise bandwidths, the slow fluctuations in the noise have a deleterious effect on detection (Bos & de Boer, 1966; Patterson & Henning, 1977).

III. ESTIMATING THE SHAPE OF THE AUDITORY FILTER

Most methods for estimating the shape of the auditory filter at a given center frequency are based on the assumptions of the power-spectrum model of masking. If the masker is represented by its long-term power spectrum, $N(f)$, and the weighting function or shape of the auditory filter is $W(f)$, then the power-spectrum model is expressed by

$$P_s = K \int_0^\infty W(f)N(f)\, df, \qquad (3)$$

where P_s is the power of the signal at threshold. By manipulating the masker spectrum, $N(f)$, and measuring the corresponding changes in P_s, it is possible to derive the filter shape, $W(f)$.

The masker chosen to measure the auditory filter shape should be such that the assumptions of the power-spectrum model are not strongly violated. A number of factors affect this choice. If the masker is composed of one or more sinusoids, beats between the signal and masker (see Chapter 1) may provide a cue to the presence of the signal. This makes sinusoids unsuitable as maskers for estimating the auditory filter shape, since the salience of beats changes as the masker frequency is altered; this violates the assumption of the power-spectrum model that threshold corresponds to a constant signal-to-masker ratio at the output of the auditory filter.

In general, noise maskers are more suitable than sinusoids for estimating the auditory filter shape, because noises have inherent amplitude fluctuations that make beats much less effective as a cue. However, for narrowband noises, which have relatively slow fluctuations, temporal interactions between the signal and masker may still be audible. In addition, the slow fluctuations may strongly influence the detectability of the signal in a way that depends on the difference between the center frequency of the masker and the frequency of the signal (Buus, 1985; Moore & Glasberg, 1987a). For these reasons, the assumptions of the power-spectrum model are best satisfied using reasonably broadband noise maskers.

A second important consideration in choosing a noise masker for measuring auditory filter shapes is that the filter giving the highest signal-to-masker ratio is not necessarily centered at the signal frequency. For example, if the signal has a frequency of 1 kHz, and the masker spectrum consists entirely of frequencies above 1 kHz, the highest signal-to-masker ratio may occur for a filter centered below 1 kHz. The process of detecting the signal through a filter that is not centered at the signal frequency is called *off-frequency listening*. In this context, the center frequency of the filter is "off frequency." Furthermore, if the masker spectrum is concentrated primarily above or below the signal frequency, there may be a range of filter center frequencies over which the signal-to-masker ratio is sufficiently high to give useful information. Under these conditions, the observer may combine information over several auditory filters, rather than listening through a single filter as assumed by the power-spectrum model (Patterson & Moore, 1986; Moore, Glasberg, & Simpson, 1992; for a similar concept applied to intensity discrimination, see Chapter 4).

A. Psychophysical Tuning Curves

The measurement of psychophysical tuning curves (PTCs) involves a procedure that is analogous in many ways to the determination of a neural tuning curve (Chistovich, 1957; Small, 1959); see Chapter 3. The signal is fixed in level, usually at a very low level, say, 10 dB SL. The masker can be either a sinusoid or a narrow band of noise, but a noise is generally preferred, for the reasons given earlier.

For each of several masker center frequencies, the level of the masker needed just to mask the signal is determined. Because the signal is at a low level it is assumed that it will produce activity primarily in one auditory filter. It is assumed further that, at threshold, the masker produces a constant output from that filter, in order to mask the fixed signal. Thus the PTC indicates the masker level required to produce a fixed output from the auditory filter as a function of frequency. Normally, a filter characteristic is determined by plotting the output from the filter for an input varying in frequency and fixed in level. However, if the filter is linear, the same result can be obtained by plotting the input required to give a fixed output. Thus, if linearity is assumed, the shape of the auditory filter can be obtained simply by inverting the PTC. Examples of some PTCs are given in Figure 2; the data are taken from Vogten (1974).

It has been assumed so far that only one auditory filter is involved in the determination of a PTC. However, there is now good evidence that off-

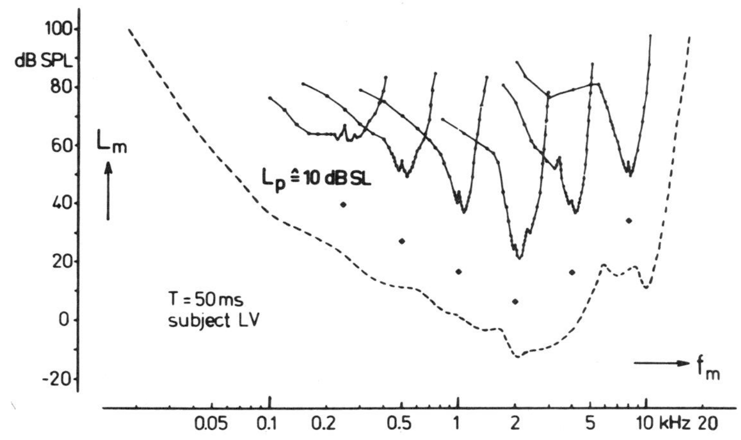

FIGURE 2 Psychophysical tuning curves determined in simultaneous masking, using sinusoidal signals at 10 dB SL. For each curve, the solid diamond below it indicates the frequency and the level of the signal. The masker was a sinusoid that had a fixed starting phase relationship to the brief, 50 ms, signal. The masker level, L_m, required for threshold is plotted as a function of masker frequency, f_m, on a logarithmic scale. The dashed line shows the absolute threshold for the signal. (From Vogten, 1974, by permission of the author.)

frequency listening can influence PTCs. When the masker frequency is above the signal frequency, the highest signal-to-masker ratio occurs for a filter centered below the signal frequency. Conversely, when the masker frequency is below the signal frequency, the highest signal-to-masker ratio occurs for a filter centered above the signal frequency. In both these cases, the masker level required for threshold is higher than would be the case if off-frequency listening did not occur. When the masker frequency equals the signal frequency, the signal-to-masker ratio is similar for all auditory filters that are excited and off-frequency listening is not advantageous. The overall effect of off-frequency listening is that the PTC has a sharper tip than would be obtained if only one auditory filter were involved (Johnson-Davies & Patterson, 1979; O'Loughlin & Moore, 1981a, 1981b).

One way to limit off-frequency listening is to add to the masker a fixed, low-level noise with a spectral notch centered at the signal frequency (O'Loughlin & Moore, 1981a; Moore, Glasberg, & Roberts, 1984; Patterson & Moore, 1986). Such a masker should make it disadvantageous to use an auditory filter whose center frequency is shifted much from the signal frequency. The effect of using such a noise, in addition to the variable narrowband masker, is illustrated in Figure 3. The main effect is to broaden the tip of the PTC; the slopes of the skirts are relatively unaffected.

A final difficulty in using PTCs as a measure of frequency selectivity is connected with the nonlinearity of the auditory filter. Evidence will be

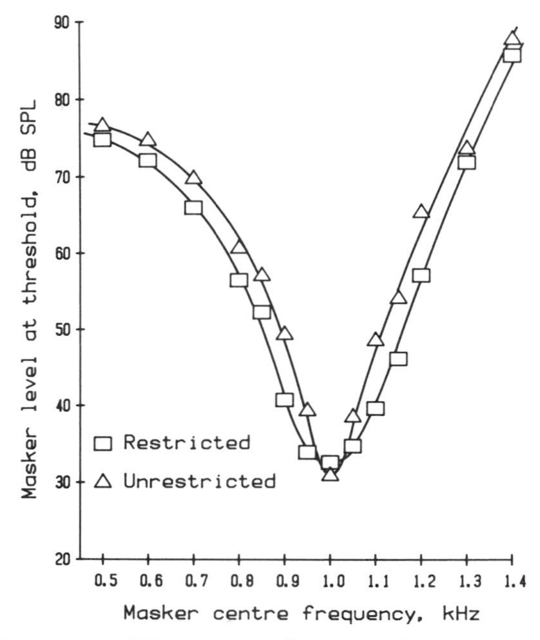

FIGURE 3 Comparison of PTCs where off-frequency listening is not restricted (triangles) and where it is restricted using a low-level notched noise centered at the signal frequency (squares). (Data from Moore et al., 1984.)

presented later indicating that the auditory filter is not strictly linear, but changes its shape with level. The shape seems to depend more on the level at the input to the filter than on the level at the output (although this is still a matter of some debate, as will be discussed later in this chapter). However, in determining a PTC, the input is varied while the output is held (roughly) constant. Thus, effectively, the underlying filter shape changes as the masker frequency is altered. This can give a misleading impression of the shape of the auditory filter; in particular, it leads to an underestimation of the slope of the lower skirt of the filter and an overestimation of the slope of the upper skirt (Verschuure, 1981a, 1981b; Moore & O'Loughlin, 1986).

B. The Notched-Noise Method

To satisfy the assumptions of the power-spectrum model, it is necessary to use a masker that limits the amount by which the center frequency of the filter can be shifted (off-frequency listening) and that limits the range of filter center frequencies over which the signal-to-masker ratio is sufficiently high to be useful. This can be achieved using a noise masker with a spectral notch around the signal frequency. For such a masker, the highest signal-to-masker ratio occurs for a filter that is centered reasonably close to the signal frequency, and performance is not improved (or is improved very little) by combining information over filters covering a range of center frequencies (Patterson, 1976; Patterson & Moore, 1986; Moore et al., 1992). The filter shape can then be estimated by measuring signal threshold as a function of the width of the notch.

For moderate noise levels, the auditory filter is almost symmetrical on a linear frequency scale (Patterson, 1974, 1976; Patterson & Nimmo-Smith, 1980; Moore & Glasberg, 1987b). Hence, the auditory filter shape can be estimated using a notched-noise masker with the notch placed symmetrically about the signal frequency. The method is illustrated in Figure 4. For a masker with a notch width of $2\Delta f$, and a center frequency f_c, Eq. (3) becomes

$$P_s = KN_0 \int_0^{f_c-\Delta f} W(f)\,\mathrm{d}f + KN_0 \int_{f_c+\Delta f}^{\infty} W(f)\,\mathrm{d}f, \tag{4}$$

where N_0 is the power spectral density of the noise in its passbands. The two integrals on the right-hand side of Eq. (4) represent the respective areas in Figure 4 where the lower and upper noise bands overlap the filter. Because both the filter and the masker are symmetrical about the signal frequency, these two areas are equal. Thus, the function relating P_s to the width of the notch provides a measure of the integral of the auditory filter. Hence, the value of $W(f)$ at a given deviation Δf from the center frequency is given by the slope of the threshold function at a notch width of $2\Delta f$.

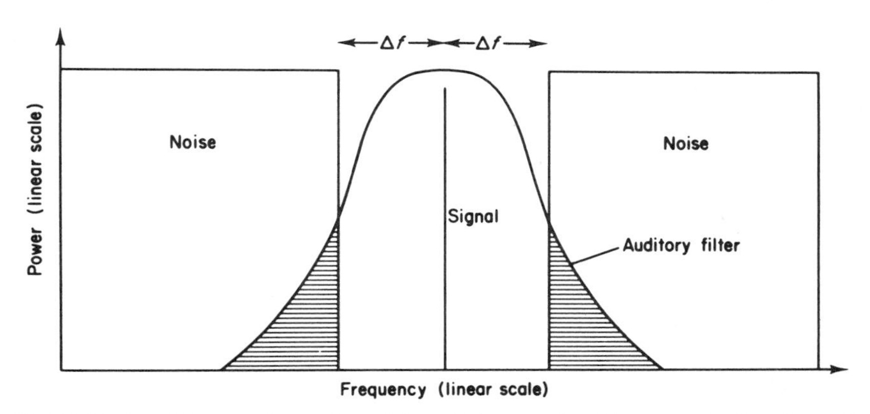

FIGURE 4 Schematic illustration of the technique used by Patterson (1976) to determine the shape of the auditory filter. The threshold of the sinusoidal signal is measured as a function of the width of a spectral notch in the noise masker. The amount of noise passing through the auditory filter centered at the signal frequency is proportional to the shaded areas.

When the auditory filter is asymmetric, as it is at high masker levels (see later), the filter shape can still be measured using a notched-noise masker if some reasonable assumptions are made and if the range of measurements is extended to include conditions where the notch is placed asymmetrically about the signal frequency. It is necessary first to assume that the auditory filter shape can be approximated by a simple mathematical expression with a small number of free parameters. Patterson, Nimmo-Smith, Weber, and Milroy (1982) suggested a family of such expressions, all having the form of an exponential with a rounded top, called *roex* for brevity. The simplest of these expressions was called the *roex(p) filter shape*. It is convenient to measure frequency in terms of the absolute value of the deviation from the center frequency of the filter, f_c, and to normalize this frequency variable by dividing by the center frequency of the filter. The new frequency variable, g, is

$$g = |f - f_c|/f_c. \tag{5}$$

The roex(p) filter shape is then given by

$$W(g) = (1 + pg)\exp(-pg), \tag{6}$$

where p is a parameter that determines both the bandwidth and the slope of the skirts of the auditory filter. The higher the value of p, the more sharply tuned is the filter. The equivalent rectangular bandwidth (ERB) is equal to $4f_c/p$ (see Chapter 1 for a definition of the ERB). When the filter is assumed to be asymmetric, then p is allowed to have different values on the two sides of the filter: p_l for the lower branch and p_u for the upper branch. The ERB in this case is $2f_c/p_l + 2f_c/p_u$.

Having assumed this general form for the auditory filter shape, the values of p_l and p_u for a particular experiment can be determined by rewriting Eq. (4) in terms of the variable g and substituting the preceding expression for W; the value of p_l is used for the first integral, and the value of p_u for the second. The equation can then be solved analytically; for full details see Patterson et al. (1982) and Glasberg, Moore, and Nimmo-Smith (1984a). Starting values of p_l and p_u are assumed, and the equation is used to predict the threshold for each condition (for notches placed both symmetrically and asymmetrically about the signal frequency). The center frequency of the filter is allowed to shift for each condition so as to find the center frequency giving the highest signal-to-masker ratio; this center frequency is assumed in making the prediction for that condition. Standard least-squares minimization procedures are then used to find the values of p_l and p_u that minimize the mean-squared deviation between the obtained and predicted values. The minimization is done with the thresholds expressed in decibels. Full details are given in Patterson and Nimmo-Smith (1980), Glasberg et al. (1984a), Patterson and Moore (1986) and Glasberg and Moore (1990).

The roex(p) filter shape is usually quite successful in predicting the data from notched-noise experiments, except when the thresholds cover a wide range of levels or when the masked thresholds approach the absolute threshold. In such cases there is a decrease in the slope of the function relating threshold to notch width, a decrease that is not predicted by the roex(p) filter shape. This can be accommodated in two ways. The first involves limiting the dynamic range of the filter, using a second parameter, r. This gives the roex(p,r) filter shape of Patterson et al. (1982):

$$W(g) = (1 - r)(1 + pg)\exp(-pg) + r. \tag{7}$$

As before, p can have different values for the upper and lower branches of the filter. However, the data can generally be well predicted using the same value of r for the two sides of the filter (Tyler, Hall, Glasberg, Moore, & Patterson, 1984; Glasberg & Moore, 1986). The method of deriving filter shapes using this expression is exactly analogous to that described earlier.

When the noise level used is relatively high and a large range of notch widths is used, there may be systematic deviations of the data from values predicted by the roex(p,r) model. In such cases, a better fit to the data can be obtained using a model in which the slope of the filter is assumed to decrease once its attenuation exceeds a certain value. This is achieved using the roex(p,w,t) model suggested by Patterson et al. (1982). The filter is assumed to be the sum of two exponentials, both of which are rounded:

$$W(g) = (1 - w)(1 + pg)\exp(-pg) + w(1 + tg)\exp(-tg) \tag{8}$$

The parameter t determines the slope of the filter at large deviations from the center frequency, and the parameter w determines the point at which the shallower "tail" takes over from the steeper central passband. In principle, all three parameters, p, w, and t, could be different for the two sides of the filter, giving six free parameters all together. In practice, it has been found that the results can be well fitted by assuming that one side of the filter is a "stretched" version of the other side (Patterson & Nimmo-Smith, 1980; Glasberg, Moore, Patterson, & Nimmo-Smith, 1984b). In this case, w is assumed to be the same for the two sides of the filter, and the ratio p/t is assumed to be the same for the two sides of the filter. This reduces the number of free parameters to four.

One limitation of the notched-noise method occurs when the auditory filter is markedly asymmetric, as it is, for example, at high sound levels. In such cases, the method does not define the sharper side of the filter very well. As a rule of thumb, when the value of p for one side of the filter is more than twice that for the other, the slope of the steeper side is very poorly determined.

A second potential problem with the method is that components within the upper band of noise may interact to produce combination products whose frequencies lie within the notch in the noise. Such combination products are produced by nonlinear processes within the cochlea, and they occur even when the input is at low to moderate sound levels (Greenwood, 1971; Smoorenburg, 1972a, 1972b). The effect of this is that the upper band of noise may produce more masking than would be the case if no combination products were present. This can result in a derived filter shape with a shallower upper skirt. However, the effect on the derived filter shape is usually small (Moore, Glasberg, van der Heijden, Houtsma, & Kohlrausch, 1995).

C. The Rippled-Noise Method

Several researchers have estimated auditory filter shapes using *rippled noise,* sometimes also called *comb-filtered noise,* as a masker. This is produced by adding white noise to a copy of itself that has been delayed by T seconds. The resulting spectrum has peaks spaced at $1/T$ Hz, with minima in between. When the delayed version of the noise is added to the original in phase, the first peak in the spectrum of the noise occurs at 0 Hz; this noise is referred to as *cosine+*. When the polarity of the delayed noise is reversed, the first peak is at 0.5/T Hz; this is referred to as *cosine−*. The sinusoidal signal is usually fixed in frequency, and the values of T are chosen so that the signal falls at either a maximum or minimum in the masker spectrum; the signal threshold is measured for both cosine+ and cosine− noise for various ripple densities (different values of T).

The auditory filter shape can be derived from the data either by approximating the auditory filter as a Fourier series (Houtgast, 1977; Pick, 1980) or by a method similar to that described for the notched-noise method (Glasberg et al., 1984a; Patterson & Moore, 1986). The filter shapes obtained in this way are generally similar to those obtained using the notched-noise method, although they tend to have a slightly broader and flatter top (Glasberg et al., 1984a). The method seems to be quite good for defining the shape of the tip of the auditory filter, but it does not allow the auditory filter shape to be measured over a wide dynamic range.

D. Allowing for the Transfer Function of the Outer and Middle Ear

The transfer function of the outer and middle ear varies markedly with frequency, particularly at very low and high frequencies. Clearly this can have a significant influence on measures of frequency selectivity. For example, if one of the bands of noise in a notched-noise experiment is very low or high in center frequency, it will be strongly attenuated by the middle ear and so will not do much masking. It is possible to conceive of the auditory filter shape as resulting from the overall response properties of the outer and middle ear and the cochlea. However, it is theoretically more appealing to conceive of the auditory filter as resulting from processes occurring after the outer and middle ear. The effect of the outer and middle ear can be thought of as a fixed frequency-dependent attenuation applied to all stimuli before auditory filtering takes place.

If this is the case, then the frequency-dependent attenuation should be taken into account in the fitting procedure for deriving filter shapes. Essentially, the spectra of the stimuli at the input to the cochlea have to be calculated by assuming a certain form for the frequency-dependent transfer. The fitting procedure then has to work on the basis of these "corrected" spectra. In practice, this means that the integral in Eq. (3) cannot be solved analytically, but has to be evaluated numerically. Glasberg and Moore (1990) have considered several possible types of "correction." One is appropriate for stimuli presented in a free field (e.g., via a loudspeaker in an anechoic chamber) or via earphones designed to have a free-field response, such as the Sennheiser HD414 or the Etymotic Research ER4. Another is appropriate for earphones designed to give a flat response at the eardrum, such as the Etymotic Research ER2. In both cases, the "correction" may be modified to take into account the specific properties of the transducers used. Glasberg and Moore (1990) list a computer program for deriving auditory filter shapes from notched-noise data that includes the option of using "corrections" to allow for the transfer function of the outer and middle ear.

E. An Example of Measurement of the Auditory Filter Shape

Figure 5 shows an example of data obtained using the notched-noise method, and of the filter shape obtained. The data are for a normally hearing subject and a signal frequency of 200 Hz. In the top panel, signal thresholds are plotted as a function of the width of the spectral notch in the noise masker. Asterisks indicate conditions where the spectral notch was placed symmetrically about the signal frequency; the notch width, Δ, is specified as the deviation of each edge of the notch from the signal frequency, divided by the signal frequency. The left-pointing arrows indicate conditions where the lower edge of the notch was 0.2 units farther from the signal frequency than the upper edge. The right-pointing arrows indicate conditions where the upper edge of the notch was 0.2 units farther from the signal frequency than the lower edge. Moving the lower edge of the notch father from the signal frequency has a greater effect than moving the upper edge farther from the signal frequency.

The lines in the top panel are the fitted values derived from the roex(p,r) model, as described by Glasberg and Moore (1990). The model fits the data well. The derived filter shape is shown in the bottom panel. The filter is

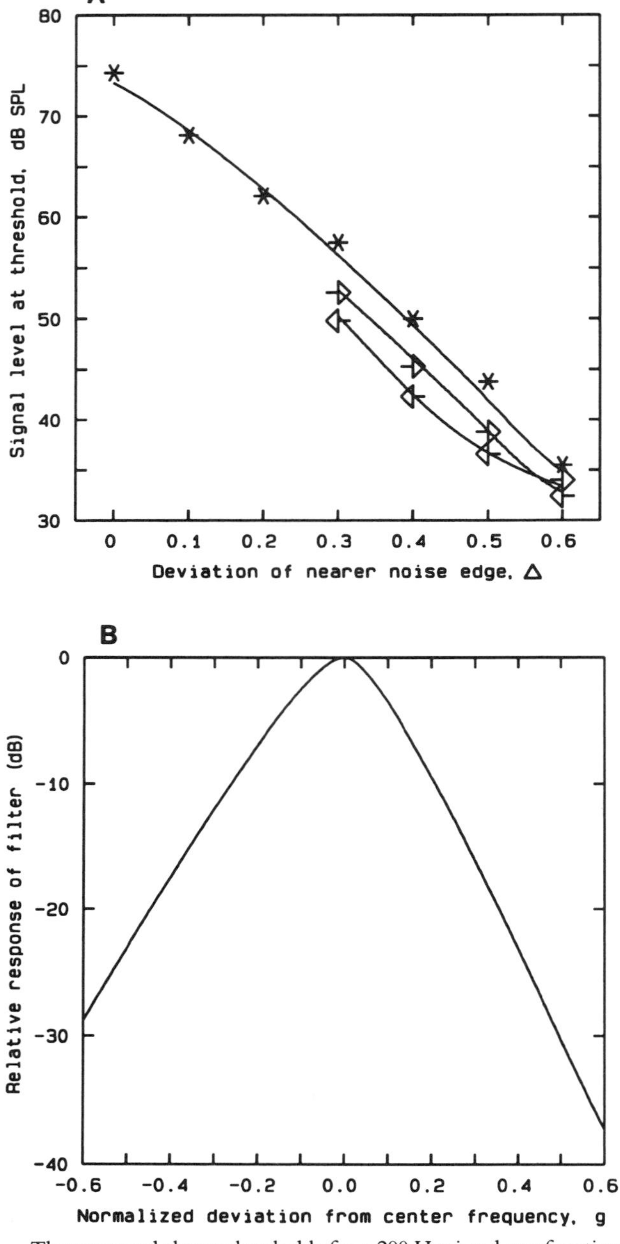

FIGURE 5 The top panel shows thresholds for a 200 Hz signal as a function of the width of a notch in a noise masker. The value on the abscissa is the deviation of the nearer edge of the notch from the signal frequency, divided by the signal frequency, represented by the symbol Δ. Asterisks (*) indicate conditions where the notch was placed symmetrically about the signal frequency. Right-pointing arrows indicate conditions where the upper edge of the notch was

somewhat asymmetric, with a shallower lower branch. The ERB is 48 Hz, which is typical at this center frequency.

IV. SUMMARY OF THE CHARACTERISTICS OF THE AUDITORY FILTER

A. Variation with Center Frequency

Moore and Glasberg (1983b) presented a summary of experiments measuring auditory filter shapes using symmetric notched-noise maskers. All of the data were obtained at moderate noise levels and were analyzed using the roex(p,r) filter shape. Glasberg and Moore (1990) updated that summary, including results that extend the frequency range of the measurements and data from experiments using asymmetric notches. The ERBs of the filters derived from the data available in 1983 are shown as asterisks in Figure 6. The dashed line shows the equation fitted to the data in 1983. Other symbols show ERBs estimated in more recent experiments, as indicated in the figure.

The solid line in Figure 6 provides a good fit to the ERB values over the whole frequency range tested. It is described by the following equation:

$$\mathrm{ERB} = 24.7(4.37F + 1), \tag{9}$$

where F is center frequency in kHz. This equation is a modification of one originally suggested by Greenwood (1961b) to describe the variation of the CB with center frequency. He based it on the assumption that each CB corresponds to a constant distance along the basilar membrane. Although the constants in Eq. (9) differ from those given by Greenwood, the form of the equation is the same as his. Each ERB corresponds to a distance of about 0.89mm on the basilar membrane.

It should be noted that the function specified by Eq. (9) differs somewhat from the "traditional" critical band function (Zwicker, 1961), which flattens off below 500 Hz at a value of about 100 Hz. The traditional function was obtained by combining data from a variety of experiments. However, the data were sparse at low frequencies, and the form of the function was strongly influenced by measures of the critical ratio. As described earlier, the critical ratio does not provide a good estimate of the CB, particularly at low frequencies. It seems clear that the CB does continue to decrease below 500 Hz.

0.2 units farther away from the signal frequency than the lower edge. Left-pointing arrows indicate conditions where the lower edge of the notch was 0.2 farther away than the upper edge. The fact that the left-pointing arrows are markedly below the right-pointing arrows indicates that the filter is asymmetric. The bottom panel shows the auditory filter shape derived from the data. (From Moore et al., 1990.)

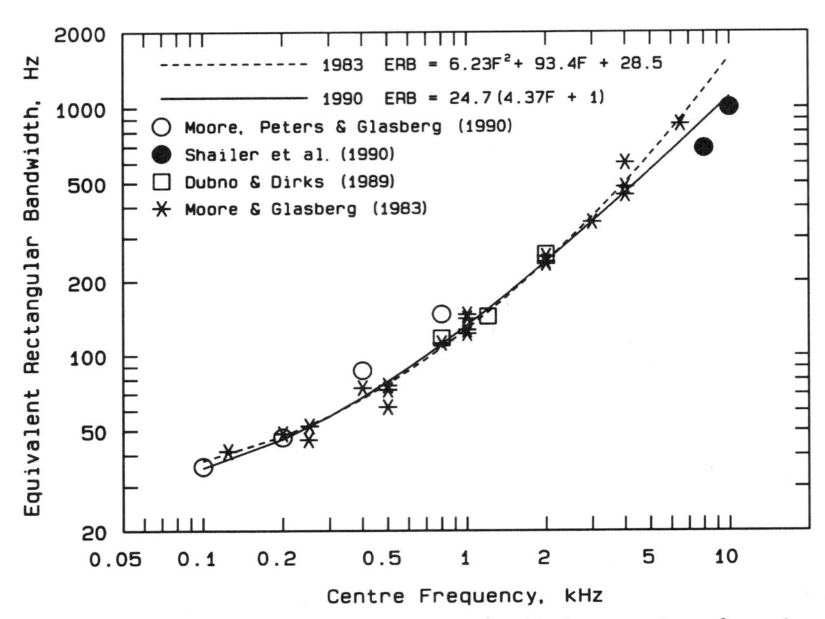

FIGURE 6 Estimates of the auditory filter bandwidth from a variety of experiments, plotted as a function of center frequency. The dashed line represents the equation suggested by Moore and Glasberg (1983b). The solid line represents the equation suggested by Glasberg and Moore (1990). (Adapted from Glasberg & Moore, 1990.)

It is sometimes useful to plot experimental data and theoretical functions on a frequency-related scale based on units of the CB or ERB of the auditory filter. A traditional scale of this type is the Bark scale (Zwicker & Terhardt, 1980) where the number of Barks is indicated by the symbol z. A good approximation to the traditional Bark scale is

$$z = [26.8/(1 + 1.96/F)] - 0.53 \tag{10}$$

(Traunmüller, 1990). A scale based on the ERB of the auditory filter, derived from Eq. (9), is

$$\text{Number of ERBs, } E = 21.4 \log_{10}(4.37F + 1) \tag{11}$$

Auditory filter bandwidths for young, normally hearing subjects vary relatively little across subjects; the standard deviation of the ERB is typically about 10% of its mean value (Moore, 1987; Moore et al., 1990). However, the variability tends to increase at very low frequencies (Moore et al., 1990) and at very high frequencies (Patterson et al., 1982; Shailer, Moore, Glasberg, Watson, & Harris, 1990).

B. Variation with Level

If the auditory filter were linear, then its shape would not vary with the level of the noise used to measure it. Unfortunately, this is not the case. Moore and Glasberg (1987b) presented a summary of measurements of the auditory filter shape using maskers with notches placed asymmetrically about the signal frequency. They concluded that the lower skirt of the filter becomes less sharp with increasing level, while the higher skirt becomes slightly steeper. Glasberg and Moore (1990) reanalyzed the data from the studies summarized in that paper, but using a modified fitting procedure including "corrections" for the transfer function of the middle ear. They also examined the data presented in Moore et al. (1990) and Shailer et al. (1990). The reanalysis led to the following conclusions:

1. The auditory filter for a center frequency of 1 kHz is roughly symmetric on a linear frequency scale when the level of the noise is approximately 51 dB/ERB. This corresponds to a noise spectrum level of about 30 dB. The auditory filters at other center frequencies are approximately symmetric when the effective input levels to the filters are equivalent to the level of 51 dB/ERB at 1 kHz (after making allowance for changes in relative level produced by passage of the sound through the outer and middle ear).

2. The low-frequency skirt of the auditory filter becomes less sharp with increasing level. The variation can be described in terms of the parameter p_l. Let X denote the effective input level in dB/ERB. Let $p_{l(X)}$ denote the value of p_l at level X. Then,

$$p_{l(X)} = p_{l(51)} - 0.38(p_{l(51)}/p_{l(51,1k)})(X - 51), \tag{12}$$

where $p_{l(51)}$ is the value of p at that center frequency for an effective input noise level of 51 dB/ERB and $p_{l(51,1k)}$ is the value of p_l at 1 kHz for an input level of 51 dB/ERB.

3. Changes in slope of the high-frequency skirt of the filter with level are less consistent. At medium center frequencies (1–4 kHz) there is a trend for the slope to increase with increasing level, but at low center frequencies there is no clear trend with level, and the filters at high center frequencies show a slight decrease in slope with increasing level.

These statements are based on the assumption that, although the auditory filter is not linear, it may be considered as approximately linear at any given noise level. Furthermore, the sharpness of the filter is assumed to depend on the input level to the filter, not the output level. This issue is considered further later. Figure 7 illustrates how the shape of the auditory filter varies with input level for a center frequency of 1 kHz.

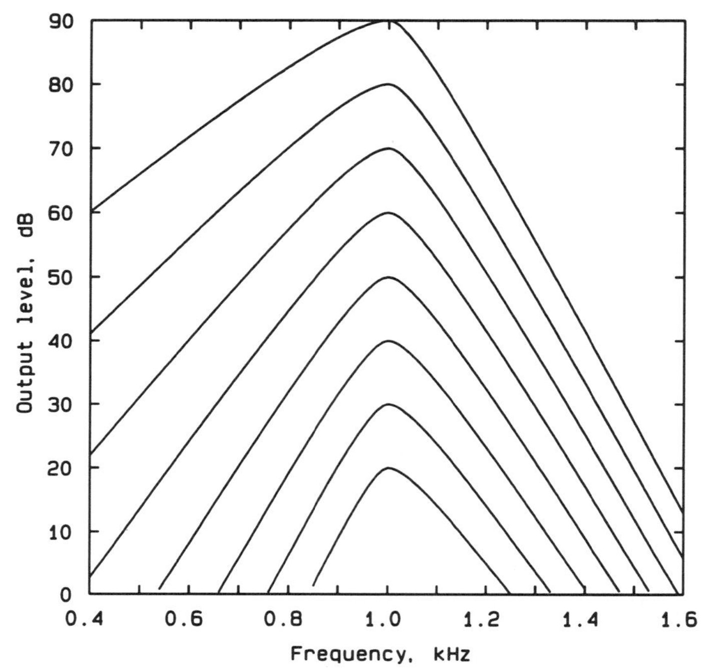

FIGURE 7 The shape of the auditory filter centered at 1 kHz, plotted for input sound levels ranging from 20 to 90 dB SPL/ERB. The level at the output of the filter is plotted as a function of frequency. On the low-frequency side, the filter becomes progressively less sharply tuned with increasing sound level. On the high-frequency side, the sharpness of tuning increases slightly with increasing sound level. At moderate sound levels the filter is approximately symmetric on the linear frequency scale used.

As mentioned earlier, the notched-noise method does not give a precise estimate of the slope of the steeper side of the filter when the filter is markedly asymmetric. This is a particular problem at high sound levels, where the lower branch becomes very shallow. Thus, at high levels, there may well be significant errors in the estimates of the sharpness of the high-frequency side of the filter.

V. MASKING PATTERNS AND EXCITATION PATTERNS

In the experiments described so far, the frequency of the signal was held constant, while the masker was varied. These experiments are most appropriate for estimating the shape of the auditory filter at a given center frequency. However, many of the early experiments on masking did the opposite; the signal frequency was varied while the masker was held constant.

Wegel and Lane (1924) reported the first systematic investigation of the masking of one pure tone by another. They determined the threshold of a signal with adjustable frequency in the presence of a masker with fixed frequency and intensity. The function relating masked threshold to the signal frequency is known as a *masking pattern,* or sometimes as a *masked audiogram.* The results of Wegel and Lane were complicated by the occurrence of beats when the signal and masker were close together in frequency. To avoid this problem, later experimenters (Egan & Hake, 1950; Fastl, 1976a) have used a narrow band of noise as either the signal or the masker.

The masking patterns obtained in these experiments show steep slopes on the low-frequency side, of between 80 and 240 dB/octave for pure tone masking and 55–190 dB/octave for narrowband noise masking. The slopes on the high-frequency side are less steep and depend on the level of the masker. A typical set of results is shown in Figure 8. Notice that on the high-frequency side the slopes of the curves tend to become shallower at high levels. Thus, if the level of a low-frequency masker is increased by, say, 10 dB, the masked threshold of a high-frequency signal is elevated by more than 10 dB; the amount of masking grows nonlinearly on the high-frequency side. This has been called the *upward spread of masking.*

The masking patterns do not reflect the use of a single auditory filter. Rather, for each signal frequency the listener uses a filter centered close to the signal frequency. Thus the auditory filter is shifted as the signal frequency is altered. One way of interpreting the masking pattern is as a crude

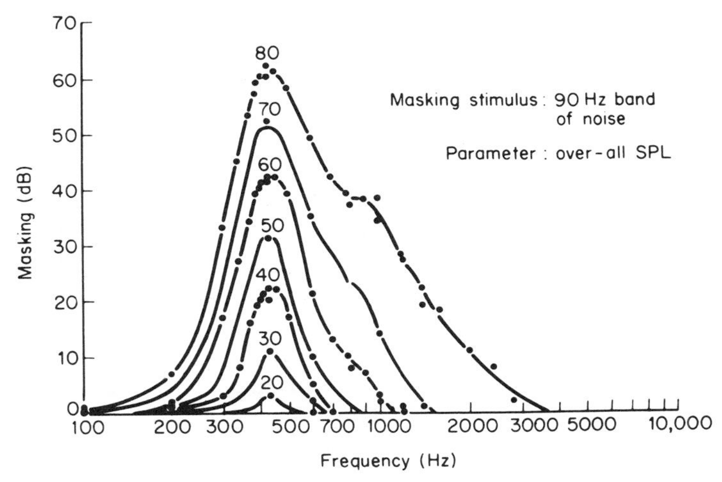

FIGURE 8 Masking patterns (masked audiograms) for a narrow band of noise centered at 410 Hz. Each curve shows the elevation in threshold of a sinusoidal signal as a function of signal frequency. The overall noise level for each curve is indicated in the figure. (Data from Egan & Hake, 1950.)

indicator of the excitation pattern of the masker. The excitation pattern of a sound is a representation of the activity or excitation evoked by that sound as a function of characteristic frequency (Zwicker, 1970). In the case of a masking pattern, one might assume that the signal is detected when the excitation it produces is some constant proportion of the excitation produced by the masker in the frequency region of the signal. Thus the threshold of the signal as a function of frequency is proportional to the masker's excitation level. The masking pattern should be parallel to the excitation pattern of the masker, but shifted vertically by a small amount. In practice, the situation is not so straightforward, since the shape of the masking pattern is influenced by factors such as off-frequency listening and the detection of combination tones produced by the interaction of the signal and the masker (Greenwood, 1971).

A. Relationship of the Auditory Filter to the Excitation Pattern

Moore and Glasberg (1983b) have described a way of deriving the shapes of excitation patterns using the concept of the auditory filter. They suggested that the excitation pattern of a given sound can be thought of as the output of the auditory filters as a function of their center frequency. This idea is illustrated in Figure 9. The upper portion of the figure shows auditory filter shapes for five center frequencies. Each filter is symmetrical on the linear frequency scale used, but the bandwidths of the filters increase with increasing center frequency, as illustrated in Figure 6. The dashed line represents a 1 kHz sinusoidal signal whose excitation pattern is to be derived. The lower panel shows the output from each filter in response to the 1 kHz signal, plotted as a function of the center frequency of each filter; this is the desired excitation pattern.

To see how this pattern is derived, consider the output from the filter with the lowest center frequency. This has a relative output in response to the 1 kHz tone of about −40 dB, as indicated by point *a* in the upper panel. In the lower panel, this gives rise to the point *a* on the excitation pattern; the point has an ordinate value of −40 dB and is positioned on the abscissa at a frequency corresponding to the center frequency of the lowest filter illustrated. The relative outputs of the other filters are indicated, in order of increasing center frequency, by points *b* to *e*, and each leads to a corresponding point on the excitation pattern. The complete excitation pattern was actually derived by calculating the filter outputs for filters spaced at 10 Hz intervals. In deriving the excitation pattern, excitation levels were expressed relative to the level at the tip of the pattern, which was arbitrarily labeled 0 dB. To calculate the excitation pattern for a 1 kHz tone with a level of, say, 60 dB, the level at the tip would be labeled 60 dB, and all other excitation levels would correspondingly be increased by 60 dB.

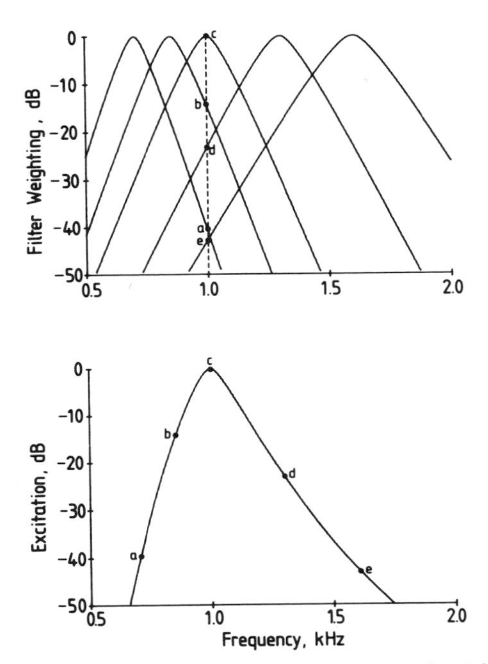

FIGURE 9 An illustration of how the excitation pattern of a 1 kHz sinusoid can be derived by calculating the outputs of the auditory filters as a function of their center frequency. The top half shows five auditory filters, centered at different frequencies, and the bottom half shows the calculated excitation pattern. (From Moore & Glasberg, 1983b.)

Note that, although the auditory filters were assumed to be symmetric on a linear frequency scale, the derived excitation pattern is asymmetric. This happens because the bandwidth of the auditory filter increases with increasing center frequency. As pointed out by Patterson (1974), the increase in auditory filter bandwidth with frequency can also explain why masking patterns are asymmetric when the auditory filter itself is roughly symmetric.

B. Changes in Excitation Patterns with Level

One problem in calculating excitation patterns from filter shapes is how to deal with the level dependence of the auditory filter. It seems clear that the shape of the auditory filter does change with level, the major change being a decrease in sharpness of the low-frequency side with increasing level. However, to determine the effect of this on excitation patterns, it is necessary to decide exactly what aspect of level determines the filter shape.

As one approach to this problem, Moore and Glasberg (1987b) considered whether the shape of the auditory filter depends primarily on the level of the input to the filter or on the level of the output of the filter. This way of posing the problem may well be over-simplistic, especially when the input is a complex sound. However, the question may be a reasonable one for a simple stimulus such as a sinusoid.

To examine this question, Moore and Glasberg (1987b) calculated excitation patterns for sinusoids assuming that the shape of the auditory filter depended either on the input level to the filter or the output level from the filter. They assumed that the filter had the form of the roex(p) filter described earlier. An example of the results is shown in Figure 10, for a 1 kHz sinusoid at levels ranging from 20 to 90 dB SPL. For the left-hand panels the output level of each filter was assumed to determine its shape. For the right-hand panels the input level was assumed to determine the shape.

As described previously, the shapes of excitation patterns for narrowband stimuli as a function of level can be determined approximately from their masking patterns. The patterns shown in the right-hand panels of Figure 10 closely resemble masked audiograms at similar masker levels, whereas those in the left-hand panels are very different in form and do not show the classic "upward spread of masking." Moore and Glasberg (1987b) concluded that the critical variable determining the auditory filter shape is the input level to the filter.

Rosen, Baker, and Kramer (1992) have taken the opposite viewpoint, arguing that the sharpness and asymmetry of the auditory filter are determined by the level at the output of the auditory filter. Their argument is based on an analysis of data from a notched-noise experiment conducted using three fixed masker spectrum levels and three fixed signal levels, with a signal frequency of 2000 Hz. Although their data can be fitted well on the assumption that the shape of the auditory filter is determined by its output level, this assumption leads to problems with other types of stimuli. One such problem was noted earlier: excitation patterns for narrowband stimuli calculated using this assumption have the "wrong" shape. In addition, the equations given by Rosen et al. to define the variation of the filter parameters (such as p_l and p_u) with level lead to substantial errors in predicting the masking produced by sounds such as low-frequency narrow bands of noise. Thus, even though their conclusion is consistent with the restricted data set analyzed, it does not seem to be generally applicable.

At present, it appears that the data from a range of experiments are best accounted for on the assumption that the shape of the auditory filter is controlled by the level at its input, rather than the level at its output. Unfortunately, the situation cannot be as simple as this. For sounds with complex broadband spectra, it seems likely that only components that produce a significant output from a given auditory filter have any influence in determining the shape of that filter. Moore and Glasberg (1987b) suggested that

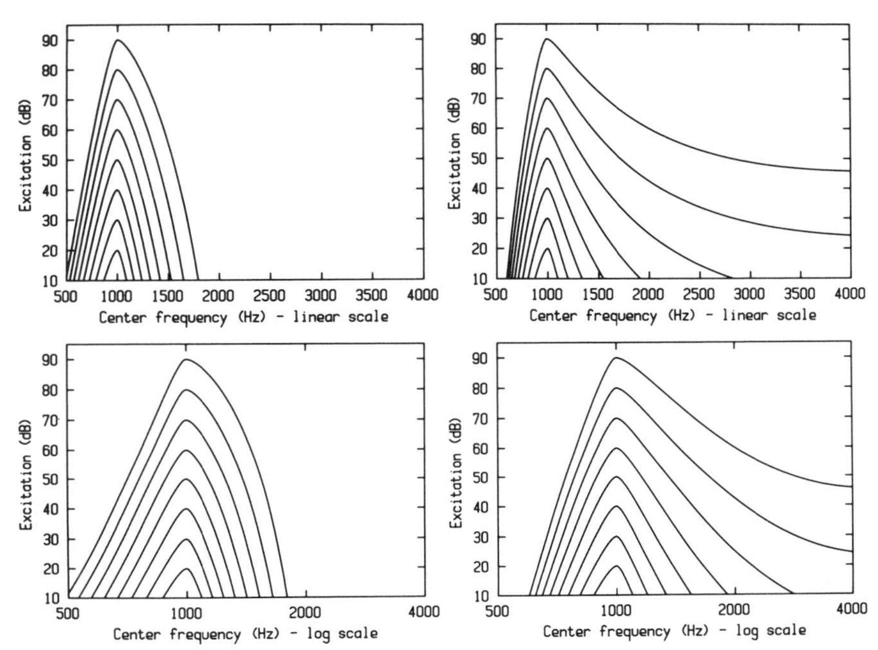

FIGURE 10 Excitation patterns calculated according to the procedure described in the text for 1 kHz sinusoids ranging in level from 20 to 90 dB SPL in 10 dB steps. The frequency scale is linear for the upper panels and logarithmic for the lower panels. The left panels show excitation patterns calculated on the assumption that the level at the output of the auditory filter is the variable determining its shape. The right panels show excitation patterns calculated assuming that the level at the input to the filter determines its shape. (From Moore & Glasberg, 1987b.)

the shape of the auditory filter may be determined primarily by the input level of the component that produces the greatest output from the filter. When the input spectrum is continuous or contains closely spaced components, they suggested that the power of the components should be summed with a range of 1 ERB (around the frequencies of the components in question) to determine the effective input level.

Further analysis of these ideas, by Brian Glasberg and myself (unpublished results), reveals a significant problem. When the input is composed of a few discrete sinusoidal components with different levels, excitation patterns calculated in this way can show discontinuities. This happens because the component producing the greatest output from the auditory filters changes as the center frequency of the filters changes. At center frequencies, where a change in the dominant component occurs, there is an abrupt jump in the calculated sharpness of the filter, and this leads to a discontinuity in the excitation pattern. It seems very unlikely that such discontinuities would occur in the auditory system.

The program for calculating excitation patterns published by Glasberg and Moore (1990) is based on a different set of assumptions, although these assumptions were not explicitly stated in that paper. The assumptions are as follows. Each component (or group of components if several lie within 1 ERB) gives rise to an excitation pattern whose spread is determined by the level of the component (or group). Thus, the extent to which a given auditory filter is excited by the component is determined by the input level of the component. This idea is similar to that proposed by Zwicker (Zwicker & Feldtkeller, 1967; Zwicker, 1970). The simplest way to calculate the effective level at a given center frequency is to sum the powers of components within ±0.5 ERB of that frequency. Zwicker used a similar approach and referred to the resulting quantity as *psychoacoustical incitation*. This summed power determines the spread of excitation from that frequency. The assumption that component powers are summed within a rectangular ERB (around the frequencies of the components) is unrealistic, and it is probably more satisfactory to perform the summation using a rounded-exponential weighting function.

In cases where the stimulus is complex, with components spread over several ERBs, it is assumed that excitation patterns arising from components lying in different ERBs are summed in terms of linear power. Again, Zwicker (Zwicker & Feldtkeller, 1967; Zwicker, 1970) made a similar assumption, and the assumption is implicit in some models for calculating loudness and for predicting intensity discrimination performance (see Chapter 4).

There is a consequence of these assumptions that may, at first sight, appear paradoxical. A given auditory filter may have several different sharpnesses (for example, several values of p_l) simultaneously. The value of p_l is calculated separately for each group of components (summed within an ERB around the frequencies of the components). However, this is not so strange in terms of excitation patterns. The program by Glasberg and Moore (1990) works by calculating excitation patterns from filter shapes, but, for the auditory system, the excitation pattern may be "primary" in some sense. If the spread of the excitation pattern produced by a given component is determined by the level of that component, then the effective shapes of the auditory filters excited by that component have to vary depending on the level of the component. And if two components are present simultaneously (separated by more than 1 ERB), each will give rise to an excitation pattern whose spread is determined by the level of the respective component.

Many of these problems arise from the fact that auditory filtering is inherently nonlinear, but the models used are based on quasi-linear filtering.

More appropriate models may lead to a better understanding of the factors controlling the selectivity of the auditory filters.

VI. THE ADDITIVITY OF MASKING AND EXCESS MASKING

Many years ago, Green (1967) measured the masking of a gated sinusoidal signal produced by a continuous sinusoidal masker of the same frequency and, separately, by a broadband continuous noise. He adjusted the levels of the two maskers so that they produced equal amounts of masking. He then measured the amount of masking produced by combining the two equally effective maskers. If the threshold of the signal were determined simply by the power of the masker at the output of the auditory filter centered at the signal frequency, the combined maskers should produce 3 dB more masking than either masker alone. In fact, the amount of extra masking produced by the combined masker was usually markedly greater than the expected 3 dB. The amount of masking above 3 dB is sometimes referred to as *excess* masking.

In the last two decades, many cases of excess masking have been reported; the amount of masking produced by two maskers that are equally effective when presented individually is often more than 3 dB greater than the masking produced by each masker alone (Lutfi, 1983, 1985; Humes & Jesteadt, 1989). Two general approaches have been taken to explain the excess masking.

In one approach, it is assumed that the detection cues used by the subject differ for the two individual maskers (Bilger, 1959; Green, 1967; Moore, 1985). For example, if one masker is a continuous broadband noise, as in Green's (1967) experiment, the detection cue may be the additional energy produced by the signal at the output of the auditory filter centered on the signal frequency. If the other masker is a continuous sinusoid, the detection cue may be a fluctuation in the envelope of the auditory filter output. This cue is very effective, so when the two maskers produce equal amounts of masking, the sinusoid produces a considerably greater output from the auditory filter than the noise. When the two maskers are combined, the noise introduces random fluctuations in amplitude that make it difficult to use the fluctuation cue previously employed with the sinusoidal masker. However, the energy cue previously used with the noise masker is also less effective, because the sinusoidal masker considerably increases the energy at the filter output. Hence, considerable excess masking occurs. According to this type of explanation, excess masking occurs when the detection processes or cues following the auditory filter are different for the two maskers used and when each masker renders less effective the cue used with the other masker.

A second type of explanation has been proposed by Lutfi (1983, 1985) and Humes and Jesteadt (1989). These researchers have suggested that the effects of the two maskers are summed after each has undergone a compressive nonlinear transformation. A model of this type was originally proposed by Penner and Shiffrin (1980) to account for the excess masking obtained with pairs of maskers that do not overlap in time (see Chapter 6, Section VIII of this chapter, and Oxenham & Moore, 1994), but Lutfi and Humes and Jesteadt have argued that a similar model can be applied to maskers that do overlap in time.

In the model of Humes and Jesteadt (1989), the compressed internal effect of each masker is given by the following transform:

$$i_{mt} = (10^{(mt/10)})^{\alpha} - (10^{(q/10)})^{\alpha} \tag{13}$$

where i_{mt} reflects the internal effect of the masker, mt is the masked threshold of the signal, q is the absolute threshold of the signal, and α is a parameter that is adjusted to fit the data. The value of α reflects the amount of compression: the smaller the value of α, the greater is the compression. The effects of combined maskers are assumed to be summed after the individual maskers have been subjected to the nonlinear compressive transform of Eq. (13). An inverse transform is then applied to the sum to predict the signal threshold for the combined maskers. If $\alpha < 1$, excess masking is predicted. If $\alpha = 1$, there is no compression, and no excess masking is predicted.

The basic concept behind this type of model is that the excess masking arises from a fundamental physiological property of the auditory system: it is assumed that all stimuli are subject to a compressive nonlinearity in the peripheral auditory system. The amount of excess masking should be determined by the characteristics of this compressive nonlinearity, and excess masking should always occur.

Although the model proposed by Humes and Jesteadt can account for a large body of experimental data, several problems can be identified with this approach. First, it is assumed that the two maskers are compressed independently. It is hard to see how this could happen for two maskers that are presented simultaneously. Consider Green's (1967) experiment described earlier. Presumably, the masking produced by both the noise masker and the sinusoidal masker depends on the extent to which those maskers produce an output from the auditory filter centered at the signal frequency. If the two maskers are presented simultaneously and excite the same auditory filter, how can the effects of the two maskers be subjected to independent compressive nonlinearities?

A second problem arises from the assumption that the amount of excess masking is determined by the characteristics of the compressive nonlinearity, and that this nonlinearity reflects a physiological property of the peripheral auditory system. If this were the case, then the form of the nonlinearity (the value of α) needed to fit the data should be similar regardless of the specific combination of maskers used. In fact, the required value of α varies markedly across different data sets (Humes & Jesteadt, 1989). This is not consistent with the idea that the excess masking arises from a fixed physiological property of the auditory system.

A third problem is raised by a study of Oxenham and Moore (1995) on excess masking in subjects with cochlear hearing loss. It is known that the compressive nonlinearity on the basilar membrane is reduced in such subjects (see Chapter 2). Hence, if the model of Humes and Jesteadt (1989) were correct, excess masking should be reduced. In an experiment similar to that of Green (1967), Oxenham and Moore showed that excess masking in simultaneous masking was similar for normally hearing subjects and subjects with cochlear hearing loss. However, excess masking produced by combined forward and backward masking was absent in subjects with cochlear hearing loss, while it was marked in normally hearing subjects. These results suggest that compressive nonlinearities in the peripheral auditory system can account for excess masking in nonsimultaneous masking, but not in simultaneous masking.

A fourth problem is that the model predicts that excess masking will always occur. This is not the case. For example, Moore (1985) determined masking functions (signal threshold versus masker level) separately for two narrowband noise maskers, one centered at 1.4 kHz and the other at 1.6 kHz. The signal was always a 2 kHz sinusoid. The masking functions were used to select pairs of maskers (i.e., the two bands of noise presented together) that would be equally effective if presented individually. When the two bands of noise had independent envelopes, excess masking occurred, as predicted by the model of Humes and Jesteadt. However, when the two bands of noise had the same envelope (i.e., the bands were comodulated; see Chapter 7), no excess masking occurred; the combined maskers produced 3 dB more masking than each masker individually. This is not consistent with the model of Humes and Jesteadt. Green (1967) and Bilger (1959) also reported cases where no excess masking occurred. In my opinion, these results are sufficient to demonstrate that the model cannot be correct.

Moore (1985) explained the pattern of his results in the following way. When a single narrowband noise masker is used, subjects exploit the envelope fluctuations in the masker to improve signal detection. They do this by detecting the signal in minima of the masker envelope and by comparing temporal patterns of modulation across different auditory filters (Buus, 1985; Moore & Glasberg, 1987a). When the signal is absent, the pattern of modulation is similar at the outputs of all auditory filters. When the signal is present, the pattern of modulation at the output of filters tuned close to the signal frequency differs from that in the remaining filters. This across-filter disparity provides a detection cue; see Chapter 7 for further discussion of this topic. When two uncorrelated narrowband noise maskers are combined, this cue is disrupted and excess masking occurs. When two narrowband maskers with the same envelope are combined, the cue is preserved, and the combined masking is correctly predicted by a linear power summation of the effects of the two maskers.

In summary, it seems that most cases of excess masking in simultaneous masking can be explained by a detailed consideration of the detection cues available to and used by the subjects for the individual maskers and for the combined maskers. Alternative models, assuming that the effects of the individual maskers are subject to a compressive nonlinearity before the effects are combined, lead to a number of conceptual difficulties, and cannot account for cases where excess masking does not occur.

Finally, a general problem in the study of excess masking should be noted. It is very hard to formulate a set of rules defining whether a given stimulus should be described as one masker or as two (or even more). For example, should a band of noise extending from 500 to 1000 Hz be described as a single masker or as two maskers, one extending from 500 to 750 Hz, and the other from 750 to 1000 Hz? In many cases, the definition of what constitutes one masker or two maskers appears completely arbitrary. This problem is discussed by Humes and Jesteadt (1989) and by Humes, Lee, and Jesteadt (1992). The latter proposed that maskers that do not overlap within the critical band centered on the signal frequency should be treated as two separate maskers; in such cases they predicted that excess masking should occur. On the other hand, they suggested that for simultaneous maskers with spectral overlap within the critical band centered on the signal frequency there is effectively only one masker, and no excess masking should occur. While this rule is consistent with the data presented by Humes et al. (1992), it clearly does not always work. For example, Green (1967) showed that excess masking could occur for spectrally overlapping maskers; and Moore (1985) showed that linear additivity of masking could occur for pairs of maskers that did not overlap spectrally either with each other or with the signal.

VII. PHENOMENA REFLECTING THE INFLUENCE OF AUDITORY FILTERING

Many aspects of auditory perception are affected by auditory filtering. Some of these are considered in other chapters, especially Chapters 4 and 8. This section describes just a few examples of these phenomena.

A. The Threshold of Complex Sounds

Gässler (1954) measured the threshold for detecting multicomponent complexes consisting of evenly spaced sinusoids. The complexes were presented both in quiet and in a special background noise, chosen to give the same masked threshold for each component in the signal. As the number of components in a complex was increased, the threshold, specified in terms of total energy, remained constant until the overall spacing of the tones reached a certain bandwidth, the CB for threshold. Thereafter the threshold increased by about 3 dB per doubling of bandwidth. The CB for a center frequency of 1 kHz was estimated to be about 180 Hz. These results were interpreted as indicating that the energies of the individual components in a complex sound will sum, in the detection of that sound, provided the components lie within a CB. When the components are distributed over more than one CB, detection is based on the single band giving the highest detectability.

Other data are not in complete agreement with those of Gässler. Indeed, most subsequent experiments have failed to replicate Gässler's results. For example, Spiegel (1981) measured the threshold for a noise signal of variable bandwidth centered at 1 kHz in a broadband background noise masker. The threshold for the signal as a function of bandwidth did not show a break point corresponding to the CB, but increased monotonically as the bandwidth increased beyond 50 Hz. The slope beyond the CB was close to 1.5 dB per doubling of bandwidth. Higgins and Turner (1990) have suggested that the discrepancy may be explained by the fact that Gässler widened the bandwidth, keeping the upper edge of the complex fixed in frequency, while Spiegel used stimuli with a fixed center frequency. However, other results clearly show that the ear is capable of combining information over bandwidths much greater than the CB (Buus et al., 1986; Langhans & Kohlrausch, 1992). For example, Buus et al. (1986) showed that multiple widely spaced sinusoidal components were more detectable than any of the individual components.

These results should not be interpreted as evidence against the concept of the auditory filter. They do indicate, however, that detection of complex signals may not be based on the output of a single auditory filter. Rather, information can be combined across filters to improve performance.

B. Sensitivity to the Relative Phase

An amplitude modulated (AM) sinewave with modulation index m and a frequency modulated (FM) sinewave with modulation index β may each be considered as composed of three sinusoidal components, corresponding to the carrier frequency and two sidebands (an FM wave actually contains many components but for small modulation indices only the first two sidebands are important); see Chapter 1. When the modulation indices are numerically equal ($m = \beta$) and the carrier frequencies and modulation frequencies are the same, the components of an AM wave and an FM wave are identical in frequency and amplitude, the only difference between them being in the relative phase of the components. If, then, the two types of wave are perceived differently, the difference is likely to arise from a sensitivity to the relative phase of the components.

Zwicker (1952), Schorer (1986), and Sek (1994) have measured one aspect of the perception of such stimuli, namely, the just-detectable amounts of amplitude or frequency modulation, for various rates of modulation. They found that, for high rates of modulation, where the frequency components were widely spaced, the detectability of FM and AM was equal when the components in each type of wave were of equal amplitude ($m = \beta$). However, for low rates of modulation, when all three components fell within a narrow frequency range, AM could be detected when the relative levels of the sidebands were lower than for a wave with a just-detectable amount of FM ($m < \beta$). This is illustrated in the upper panel of Figure 11. Thus, for small frequency separations of the components, subjects appear to be sensitive to the relative phases of the components, while for wide frequency separations they are not.

If the threshold for detecting modulation is expressed in terms of the modulation index, m or β, the ratio β/m decreases as the modulation frequency increases and approaches an asymptotic value of unity. This is illustrated in the lower panel of Figure 11. The modulation frequency at which the ratio first becomes unity is called the *critical modulation frequency* (CMF). Zwicker (1952) and Schorer (1986) suggested that the CMF corresponded to half the value of the CB; essentially, the CMF was assumed to be reached when the overall stimulus bandwidth reached the CB. If this is correct, then the CMF may be regarded as providing an estimate of the CB at the carrier frequency.

Further analysis suggests that this interpretation of the results may not be completely correct. The CMF appears to correspond to the point where one of the sidebands in the spectrum first becomes detectable; usually the lower sideband is more detectable than the upper one (Hartmann & Hnath, 1982; Moore & Sek, 1992). The threshold for detecting a sideband depends more on the selectivity of auditory filters centered close to the frequency of the sideband than on the selectivity of the auditory filter centered on the carrier frequency. Furthermore, for low carrier frequencies, the upper sideband may be more detectable than the lower sideband (Sek & Moore, 1994). The change in the most detectable sideband with carrier frequency can account for the finding that the ERB decreases more with decreasing center frequency than does the CMF. It also makes the CMF unsuitable as a direct measure of the CB. Additionally, it should be noted that the detectability of a sideband may be influenced by factors not connected with frequency selectivity, such as the efficiency of the detection process following auditory filtering. This efficiency may well vary with center frequency, just as it does for the detection of tones in notched noise. Thus, like the critical ratio described earlier, the CMF does not provide a direct measure of the CB.

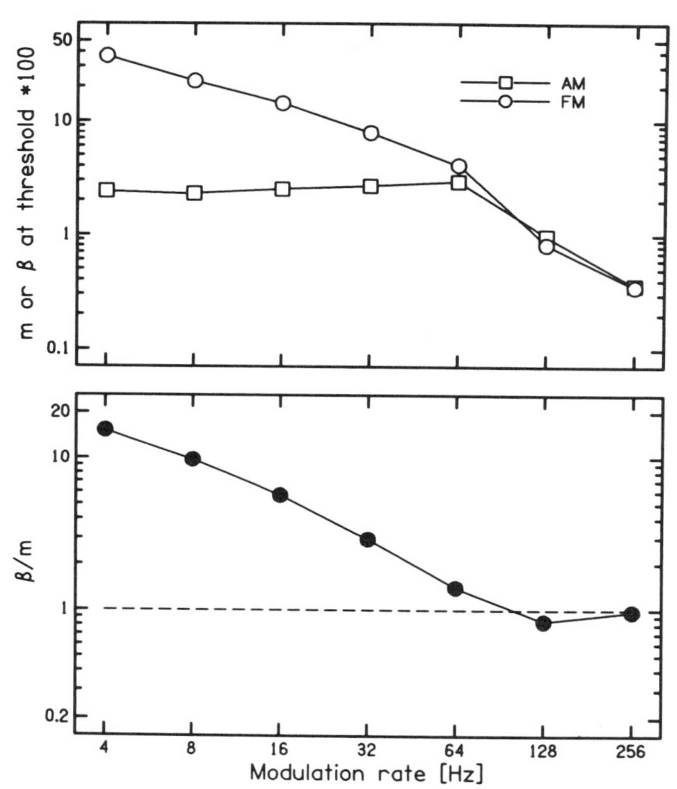

FIGURE 11 The upper panel shows thresholds for detecting sinusoidal amplitude modulation (squares) or frequency modulation (circles) of a 1 kHz carrier, plotted as a function of modulation rate. The thresholds are expressed in terms of the modulation indices, m and β, respectively (the indices are multiplied by 100 to give convenient numbers). The lower panel shows the ratio β/m, plotted on a logarithmic scale as a function of modulation rate. (Data from Sek, 1994, with permission of the author.)

It also appears to be incorrect to assume that changes in the relative phase of the components in a complex sound are detectable only when those components lie within a CB. In cases where all components are well above threshold, subjects can detect phase changes between the components in complex sounds in which the components are separated by considerably more than a CB (Craig & Jeffress, 1962; Blauert & Laws, 1978; Patterson, 1987). The detection of these phase changes may depend partly on the ability to compare the time patterns at the outputs of different auditory filters; see Chapter 6 for further information on this topic.

C. The Audibility of Partials in Complex Tones

According to Ohm's (1843) Acoustical Law, the ear is able to hear pitches corresponding to the individual sinusoidal components in a complex sound. In other words, we can "hear out" the individual partials. For periodic complex sounds, the most prominent pitch is usually the residue pitch or virtual pitch associated with the sound as a whole, and we are not normally aware of hearing pitches corresponding to individual partials; see Chapter 8. Nevertheless, such pitches can be heard if attention is directed appropriately (Helmholtz, 1863).

Plomp (1964) and Plomp and Mimpen (1968) used complex tones with 12 sinusoidal components to investigate the limits of this ability. The listener was presented with two comparison tones, one of which was of the same frequency as a partial in the complex; the other lay halfway between that frequency and the frequency of the adjacent higher or lower partial. The listener was allowed to switch freely between the complex tone and the comparison tones and was required to decide which of the two comparison tones coincided with the partial in the complex tone. The score (varying between 50 and 100%) was used as an index of how well the partial could be heard out from the complex tone. For harmonic complex tones, only about the first five to seven harmonics could be heard out.

Plomp and Mimpen (1968) suggested that a component can be heard out only when its frequency separation from adjacent components exceeds the CB. The spacing of the components in a harmonic complex is uniform on a linear frequency scale, but, relative to the CB, the upper harmonics are more closely spaced than the lower harmonics. Harmonics above about the eighth are separated by less than a CB and cannot be heard out. Results for two subjects using complex tones where the frequency ratios between components were "compressed" relative to a harmonic complex gave basically the same result; only the lower components could be heard out.

The concept that the CB was the main factor limiting the audibility of partials in complex tones was questioned by Soderquist (1970). He used a task similar to that of Plomp (1964), but compared the results for musicians and nonmusicians. He found that musicians performed markedly better than nonmusicians. A possible explanation for this is that musicians have narrower CBs than nonmusicians. However, Fine and Moore (1993) estimated auditory filter bandwidths in musicians and nonmusicians, using the notched-noise method, and found that ERBs did not differ for the two groups. An alternative possibility is that performance depends on some factor or factors other than the CB.

Some aspects of the data of Plomp and Mimpen (1968) also suggest the involvement of factors other than the CB. The frequency difference between adjacent harmonics required to hear them separately was somewhat greater than traditional CB values (Zwicker & Terhardt, 1980) above 1000 Hz and was distinctly smaller below 1000 Hz. Recent estimates of the ERB of the auditory filter, described in Section IV.A, are smaller than traditional CB values at low frequencies and more consistent with the data of Plomp and Mimpen. Nevertheless, some discrepancy remains. The relative value of the CB or the ERB (i.e., bandwidth divided by center frequency) increases as the center frequency decreases below about 1000 Hz. As a consequence, the number of resolvable harmonics in a harmonic complex tone would be expected to decrease at low fundamental frequencies. In fact, the data of Plomp and Mimpen show that the number of resolvable harmonics increases as the fundamental frequency decreases below 250 Hz.

Moore and Ohgushi (1993) examined the ability of musically trained subjects to hear out individual partials in complex tones with partials uniformly spaced on a scale related to the ERB of the auditory filter. ERB spacings of 0.75, 1.0, 1.25, 1.5, and 2 were used, and the central component always had a frequency of 1000 Hz. On each trial, subjects heard a pure tone (the "probe") followed by a complex tone. The probe was close in frequency to one of the partials in the complex, but was mistuned downward by 4.5% on half the trials (at random) and mistuned upward by 4.5% on the other half. The task of the subject was to indicate whether the probe was higher or lower in frequency than the nearest partial in the complex. The partial that was "probed" varied randomly from trial to trial. If auditory filtering were the only factor affecting performance on this task, then scores for a given ERB spacing should be similar for each component in the complex sound.

Scores for the highest and lowest components in the complexes were generally high for all components spacings, although they worsened somewhat for ERB spacings of 0.75 and 1.0. Scores for the inner components were close to chance level at 0.75 ERB spacing, and improved progressively as the ERB spacing was increased from 1 to 2 ERBs. For ERB spacings of 1.25 or less, the scores did not change smoothly with component frequency; marked irregularities were observed, as well as systematic errors. Moore and Ohgushi suggested that these resulted from irregularities in the transmission of sound through the middle ear; such irregularities could change the relative levels of the components, making some components more prominent than others and therefore easier to hear out.

Performance for the inner components tended to be worse for component frequencies above 1000 Hz than below 1000 Hz. This is consistent with the pattern of results found by Plomp and Mimpen (1968) and indicates that some factor other than auditory filtering influences the audibility of partials in complex tones.

Moore and Ohgushi suggested that the pitches of individual components may be partly coded in the time patterns of neural activity (phase locking) in the auditory nerve, as has also been suggested by previous researchers (Ohgushi, 1978, 1983; Srulovicz & Goldstein, 1983; Moore, Glasberg, & Shailer, 1984; Moore, 1989; Hartmann, McAdams, & Smith, 1990; Moore & Glasberg, 1990; see Chapter 3). Phase locking is more precise below 1000 Hz than above 1000 Hz. This can explain why, for partials uniformly spaced on an ERB scale, the identification of partials was better for components with frequencies below 1000 Hz than above 1000 Hz.

The influence of phase locking can also explain the superior identification of the lowest and highest components in the complex tones. Generally, neurons with characteristic frequencies (CFs) close to the frequency of a given partial will phase lock to that partial, provided the frequency separation of partials is sufficient and the frequency of the partial is not too high. For an inner partial in a complex tone, the pattern of phase locking in neurons with CFs close to the frequency of the partial will be disturbed by the partials on either side, making it more difficult to extract the pitch of that partial, especially when the components are closely spaced. In contrast, neurons tuned just below the lower edge frequency or just above the higher edge frequency will show a pattern of phase locking that is less disturbed by the other components. A similar explanation can be offered for Plomp's (1964) finding that the partials in a two-tone complex could be heard out for smaller frequency separations than were found for multitone complexes.

In summary, it seems clear that auditory filtering plays a strong role in limiting the ability to hear out partials in complex tones. However, it is probably not the only factor involved. Specifically, the pitches of individual partials may partly be coded in the patterns of phase locking of neurons in the auditory nerve. This coding is more accurate at low frequencies than at high.

VIII. NONSIMULTANEOUS MASKING

Simultaneous masking describes situations where the masker is present for the whole time that the signal occurs. Masking can also occur when a brief signal is presented just before or after the masker; this is called *nonsimultaneous masking*. Two basic types of nonsimultaneous masking can be distinguished: (1) *backward masking,* in which the signal precedes the masker (also known as *prestimulatory masking*); and (2) *forward masking,* in which the signal follows the masker (also known as *poststimulatory masking*).

Although many studies of backward masking have been published, the phenomenon is poorly understood. The amount of backward masking obtained depends strongly on how much practice the subjects have received, and practiced subjects often show little or no backward masking (Miyazaki & Sasaki, 1984; Oxenham & Moore, 1994, 1995). The larger masking effects found for unpracticed subjects may reflect some sort of "confusion" of the signal with the masker. In contrast, forward masking can be substantial even in highly practiced subjects. The main properties of forward masking are as follows:

1. Forward masking is greater the nearer in time to the masker that the signal occurs. This is illustrated in the left panel of Figure 12. When the delay *D* of the signal after the end of the masker is plotted on a logarithmic scale, the data fall roughly on a straight line. In other words, the amount of forward masking, in dB, is a linear function of log(*D*).

2. The rate of recovery from forward masking is greater for higher masker levels. Thus, regardless of the initial amount of forward masking, the masking decays to 0 after 100–200 ms.

3. Increments in masker level do not produce equal increments in amount of forward masking. For example, if the masker level is increased by 10 dB, the masked threshold may increase by only 3 dB. This contrasts with simultaneous masking, where, at least for wideband maskers, the threshold usually corresponds to a constant signal-to-masker ratio. This effect can be quantified by plotting the signal threshold as a function of masker level. The resulting function is called a *growth of masking* function.

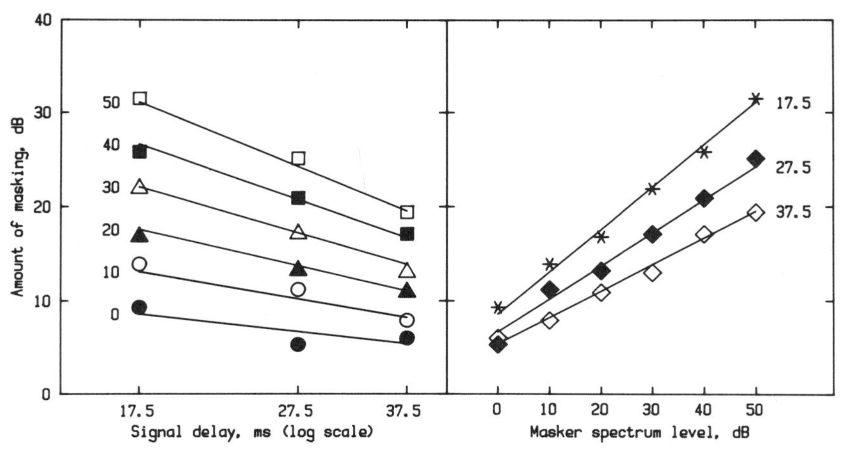

FIGURE 12 The left panel shows the amount of forward masking of a brief 2 kHz signal, plotted as a function of the time delay of the signal after the end of the noise masker. Each curve shows results for a different noise spectrum level (10–50 dB). The results for each spectrum level fall on a straight line when the signal delay is plotted on a logarithmic scale, as here. The right panel shows the same thresholds plotted as a function of masker spectrum level. Each curve shows results for a different signal delay time (17.5, 27.5, or 37.5 ms). Note that the slopes of these growth of masking functions decrease with increasing signal delay. (Adapted from Moore and Glasberg, 1983a.)

Several such functions are shown in the right panel of Figure 12. In simultaneous masking such functions would have slopes close to 1. In forward masking the slopes are less than 1, and the slopes decrease as the value of D increases.

4. The amount of forward masking increases with increasing masker duration for durations up to at least 20 ms. The results for greater masker durations vary somewhat across studies. Some studies show an effect of masker duration for durations up to 200 ms (Kidd & Feth, 1982), while others show little effect for durations beyond 50 ms (Fastl, 1976b).

The mechanisms underlying forward masking are not clear. It could be explained in terms of a reduction in sensitivity of recently stimulated neurons or in terms of a persistence in the pattern of neural activity evoked by the masker. Both points of view can be found in the literature. In addition, the response of the basilar membrane to the masker takes a certain time to decay, and for small intervals between the signal and the masker this may result in forward masking (Duifhuis, 1973); see Chapter 6 for further discussion of these issues.

IX. EVIDENCE FOR LATERAL SUPPRESSION FROM NONSIMULTANEOUS MASKING

Measurements of basilar membrane motion (Ruggero, 1992), or in single neurons (Arthur, Pfeiffer, & Suga, 1971), show that the response to a tone of a given frequency can sometimes be suppressed by a tone with a different frequency, a phenomenon known as *two-tone suppression;* see Chapters 2 and 3. For other complex signals, similar phenomena occur and are given the general name *lateral suppression* or *suppression*. This can be characterized in the following way. Strong activity at a given characteristic frequency can suppress weaker activity at adjacent CFs. In this way, peaks in the excitation pattern are enhanced relative to adjacent dips. The question now arises as to why the effects of suppression are not usually seen in experiments on simultaneous masking.

Houtgast (1972) has argued that simultaneous masking is not an appropriate tool for detecting the effects of suppression. In simultaneous masking, the masking stimulus and the signal are processed simultaneously in the same channel (the same auditory filter). Thus any suppression in that channel will affect the neural activity caused by both the signal and the masker. In other words, the signal-to-masker ratio in a given frequency region will be unaffected by suppression, and thus the threshold of the signal will remain unaltered.

Houtgast suggested that this difficulty could be overcome by presenting the masker and the signal successively, for example, by using forward masking. If suppression does occur, then its effects will be seen in forward masking provided (1) in the chain of levels of neural processing, the level at which the suppression occurs is not later than the level at which most of the forward masking effect arises; and (2) the suppression built up by the masker has decayed by the time that the signal is presented (otherwise the problems described for simultaneous masking will be encountered).

Following the pioneering work of Houtgast (1972, 1973, 1974), many workers have reported that there are systematic differences between the results obtained using simultaneous and nonsimultaneous masking techniques. An extensive review is provided by Moore and O'Loughlin (1986). One major difference is that nonsimultaneous masking reveals effects that can be directly attributed to suppression. A good demonstration of this involves a psychophysical analog of neural two-tone suppression. Houtgast (1973, 1974) measured the threshold for a 1 kHz signal and a 1 kHz nonsimultaneous masker. He then added a second tone to the masker and measured the threshold again. He found that sometimes the addition of this second tone produced a reduction in the threshold, and he attributed this to a suppression of the 1 kHz component in the masker by the second component. If the 1 kHz component is suppressed, then there will be less activity in the frequency region around 1 kHz, producing a drop in the threshold for detecting the signal. The second tone was most effective as a "suppressor" when it was somewhat more intense than the 1 kHz component and above it in frequency. Similar results have been obtained by Shannon (1976).

Under some circumstances, the reduction in threshold (unmasking) produced by adding one or more extra components to a masker can be partly explained in terms of additional cues provided by the added components, rather than in terms of suppression. Specifically, in forward masking the added components may reduce "confusion" of the signal with the masker by indicating exactly when the masker ends and the signal begins (Moore, 1980, 1981; Moore & Glasberg, 1982a, 1985; Neff, 1985; Moore & O'Loughlin, 1986). This may have led some researchers to overestimate the magnitude of suppression as indicated in nonsimultaneous masking experiments. However, it seems clear that not all unmasking can be explained in this way.

X. THE ENHANCEMENT OF FREQUENCY SELECTIVITY REVEALED IN NONSIMULTANEOUS MASKING

A second major difference between simultaneous and nonsimultaneous masking is that the frequency selectivity revealed in nonsimultaneous masking is greater than that revealed in simultaneous masking. A well-studied example of this is the psychophysical tuning curve. PTCs determined in forward masking are typically sharper than those obtained in simultaneous masking (Moore, 1978). An example is given in Figure 13. The difference is

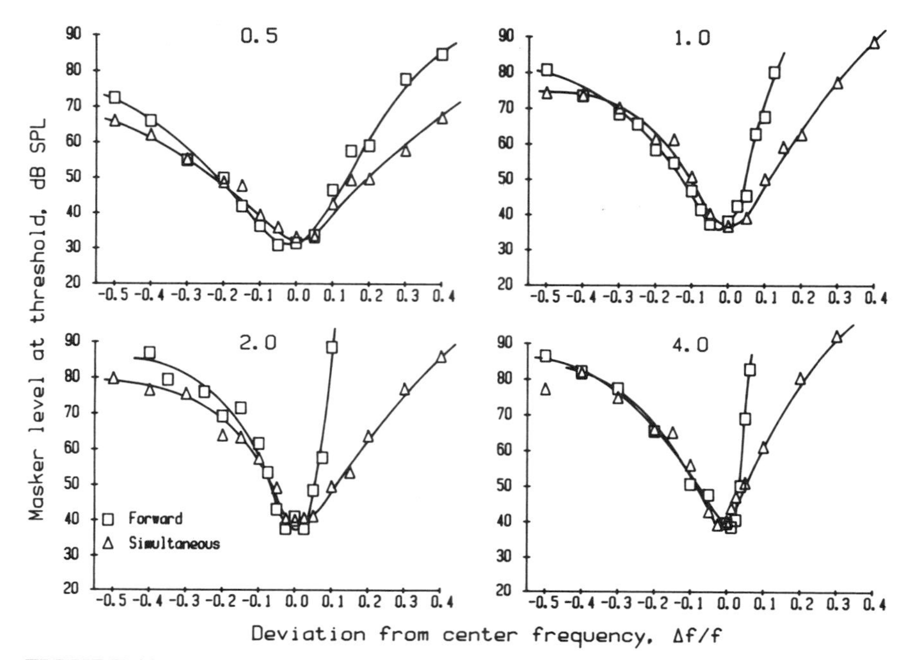

FIGURE 13 Comparison of psychophysical tuning curves determined in simultaneous masking (triangles) and forward masking (squares). The masker frequency is plotted as deviation from the center frequency divided by the center frequency ($\Delta f/f$). The center frequency is indicated in kHz in each panel. A low-level notched noise was gated with the masker to provide a consistent detection cue in forward masking and to restrict off-frequency listening. (From Moore et al., 1984.)

particularly marked on the high-frequency side of the tuning curve. According to Houtgast (1974) this difference arises because the internal representation of the masker (its excitation pattern) is sharpened by a suppression process, with the greatest sharpening occurring on the low-frequency side. In simultaneous masking, the effects of suppression are not seen, because any reduction of the masker activity in the frequency region of the signal is accompanied by a similar reduction in signal-evoked activity. In other words, the signal-to-masker ratio in the frequency region of the signal is unaffected by the suppression. In forward masking, on the other hand, the suppression does not affect the signal. For maskers with frequencies above that of the signal, the effect of suppression is to sharpen the excitation pattern of the masker, resulting in an increase of the masker level required to mask the signal. Thus the suppression is revealed as an increase in the slopes of the PTC.

An alternative explanation is that, in simultaneous masking, the low-level signal may be suppressed by the masker, so that it falls below absolute threshold. The neural data indicate that tones falling outside of the region bounded by the neural tuning curve can produce suppression. Thus, the PTC in simultaneous masking might map out the boundaries of the more broadly tuned suppression region (Delgutte, 1988).

It remains unclear which of these two explanations is correct. Moore and Glasberg (1982b) concluded on the basis of a psychophysical experiment that the first explanation was correct, and a physiological experiment by Pickles (1984) supported this view. However, Delgutte (1990) has presented physiological evidence suggesting that simultaneous masking by intense low-frequency tones (upward spread of masking) is due largely to suppression rather than spread of excitation.

Several other methods of estimating frequency selectivity have indicated sharper tuning in nonsimultaneous masking than in simultaneous masking. For example, auditory filter shapes estimated in forward masking using a notched-noise masker, have smaller bandwidths and greater slopes than those estimated in simultaneous masking (Moore & Glasberg, 1981; Moore, Poon, Bacon, & Glasberg, 1987). This encourages the belief that a general consequence of suppression is an enhancement of frequency selectivity.

XI. SUMMARY

The peripheral auditory system contains a bank of bandpass filters, the auditory filters, with center frequencies spanning the audible range. The basilar membrane appears to provide the initial basis of the filtering process. The auditory filter can be thought of as a weighting function that characterizes frequency selectivity at a particular center frequency. The shape of the auditory filter at a given center frequency can be estimated using the notched-noise masking technique and the assumptions of the power-spectrum model. Its bandwidth for frequencies above 1 kHz is about 10–17% of the center frequency. At moderate sound levels the auditory filter is roughly symmetric on a linear frequency scale. At high sound levels the low-frequency side of the filter becomes less steep than the high-frequency side. The shape of the auditory filter appears to depend mainly on the level at the input to the filter.

When two maskers are combined, the resulting masking is sometimes greater than predicted from linear summation of the individual effects of the maskers. One explanation for this excess masking is that the individual maskers are subject to a compressive nonlinearity before their effects are combined. However, a more plausible explanation can be given in terms of the detection cues used with the individual maskers and the combined maskers.

The excitation pattern of a given sound represents the distribution of activity evoked by that sound as a function of the characteristic frequency of the neurons stimulated. In psychophysical terms, the excitation pattern can be defined as the output of each auditory filter as a function of its center frequency. The shapes of excitation patterns for sinusoids or narrowband noises are similar to the masking patterns of narrowband noises.

The critical bandwidth is related to the bandwidth of the auditory filter. It is revealed in experiments on masking, loudness, absolute threshold, phase sensitivity, and the audibility of partials in complex tones. However, factors other than auditory filtering play a role in many of these experiments, so that they often do not provide a "direct" measure of the bandwidth of the auditory filter.

Houtgast and others have shown that nonsimultaneous masking reveals suppression effects similar to the suppression observed in primary auditory neurons. This suppression is not revealed in simultaneous masking, possibly because suppression at a given CF does not affect the signal-to-masker ratio at that CF. One result of suppression is an enhancement in frequency selectivity.

Acknowledgments

I thank Joseph Alcántara, Tom Baer, Brian Glasberg, Andrew Oxenham, Roy Patterson, Robert Peters, Aleksander Sek, Michael Shailer, and Michael Stone for helpful comments on an earlier version of this chapter. I also thank Aleksander Sek for producing Figure 11.

References

Arthur, R. M., Pfeiffer, R. R., & Suga, N. (1971). Properties of "two-tone inhibition" in primary auditory neurones. *Journal of Physiology, 212,* 593–609.

Bernstein, R. S., & Raab, D. H. (1990). The effects of bandwidth on the detectability of narrow- and wideband signals. *Journal of the Acoustical Society of America, 88,* 2115–2125.

Bilger, R. (1959). Additivity of different types of masking. *Journal of the Acoustical Society of America, 31,* 1107–1109.

Blauert, J., & Laws, P. (1978). Group delay distortions in electroacoustical systems. *Journal of the Acoustical Society of America, 63,* 1478–1483.

Bos, C. E., & de Boer, E. (1966). Masking and discrimination. *Journal of the Acoustical Society of America, 39,* 708–715.

Buus, S. (1985). Release from masking caused by envelope fluctuations. *Journal of the Acoustical Society of America, 78,* 1958–1965.

Buus, S., Schorer, E., Florentine, M., & Zwicker, E. (1986). Decision rules in detection of simple and complex tones. *Journal of the Acoustical Society of America, 80,* 1646–1657.

Chistovich, L. A. (1957). Frequency characteristics of masking effect. *Biophysics, 2,* 743–755.

Craig, J. H., & Jeffress, L. A. (1962). Effect of phase on the quality of a two-component tone. *Journal of the Acoustical Society of America, 34,* 1752–1760.

Delgutte, B. (1988). Physiological mechanisms of masking. In H. Duifhuis, J. W. Horst, & H. P. Wit (Eds.), *Basic issues in hearing* (pp. 204–214). London: Academic Press.

Delgutte, B. (1990). Physiological mechanisms of psychophysical masking: Observations from auditory-nerve fibers. *Journal of the Acoustical Society of America, 87,* 791–809.

Duifhuis, H. (1973). Consequences of peripheral frequency selectivity for nonsimultaneous masking. *Journal of the Acoustical Society of America, 54,* 1471–1488.

Egan, J. P., & Hake, H. W. (1950). On the masking pattern of a simple auditory stimulus. *Journal of the Acoustical Society of America, 22,* 622–630.

Fastl, H. (1976a). Masking patterns of subcritical versus critical band maskers at 8.5 kHz. *Acustica, 34,* 167–171.

Fastl, H. (1976b). Temporal masking effects. I. Broad band noise masker. *Acustica, 35,* 287–302.

Fine, P. A., & Moore, B. C. J. (1993). Frequency analysis and musical ability. *Music Perception, 11,* 39–53.

Fletcher, H. (1940). Auditory patterns. *Reviews of Modern Physics, 12,* 47–65.

Gässler, G. (1954). Über die Hörschwelle für Schallereignisse mit verschieden breitem Frequenzspektrum. *Acustica, 4,* 408–414.

Glasberg, B. R., & Moore, B. C. J. (1986). Auditory filter shapes in subjects with unilateral and bilateral cochlear impairments. *Journal of the Acoustical Society of America, 79,* 1020–1033.

Glasberg, B. R., & Moore, B. C. J. (1990). Derivation of auditory filter shapes from notched-noise data. *Hearing Rresearch, 47,* 103–138.

Glasberg, B. R., Moore, B. C. J., & Nimmo-Smith, I. (1984a). Comparison of auditory filter shapes derived with three different maskers. *Journal of the Acoustical Society of America, 75,* 536–544.

Glasberg, B. R., Moore, B. C. J., Patterson, R. D., & Nimmo-Smith, I. (1984b). Dynamic range and asymmetry of the auditory filter. *Journal of the Acoustical Society of America, 76,* 419–427.

Green, D. M. (1967). Additivity of masking. *Journal of the Acoustical Society of America, 41,* 1517–1525.

Greenwood, D. D. (1961a). Auditory masking and the critical band. *Journal of the Acoustical Society of America, 33,* 484–501.

Greenwood, D. D. (1961b). Critical bandwidth and the frequency coordinates of the basilar membrane. *Journal of the Acoustical Society of America, 33,* 1344–1356.

Greenwood, D. D. (1971). Aural combination tones and auditory masking. *Journal of the Acoustical Society of America, 50,* 502–543.

Hall, J. W., Haggard, M. P., & Fernandes, M. A. (1984). Detection in noise by spectro-temporal pattern analysis. *Journal of the Acoustical Society of America, 76,* 50–56.

Hamilton, P. M. (1957). Noise masked thresholds as a function of tonal duration and masking noise bandwidth. *Journal of the Acoustical Society of America, 29,* 506–511.

Hartmann, W. M., & Hnath, G. M. (1982). Detection of mixed modulation. *Acustica, 50,* 297–312.

Hartmann, W. M., McAdams, S., & Smith, B. K. (1990). Hearing a mistuned harmonic in an otherwise periodic complex tone. *Journal of the Acoustical Society of America, 88,* 1712–1724.

Helmholtz, H. L. F. von (1863). *Die Lehre von den Tonempfindungen als physiologische Grundlage für die Theorie der Musik.* Braunschweig: F. Vieweg.

Higgins, M. B., & Turner, C. W. (1990). Summation bandwidths at threshold in normal and hearing-impaired listeners. *Journal of the Acoustical Society of America, 88,* 2625–2630.

Houtgast, T. (1972). Psychophysical evidence for lateral inhibition in hearing. *Journal of the Acoustical Society of America, 51,* 1885–1894.

Houtgast, T. (1973). Psychophysical experiments on "tuning curves" and "two-tone inhibition." *Acustica, 29,* 168–179.

Houtgast, T. (1974). Lateral suppression in hearing. Ph.D. thesis, Free University of Amsterdam.

Houtgast, T. (1977). Auditory-filter characteristics derived from direct-masking data and pulsation-threshold data with a rippled-noise masker. *Journal of the Acoustical Society of America, 62,* 409–415.

Humes, L. E., & Jesteadt, W. (1989). Models of the additivity of masking. *Journal of the Acoustical Society of America, 85,* 1285–1294.

Humes, L. E., Lee, L. W., & Jesteadt, W. (1992). Two experiments on the spectral boundary conditions for nonlinear additivity of simultaneous masking. *Journal of the Acoustical Society of America, 92,* 2598–2606.

Johnson-Davies, D., & Patterson, R. D. (1979). Psychophysical tuning curves: Restricting the listening band to the signal region. *Journal of the Acoustical Society of America, 65,* 675–770.

Kidd, G., & Feth, L. L. (1982). Effects of masker duration in pure-tone forward masking. *Journal of the Acoustical Society of America, 72,* 1384–1386.

Kohlrausch, A. (1988). Masking patterns of harmonic complex tone maskers and the role of the inner ear transfer function. In H. Duifhuis, J. W. Jorst, & H . P. Wit (Eds.), *Basic issues in hearing* (pp. 339–346). London: Academic Press.

Langhans, A., & Kohlrausch, A. (1992). Spectral integration of broadband signals in diotic and dichotic masking experiments. *Journal of the Acoustical Society of America, 91,* 317–326.

Lutfi, R. A. (1983). Additivity of simultaneous masking. *Journal of the Acoustical Society of America, 73,* 262–267.

Lutfi, R. A. (1985). A power-law transformation predicting masking by sounds with complex spectra. *Journal of the Acoustical Society of America, 77,* 2128–2136.

Miyazaki, K., & Sasaki, T. (1984). Pure-tone masking patterns in nonsimultaneous masking conditions. *Japanese Psychological Research, 26,* 110–119.

Moore, B. C. J. (1978). Psychophysical tuning curves measured in simultaneous and forward masking. *Journal of the Acoustical Society of America, 63,* 524–532.

Moore, B. C. J. (1980). Detection cues in forward masking. In G. van den Brink & F. A. Bilson (Eds.), *Psychophysical, physiological and behavioural studies in hearing* (pp. 222–229). Delft: Delft University Press.

Moore, B. C. J. (1981). Interactions of masker bandwidth with signal duration and delay in forward masking. *Journal of the Acoustical Society of America, 70,* 62–68.

Moore, B. C. J. (1985). Additivity of simultaneous masking, revisited. *Journal of the Acoustical Society of America, 78,* 488–494.

Moore, B. C. J. (1987). Distribution of auditory-filter bandwidths at 2 kHz in young normal listeners. *Journal of the Acoustical Society of America, 81,* 1633–1635.

Moore, B. C. J. (1988). Dynamic aspects of auditory masking. In G. Edelman, W. Gall, & W. Cowan (Eds.), *Auditory function: Neurobiological bases of hearing* (pp. 585–607). New York: Wiley.

Moore, B. C. J. (1989). *An Introduction to the Psychology of Hearing,* 3rd ed. London: Academic Press.

Moore, B. C. J., & Glasberg, B. R. (1981). Auditory filter shapes derived in simultaneous and forward masking. *Journal of the Acoustical Society of America, 70,* 1003–1014.

Moore, B. C. J., & Glasberg, B. R. (1982a). Contralateral and ipsilateral cueing in forward masking. *Journal of the Acoustical Society of America, 71,* 942–945.

Moore, B. C. J., & Glasberg, B. R. (1982b). Interpreting the role of suppression in psychophysical tuning curves. *Journal of the Acoustical Society of America, 72,* 1374–1379.

Moore, B. C. J., & Glasberg, B. R. (1983a). Growth of forward masking for sinusoidal and noise maskers as a function of signal delay: Implications for suppression in noise. *Journal of the Acoustical Society of America, 73,* 1249–1259.

Moore, B. C. J., & Glasberg, B. R. (1983b). Suggested formulae for calculating auditory-filter bandwidths and excitation patterns. *Journal of the Acoustical Society of America, 74,* 750–753.

Moore, B. C. J., & Glasberg, B. R. (1985). The danger of using narrowband noise maskers to measure suppression. *Journal of the Acoustical Society of America, 77,* 2137–2141.

Moore, B. C. J., & Glasberg, B. R. (1987a). Factors affecting thresholds for sinusoidal signals in narrow-band maskers with fluctuating envelopes. *Journal of the Acoustical Society of America, 82,* 69–79.

Moore, B. C. J., & Glasberg, B. R. (1987b). Formulae describing frequency selectivity as a function of frequency and level and their use in calculating excitation patterns. *Hearing Research, 28,* 209–225.

Moore, B. C. J., & Glasberg, B. R. (1990). Frequency discrimination of complex tones with overlapping and non-overlapping harmonics. *Journal of the Acoustical Society of America, 87,* 2163–2177.

Moore, B. C. J., Glasberg, B. R., & Roberts, B. (1984). Refining the measurement of psychophysical tuning curves. *Journal of the Acoustical Society of America, 76,* 1057–1066.

Moore, B. C. J., Glasberg, B. R., & Shailer, M. J. (1984). Frequency and intensity difference limens for harmonics within complex tones. *Journal of the Acoustical Society of America, 75,* 550–561.

Moore, B. C. J., Glasberg, B. R., & Simpson, A. (1992). Evaluation of a method of simulating reduced frequency selectivity. *Journal of the Acoustical Society of America, 91,* 3402–3423.

Moore, B. C. J., Glasberg, B. R., van der Heijden, M., Houtsma, A. J. M., & Kohlrausch, A. (1995). Comparison of auditory filter shapes obtained with notched-noise and noise-tone maskers. *Journal of the Acoustical Society of America, 97,* 1175–1182.

Moore, B. C. J., & Ohgushi, K. (1993). Audibility of partials in inharmonic complex tones. *Journal of the Acoustical Society of America, 93,* 452–461.

Moore, B. C. J., & O'Loughlin, B. J. (1986). The use of nonsimultaneous masking to measure frequency selectivity and suppression. In B. C. J. Moore (Ed.), *Frequency selectivity in hearing* (pp. 179–250). London: Academic Press.

Moore, B. C. J., Peters, R. W., & Glasberg, B. R. (1990). Auditory filter shapes at low center frequencies. *Journal of the Acoustical Society of America, 88,* 132–140.

Moore, B. C. J., Poon, P. W. F., Bacon, S. P., & Glasberg, B. R. (1987). The temporal course of masking and the auditory filter shape. *Journal of the Acoustical Society of America, 81,* 1873–1880.

Moore, B. C. J., & Sek, A. (1992). Detection of combined frequency and amplitude modulation. *Journal of the Acoustical Society of America, 92,* 3119–3131.

Moore, B. C. J., Shailer, M. J., Hall, J. W., & Schooneveldt, G. P. (1993). Comodulation masking release in subjects with unilateral and bilateral cochlear hearing impairment. *Journal of the Acoustical Society of America, 93,* 435–451.

Neff, D. L. (1985). Stimulus parameters governing confusion effects in forward masking. *Journal of the Acoustical Society of America, 78,* 1966–1976.

Ohgushi, K. (1978). On the role of spatial and temporal cues in the perception of the pitch of complex tones. *Journal of the Acoustical Society of America, 64,* 764–771.

Ohgushi, K. (1983). The origin of tonality and a possible explanation of the octave enlargement phenomenon. *Journal of the Acoustical Society of America, 73,* 1694–1700.

Ohm, G. S. (1843). Über die Definition des Tones, nebst daran geknüpfter Theorie der Sirene und ähnlicher tonbildender Vorrichtungen. *Annalen der Physik und Chemie, 59,* 513–565.

O'Loughlin, B. J., & Moore, B. C. J. (1981a). Improving psychoacoustical tuning curves. *Hearing Research, 5,* 343–346.

O'Loughlin, B. J., & Moore, B. C. J. (1981b). Off-frequency listening: Effects on psychoacoustical tuning curves obtained in simultaneous and forward masking. *Journal of the Acoustical Society of America, 69,* 1119–1125.

Oxenham, A. J., & Moore, B. C. J. (1994). Modeling the additivity of nonsimultaneous masking. *Hearing Research, 80,* 105–118.

Oxenham, A. J., & Moore, B. C. J. (1995). Additivity of masking in normally hearing and hearing-impaired subjects. *Journal of the Acoustical Society of America* (in press).

Patterson, R. D. (1974). Auditory filter shape. *Journal of the Acoustical Society of America, 55,* 802–809.

Patterson, R. D. (1976). Auditory filter shapes derived with noise stimuli. *Journal of the Acoustical Society of America, 59,* 640–654.

Patterson, R. D. (1987). A pulse ribbon model of monaural phase perception. *Journal of the Acoustical Society of America, 82,* 1560–1586.

Patterson, R. D., & Henning, G. B. (1977). Stimulus variability and auditory filter shape. *Journal of the Acoustical Society of America, 62,* 649–664.

Patterson, R. D., & Moore, B. C. J. (1986). Auditory filters and excitation patterns as representations of frequency resolution. In B. C. J. Moore (Ed.), *Frequency selectivity in hearing* (pp. 123–177). London: Academic Press.

Patterson, R. D., & Nimmo-Smith, I. (1980). Off-frequency listening and auditory filter asymmetry. *Journal of the Acoustical Society of America, 67,* 229–245.

Patterson, R. D., Nimmo-Smith, I., Weber, D. L., & Milroy, R. (1982). The deterioration of hearing with age: Frequency selectivity, the critical ratio, the audiogram, and speech threshold. *Journal of the Acoustical Society of America, 72,* 1788–1803.

Penner, M. J., & Shiffrin, R. M. (1980). Nonlinearities in the coding of intensity within the context of a temporal summation model. *Journal of the Acoustical Society of America, 67,* 617–627.

Pick, G. (1980). Level dependence of psychophysical frequency resolution and auditory filter shape. *Journal of the Acoustical Society of America, 68,* 1085–1095.

Pickles, J. O. (1984). Frequency threshold curves and simultaneous masking functions in single fibers of the guinea pig auditory nerve. *Hearing Research, 14,* 245–256.

Plomp, R. (1964). The ear as a frequency analyzer. *Journal of the Acoustical Society of America, 36,* 1628–1636.

Plomp, R., & Mimpen, A. M. (1968). The ear as a frequency analyzer, II. *Journal of the Acoustical Society of America, 43,* 764–767.

Rosen, S., Baker, R. J., & Kramer, S. (1992). Characterizing changes in auditory filter bandwidth as a function of level. In Y. Cazals, K. Horner, & L. Demany (Eds.), *Auditory physiology and perception* (pp. 171–177). Oxford: Pergamon Press.

Ruggero, M. A. (1992). Responses to sound of the basilar membrane of the mammalian cochlea. *Current Opinion in Neurobiology, 2,* 449–456.

Scharf, B. (1970). Critical bands. In J. V. Tobias (Ed.), *Foundations of modern auditory theory* (pp. 157–202). New York: Academic Press.

Schooneveldt, G. P., & Moore, B. C. J. (1989). Comodulation masking release (CMR) as a function of masker bandwidth, modulator bandwidth and signal duration. *Journal of the Acoustical Society of America, 85,* 273–281.

Schorer, E. (1986). Critical modulation frequency based on detection of AM versus FM tones. *Journal of the Acoustical Society of America, 79,* 1054–1057.

Sek, A. (1994). Modulation detection thresholds and critical modulation frequency based on random amplitude and frequency changes. *Journal of the Acoustical Society of Japan (E), 15,* 67–75.

Sek, A., & Moore, B. C. J. (1994). The critical modulation frequency and its relationship to auditory filtering at low frequencies. *Journal of the Acoustical Society of America, 95,* 2606–2615.

Shailer, M. J., Moore, B. C. J., Glasberg, B. R., Watson, N., & Harris, S. (1990). Auditory filter shapes at 8 and 10 kHz. *Journal of the Acoustical Society of America, 88,* 141–148.

Shannon, R. V. (1976). Two-tone unmasking and suppression in a forward masking situation. *Journal of the Acoustical Society of America, 59,* 1460–1470.

Small, A. M. (1959). Pure-tone masking. *Journal of the Acoustical Society of America, 31,* 1619–1625.

Smoorenburg, G. F. (1972a). Audibility region of combination tones. *Journal of the Acoustical Society of America, 52,* 603–614.

Smoorenburg, G. F. (1972b). Combination tones and their origin. *Journal of the Acoustical Society of America, 52,* 615–632.

Soderquist, D. R. (1970). Frequency analysis and the critical band. *Psychonomic Science, 21,* 117–119.

Spiegel, M. F. (1981). Thresholds for tones in maskers of various bandwidths and for signals of various bandwidths as a function of signal frequency. *Journal of the Acoustical Society of America, 69,* 791–795.

Srulovicz, P., & Goldstein, J. L. (1983). A central spectrum model: a synthesis of auditory-nerve timing and place cues in monaural communication of frequency spectrum. *Journal of the Acoustical Society of America, 73,* 1266–1276.

Traunmüller, H. (1990). Analytical expressions for the tonotopic sensory scale. *Journal of the Acoustical Society of America, 88,* 97–100.

Tyler, R. S., Hall, J. W., Glasberg, B. R., Moore, B. C. J., & Patterson, R. D. (1984). Auditory filter asymmetry in the hearing impaired. *Journal of the Acoustical Society of America, 76,* 1363–1368.

Verschuure, J. (1981a). Pulsation patterns and nonlinearity of auditory tuning. I. Psychophysical results. *Acustica, 49,* 288–295.

Verschuure, J. (1981b). Pulsation patterns and nonlinearity of auditory tuning. II. Analysis of psychophysical results. *Acustica, 49,* 296–306.

Vogten, L. L. M. (1974). Pure-tone masking: A new result from a new method. In E. Zwicker & E. Terhardt (Eds.), *Facts and models in hearing* (pp. 142–155). Berlin: Springer-Verlag.

Wegel, R. L., & Lane, C. E. (1924). The auditory masking of one sound by another and its probable relation to the dynamics of the inner ear. *Physical Review, 23,* 266–285.

Zwicker, E. (1952). Die Grenzen der Hörbarkeit der Amplitudenmodulation und der Frequenzmodulation eines Tones. *Acustica, 2,* 125–133.

Zwicker, E. (1961). Subdivision of the audible frequency range into critical bands (Frequenzgruppen). *Journal of the Acoustical Society of America, 33,* 248.

Zwicker, E. (1970). Masking and psychological excitation as consequences of the ear's frequency analysis. In R. Plomp & G. F. Smoorenburg (Eds.), *Frequency analysis and periodicity detection in hearing* (pp. 376–394). Leiden: Sijthoff.

Zwicker, E., & Feldtkeller, R. (1967). *Das Ohr als Nachtrichtenempfänger.* Stuttgart: Hirzel-Verlag.

Zwicker, E., & Terhardt, E. (1980). Analytical expressions for critical band rate and critical bandwidth as a function of frequency. *Journal of the Acoustical Society of America, 68,* 1523–1525.

CHAPTER 6

Temporal Integration and Temporal Resolution

David A. Eddins
David M. Green

I. INTRODUCTION

In trying to construct a mechanistic theory of how the auditory system functions, theorists are forced to speculate about certain dynamic features and to assume certain temporal parameters. Were the system a very simple one, then different experiments should provide similar estimates of these temporal parameters. Such has not been the case. "Time constants" estimated from different experimental tasks range over three orders of magnitude, from 250 μs to 200,000 μs. This range far exceeds experimental error and clearly indicates that something is amiss.

Theorists disagree on how the predicament should be resolved. One approach is to reject the theories as ad hoc (de Boer, 1985). De Boer describes our present theories as designed to explain a limited set of data. The inconsistency of estimates reflects the parochial nature of the approach. When a better "super theory" is achieved, the discrepancies will be resolved.

A different approach is the one taken in this chapter. It denies that the auditory system is a simple one and therefore expects different temporal estimates in different experimental tasks. Different modes of processing the auditory information are required by the different experimental tasks. The system can be slow, and will appear so, when integration of information

Hearing

over a long period of time may be beneficial in detecting a weak signal in noise. The system can also be fast, and will appear so, when trying to avoid the effects of forward and backward masking. The problem with this view is that it has the potential to be completely arbitrary, invoking different processing modes for each experimental situation or result. The princple that prevents such capriciousness is the assumption that the auditory system is configured to optimize the detection of a signal. Thus, it is claimed that the time constant used in any experimental task is the one that maximizes the objective in that task.

In terms of the physiology of the process, such a premise amounts to the assumption that different parts of the nervous system process auditory information in different ways. In any particular psychophysical task, the physiological information is selected to maximize the performance in that task. The different time constants are associated with different neural substrates. Visual theorists would not assume that the eyes are always fixed straight ahead, nor should auditory theorists assume that only one auditory process is used to transform the incoming waveform.

In keeping with this approach, we organized this chapter by dividing auditory temporal processing into two broad topics; namely, temporal integration and temporal resolution. Temporal integration is the process described in time–intensity trades, where we study how increasing the duration of a signal makes it easier to detect. Historically, this is the oldest area of study in the field of auditory temporal phenomena. In this review, we will cover the older approaches and some recent theorizing. The major preoccupation of current research has been on the opposite side of the temporal dimension; namely, temporal acuity. Acuity represents the fastest auditory processing that can occur within a given experimental task. Often such speed is required to avoid the dual effects of forward and backward masking. In effect, temporal acuity investigates the limits of auditor inertia. The topic of temporal acuity has produced a number of subfields that investigate the central topic in their own particular manners. These subfields include gap and modulation detection. Both have produced a substantial body of empirical data and some attempts to construct theories that bridge the different subfields. The chapter concludes with a brief summary of the empirical results from the various areas.

II. TEMPORAL INTEGRATION

A. Classical Theory

When investigators first studied the ability of observers to detect weak auditory signals in quiet the signals were continuous and the intensity or level of the signal was adjusted until it was just detectable. As electronic technology developed, investigators could present the signal for controlled durations, and it was obvious that shorter duration signals required more intensity than longer duration signals to be just detectable. We describe the signal duration and signal intensity needed to achieve a just-detectable signal as the time–intensity trade. Before reviewing the empirical evidence on this topic, which has been explored in a number of different experimental settings, let us begin with a consideration of theory. Why are short signals harder to hear than longer signals? Although the empirical relation is both obvious and evident, what are the theoretical reasons for this relation?

The usual reason given for the time–intensity trade is some kind of accumulation or integration process. The simplest and most general way to express this idea is with the convolution integral. Suppose $x(t)$ was the input to some accumulator or integration process. For the present, consider $x(t)$ to be any real-valued waveform. If we weight that input by a function $h(t)$ and add up the weighted values, we obtain the output, $y(t)$, according to the following formula:

$$y(t) = \int_{-\infty}^{t} h(t - \tau)x(\tau)\, d\tau \tag{1a}$$

Note that the argument for the weighting function h runs backward in time—it looks over the recent past of x—and weights the past value of $x(\tau)$ by $h(t - \tau)$. Although the integral runs over all past time $\{-\infty$ to $t\}$, usually $h(t)$ is essentially 0 for large values of the argument, and thus only those events close to the present time and within the nonzero portion of $h(t)$ materially affect the output $y(t)$. An equivalent form for the convolution integral is obtained by substituting $z = t - \tau$; in that case, the equation becomes

$$y(t) = \int_{0}^{\infty} x(t - z)h(z)\, dz \tag{1b}$$

This is a form often seen in textbooks. It weights the past of x by the function h and integrates over all (past) time. To fix our understanding of this idea, let us consider the simplest form for $h(t)$,

$$h(t) = \begin{matrix} 1 & 0 < t < \tau \\ 0 & \text{elsewhere} \end{matrix} \tag{2}$$

This weighting function when used in Eq. (1a) and (1b) simply adds up, with equal weight, all of $x(t)$ occurring within τ of the present. Note that $h(t)$ is assumed to be 0 for negative arguments. This prevents an output from occurring before an input occurs (recall that h is time reversed in the convolution integral). We call filters that work this way *causal filters*. Without this assumption, $h(t)$ would produce outputs before the input occurs, a feature not usually considered desirable in scientific analysis.

Clearly, such a system will produce a time–intensity trade. Suppose the input is a fixed quantity, call it S, that is monotonic with the intensity of the signal I and lasts for the duration of the signal D. If we integrate or add the quantity S for a period of time τ, then we will produce an amount $S\tau$. Assume the detection threshold occurs when $y(t) = C$. If $D < \tau$, the output of this process is SD. To achieve a threshold quantity, the product SD must equal C so that $S^{-1} \sim D$. Because S is monotonic with signal intensity, we find as D decreases I must be increased to achieve a threshold quantity. The simplest assumption is that S is proportional to I, $S \sim I$, in which case $I^{-1} \sim D$, and the threshold for the signal should decrease 10 dB per decade change in duration. The solid curve with negative slope in Figure 1 shows the expected results for a fixed integrator, with $\tau = 0.5$ s. If $D > \tau$, the output is $S\tau$ for any duration signal, and then S is independent of signal duration D—yielding the line of zero slope for the longer durations in Figure 1. The other curves in Figure 1 will be discussed shortly. This is the basic idea of an integration process and why it produces a time–intensity trade.

It is essentially the idea introduced by Munson (1947), Plomp and Bouman (1959), and Zwislocki (1960) in their seminal papers. They all used an exponential function for $h(t)$:

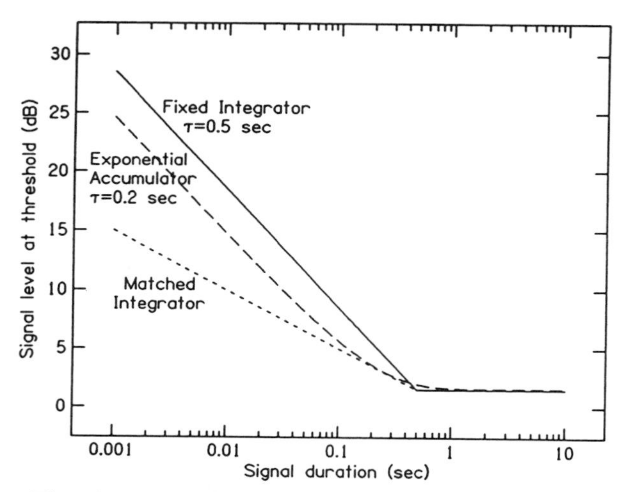

FIGURE 1 Time–intensity trades for several integration models. The fixed integrator has a time window of 0.5 s. The matched integrator has a window matched to the signal duration with a maximum value of 0.5 s. The exponential integrator with a time constant of 0.2 s is shown for comparison [see Eq. (3)]. In all cases, it is assumed that a quantity proportional to waveform power is integrated.

$$h(t) = \begin{matrix} e^{-t/\tau} & t > 0 \\ 0 & \text{elsewhere} \end{matrix} \tag{3}$$

This integrator weights past time by an amount $e^{-t/\tau}$, so that events more than 3τ in the past are greatly attenuated and have little impact on the present output. The time–intensity trade for an exponential integrator is also shown in Figure 1; the time constant is 0.2 s. As can be seen, over most time the expected trade is 10 dB per decade, as it was with the fixed integrator. We now pause in our discussion of theory to briefly review the available empirical data concerning time–intensity trading ratios.

B. Time–Intensity Trades

1. Sinusoids in Quiet

The most recent results on temporal integration of sinusoidal signals at absolute threshold are found in a study by Florentine, Fastl, and Buus (1988). They present new data using six normal-hearing listeners, as well as a thorough review of previous data. They studied four frequencies, 250, 1000, 4000, and 14,000 Hz, and varied the duration of the signals from about 5 cycles to 500 ms. The subjects listened for the signals in quiet using a two-interval, forced-choice adaptive task. The data are approximated nicely by a straight line in a plot of signal thresholds in decibels versus the logarithm of signal duration. The slope of the line is about $-3/4$ for the three lower frequencies and perhaps slightly smaller at the highest frequency. They also plotted the results obtained by previous investigators of this problem including Fastl (1977), Olsen and Carhart (1966), Hughes (1946), Watson and Gengel (1969), and Zwicker and Wright (1963). Data from the earlier studies have been fitted by a variety of different functions. Florentine et al. (1988; see their Fig. 2) demonstrate that all the data show essentially the same time–intensity trade; namely, signal intensity decreases as a power function of signal duration with an exponent of about $-3/4$. The only area of minor discrepancy concerns the slope of the time–intensity trade for higher frequency signals. Zwicker and Wright (1963), in particular, found a shallower slope for the higher frequencies.

2. Sinusoids in Noise

Plomp and Bouman's (1959) study remains the most comprehensive on how signal duration influences the detectability of a sinusoidal signal partially masked by noise. They used six signal frequencies, ranging from 250 to 8000 Hz, and varied the signal duration between about 5 cycles and 5000 ms. Their data are well approximated by accumulations produced by a simple exponential integrator with a time constant of about 200–300 ms.

Thus, for durations less than about 300 ms, the time–intensity trade is nearly a straight line with unit slope when the threshold in decibels is plotted against the logarithm of duration (see Figure 1). The threshold is independent of duration (zero slope) for the longer durations. Any departure from this simple summary occurs at the very shortest duration, where the thresholds may rise more than this model predicts. This function also provides a reasonably good summary of most of the previous studies in the area including Blodgett, Jeffress, and Taylor (1958), Garner and Miller (1947), Garner (1947), Green, Birdsall, and Tanner (1957), Hamilton (1957), Simon (1963), and Zwicker and Wright (1963).

3. Noise in Quiet

We could find only two studies on the issue of how the duration of a noise burst affects its detectability when presented in quiet. Penner (1978) shows data for three subjects at durations from 0.1 to 1000 ms. The data fall along a straight line when plotting thresholds in decibels against the logarithm of duration with a slope of about −0.73. In an earlier study, Garner (1947) found the slope to be about −0.8. Thus, it appears that a noise in quiet produces a time–intensity trade that is similar to that produced by a sinusoid in quiet.

4. Noise in Noise

There are more data on the time–intensity trade for a signal which itself is a noise burst added to a masking noise. Green (1960) presents a theoretical account of the detection of such signals using a simple energy model. He shows that such a model predicts a square-root trade between threshold signal power and signal duration (slope of −0.5). He presents data at five durations ranging from 3 to 300 ms that fall almost exactly along the theoretical line. Campbell (1963) measured the detectability of noise bursts for durations ranging from 1 to 1000 ms. He found the threshold decreased with the expected slope of 5 dB per decade for durations from 1 to about 100 ms. The threshold continued to decrease to 1000 ms, but the rate of improvement was much slower. Raab, Osman, and Rich (1963) also presented data showing that the threshold of a noise burst decreases 5 dB per decade change in duration from 5 to about 200 ms. They measured at four different sound pressure levels (20 to 75 SPL). Penner (1978) studied durations over a wider range (from 0.1 to 1000 ms) and showed that the time–intensity trade is essentially the same as for a noise burst in quiet—a slope of −0.69 when plotted on log-log coordinates. Penner's data for the longer durations (3 to 1000 ms) fall very close to the slope of −0.5. The reason for the discrepancy at the very short durations is not known.

5. Increment in a Pulsed Sinusoid

Our last time–intensity trade concerns how the duration of the sound affects the just-detectable increment in a sinusoidal signal. This was first studied by Henning (1970) and has recently been the subject of a more extensive series of measurements by Florentine (1986). She measured the difference limen ($\Delta L = 20 \log[(\Delta p + p)/p]$), where p is the pressure of the standard and Δp is the increment in pressure created by adding the signal to the standard. She used three different frequencies, 250, 500, and 8000 Hz, at three different levels, 40, 65, and 85 SPL, with durations ranging from 2 to 2000 ms. The time–intensity trade is nicely summarized by a straight line with slope of about $-1/4$ (-0.27 ± 0.06) if ΔL is plotted against the logarithm of the signal duration. An odd feature of the data is that ΔL does not appear to reach a clear asymptotic value even for the longest duration. The consistent decline of ΔL with duration up to the longest duration is especially apparent for the higher frequencies and the higher levels for the standard. The relatively shallow slope for the time–intensity trade is also inconsistent with simple statistical integration models, which will be discussed shortly. They predict that the slope should be equal to -0.5.

We now return to a more detailed consideration of theory, including the more modern theory.

C. The Presence of Noise

Although the preceding analysis provides a good summary of the situation when integrating a noiseless signal, the presence of noise complicates the analysis of the temporal integration process. Many of the problems are not explicitly discussed in any of the current models, and our analysis will focus on these problems in the following discussion. Noise not only produces an added burden of complexity because one must discriminate noise from signal plus noise, it also raises questions about the precise way in which the integrator is used. Two important issues are whether or not the integrator is matched to the signal duration and when to turn the accumulation process on or off.

To understand the theoretical problems produced by noise, consider the time traces shown in Figure 2 for a simple exponential integrator. The time traces show what happens when we actually simulate such an accumulation process. The bottom line in Figure 2 shows the weighting function $h(t)$; in this example, it is exponential with a time constant of about 100 ms. The next time line shows the signal $s(t)$, two sinusoidal bursts each 80 ms in duration but differing in intensity. The middle time line is the noise $n(t)$. Above the time line for the noise is the square of the signal plus noise. This

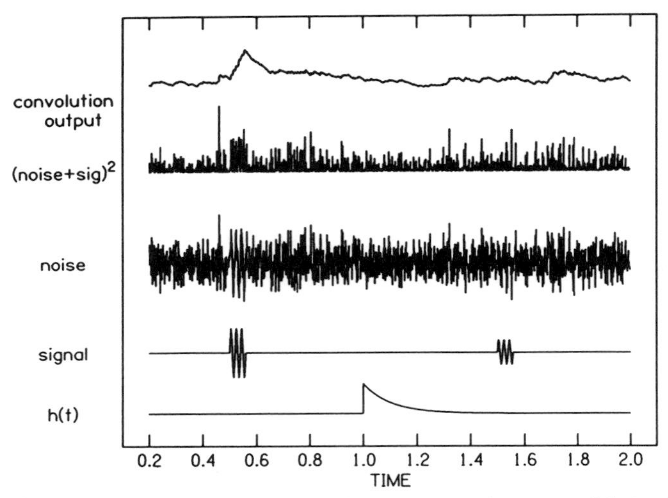

FIGURE 2 A simulation of an exponential integrator with a sinusoidal signal and noise present at its input. The time traces from bottom to top are the impulse response of the integrator $h(t)$, the sinusoidal signal, the noise waveform, the square of the signal plus noise, the output of the convolution integral, Eq. (1).

quantity is always positive and can therefore be accumulated to determine if the signal produces a greater amount of this quantity. The output of the convolution integral, Eq. (1), is the top time line in the figure—marked *convolution output*. The maximum value for the output occurs near the end of the largest signal burst. Other local maxima occur; some caused by noise alone. The example illustrates the advantage of knowing when the signal might occur.

If we know the exact time the signal might occur, a statistic useful in deciding about the presence of a signal is the output of the integrator at the time the signal terminates. Because noise is a random process, the integrated quantity, in this example the squared values of the waveform, is also a random variable. There will be a distribution of convolution outputs having different means, depending on whether or not the signal is present in the noise. The quality of such discriminations is determined by the difference in the means of the two distributions compared to the variability. Figure 3 illustrates this situation and defines the detection index d' as the difference in means divided by the standard deviation of the noise process. For the purpose of this discussion, we assume that the variance of the integrated random variable (the square of the waveform) is proportional to the mean. This is characteristic of a number of statistical processes such as the chi-squared distribution or the Poisson. We also assume the noise has bandwidth W (cycles per second), so that a noise waveform lasting T seconds has $2WT$

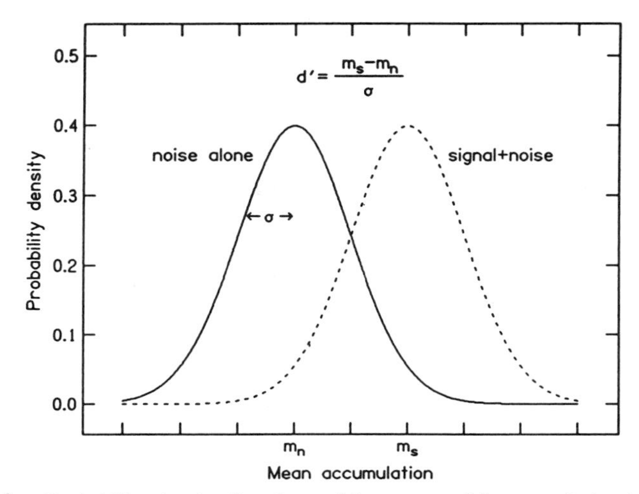

FIGURE 3 Probability density functions of the output of the convolution integral given noise alone and signal plus noise. The detectability index *d'* is the separation of the two means divided by the standard deviation of each distribution.

degrees of freedom (see Green & Swets, 1966). This assumption will simplify the ensuing discussion.

Another issue of importance in analyzing an integrator in the presence of noise is whether the period of integration is matched to the signal duration or whether it is a fixed quantity. Let us analyze both assumptions. To begin, assume the signal duration is less than the maximum integration time; that is, $D < \tau$.

For D $<$ τ. In Case I, integration time is *fixed* and independent of signal duration. Mean (noise) $= W\tau N$, mean (signal plus noise) $= WDS + W\tau N$, and $\sigma = (W\tau)^{1/2}N$:

$$d' = WDS\,/N(W\tau)^{1/2} \tag{4}$$

In Case II, integration time is *matched* to the signal duration $\tau = D$. Mean (noise) $= WDN$, mean (signal plus noise) $= WD(S + N)$, and $\sigma = (WD)^{1/2}N$:

$$d' = (WD)^{1/2}\,S/N \tag{5}$$

For $D > \tau$. If $D > \tau$, then there is no difference between the two models. In either case, it is easy to verify that

$$d' = (W\tau)^{1/2}\,S/N \tag{6}$$

so that detection is independent of signal duration D for either the matched or fixed integration time model.

To obtain the time–intensity trades for these models, we set $d' = 1$ and assume some relationship between signal intensity I and the quantity S of the model. Figure 1 shows the results, assuming S is proportional to I, for both the matched and fixed integrator. The matched integrator shows less dependence on signal duration than the fixed integrator, the threshold decreasing 5 dB per decade change in duration compared with the 10 dB per decade change for the fixed integrator. Note also that the fixed integrator shows exactly the same variation with signal duration as the older, classical model that ignored the effects of noise.

Before leaving our theoretical discussion of temporal integrators, we mention one more integration model, the statistical integrator. Suppose we observe some temporal process and calculate m quantities similar to d'. We then combine these m observations into a final statistic that is used to decide whether or not the signal is present. If the observations are independent and the individual values of d' are all equal, it is easy to show that the quality of the final statistic, call it d'_{tot}, is equal to $(m)^{1/2}d'$. This has been called the *independent looks model*. In the context of the present discussion, because the number of looks is proportional to the signal duration D, it will also predict that d' will increase proportionally to $D^{1/2}$.

The preceding discussion contains all the elements of a complete analysis of the problem of detecting signals in noise. Existing models generally ignore one or more of these elements. Often only the means of the distributions of noise and signal plus noise are derived; the variability is ignored. A fixed integration time is also usually assumed, thereby ignoring the issue of whether the integration time is matched to the signal. For the most part, it is assumed that the decision statistic (the quantity used to decide whether signal or noise alone was present) is the integrator value achieved at the time the signal terminates. This is because, on average, the integrator output will be at a maximum at that time.

D. Recent Theories

There has been comparatively little theory about the time–intensity trade since the seminal papers of the 1940s and 1950s. Zwislocki (1969) has written an extensive theoretical treatment of the temporal summation of loudness. The basis of his theory is derived from neurophysiological recordings, but it provides an adequate description of much psychoacoustic data. The theory assumes a linear integration process with an exponential window h, having a time constant of about 200 ms. Nonlinear processes also modify this mechanism, so that the integrator appears to be somewhat faster ($\tau \approx$ 100 ms) at higher sound pressure levels. The basic problem with this and other classical theories is that they all predict a time–intensity trade of about −10 dB per decade of duration, as shown in Figure 1. As we have seen in reviewing the empirical evidence, the time–intensity trade is often somewhat smaller than this value, −6 to −8 dB per decade being more typical values for some tasks. How can we account for the smaller trading ratios?

One simple approach is to assume a nonlinear relation between stimulus intensity I and the quantity accumulated by the integrator, which we call S. Such an assumption seems entirely plausible. Suppose the apparent magnitude or loudness of the stimulus is integrated rather than the stimulus intensity. Loudness is thought to grow as a compressive power function of stimulus intensity with an exponent of about 0.3 when expressed relative to stimulus intensity (loudness $\approx$ intensity$^{0.3}$, Stevens, 1975; see Chapter 4). The problem is that we find a quite different exponent when we use a compressive power function to account for the time–intensity data. To predict a time–intensity trade of −6 dB per decade, we must assume $S = I^{\beta}$, where $\beta = 0.6$, rather than the generally accepted value of 0.3. One ingenious way around this dilemma is suggested by Penner (1978). She assumes a nonlinear mapping of intensity to some internal quantity, but also suggests constructing a weighting function $h(t)$ to produce the desired time–intensity trade. The $h(t)$ derived from these assumptions is

$$h(t) = \begin{array}{ll} S_0 a\beta t^{a\beta-1} & t \geq 1 \\ S_0 & 0 < t < 1 \\ 0 & t < 0 \end{array} \tag{7}$$

S_0 is a constant, and a convenient choice is $1/a\beta$. There are then actually three free parameters in this model, a, β, and the choice of units for time t. The time–intensity trade suggested by this model is (−10·a dB per decade) and is independent of the value of β. In her article, Penner shows that a value of $a = 0.7$ gives a good fit to some time–intensity data, and she can still assume the generally accepted value for β; namely, $\beta = 0.3$. With these parameters, Penner claims that a much smaller apparent time constant results if one compares the $h(t)$ of Eq. (7) with, for example, an exponential time constant, as in Eq. (3). If viewed in the time domain, there certainly is some resemblance her $h(t)$ with $a = 0.7$ and an exponential window with a time constant of only 5 ms. A major difference is that the exponential window (Eq. 3) is essentially 0 after 15 ms, whereas Penner's $h(t)$ has not diminished to 5% of its peak value response in 54 ms. Such comparisons of different values of $h(t)$ should be viewed with caution.

Another way to compare temporal windows is in the frequency domain. Because the accumulation process is a linear operator, we can also view it as a process that differentially weights different input frequencies. This is achieved by taking the magnitude of the Fourier transform of $h(t)$. When this transformation is performed, the 3 dB point of the frequency response of the filter corresponding to $h(t)$ is often cited as a parameter to indicate the speed of the system. A large frequency corresponding to the 3 dB value is indicative of a fast system. For the 5 ms exponential window, the 3 dB point is simply $1/(2\pi\tau) = 32$ Hz. For Penner's $h(t)$, the frequency response is hard to derive analytically, but the 3 dB frequency response of that window can be calculated from the Fourier transform of a discrete approximation to Eq. (7). With the constant $a = 0.742$, the 3 dB point is about 0.36 Hz. Such a 3 dB point is produced by an exponential window with a time constant of 435 ms. Thus, if compared in the time domain, Penner's $h(t)$ is fast, but if compared in the frequency domain, it is very slow. The only safe conclusion is that the exponential window and Penner's $h(t)$ are very different.

III. TEMPORAL ACUITY

A. Introduction

There are several ways to pose the question considered in this section. The simplest is to ask, How fast is the ear? How much time does the ear take to process auditory information? To answer this question experimentally, one varies the delay between two auditory events to determine the minimal time delay that can be discriminated. In essence, the focus is the converse of the issue studied in temporal integration. There we wanted to know the duration over which the ear could integrate or collect information. Here we want to know the minimum time interval within which different acoustic events can be distinguished. The question appears simple, but a variety of potentially confounding cues are often produced by changing the starting time or duration of acoustic events. Some are of interest as temporal parameters; some are not. We will consider five different approaches to these temporal issues, describe the procedures used in the experiments, and summarize the results.

B. Temporal Order

Historically, the earliest studies of temporal acuity involved the perception of temporal order. Hirsh (1959) used two tones of different frequency but of nearly equal (500 ms) duration. He started one tone slightly before the other in time, and asked the listener which tone started first? Because chance is 50% correct in this task, he determined the difference in starting time, Δt, that would produce 75% correct responses as his threshold for temporal order. The threshold was about 20–30 ms for a variety of different pairs of sounds. Surprisingly, the threshold did not seem to change as he varied the frequency difference between the two tones or even if he used a noise burst and a tone as his stimulus pair.

In a later study, Hirsh and Sherrick (1961) showed that this 20–30 ms value was needed to discriminate reliably which of two lights came on first.

Indeed, the threshold value, Δt, was about the same when discriminating whether a light flash preceded a tone burst or the reverse. Another type of temporal order judgment was studied by Warren, Obusek, Farmer, and Warren (1969), who asked subjects to indicate the temporal sequence of four distinct sounds that were played repetitively (. . . abcdabcdabcd . . .). They found, for nonmusical or nonspeech sounds, rates of repetition as low as 5/s were needed to correctly distinguish the order of the sequences (200 ms/element). It appears that these repeating sequences produce problems for the listener in coding the auditory events, rather than taxing the inertial aspects of the auditory system. Listeners can discriminate among the different sequences, but have difficulty in mapping the perceived sequence to the required output code. Nickerson and Freeman (1974) reviewed the area in some detail. They showed that one subject, after receiving extensive training, could reliably report the order of a four-tone sequence at rates as high as 500/s (2 ms/element).

With Hirsh's temporal order experiment, it was also found that extensive practice produced better discrimination performance. Also, it is important to use a relatively long duration for the sounds of the pair of stimuli to be discriminated. If brief sounds are used, for example, 10 ms, then more subtle changes in timbre become evident as the delay between the starting times is varied, and very small thresholds for temporal order can be obtained. These changes in timbre can probably be best understood by considering the long-term power spectra of the stimulus pairs.

The most dramatic demonstration of a spectral artifact producing extremely good temporal discrimination scores comes from an experiment by Leshowitz (1971). He asked what was the smallest temporal gap one could detect between two very brief pulses? Each pulse was produced by a 10 μs rectangular wave applied to the earphone. The listeners heard two sounds in succession. In the standard sound, the two pulses followed each other with no gap, $\Delta t = 0$; in the other sound, a gap of Δt was introduced between the two pulses. Leshowitz's listeners could reliably discriminate gaps in the 5–10 μs range. He argued that the basis for this discrimination were slight (1–3 dB) differences in the high frequencies of the audio spectrum (>10,000 Hz). The introduction of a high-pass noise destroyed the ability to make these discriminations, although the clicks were still clearly audible. As Leshowitz suggests, this procedure has little to do with temporal acuity; rather, it studied the ear's ability to detect changes in power spectra of very short pulses, especially changes in the high-frequency part of those spectra.

C. Phase Detection

Leshowitz's results underscore the need for experimenters to hold the long-term power spectrum of the stimuli constant when designing stimuli to probe the limits of temporal discrimination. To rephrase this injunction, what we need to do is produce stimulus pairs that have the same overall power spectrum. If the listener can discriminate between such pairs of stimuli, then the discrimination cannot be on the basis of amplitude or intensity differences that occur at some frequency in the spectrum. The ability to discriminate such pairs must depend on the order of temporal events within the brief interval. A technical way to describe this requirement is to say that the stimulus pairs can differ only in their phase spectra.

The simplest sounds to satisfy this constraint are a pair of clicks, one click slightly larger in amplitude than the other. The discrimination task is to determine the order of the clicks within the pair, large small or small large. We construct this stimulus pair by playing the waveform either forward or backward in time, that is, either $f(t)$ or $f(-t)$. It is easy to show that the power spectra of such waveforms are identical (see Appendix). When viewed from the frequency domain, the only difference between $f(t)$ and $f(-t)$ is in the phase spectrum.

Obviously, if we separate the clicks by tens of milliseconds, the task is very easy. As the total duration of the sound decreases, however, the discrimination becomes increasingly difficult. When the time taken for the two clicks is less than about 2 ms, the ability to discriminate the temporal order fails (Ronken, 1970). We do not claim that the listeners calculate the phase spectrum and discriminate on the basis of such calculations. Rather the short-term power spectrum is the most likely basis for the discrimination. Consider another stimulus pair consisting of a 1000-Hz and a 2000-Hz sinusoid, as used by Wier and Green (1975). One stimulus consisted of a single cycle of the 1000 Hz sinusoid followed by two cycles of the 2000 Hz sinusoid, $f(t)$. The other stimulus was that waveform reversed in time, $f(-t)$. The basis for such discriminations is pitch; the forward stimuli of 1000–2000 Hz sounded higher in pitch than the backward version of 2000–1000 Hz. The listeners in that experiment could easily discriminate (90% correct) whether a stimulus whose total duration was 2 ms was played forward or backward in time. Thus, the inertial properties of the auditory system were fast enough, so that the listeners could discriminate whether the energy bursts were high–low or low–high. When discrimination of temporal direction cannot be made, then the inertial properties of the auditory system are causing different temporal events to be combined, and temporal order information is lost.

While reversing the waveform in time is a simple way to hold the power spectrum of the stimulus constant, a more sophisticated approach can also be used. Huffman (1962) provided a general algorithm for generating a class of waveforms. All members of the class are zero except for an interval of time T seconds long. All members of the class have the same energy spectrum; that is, the spectra are the same when calculated over the entire interval T. Members of the class differ only in their phase spectrum. As used in psychoacoustic experiments, T is small and the Huffman sequence sounds like a brief transient or click. One sequence may have energy delayed at some frequency region while another sequence delays energy at a different frequency region. In this way, pairs of Huffman sequences can be discriminated if the analysis process is faster than T, the total duration of the sequence. If a listener can discriminate between a pair of stimuli selected from the class, we infer that the auditory system can somehow order events within the total duration of the waveform T. Experiments using these waveforms were first pursued by Patterson and Green (1970). Additional results were reported by Green (1973). A simple summary of these results is that discrimination of such waveforms fails when the duration is less than about 2 ms (Green, 1971).

Some recent work by Henning and Gaskell (1981) has challenged the 2 ms limit. They used clicks of different amplitude, much like those of Ronken (1970). They showed that discrimination was possible, after very extensive practice, at total durations of 200 μs! The only obvious stimulus differences in the two experiments are the durations of the clicks. Ronken's clicks were 250 μs in duration, whereas Henning and Gaskell's clicks were 20 μs. We do not know whether this difference in the stimuli or the degree of training explains the difference between Henning and Gaskell's results and the results of many previous experiments.

D. Temporal Gap Detection

The detection of a gap in an otherwise continuous sound has been a much used paradigm in the study of temporal acuity. Before describing that experimental procedure, let us review the basic ideas behind this approach as outlined in Plomp's (1964) classic study. He was interested in measuring the decay of auditory sensation. To measure this decay, he measured the minimum silent interval Δt between two broadband noise pulses. This procedure is shown schematically in Figure 4. According to Plomp, the sensation of the first pulse will continue for some time after the cessation of the pulse, as shown in the lower portion of the figure. To detect the silent interval, the sensation introduced by the second pulse must cause an increase in sensation by an amount ΔS.

Plomp used a two-interval, forced-choice (2IFC) paradigm in which the signal interval consisted of two noise pulses separated by Δt, and the standard interval consisted of two noise pulses without a separation. The task of the subject was to identify which interval contained the silent interval Δt. The level of the second pulse and the just-detectable silent interval are shown for one listener in Figure 5. The parameter of the curve is the level of the first pulse. The value of the just-detectable silent interval increases as the

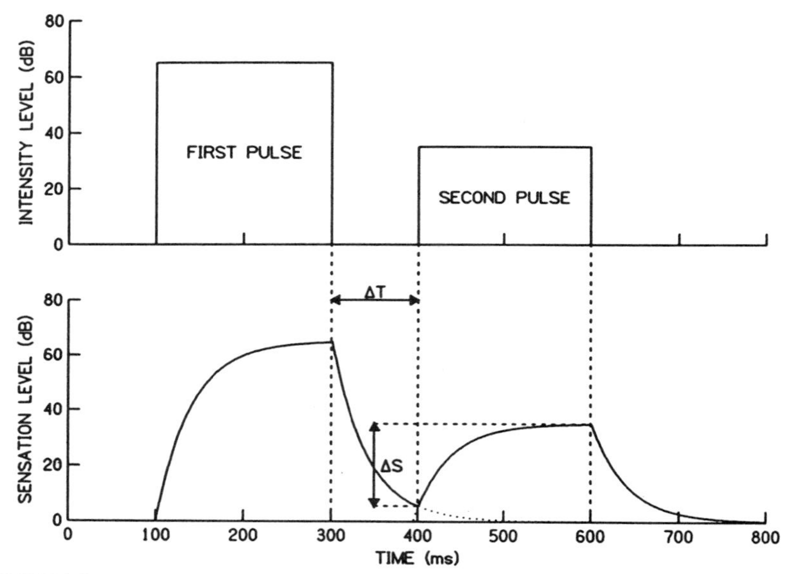

FIGURE 4 Schematic diagram of the stimulus level (upper portion) and the sensation level evoked by that stimulus (lower portion) as a function of time, as proposed by Plomp for the decay of auditory sensation. (Plomp, 1964.)

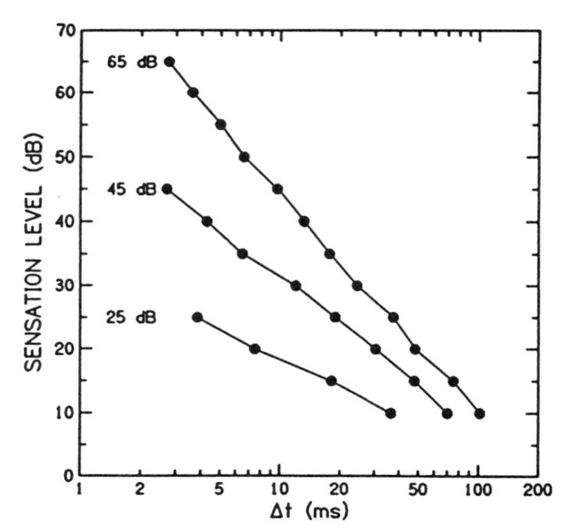

FIGURE 5 The minimum detectable silent interval is plotted along the abscissa. The sensation level of the second pulse that gives rise to that threshold silent interval is plotted along the ordinate. The sensation level of the first pulse is indicated in the figure. The data are for one subject. (Plomp, 1964.)

sensation level of the second pulse decreases, reaching its largest value of about 200 ms when the second pulse has 0 sensation level. This is true for all three levels of the first pulse. Note that time is expressed in *logarithmic* units. The data show a linear relation between the sensation level of the second pulse, in dB, and the logarithm of time. This is quite different from the results we would expect on the basis of simple exponential decay. If the decay were exponential, there would be a linear relation between the sensation level of the second pulse, and time would be expressed in *linear* units. Plomp's results violate the expectations of a simple exponential decay and suggest the operation of a strongly nonlinear system.

A simple adaptation of these ideas has become one of the more popular methods for estimating temporal acuity. When the levels of the first and second noise pulses are made equal, we simply measure the minimum temporal gap that can be detected—what we commonly call *temporal-gap detection*. This procedure has been used with a variety of different sounds. Let us now review some empirical data.

1. Broadband Noise

Plomp (1964) measured a gap-detection threshold of about 3 ms for noise pulses of equal level, provided that the sensation level of the noise exceeded about 30 dB. Gap-detection thresholds increased for sensation levels below 30 dB. Numerous studies have since confirmed these results (Penner, 1977; Irwin, Hinchcliff, & Kemp, 1981; Fitzgibbons, 1983; Florentine & Buus, 1983; Shailer & Moore, 1983; Forrest & Green, 1987; Formby & Muir, 1988; Green & Forrest, 1989).

Remarkably, gap detection thresholds are relatively insensitive to the manipulation of many stimulus parameters. We will briefly mention how several parameters influence gap-detection thresholds for broadband noise. The independence of gap threshold and stimulus level (above 30 dB SL) is consistent with the fact that gap-detection thresholds increase only slightly when the overall stimulus level is randomized on each stimulus presentation over a range of 25 dB about a median stimulus level (Forrest & Green, 1987; Formby & Muir, 1989).

Another potential influence on gap detection is the total noise duration. Forrest and Green (1987) reported that changes in the total stimulus duration from 5 to 400 ms had little effect on gap-detection thresholds, which ranged from 1.6 to 2.6 ms with a minimum threshold at a duration of about 50 ms. They held the overall duration constant, and randomized the intensity level from interval to interval so that a difference in overall loudness would not be a potential artifact. Formby and Muir (1989) held intensity level constant as well as the duration of the noise that marked the gap. Thus, the total stimulus interval increased in time by an amount equal to the duration of the gap. They found that total stimulus duration was used as a potential cue to the presence of a gap, at least for long gap durations. Penner (1977) also reported nearly equal gap thresholds for durations from 20 to 400 ms. Below 20 ms, however, she reported improved gap-detection thresholds. In that study, the signal interval was increased in duration by the length of the gap so the results probably indicate that the listeners were detecting changes in the overall duration of the sound.

Forrest and Green (1987) showed that gap detection is largely independent of the temporal position of the gap within the noise. They also reported that the psychometric function for gap detection is nearly linear when *d'* is plotted against gap duration. The range of the psychometric function for a completely silent interval is about 2 ms and is shallower for partially filled gaps.

We conclude that gap-detection thresholds are relatively insensitive to changes in level (above 30 dB SL), total duration, and temporal position.

2. Narrowband Noise

A natural question to ask is whether the temporal parameters are different for different frequency channels. This question is especially appropriate for a tonotopic system such as the human auditory system. To investigate auditory temporal acuity at specific frequency locations, gap detection has been measured with narrowband noise and sinusoids. With narrowband signals, changes in the power spectrum produced by the introduction of a silent interval create a potential artifact. Are the listeners hearing the temporal gap or the change in the power spectrum?

Two methods have been used to avoid detection based on the spectral rather than the temporal cues. First, gap-detection thresholds may be measured in the presence of broadband maskers having a spectral notch located at the center frequency of the stimulus band (e.g., Fitzgibbons & Wightman, 1982). Alternatively, we can take a broadband noise, introduce the silent interval, and then filter the resulting waveform (e.g., Grose, Eddins, & Hall, 1989). Gap-detection thresholds are nearly the same using either the notch-noise masking technique or the technique of filtering after the temporal gap (Eddins, Hall, & Grose, 1992).

As with broadband stimuli, gap-detection thresholds in narrowband noise decrease with increasing stimulus level to about 30 dB sensation level and remain nearly constant for higher levels (Shailer & Moore, 1983; Buus & Florentine, 1985; Fitzgibbons & Gordon-Salant, 1987).

A clear principle is that the larger is the bandwidth of the noise, the easier it is to hear the gap. Figure 6 shows the results of seven investigations that varied bandwidth over a significant range. These results are also consistent with a number of other studies where bandwidth was a parameter (Fitzgib-

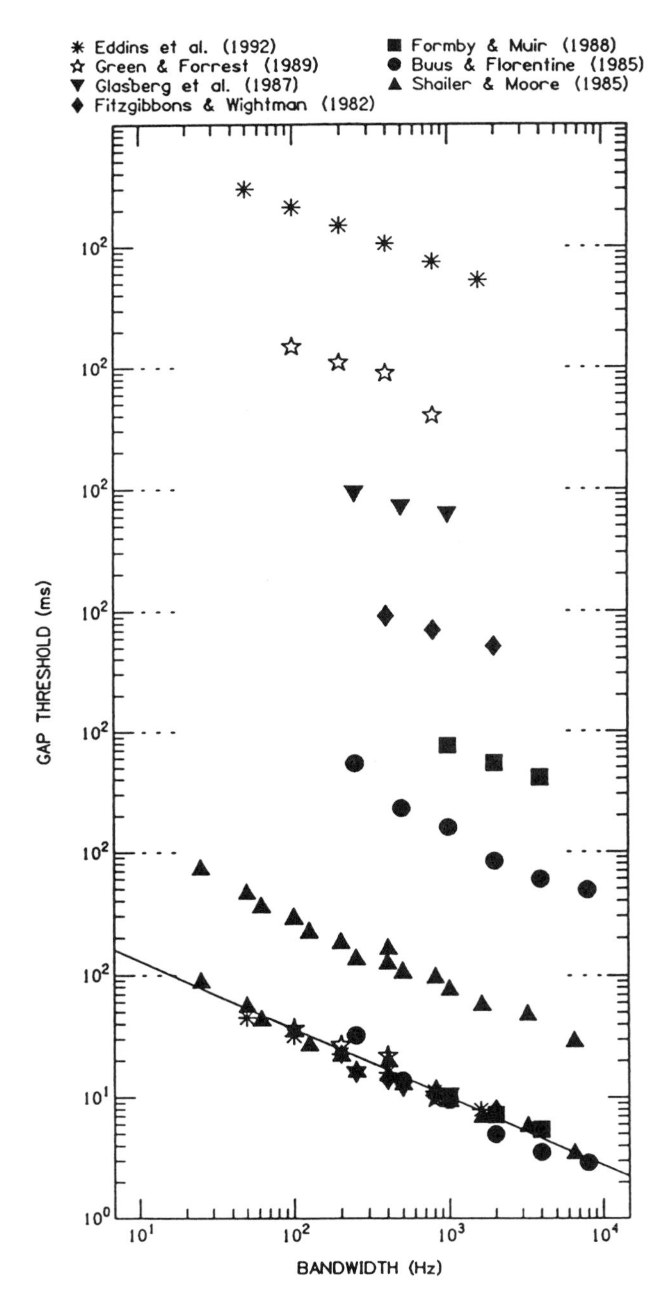

FIGURE 6 Gap detection thresholds as a function of noise bandwidth for seven investigations. The successive ordinates for each data set are offset by a factor of 10 to faciliate comparison. In the lower portion, the data for each investigation are normalized and shown with the best fitting (least-squares) line. The slope of this line is −0.55, close to the slope predicted by a square root of bandwidth rule (−0.5).

bons, 1983, 1984; Florentine & Buus, 1983; Shailer & Moore, 1983, 1985; Buus & Florentine, 1985; De Filippo & Snell, 1986; Formby & Muir, 1988; Glasberg, Moore, & Bacon, 1987; Moore & Glasberg, 1988). In the lower portion of the figure, the data are normalized so that the predicted threshold at 1000 Hz is about 10 ms. The slope of the best fitting line, in the least-squares sense, is −0.55. That is, gap threshold varies roughly as the reciprocal of the square root of bandwidth.

Having demonstrated the importance of stimulus bandwidth for gap detection, the question remains whether stimulus center frequency influences gap detection. Most of the points plotted in Figure 6, even those within a single study, were measured at different center frequencies, which we have ignored in plotting the figure. The first systematic study of bandwidth at different frequency regions was conducted by Eddins et al. (1992). Figure 7 shows their data. Bands of noise having three different upper-frequency cutoffs, 600, 2200, and 4400 Hz, were used. For each frequency region, the bandwidth of the noise was varied, as indicated in the figure. Gap-detection thresholds did not differ across the three cutoff frequencies. The same effects of bandwidth were seen at each frequency region; the thresholds vary as the reciprocal of the square root of bandwidth.

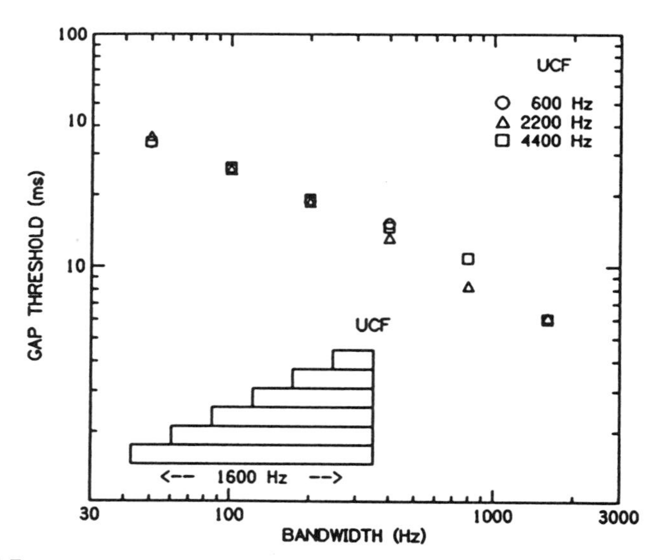

FIGURE 7 Gap detection thresholds as a function of noise bandwidth. The insert indicates increasing noise bandwidth for a constant upper-cutoff frequency, which was either 600, 2200, or 4400 Hz, as indicated by the symbols. (The data are taken from Eddins, Hall, and Grose, 1992.)

3. Sinusoids

Gap-detection thresholds for sinusoidal stimuli have also been measured. Potential spectral artifacts are, however, an even greater concern with sinusoidal stimuli because the spectrum is nonvarying, and any change in the spectral pattern of the stimulus is certain evidence that the gap is present. Shailer and Moore (1987), using a notched-noise masker to limit audibility of the spectrum to a narrow range, measured psychometric functions for gap detection for sinusoidal signals at frequencies of 200, 400, 1000, and 2000 Hz. Resulting gap thresholds were about 5 ms (75% correct) for their "preserved phase" condition, in which the signal phase after the gap was the same as if no interruption were present. In that case, the thresholds were largely independent of frequency region. The range of the psychometric function was roughly 3 to 5 ms. Moore and Glasberg (1988) measured gap detection for sinusoids of 500, 1000, and 2000 Hz using an adaptive technique. The obtained gap thresholds were about 4 ms and showed no consistent change with center frequency. The thresholds reported by Moore and Glasberg were slightly better than those of Shailer and Moore, probably the result of a greater signal-to-masker ratio (30 dB versus 15 dB).

Green and Forrest (1989) measured gap detection for multitonal complexes consisting of 21 components spaced at equal intervals in logarithmic frequency. Because of the spacing among the components, the sensation level of any single component was only about 9 dB; thus, spectral artifacts were probably inaudible. For this sensation level, gaps were easiest to hear at the higher component frequencies, 5–10 ms above 2000 Hz as opposed to 40 ms at 235 Hz. When the sensation level was raised to 20 dB, roughly comparable to the sensation levels of Shailer and Moore (1987), the results were similar to those found for single tones in a notched noise and were nearly independent of component frequency.

4. Theory

The dependence of the gap-detection threshold on stimulus bandwidth is consistent with any account that considers the fluctuations in the noise. The mean-to-sigma ratio of the noise power is proportional to the square root of the noise bandwidth. As the noise bandwidth increases, therefore, the relative fluctuations of the noise power decrease and the change in power produced by the gap becomes easier to detect. The exact way this will affect detection depends on the specific model. For an energy-detection model (Green & Swets, 1966), the detection index is

$$d' = \sqrt{WT}\,\frac{S}{N} \tag{8}$$

where W represents the bandwidth of the stimulus, T represents the duration of the gap, N is noise power, and S is the *difference* in power between the noise power, N, and the noise power in the gap. Note that the energy detector model predicts performance independent of center frequency, because the bandwidth W of that model is the bandwidth of the noise process, not the receiver (Green, 1960). Gap detection can be viewed as detecting a decrement in the noise, and thus, for constant d', the slope of the gap detection-noise bandwidth function should be about -0.5, which is reasonably close to the empirical slope of -0.55.

The empirical results deserve some comment. It is widely accepted that the width of the auditory filter increases with increasing frequency (see Chapter 5). Therefore, we might expect better temporal acuity at the higher frequencies. The majority of available data suggest that gap-detection thresholds are independent of center frequency. This is true whether the stimulus bandwidth is narrower or wider than the auditory filter. Several authors have suggested that information from the output of several auditory filters may be combined at a more central location, thus overcoming the limiting effects of peripheral filtering (Green, 1960; De Boer, 1966; Schacknow & Raab, 1976; Viemeister, 1979; Shailer & Moore, 1983, 1985; Bacon & Viemeister, 1985; Formby & Muir, 1988; Grose et al., 1989; Eddins et al., 1992). We will see later that at least two models for the detection of amplitude-modulated noise also assume that the bandwidth of the initial filtering stage is much wider than any single critical bandwidth.

A second explanation for the improvement in gap threshold with increasing bandwidth argues that an increase of the stimulus bandwidth increases the number of across-channel comparisons. If the presence of a temporal gap were signaled by several independent channels, then the gap-detection threshold would decrease. Green and Forrest (1989) showed that gaps were easier to hear when multiple components in their 21-component complex were gated off and on synchronously. When multiple silent intervals were nonsynchronously placed on different frequency components, gap thresholds were typically intermediate between those for single and multiple (synchronous) components. The improvement in detectability falls well below the square root of n rule.

Grose et al. (1989) measured gap detection for three conditions chosen to separate the effects of bandwidth and synchronous comparisons across frequency. In their baseline conditions, gap detection was measured for a noise having a bandwidth of 1600 Hz (600–2200 Hz) and a 20 Hz low-pass noise multiplied by a 2180 Hz sinusoid, yielding a 40 Hz noise band centered on 2180 Hz, the components above 2180 Hz mirroring those below 2180 Hz. According to a square-root-of-bandwidth argument, the gap thresholds should be different by a factor of about 9 (the ratio of the number of independent components in each noise process is 1600/20, hence, $(80)^{1/2} \approx 9$). The thresholds were 8 ms for the wideband noise and 76 ms for the narrowband noise (a factor of 9.5). The critical measurement was the threshold for the wideband noise amplitude modulated by the narrow, 20 Hz, noise. Gap thresholds were intermediate for this condition, about 31 ms.

The fluctuations in the envelope for this wideband multiplied noise are dominated by the much narrower 20 Hz noise. On the basis of noise fluctuation, then, we would expect little difference between the threshold for the 20 Hz multiplied noise (76 ms) and the 1600 Hz noise with a 20 Hz envelope (31 ms). If, however, we assume the listeners compared fluctuations over different frequency channels, then thresholds should reflect the increase in the number of channels that are stimulated. Roughly, ten critical bands are stimulated by the 1600 Hz noise band, compared with one critical band for the 20 Hz multiplied noise centered at 2180 Hz. Using the square root of the number of multiple looks, we would expect a difference in threshold of $[(10/1)^{1/2} = 3.2]$. The obtained thresholds differ by a factor of 2.5. As was true in the study of Green and Forrest (1989), the improvement falls short of the square root of n rule.

Clearly, an increase in noise fluctuations, whether created by modulation or by changing the noise bandwidth, has a detrimental effect on gap detection; the amount of threshold increase is proportional to the square root of the change in bandwidth. Synchronous comparisons of temporal gaps across frequency also improve gap detection. It appears that fluctuation is somewhat more important than cross-channel synchrony in gap detection.

A quite different attempt to summarize much of the data associated with temporal acuity is the temporal weighting function (Moore, Glasberg, Plack, & Biswas, 1988; Plack & Moore, 1990). This approach is empirical in origin and modeled after the measurement of auditory filter shape pioneered by Patterson (1974). It is not a process model in the sense of describing how the signal is detected. Rather, the idea is to characterize how masking is related to the temporal proximity of the signal and noise. To estimate the temporal weighting function, a brief sinusoidal signal was placed between two successive noise bursts. They then determined how the detectability of the signal depended on the temporal relationships of the signal and the two noise bursts. These experimental manipulations provide an assessment of the relative importance of the temporal sections of the noise bursts that preceed and succeed the signal. The approach is similar in spirit to the determination of the auditory filter shape using notched noise, except the manipulations are carried out in the time domain rather than the frequency domain (see Chapter 5). The resulting empirical function is a weighting function $w(t)$. This function bears some resemblance to $h(t)$ of the convolution integral, Eq. (1). The difference is that $w(t)$ weights the *average power of the noise,* not the instantaneous square of the noise waveform. The empirically determined weighting function [10 log $w(t)$] is shown in Figure 8 for a 5 ms, 2000 Hz signal at two different noise levels. Robinson and Pollack (1973) and Penner and her colleagues (Penner, Robinson, & Green, 1972; Penner and Cudahy, 1973) attempted to determine such a weighting function, but their efforts were not as comprehensive as those described by Moore et al. (1988).

Several points are worth noting about this approach. Clearly, the weighting function $w(t)$ should not be considered the $h(t)$ of the convolution integral. It is not a temporal filter, because if the weighting function is interpreted as an impulse response it has finite values for negative times—it would be both predictive and noncausal. The authors have argued that the noncausal property of the weighting function could be avoided by moving the 0 point on the time axis to some positive value, for example to 50 ms in Figure 8. In effect, this would amount to assuming a 50 ms delay in the nervous system before $w(t)$ is applied. The exact amount of this delay cannot be determined from the data, so that any value is arbitrary and the authors of the theory have presented the weighting function centered at zero delay. The negative values of Figure 8 arise because a noise occurring after a brief signal causes masking—we call the phenomenon *backward masking,* but it is little understood. In our opinion, the weighting function $w(t)$ should be considered a parsimonious summary of a large number of different experi-

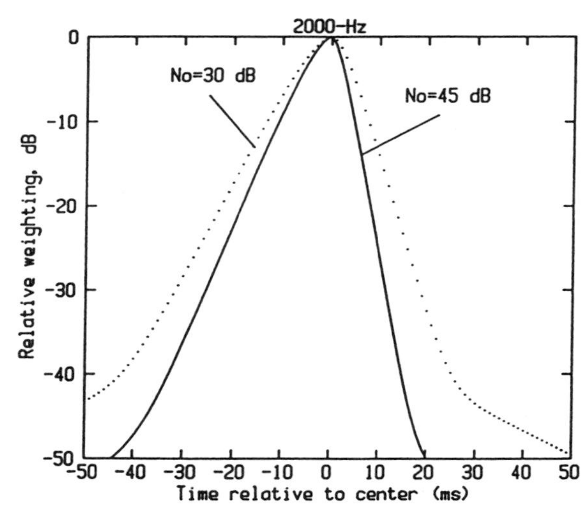

FIGURE 8 The shape of the empirically determined weighting function $w(t)$ derived from the thresholds for detecting a 2000 Hz, 5 ms signal located in time between two bursts of noise having a spectrum level of 45 (solid curve) and 30 dB (dotted curve). (From Moore, Glasberg, Plack, and Biswas, 1988.)

ments. A considerable amount of data can be successfully predicted with only four free parameters. Three of the parameters are related to the shape of the weighting function. The fourth parameter reflects the asymmetry between forward and backward masking—the shape of the weighting function is very different for positive and negative times. The same weighting function works for two different frequency regions, 500 and 2000 Hz. Furthermore, the shape of the function is not greatly altered with overall changes in noise level, as Figure 8 demonstrates.

An implicit assumption of this approach is that the threshold of the signal is determined by the additive effects of noise in the temporal vicinity of the signal. Although the assumption is not specifically tested in the Moore et al. paper, the success of this approach suggests that forward and backward masking may be treated as combining according to simple power addition.

Direct empirical measurements of masker additivity strongly suggest that the opposite is true. Penner (1980) measured the audibility of a click surrounded by two noise bursts. The additional masking produced by the combination of two equally effective maskers was 8 dB rather than the expected 3 dB. The earlier work by Robinson and Pollack (1973) showed additional masking values as large as 10 dB beyond power addition, depending on the position of the forward and backward masker. Penner and Schiffrin (1980) developed a model of temporal summation and intensity coding. Compression of the stimulus intensity is a central part of this theory, thereby accounting for the highly nonlinear way in which two different maskers combine (see Chapter 5 for an explanation of the effects of compression on the additivity of masking). Jesteadt and Wilke (1982) and Jesteadt, Weber, and Wilke (1982) have also demonstrated nonlinear effects in the combination of two forward and simultaneous maskers. Lutfi (1983) and Humes and Jesteadt (1989) have also assumed power-law compression of stimulus intensity to account for the highly nonlinear way in which different maskers combine. Lutfi (1990) has recently proposed an informational masking account of why simple power summation often fails to account for masking produced by the combination of two maskers.

Moore et al. (1988) suggested that such apparent nonlinear effects might arise because the temporal position of the window depends on the masker type. Specifically they assume that the observer can listen before or after the signal (off-time listening) to maximize the signal-to-noise ratio. For the forward masking case, the temporal window should be positioned some time after the signal occurs. For the backward masking case, the temporal window should be positioned some time before the signal occurs for the same reason. When both maskers are used, the advantage of either off-time listening is reduced, and an increase in threshold would be expected. It remains unclear whether this increase would be sufficient to account for the nonlinear way in which forward and backward masking combine.

E. Amplitude-Modulation Detection

1. Broadband Noise

Another method used to estimate temporal acuity is to measure the ability to detect amplitude modulation of a noise waveform. Traditionally, a sinusoidal modulation is imposed of the form

$$x(t) = [1 + m \cos(2\pi f_m t)]n(t) \tag{9}$$

where $n(t)$ is the noise waveform, f_m is the rate of modulation, and m is the depth of modulation. We define the sensitivity to modulation as the value of $20 \log(1/m)$ necessary just to distinguish an amplitude-modulated stimulus from an unmodulated stimulus. Measuring the modulation threshold for several frequencies of modulation, we obtain a measure of the ability to follow changes in amplitude as a function of the rate of amplitude change. A modulation transfer function obtained in this manner is shown in Figure 9 (adapted from Viemeister, 1979).

The analogy between such a function and the systems analysis technique is obvious. It must be used with some caution, as was discussed by Rodenburg (1972) and Viemeister (1977). The Fourier transform of the modulation transfer function can be likened to the temporal weighting function of a convolution integral $h(t)$ [see Eq. (1)]. The modulation transfer function resembles a low-pass filter, and if a simple first-order filter is assumed, then $\tau = 1/(2\pi f_c)$, where f_c is the 3 dB point of the low-pass filter. Nonlinearities in the auditory system will restrict the generality of this approach. Evidence for such nonlinearity was cited earlier (see Figure 5). Additional problems for this simple approach are evidenced by the attenuation slope of the transfer function, which is closer to 3 dB per octave than the 6 dB per octave expected with a first-order system. Despite these problems, the linear systems approach has proven useful; the interested reader is referred to Viemeister (1979).

Now, let us consider empirical results from several investigations. The modulation thresholds shown in Figure 9 are roughly constant for modulation frequencies below about 50 Hz. Above 50 Hz, sensitivity to modulation decreases at a rate of about 3 dB/octave up to 1000 Hz. Assuming a half-power cutoff frequency, f_c, of 50 Hz, the derived time constant would then be about 3 ms. Time constants of 2–3 ms are consistent with many estimates derived from modulation transfer functions for broadband noise (Rodenburg, 1972, 1977; Viemeister, 1977; Bacon & Viemeister, 1985; Formby, 1985; Forrest & Green, 1987; Formby & Muir, 1988). Furthermore, the form of the modulation transfer function is largely independent of stimulus level (Viemeister, 1979).

Several authors have commented on the attenuation slope of the transfer function. From a simple low-pass filter we would expect an attenuation

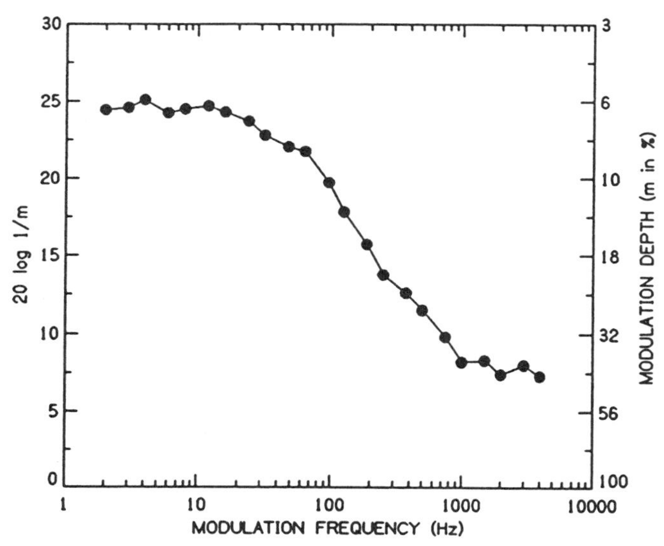

FIGURE 9 Modulation detection thresholds for broadband noise. Sensitivity to modulation (20 log 1/m) is plotted as a function of modulation frequency (2 to 4000 Hz). (Adapted from Viemeister, 1979.)

slope of −6 dB per octave. Although Rodenburg reported a slope of nearly −6 dB per octave, investigators have typically reported a slope of about −3 dB per octave (Viemeister, 1977, 1979; Bacon & Viemeister, 1985; Formby, 1985; Forrest & Green, 1987; Formby & Muir, 1988; Eddins, 1993). In general, this problem has been ignored, and a model using a simple low-pass filter has been used that accurately predicts the empirical results from several investigations of temporal acuity.

2. Narrowband Noise

As with gap detection, we are interested in whether temporal acuity as measured by modulation detection differs for different frequency channels. Again, a basic principle is that amplitude modulation is easier to detect for larger noise bandwidths. This was first shown by Rodenburg (1972) and has since been demonstrated by several investigators (Viemeister, 1979; van Zanten, 1980; Formby & Muir, 1988). In general, as the noise bandwidth increases, the half-power bandwidth of the transfer function shifts to higher modulation frequencies and corresponding time constants decrease. Furthermore, as the noise bandwidth increases, the sensitivity to modulation improves across all modulation frequencies. In each of the studies mentioned, increases in noise bandwidth accompanied increases in center frequency, limiting the separation of these two parameters.

More recent work by Eddins (1993) indicates that the attenuation characteristics of modulation transfer functions for narrowband noise are relatively invariant with changes in frequency region and are almost entirely determined by stimulus bandwidth. Figure 10 shows Eddins' data for stimulus bandwidths ranging from 200 to 1600 Hz having upper-cutoff frequencies of 600, 2200, or 4400 Hz. The data are similar in form to data for broadband stimuli (Figure 9). Each panel of the figure shows a different bandwidth. As bandwidth decreases from 1600 to 200 Hz, the cutoff frequency f_c decreases from 83 Hz to 27 Hz and the corresponding time constants increase from about 1.9 to 5.9 ms. Time constants estimated from modulation transfer functions, like gap-detection thresholds, decrease at a rate proportional to the square root of the stimulus bandwidth. The sensitivity to modulation also improves by approximately 3 dB per doubling of noise bandwidth. For a given bandwidth, the half-power point on these functions is nearly independent of center frequency. In accordance with previous results for broadband and narrowband noise, the slope of the functions was approximately −3 dB per octave beyond f_c. In summary, time constants associated with modulation transfer functions for narrowband noise decrease with increasing bandwidth and are constant across frequency region, at least to 4400 Hz.

3. Theory

In view of the many measures of temporal acuity, it would be helpful to have a model of temporal acuity to account for at least a few, if not most of these results. A common model of auditory processing involves four stages: an initial bandpass filter, a nonlinear operator, a temporal integrator, and a decision device. Viemeister (1979) used such a model to predict his modulation detection data with reasonable accuracy. Briefly, the initial bandpass filter is analogous to peripheral auditory filtering. The nonlinear operator served to convert the input waveform to a purely positive quantity; a half-wave rectifier was used by Viemeister. The temporal integrator was modeled as a low-pass filter. The final-stage decision process computed the variance of the output of the low-pass filter. The free parameters in the model were the bandwidths of the initial filter and the low-pass filter. The model predicted the form of the modulation transfer function with reasonable accuracy. An initial filter having a bandwidth of 2000 Hz and a time constant of 2.5 ms (low-pass filter bandwidth of 65 Hz) nicely fits the data obtained with wideband noise. The model requires a predetection filter that is clearly wider than a single auditory filter, implying that the auditory system somehow increases its effective bandwidth.

Forrest and Green (1987) used a model similar to Viemeister's to predict the results for both modulation and gap detection. The major difference between the two models was the decision statistic. Forrest and Green used

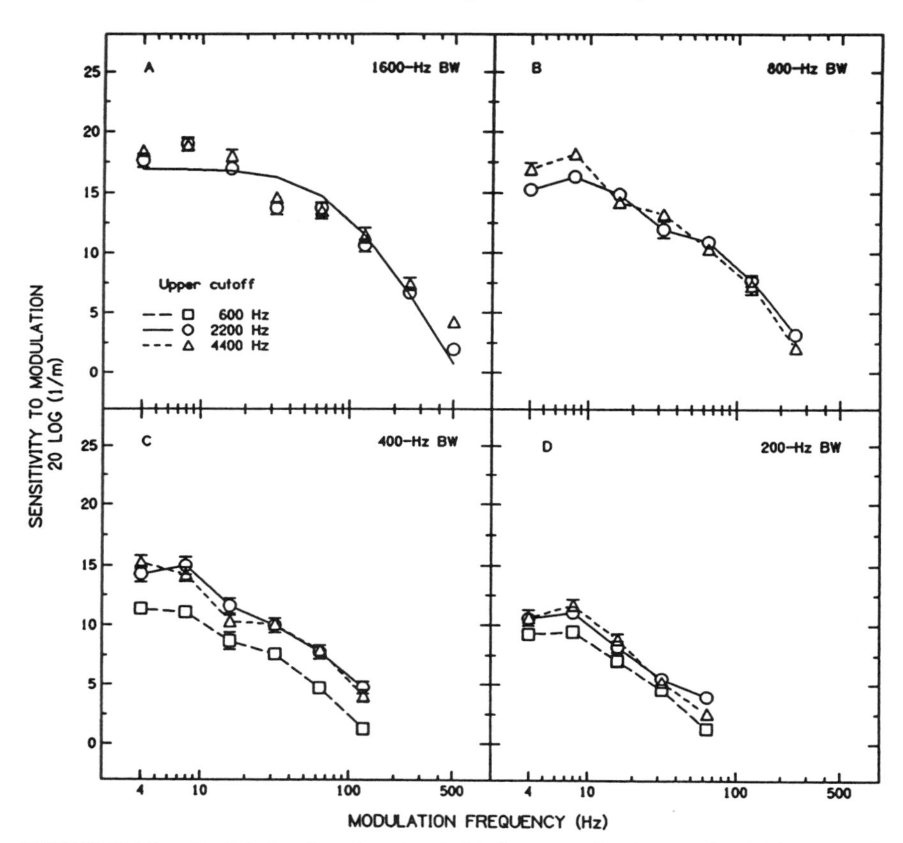

FIGURE 10 Modulation detection thresholds for narrowband noise. Sensitivity to modulation (20 log $1/m$) is plotted as a function of modulation frequency with upper-cutoff frequency as the parameter. Each panel represents the indicated bandwidth condition. The solid line in panel A is the fitted low-pass function to the data for the 2200 Hz upper cutoff (see Eddins, 1993). (The data are taken from Eddins, 1993.)

the ratio of the maximum to the minimum sample at the output of the low-pass filter, rather than its variance. The choice of the max–min statistic was partially motivated by the need for a decision statistic independent of the overall level of the stimulus, as the level was randomized from interval to interval in their gap-detection experiments. An advantage of this model is that it does predict the nearly -3 dB per octave slope of the modulation transfer functions. Both gap and modulation detection results were well approximated by the model using an initial bandpass filter 4000 Hz wide and a 3 ms time constant. Again, note the wide bandwidth of the pre-detection filter.

The models of both Viemeister and Forrest and Green demonstrated that the width of the predetection filter necessary for accurate predictions was much greater than a single auditory filter. One interpretation is that peripheral filtering does not have a strong influence on modulation or gap detection. That is not to say that peripheral filtering is not present. It is certainly possible that the auditory system makes use of information obtained from the output of many auditory filters combined at some higher neural level. Although the models discussed here were used to predict data from broadband noise, they appear to be consistent with experiments using narrowband noise. Recall that neither the gap nor modulation detection experiments employing narrowband stimuli showed obvious effects of peripheral filtering (Eddins et al., 1992; Eddins, 1993).

F. Temporal Asynchrony

In a discrimination task similar to Hirsh's temporal order judgment, Zera and Green (1993a, 1993b) measured the ability of listeners to hear temporal synchrony in multicomponent complexes. Two sounds, each of which lasted about 0.5 s, were presented in a two-alternative, forced-choice task. For one sound, the standard, all components of the complex started (or ended) at the same time. For the other sound, the asynchronous signal, one or more components began (onset condition) or ended (offset condition) at a different time from the remainder. The listener was asked to indicate which sound was asynchronous. In their experiment, for each onset condition, a similar offset condition was measured by simply reversing the waveform in time. The first major finding was that the thresholds for detecting onset asynchrony are about three to ten times smaller than those for detecting offset asynchrony. Indeed, for onset detection, an asynchrony of 1 ms is almost always detectable in any condition and for any component frequency. In some conditions, the listeners could hear 0.1 ms of asynchrony.

Several different kinds of asynchrony should be distinguished because they influence the ability to detect such changes. The asynchronous components can begin before or after the remainder of the components. As one might expect on the basis of masking, it was generally easier to hear asynchrony if the asynchronous components occurred alone: either before the other components had started in the onset condition, or after the others had terminated in the offset condition. Another factor that strongly influenced the results was the frequency of the asynchronous component and the composition of the complex. Detection of asynchrony is usually easier for harmonic complexes rather than for complexes where successive frequencies have a constant frequency ratio (equal spacing in logarithmic frequency). The exception to this generalization occurs when the asynchronous component is very high in frequency. In that case, several components fall within the same critical band for the harmonic complex and asynchrony in logarithmic complexes may be easier to hear, although overall performance is very poor in such conditions.

Detection of onset asynchrony is especially acute for lower frequency components in *harmonic* complexes. For frequencies below about 2000 Hz, onset asynchrony is detectable if the odd component starts only 1/4 cycle before the remaining components of the harmonic complex. The detection of asynchrony in an *inharmonic* complex is less dependent on frequency, either for onset or offset conditions.

Detecting the asynchrony of single components in a complex is best when all the remaining components start or end at exactly the same time. In one experiment, the standard stimulus was a complex in which each successive component of the complex began a constant time interval before or after the preceding component. Thus, the time at which each component began (or ended) was a linear function of time. Additional asynchrony was created by altering the temporal position of a single component of the complex. As might be expected, it is more difficult to detect the *onset* asynchrony of a single component when the standard itself is asynchronous. On the other hand, *offset* asynchrony detection was hardly affected by the linear asynchrony of the standard sound.

There is as yet little theory to explain any of these findings. Among the more obvious questions are these: Why does harmonic frequency spacing lead to better detection of asynchrony than when the components are inharmonically spaced, and why is onset asynchrony so much easier to detect than offset asynchrony?

IV. CONCLUSIONS

A. Temporal Integration

All stimulus conditions investigated produce a time–intensity trade. The decrease in the signal level or relative signal level at the threshold is about x dB when the signal duration is increased by a factor of 10. The value of x ranges between 0.25 and 1.0, depending on the stimulus condition. The maximum duration of the temporal integrator appears to be about 100 to 200 ms, except when the task is to detect an increment in a sinusoid. In that case, it appears to be appreciably longer.

B. Temporal Acuity

The time constant for temporal acuity is appreciably shorter than the estimated value for temporal integration. The estimated time constants range from a fraction of a millisecond to 30 ms. Many of the estimates cluster at about 2 ms. The number of very different experimental techniques leading to this similar estimate is impressive. Estimates of temporal acuity are relatively stable over a broad range of stimulus conditions.

Sensitivity to either the amplitude modulation of noise or to gaps in the noise varies as the square root of the bandwidth of the stimulus. The center frequency of a narrowband noise has little influence on the estimated time constant.

Temporal asynchrony of a single component in a multicomponent complex is much easier to detect when the complex is harmonic than when it is inharmonic and when the asynchronous component occurs at the onset rather than the offset of the components.

Appendix

The Fourier transform $F(w)$ of a signal of finite duration having nonzero value only in the interval $-T$ to $+T$ is

$$F(w) = \int_{-T}^{+T} f(t)\, e^{jwt}\, dt \tag{A1}$$

where $f(t)$ is the time waveform, w is the frequency in radians per second, and j is the imaginary number $j = \sqrt{-1}$. The energy spectrum of a signal is simply the Fourier transform times its conjugate $F^*(w)$. If we represent the Fourier transform as a real function $A(w)$ and an imaginary function $B(w)$, then the energy spectrum is simply $[A(w)^2 + B(w)^2]$.

Using Euler's identity, we can rewrite Eq. (A1) in terms of the real and imaginary parts, in which case

$$A(w) = \int_{-T}^{+T} f(t) \cos(wt)\, dt \tag{A2}$$

or

$$B(w) = j \int_{-T}^{+T} f(t) \sin(wt)\, dt \tag{A3}$$

An even function has the property that $f_e(t) = f_e(-t)$, and an odd function has the property that $f_o(t) = -f_o(-t)$. Any function $f(t)$ can be considered the sum of an odd part f_o and an even part f_e. The even part is simply $[f(t) + f(-t)]/2$, and the odd part is $[f(t) - f(-t)]/2$.

Consider the energy spectrum of a waveform $f(t)$ and a time-reversed waveform $f(-t)$. It is easy to verify that the odd part of $f(t)$ yields a Fourier transform that is purely imaginary; $B(w)$. The even part of $f(t)$ yields a Fourier transform that is purely real, with the forward wave, $f(t)$, producing $A(w)$ and the time-reversed wave, $f(-t)$, producing $-A(w)$. Thus, the energy spectra of the forward and time-reversed waves are the same; both are equal to $A(w)^2 + B(w)^2$.

Acknowledgments

Preparation of this chapter was supported by the NIH and the Air Force Office of Scientific Research. We wish to thank B. C. J. Moore and Huanping Dai for their comments on an earlier draft of this chapter.

References

Bacon, S. P., & Viemeister, N. F. (1985). Temporal modulation transfer functions in normal-hearing and hearing-impaired listeners. *Audiology, 24,* 117–134.

Blodgett, H. C., Jeffress, L. A., & Taylor, R. W. (1958). Relation of masked threshold to signal-duration for various interaural phase-combinations. *American Journal of Psychology, 71,* 283–290.

Buus, S., & Florentine, M. (1985). Gap detection in normal and impaired listeners: The effect of level and frequency. In A. Michelsen (Ed.), *Time resolution in auditory systems* (pp. 159–179). New York: Springer-Verlag.

Campbell, R. A. (1963). Detection of a noise signal of varying duration. *Journal of the Acoustical Society of America, 35,* 1732–1737.

De Boer, E. (1966). Intensity discrimination of fluctuating signals. *Journal of the Acoustical Society of America, 40,* 552–560.

De Boer, E. (1985). Auditory time constants: A paradox? In A. Michelsen (Ed.), *Time resolution in auditory systems* (pp. 141–158). New York: Springer-Verlag.

De Filippo, C. L., & Snell, L. B. (1986). Detection of a temporal gap in low-frequency narrow-band signals by normal-hearing and hearing-impaired subjects. *Journal of the Acoustical Society of America, 80,* 1354–1358.

Eddins, D. A. (1993). Amplitude modulation detection of narrowband noise: effects of absolute bandwidth and frequency region. *Journal of the Acoustical Society of America, 93,* 470–479.

Eddins, D. A., Hall, J. W., & Grose, J. H. (1992). The detection of temporal gaps as a function of frequency region and absolute stimulus bandwidth. *Journal of the Acoustical Society of America, 91,* 1069–1077.

Fastl, H. (1977). Simulation of a hearing loss at long versus short test tones. *Audiology, 16,* 102–109.

Fitzgibbons, P. J. (1983). Temporal gap detection in noise as a function of frequency, bandwidth, and level. *Journal of the Acoustical Society of America, 74,* 67–72.

Fitzgibbons, P. J. (1984). Temporal gap resolution in narrow-band noises with center frequencies from 6000 to 14000 Hz. *Journal of the Acoustical Society of America, 75,* 566–569.

Fitzgibbons, P. J., & Gordon-Salant, S. J. (1987). Temporal gap resolution in listeners with high frequency sensorineural hearing loss. *Journal of the Acoustical Society of America, 81,* 133–137.

Fitzgibbons, P. J., & Wightman, F. L. (1982). Gap detection in normal and hearing-impaired listeners. *Journal of the Acoustical Society of America, 72,* 761–765.

Florentine, M. (1986). Level discrimination of tones as a function of duration. *Journal of the Acoustical Society of America, 79,* 792–798.

Florentine, M., & Buus, S. (1983). Temporal acuity as a function of level and frequency. *Proceedings of the 11th International Congress of Acoustics, 3,* 103–106.

Florentine, M., Fastl, H., & Buus, S. (1988). Temporal integration in normal hearing, cochlear impairment, and impairment simulated by masking. *Journal of the Acoustical Society of America, 84,* 195–203.

Formby, C. (1985). Differential sensitivity to tonal frequency and to the rate of amplitude modulation of broadband noise by normally-hearing listeners. *Journal of the Acoustical Society of America, 78,* 70–77.

Formby, C., & Muir, K. (1988). Modulation and gap detection for broadband and filtered noise signals. *Journal of the Acoustical Society of America, 84,* 545–550.

Formby, C., & Muir, K. (1989). Effects of randomizing signal level and duration on temporal gap detection. *Audiology, 28,* 250–257.

Forrest, T. G., & Green, D. M. (1987). Detection of partially filled gaps in noise and the temporal modulation transfer function. *Journal of the Acoustical Society of America, 82,* 1933–1943.

Garner, W. R. (1947). The effect of frequency spectrum on temporal integration of energy in the ear. *Journal of the Acoustical Society of America, 19,* 808–815.

Garner, W. R., & Miller, G. A. (1947). The masked threshold of pure tones as a function of duration. *Journal of Experimental Psychology, 37,* 293–303.

Glasberg, B. R., Moore, B. C. J., & Bacon, S. P. (1987). Gap detection and masking in hearing-impaired and normal-hearing subjects. *Journal of the Acoustical Society of America, 81,* 1546–1556.

Green, D. M. (1960). Auditory detection of a noise signal. *Journal of the Acoustical Society of America, 32,* 121–131.

Green, D. M. (1971). Temporal auditory acuity. *Psychological Review, 78,* 540–551.

Green, D. M. (1973). Temporal acuity as a function of frequency. *Journal of the Acoustical Society of America, 54,* 373–379.

Green, D. M., Birdsall, T. G., & Tanner, W. P., Jr. (1957). Signal detection as a function of signal intensity and duration. *Journal of the Acoustical Society of America, 29,* 523–531.

Green, D. M., & Forrest, T. G. (1989). Temporal gaps in noise and sinusoids. *Journal of the Acoustical Society of America, 86,* 961–970.

Green, D. M., & Swets, J. A. (1966). *Signal detection theory and psychophysics.* New York: Wiley (reprinted Huntington, NY: R. E. Krieger, 1974; Los Altos, CA: Peninsula Publishing, 1988).

Grose, J. H., Eddins, D. A., & Hall, J. W. (1989). Gap detection as a functin of stimulus bandwidth with fixed high-frequency cutoff in normal-hearing and hearing-impaired listeners. *Journal of the Acoustical Society of America, 86,* 1747–1755.

Hamilton, P. M. (1957). Noise masked threshold as a function of tonal duration and masking noise bandwidth. *Journal of the Acoustical Society of America, 29,* 506–511.

Henning, G. B. (1970). A comparison of the effects of signal duration on frequency and amplitude discrimination. In R. Plomp & G. F. Smoorenburg (Eds.), *Frequency analysis and periodicity detection in hearing* (pp. 350–361). Lieden: A. W. Sijthoff.

Henning, G. B., & Gaskell, H. (1981). Monaural phase sensitivity with Ronken's paradigm. *Journal of the Acoustical Society of America, 70,* 1669–1673.

Hirsh, I. J. (1959). Auditory perception of temporal order. *Journal of the Acoustical Society of America, 31,* 759–767.

Hirsh, I. J., & Sherrick, C. E., Jr. (1961). Perceived order in different sense modalities. *Journal of Experimental Psychology, 62,* 423–32.

Huffman, D. A. (1962). The generation of impulse-equivalent pulse trains. *IRE Transactions, IT8,* S10–S16.

Hughes, J. W. (1946). The threshold of audition for short periods of stimulation. *Proceedings of the Royal Society* [London], *B133,* 486–490.

Humes, L. E., & Jesteadt, W. (1989). Models of the additivity of masking. *Journal of the Acoustical Society of America, 85,* 1285–1294.

Irwin, R. J., Hinchcliff, L. K., & Kemp, S. (1981). Temporal acuity in normal and hearing-impaired listeners. *Audiology, 20,* 234–243.

Jesteadt, W., Weber, D. L., & Wilke, S. S. (1982). Interaction of simultaneous and forward masking for maskers and signals differing in frequency. Abstract given at the 104th meeting, Acoustical Society of America, *72,* S67.

Jesteadt, W., & Wilke, S. (1982). Interaction of simultaneous and forward masking, Abstract given at the 103rd meeting, Acoustical Society of America, *71,* S72.

Leshowitz, B. (1971). Measurement of the two-click threshold. *Journal of the Acoustical Society of America, 49,* 462–466.

Lutfi, R. A. (1983). Additivity of simultaneous masking. *Journal of the Acoustical Society of America, 73,* 262–267.

Lutfi, R. A. (1990). How much masking is informational masking? *Journal of the Acoustical Society of America, 88,* 2607–2610.

Moore, B. C. J., & Glasberg, B. R. (1988). Gap detection with sinusoids and noise in normal, impaired, and electrically stimulated ears. *Journal of the Acoustical Society of America, 83,* 1093–1101.

Moore, B. C. J., Glasberg, B. R., Plack, C. J., & Biswas, A. K. (1988). The shape of the ear's temporal window. *Journal of the Acoustical Society of America, 83,* 1102–1116.

Munson, W. A. (1947). The growth of auditory sensation. *Journal of the Acoustical Society of America, 19,* 584–591.

Nickerson, R. S., & Freeman, B. (1974). Discrimination of the order of the components of repeating tone sequences: Effects of frequency separation and extensive practice. *Perception and Psychophysics, 16,* 471–477.

Olsen, W. A., & Carhart, R. (1966). Integration of acoustic power at threshold by normal hearers. *Journal of the Acoustical Society of America, 40,* 591–599.

Patterson, J. H. (1974). Auditory filter shapes derived with noise stimuli. *Journal of the Acoustical Society of America, 55,* 640–654.

Patterson, J. H., & Green, D. M. (1970). Discrimination of transient signals having identical energy spectra. *Journal of the Acoustical Society of America, 48,* 894–905.

Penner, M. J. (1977). Detection of temporal gaps in noise as a measure of decay of auditory sensation. *Journal of the Acoustical Society of America, 61,* 552–557.

Penner, M. J. (1978). A power law transformation resulting in a class of short-term integrators that produce time-intensity trades for noise bursts. *Journal of the Acoustical Society of America, 63,* 195–201.

Penner, M. J. (1980). The coding of intensity and the interaction of forward and backward masking. *Journal of the Acoustical Society of America, 67,* 608–616.

Penner, M. J., & Cudahy, E. (1973). Critical masking interval: A temporal analog of the critical band. *Journal of the Acoustical Society of America, 65,* 1530–1534.

Penner, M. J., Robinson, C. E., & Green, D. M. (1972). The critical masking interval. *Journal of the Acoustical Society of America, 52,* 1661–1668.

Penner, M. J., & Shiffrin, R. M. (1980). Nonlinearities in the coding of intensity within the context of a temporal summation model. *Journal of the Acoustical Society of America, 67,* 617–627.

Plack, C. J., & Moore, B. C. J. (1990). Temporal window shape as a function of frequency and level. *Journal of the Acoustical Society of America, 87,* 2178–2187.

Plomp, R. (1964). Rate of decay of auditory sensation. *Journal of the Acoustical Society of America, 36,* 277–282.

Plomp, R., & Bouman, M. (1959). Relation between hearing threshold and duration for tone pulses. *Journal of the Acoustical Society of America, 31,* 749–758.

Raab, D. H., Osman, E., & Rich, E. (1963). Effect of waveform correlation and signal duration on detection of noise bursts in continuous noise. *Journal of the Acoustical Society of America, 35,* 1942–1946.

Robinson, C. E., & Pollack, I. (1973). Interaction between forward and backward masking: A measure of the integrating period of the auditory system. *Journal of the Acoustical Society of America, 53,* 1313–1316.

Rodenburg, M. (1972). Sensitivity of the auditory system to differences in intensity. Ph.D. dissertation, Medical Faculty, Rotterdam.

Rodenburg, M. (1977). Investigation of temporal effects with amplitude modulated signals. In E. F. Evans & J. P. Wilson (Eds.), *Psychophysics and physiology of hearing* (pp. 429–437). London: Academic Press.

Ronken, D. (1970). Monaural detection of a phase difference between clicks. *Journal of the Acoustical Society of America, 47,* 1091–1099.

Schacknow, P. N., & Raab, D. H. (1976). Noise-intensity discrimination: Effects of bandwidth conditions and mode of masker presentation. *Journal of the Acoustical Society of America, 60,* 893–905.

Shailer, M. J., & Moore, B. C. J. (1983). Gap detection as a function of frequency, bandwidth, and level. *Journal of the Acoustical Society of America, 74,* 467–473.

Shailer, M. J., & Moore, B. C. J. (1985). Detection of temporal gaps in bandlimited noise: Effects of variations in bandwidth and signal-to-masker ratio. *Journal of the Acoustical Society of America, 77,* 635–639.

Shailer, M. J., & Moore, B. C. J. (1987). Gap detection and the auditory filter: Phase effects using sinusoidal stimuli. *Journal of the Acoustical Society of America, 81,* 1110–1117.

Simon, G. R. (1963). The critical bandwidth level in recruiting ears and its relation to temporal summation. *Journal of Auditory Research, 3,* 109–119.

Stevens, S. S. (Ed.) (1975). *Psychophysics: Introduction to its perceptual, neural, and social prospects.* New York: Wiley.

van Zanten, G. A. (1980). Temporal modulation transfer functions for intensity modulated noise bands. In G. van den Brink & F. A. Bilsen (Eds.), *Psychophysical and behavioural studies in hearing* (pp. 206–209). Delft: Delft University Press.

Viemeister, N. F. (1977). Temporal factors in audition: A systems analysis approach. In E. F. Evans & J. P. Wilson (Eds.), *Psychophysics and physiology of hearing* (pp. 427–437). London: Academic Press.

Viemeister, N. F. (1979). Temporal modulation transfer functions based upon modulation thresholds. *Journal of the Acoustical Society of America, 66,* 1364–1380.

Warren, R. M., Obusek, C. J., Farmer, R. M., & Warren, R. P. (1969). Auditory sequence: Confusion of patterns other than speech or music. *Science, 164,* 586–587.

Watson, C. S., & Gengel, R. W. (1969). Signal duration and signal frequency in relation to auditory sensitivity. *Journal of the Acoustical Society of America, 46,* 989–997.

Wier, C. C., & Green, D. M. (1975). Temporal acuity as a function of frequency difference. *Journal of the Acoustical Society of America, 57,* 1512–1515.

Zera, J., & Green, D. M. (1993a). Detecting temporal onset and offset asynchrony in multicomponent complexes. *Journal of the Acoustical Society of America, 93,* 1038–1052.

Zera, J., & Green, D. M. (1993b). Detecting temporal asynchrony with asynchronous standards. *Journal of the Acoustical Society of America, 93,* 1571–1579.

Zwicker, E., & Wright, H. N. (1963). Temporal summation for tones in narrow-band noise. *Journal of the Acoustical Society of America, 35,* 691–699.

Zwislocki, J. (1960). Theory of temporal auditory summation. *Journal of the Acoustical Society of America, 26,* 1046–1060.

Zwislocki, J. (1969). Temporal summation of loudness: An analysis. *Journal of the Acoustical Society of America, 46,* 431–441.

CHAPTER 7

Across-Channel Processes in Masking

Joseph W. Hall III
John H. Grose
Lee Mendoza

I. INTRODUCTION

In many listening situations, the ear appears to be able to analyze energy at one frequency to the effective exclusion of energy at surrounding frequencies. Over the past 50 years, this "critical band" or "auditory filter" concept has accounted for an immense body of psychophysical masking data (see Chapter 5), and has been supported strongly by physiological data characterizing frequency selectivity at the levels of the basilar membrane and the eighth nerve (see Chapters 2 and 3). A tacit corollary of the auditory filter concept has been that, in analyzing the output of a particular auditory filter, the output of remote auditory filters are essentially inconsequential. Surprisingly, however, it has recently become clear that the auditory system is sometimes disposed to adopt strategies involving the combination of information across auditory filters, even in apparently simple tasks such as the detection of an intensity increment in a sinusoid or the detection of a pure-tone signal presented in a masking noise. Whereas some of these across-critical-band strategies can enhance performance considerably, other strategies involving the processing of information across multiple auditory filters actually lead to *poorer* performance than would be expected from the analysis of information from a single auditory filter. This chapter describes sev-

Hearing

eral of these psychoacoustic phenomena that appear to hinge upon the analysis of information across multiple auditory filters.

II. PROFILE ANALYSIS

The sensitivity of the auditory system to changes in the spectral envelope, or shape, of a complex signal has been characterized by a paradigm known as *profile analysis.* This manifestation of across-frequency processing was first noticed by Spiegel, Picardi, and Green (1981), who were investigating the effects of spectral uncertainty on the detection of an increment in level of one component of a multiple-component masker. A curious finding emerged: The detection of the increment was more difficult when the frequency of the incremented tone was predictable while the components constituting the masker were unpredictable than when the frequency of the tone was unpredictable while the components constituting the masker were predictable. A parsimonious interpretation was that, in the latter case, the auditory system was detecting the increment as a change in spectral shape from a known reference spectrum. Because the components of the masker were distributed across a number of auditory filters, the extraction of the spectral shape or profile required an across-frequency process.

In a typical profile analysis paradigm, the listener is presented with a standard complex of equal-amplitude sinusoids distributed logarithmically across some frequency range; for example, 200 to 5000 Hz (see inset to Figure 1). The signal consists of a tone matched in both frequency and phase to one of the masker components (the target) such that the addition of the signal results in an amplitude increment in the target component. To ensure that detection is based upon the change in spectral profile rather than level, the overall level of the standard complex and the standard + signal complex is randomly varied (roved) over some intensity range for each stimulus presentation. The signal threshold is expressed as the level of the signal relative to the level of the target component.

Given this basic paradigm, many stimulus parameters have been manipulated to ascertain their importance to profile analysis (Green, 1988). The results can be summarized as follows:

1. The range over which the level can be roved while maintaining "spectral shape constancy" is about 40 dB (Mason, Kidd, Hanna, & Green, 1984).

2. Performance is generally good for a flat spectrum and becomes increasingly degraded as the background spectrum is perturbed in amplitude (Kidd, Mason, & Green, 1986; Kidd, 1987). However, with a (dense) flat spectrum, performance improves slightly if the target component of the masker is about 6 dB (but less than 12 dB) higher in level than the nontarget components (Green & Kidd, 1983).

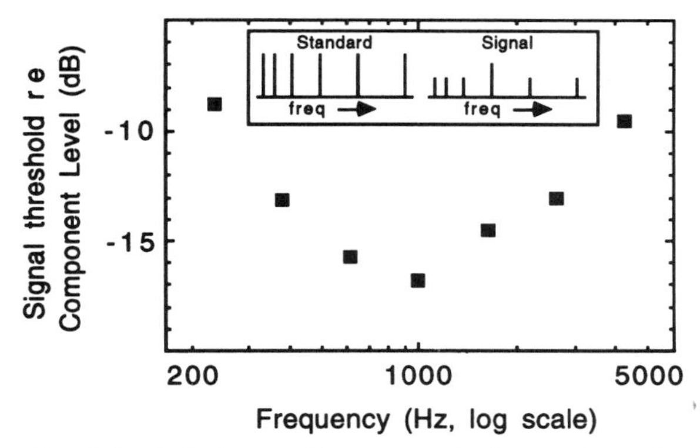

FIGURE 1 Signal threshold in a multicomponent background as a function of the frequency of the signal. (Data are redrawn from Green, Onsan, & Forrest, 1987.) Inset shows schematic spectra of a nonsignal interval and a signal interval for a two-alternative forced-choice task where the overall level is roved from interval to interval. The frequency scale on the inset is linear; the background components were uniformly spaced on a logarithmic scale.

3. Some benefit to signal detection produced by nontarget masker components can be observed if those components are presented to the contralateral ear, but such dichotic effects are more limited than the monaural effects (Bernstein & Green, 1987b; Green & Kidd, 1983).

4. If a multicomponent signal is used that, when added to the masker, results in a ripple of the spectral envelope, performance is relatively constant as long as the number of ripples across the stimulus spectrum remains below about ten. Above this, performance deteriorates. However, performance for this multicomponent signal is poorer than would be expected from the optimal combination of information from changes in level of the individual components (Bernstein & Green, 1987b; Green, Onsan, & Forrest, 1987; Richards, Onsan, & Green, 1989).

5. As the number of components in the background spectrum increases (i.e., spectral density increases), signal threshold improves up to about 11 to 21 components. Beyond this, the threshold deteriorates due to increased peripheral masking as components fall within the same auditory filter as the target (Bernstein & Green, 1987b; Green & Mason, 1985; Kidd, Mason, Uchanski, Brantley, & Shah, 1991). The optimal frequency ratio between successive components is 1.1 to 1.3 (Green, Kidd, & Picardi, 1983; Green, Mason, & Kidd, 1984).

6. For a given number of components in the background spectrum, performance improves with increasing spectral extent of the components (Green et al., 1984).

7. The signal threshold is lowest when the target component falls in the middle region of the background spectrum and increases monotonically as the position of the target component approaches either spectral edge, giving a so-called bowl-shaped threshold curve as shown in Figure 1 (Bernstein & Green, 1987a; Green & Mason, 1985; Green et al., 1987).

The final observation raises the question of whether all components in the background spectrum are equally informative to the detection process. Some insight on this issue can be gleaned from a method called *conditional-on-a-single-stimulus* (COSS) analysis (Berg, 1989). COSS analysis involves placing a small amplitude perturbation on each component of the background such that the level of that component varies slightly between the two observation intervals of a two-alternative forced-choice procedure. The degree to which this influences the probability of the listener identifying the signal is computed or each component and expressed as a weight. The response of an optimal observer will depend only on perturbations in the target component and not on the nontarget components. Thus the optimal observer will give the target component a weighting of 1.0 and all other components equal negative weightings that sum to −1.0. When COSS analysis is applied to conditions giving rise to the bowl-shaped threshold curve seen in Figure 1, it is evident that listeners perform almost optimally for conditions where the target component falls in the middle region of the background spectrum. However, when the target component falls near the spectral edges, listeners perform less than optimally (Berg & Green, 1992). Although COSS analysis accounts for the bowl-shaped threshold curve, it does not shed light on *why* listeners apparently apply different weightings, dependent upon the relative spectral location of the target. COSS analysis also holds some promise in highlighting the nature of the large individual differences observed in profile analysis tasks (Berg & Green, 1990; Kidd, 1993). The range of individual differences is unlikely to be due entirely to differences in degree of training (Kidd et al., 1991) and may reflect idiosyncratic use of the variety of subjective cues that play a role in such listening tasks, including pitch jumps and timbre changes (Versfeld & Houtsma, 1991; Feth & Stover, 1987; but see Richards et al., 1989).

Studies of profile analysis have also explored the role of dynamic aspects of the stimuli, in particular onset–offset asynchrony and amplitude modulation (AM). Because profile analysis depends on the simultaneous comparison of the amplitude of the target component with the amplitudes of the nontarget components, it is to be expected that any manipulation that promotes perceptual segregation between these regions will be detrimental to the across-frequency process (see Chapter 11). Consistent with this, the presentation of the target component at a lead–lag delay of greater than about 10 ms relative to the nontarget components greatly disrupts the detection of the signal (Green & Dai, 1992).

In a similar vein, the differential AM of the target component relative to the nontarget components has been explored. When the stimulus components are amplitude modulated at a low rate (<20–40 Hz), but the nontarget components are modulated 180° out of phase with the target component, signal threshold is greatly increased (Green & Nguyen, 1988). When the target component remains unmodulated while the nontarget components are modulated in phase (i.e., comodulated) at a variety of rates, several effects emerge:

1. Comodulation of the nontarget components greatly disrupts detection of the signal and the degree of disruption is an inverse function of modulation rate.
2. The elevation of threshold is not due simply to the presence of the spectral sidebands produced by the AM (see Chapter 1) because the replacement of these sidebands with components of the same frequency and amplitude, but with pseudorandom phases, is far less disruptive and the effect is not sensitive to modulation rate.
3. The degree of signal threshold elevation with comodulation of the nontarget components is a monotonic function of the depth of modulation.
4. The degree of threshold elevation caused by 100% modulation of the nontarget components is unaffected by the number of nontarget components.
5. When the nontarget components are spaced sufficiently to fall into separate auditory filters, the amount of threshold elevation is not dependent upon the relative phases of modulation of the nontarget components (Dai & Green, 1991).

The general conclusion is that profile analysis occurs optimally when the target and nontarget components are all unmodulated or are all comodulated. Profile analysis highlights the sensitivity of the auditory system to changes in spectral shape. Although this sensitivity can be inferred from work on timbre perception (because of the association between timbre and spectral shape), profile analysis quantifies the limits of this sensitivity. In addition, profile analysis establishes that spectral shape sensitivity is not a special case of intensity discrimination because it remains robust over such a wide range of random level variation.

The work on profile analysis has also shown that under certain conditions the presence of nontarget components can be *deleterious* to the detection of the signal, such as when the target and nontarget components are modulated out of phase—that is, incoherently—at low rates. In this vein, Green and Nguyen (1988) reported a condition where the signal to be detected consisted of sidebands around the target component, a task that is reduced to the detection of AM of the target component. They found that the detection threshold was markedly elevated when unmodulated flanking tones were present around the target component compared to when they were absent. In other words, AM detection at the target frequency was disrupted by the presence of nontarget components. This phenomenon of disruption to the processing of AM at one frequency due to the presence of sounds at another frequency has been formalized as modulation detection interference, or modulation discrimination interference, both given the acronym MDI.

III. MODULATION DETECTION OR DISCRIMINATION INTERFERENCE

Yost and Sheft (1989) posed the following question: If two widely separated carriers are fused into a single auditory object by virtue of their common AM, does this fusion preclude the independent processing of the envelope information of one of those components? They found that the detection of AM of one component deteriorated markedly when the second component was modulated at the same rates as the target component (see Figure 2). In other words, when the AM detection threshold was expressed in units of 20 L og (*m*), where *m* is the modulation depth, the threshold changed by about 12 dB when the second component was present compared to when it was absent. This reduction in performance was thought not to be due to masking in the auditory periphery, but rather to an "interference" in modulation processing at a more central stage; hence, the term *MDI*. MDI therefore is the change in AM threshold, expressed in dB, brought about by adding the modulated interfering components. Some studies have referenced the change to the AM threshold in the absence of any interfering sounds, while others have referenced it to the AM threshold in the presence of unmodulated interfering sounds.

In the basic paradigm just described, MDI has been found to be a function of the depth of modulation of the interfering or masking tone and, to some extent, the rate of modulation of the masking tone relative to that of the target. That is, the interference effect is broadly tuned for modulation rate, with maximum MDI occurring for target and maskers modulated at the same rate. MDI tapers off gradually with a slope of roughly 3–4 dB/octave when the modulation rates of the two components become increasingly dissimilar (Bacon & Konrad, 1993; Bacon & Opie, 1994; Moore, Glasberg, Gaunt, & Child, 1991; Shailer & Moore, 1993; Yost & Sheft, 1989; Yost, Sheft, & Opie, 1989). A small but consistent threshold elevation can occur for unmodulated maskers that are gated synchronously with the target component (Grose & Hall, 1993b; Moore & Jorasz, 1992; Moore & Shailer, 1992). MDI is also inversely related to the rate of (co)modulation, being maximal for lower rates (Yost et al., 1989). When the masker tone is presented to the ear contralateral to that receiving the target, MDI can still be

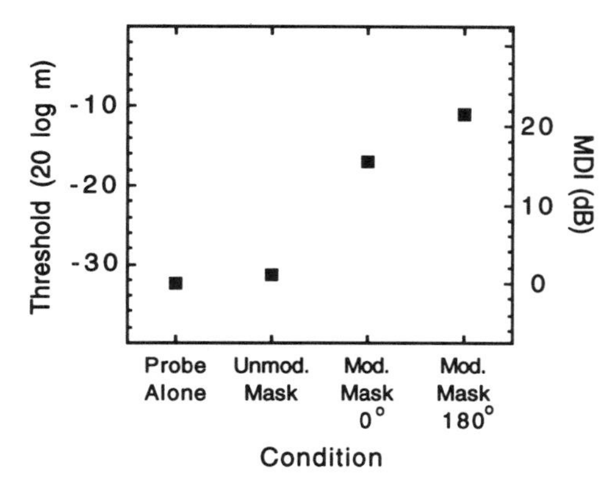

FIGURE 2 Modulation detection thresholds for a 1000 Hz probe expressed in 20 log (m), where m = modulation depth (left ordinate). The probe was presented alone, in the presence of an unmodulated 4000 Hz tone, in the presence of a 4000 Hz tone modulated in phase with the probe, and in the presence of a 4000 Hz tone modulated out of phase with the probe. The right ordinate indicates the magnitude of interference (i.e., shift in threshold relative to the probe alone). (Data are averaged points redrawn from Yost & Sheft, 1989.)

obtained (Bacon and Opie, 1994; Mendoza, Hall, & Grose, 1994b; Yost and Sheft, 1990). MDI appears to be restricted to tasks requiring judgments of envelope information: the presence of the masker tone does not affect the discrimination of frequency or intensity differences in the target component (Yost, 1992). In addition to interfering with the detection of AM of a target frequency, the presence of AM on a remote carrier has also been found to interfere with the discrimination of modulation depth (Moore & Jorasz, 1992). This point will be returned to later.

From the outset, it was noted that MDI was largely insensitive to the relative modulator phases of the two carriers (Yost & Sheft, 1989; Moore et al., 1991; Bacon & Konrad, 1993). This is also supported by studies using the paradigm of gap detection, where the processing of temporal envelope information at one frequency is greatly disrupted by the presence of even uncorrelated modulation at a remote frequency, both for monaural presentation (Grose & Hall, 1993b) and dichotic presentation (Moore, Shailer, & Black, 1993). The lack of envelope phase sensitivity is somewhat at odds with the premise that MDI reflects the fusion of components strictly on the basis of coherent AM (Yost and Sheft, 1989; Yost and Sheft, 1994). Direct tests of the "fusion by common AM" hypothesis indicate that AM detection (Mendoza, Hall, & Grose, 1994a) and AM discrimination (Moore, Sek, & Shailer, 1994) are not affected by the modulation coherence of the target and interfering sound. Another challenge to the "fusion by common AM" hypothesis is that it suggests that the listener's task in the typical AM detection paradigm should actually be quite easy. Forced to choose between two intervals that contain either the modulated masker plus the unmodulated target tone (standard interval) or the modulated masker plus the modulated target tone (signal interval), the listener should simply be able to choose the interval that sounds "fused" because the two tones in the standard interval should not be grouped together (Hall & Grose, 1991b; Moore et al., 1991). Most listeners, however, find the task difficult; auditory fusion seems to occur in both intervals suggesting that the fusion is not due simply to common modulation. An alternative possibility is that perceptual grouping is promoted by the synchronous gating of the masker and target tones. When on onset–offset asynchrony of a few tens of milliseconds is introduced into the gating pattern, MDI is greatly reduced (Hall & Grose, 1991b; Moore & Shailer, 1992). Auditory grouping by frequency proximity of the target and interferer also increases MDI: The effect of a masking sound close in frequency to the target is greater than that explainable by peripheral interactions alone (Bacon & Moore, 1993; Mendoza et al., 1994b).

MDI also occurs in the frequency modulation (FM) domain (Wilson, Hall, & Grose, 1990; Moore et al., 1991; Carlyon, 1992). For most listeners, the detection of FM of a target carrier is greatly disrupted by the simultaneous presence of FM on a distal carrier. Wilson et al. (1990) found that the amount of FM–MDI is sensitive to the relative phase of modulation between the masker and target tones. However, this appears to be due largely to the detection of distortion products (Carlyon, 1992).

MDI is not modulation-domain specific. An AM tone can interfere with the discrimination of FM depth, and an FM tone can interfere with the discrimination of AM depth (Moore et al., 1991). In such cross-domain interference, as in within-domain interference, the amount of interference is broadly tuned to modulation rate and relatively insensitive to modulation phase. However, since FM translates to AM within individual auditory channels, the apparent cross-domain interference may reduce in actuality to MDI in the AM domain; that is, when a frequency-modulated tone sweeps across an auditory filter, the output of that filter has an amplitude-modulated envelope imposed by the skirts of the filter.

Yost, Sheft, and Opie (1989), suggesting that grouping by common modulation might underlie MDI, presented a scheme whereby such grouping may be accomplished. First, the complex stimulus is passed through the bank of auditory filters in the auditory periphery. Next, the envelopes at the output of each filter are extracted. These envelopes are then grouped according to common AM rate or pattern; that is, modulation-specific channels are activated. The notion of modulation channels has been put forward by a number of investigators (Kay, 1982; Tansley & Suffield, 1983; Bacon & Grantham, 1989). According to this scheme, MDI would be the result of a modulation-specific channel being "swamped" by the predominant modulation of the AM masker with the contribution from the target modulation having little effect. A feature of this scheme is that the information regarding carrier frequency is lost. This scheme accounts for some aspects of MDI data, including the observation of Hall and Grose (1991b) that listeners often have difficulty assigning perceived modulation to its carrier. However, to deal with the relative insensitivity of MDI to modulator rate, it must be presumed that the modulation-specific channels are, in effect, not highly specific but are actually rather broadly tuned. This relative insensitivity to modulator rate and phase makes the premise that grouping by common modulation alone underlies MDI somewhat unsatisfactory. Moore and Shailer (1992) suggested that perceptual grouping is only one factor contributing to MDI because factors known to be acutely detrimental to auditory grouping (such as temporal asynchrony) do not completely abolish MDI. They proposed that a more parsimonious hypothesis would be that MDI reflects (at least) two components: carrier-specific MDI, and modulation-specific MDI. Modulation-specific MDI would result from activity within modulation channels and would be resilient to auditory segregation. Carrier-specific MDI would result from a difficulty in "hearing out" the pitch of the target carrier and would be highly susceptible to parameters that promote auditory segregation. In a similar vein, Mendoza et al. (1994a) noted that MDI is likely to reflect the contribution of multiple processes. For example, when the target and interferer tend to fuse (by virtue of common gating and frequency proximity), the listener may choose the interval that sounds more modulated. Here, the task is similar to AM discrimination. However, when the target and interferer are perceptually more distinct, the listener may choose the interval in which the target sounds modulated rather than steady. Here, the task is more similar to AM detection.

The phenomenon of MDI indicates that, when detecting the presence of modulation of a carrier or discriminating a change in depth of modulation of the carrier, the simultaneous presence of AM or FM at a remote frequency interferes with the task. However, under certain conditions the reverse can be true. When the task consists of discriminating a *decrease* in the depth of modulation in configurations where the masker consists of more than two flanking tones modulated in phase with the target, then the discrimination threshold can actually *improve* relative to the threshold when the flanking tones are absent (Moore & Jorasz, 1992). The improvement in threshold with the presence of comodulated flanking sounds has been termed *comodulation masking release* (CMR).

IV. ACROSS-CHANNEL EFFECTS BASED UPON COMODULATION

A. Comodulation Masking Release

As noted in the Introduction, the results of many masking experiments using random noise maskers conform to the idea that only the masking noise components within a relatively restricted bandwidth around the frequency of a signal contribute materially to the masking of the signal (see Chapter 5). However, it has recently been found that this model is not sufficient to explain the data when a wideband random noise masker is modulated by a low-frequency modulator. In masking experiments investigating the detectability of pure-tone signals in modulated noise, it has been found that the presence of noise components at frequencies well removed from the signal frequency can actually result in an *improvement* in detection threshold (see Figure 3). In accounting for this result, it has been noted that, for modulated masking noise, the fluctuations of energy in a given frequency region are correlated with the fluctuations in other frequency regions. This kind of noise has therefore been referred to as *comodulated noise,* and the resulting masking release has been termed *comodulation masking release* (Hall, Haggard, & Fernandes, 1984). CMR has also been demonstrated in a related paradigm (Hall et al., 1984a). In the baseline condition of this paradigm, the signal is masked by a narrow band of noise centered on the signal frequency (the on-signal noise band). In the experimental condition, one or more narrow bands of noise are added at other frequencies; if the added bands have the same envelope as the on-signal band, a masking release (CMR) occurs. CMR occurs both for monaural presentation and for dichotic presentation, where the on-signal band and signal are presented to one ear, and the flanking bands are presented to the contralateral ear (Cohen and Schubert, 1987b; Hall, Haggard, & Harvey, 1984; Schooneveldt and Moore, 1989b). Although CMR effects are relatively robust for noise bands that have common amplitude fluctuation, maskers having common FM do not tend to give masking release (Schooneveldt and Moore, 1988), although

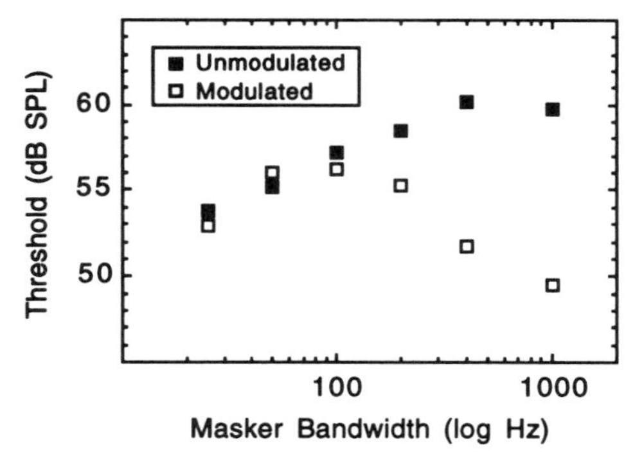

FIGURE 3 Signal threshold as a function of masker bandwidth for unmodulated noise (filled squares) and noise amplitude modulated by a lowpass-filtered noise (unfilled squares). The signal frequency, and the center frequency of the noise were both 1000 Hz. (Data redrawn from Hall, Haggard, & Fernandes, 1984.)

there is some evidence for a small effect (Grose and Hall, 1990). As in profile analysis, detection in comodulated noise appears to be based upon the pattern of outputs across a range of auditory filters, rather than only upon the output of the auditory filter centered on the signal frequency. Indeed, the CMR and profile analysis paradigms often share features. The cue of across-frequency level disparity, a main feature of profile analysis, also contributes substantially to CMR in some paradigms (Fantini & Moore, 1994).

The magnitude of CMR depends upon several variables: larger CMRs are associated with narrow masker bandwidth (Hall, Cokely, & Grose, 1988; Schooneveldt and Moore, 1987), low frequency of modulation (Hall and Grose, 1989; Schooneveldt and Moore, 1989a), high masker level (Cohen and Schubert, 1987b; Moore and Shailer, 1991), large percent of modulation (Fantini, 1991; Grose and Hall, 1989), small or no across-frequency delay in noise envelopes (Haggard, Harvey, & Carlyon, 1985; D. McFadden, 1986; Moore and Schooneveldt, 1990), and masker fringe preceding signal onset (Fantini, Moore, & Schooneveldt, 1993; D. McFadden and Wright, 1990). Perhaps the most important factor affecting the CMR magnitude is the number of comodulated noise bands present: The more bands that are present, the greater is the masking release (Carlyon, Buus, & Florentine, 1989; Haggard, Hall, & Grosse, 1990; Hall, Grose, & Haggard, 1990; Schooneveldt and Moore, 1989b). There appear to be diminishing returns once the number of bands exceeds three or four (Hall et al., 1990). The critical variable underlying the effect related to number of bands is probably the total number of quasi-independent auditory filters that contribute envelope information (Haggard et al., 1990). A corollary of this interpretation is that under circumstances where frequency selectivity is poor (and, therefore, the number of quasi-independent auditory filters is reduced), CMR should be reduced. This has received support from studies examining CMR under conditions of reduced frequency selectivity, investigating listeners having cochlear hearing impairment (Hall, Haggard, & Pillsbury; Hall and Grose, 1989; Moore, Shailer, Hall, & Schooneveldt, 1993). These studies have indicated modest, but significant, correlations of reduced frequency selectivity and small CMR. Furthermore, conditions yielding relatively poor frequency selectivity in normal-hearing listeners (Hall, Grose, & Moore, 1992) are associated with reduced CMR.

Results on the effect of across-frequency level disparity between the on-signal and flanking bands are also relevant to the issue of independence of across-channel information. If the on-signal and flanking bands are relatively close together, CMR exists when the on-signal band is 10 to 30 dB higher than the level of the flanking band(s) (Hall, 1986; McFadden, 1986; Schooneveldt and Moore, 1987). However, if frequency spacings are relatively wide or the flanking bands are presented to the contralateral ear, CMR can persist when the on-signal band is 40 to 60 dB more intense than the flanking bands (Cohen, 1991; Hall, 1987; Moore and Shailer, 1991). CMR is particularly robust to level differences in cases of dichotic stimulation where the flanking bands are maintained at high level and the on-signal band is reduced in level (Moore and Shailer, 1991). Independence of envelope information for bands of unequal level may be better preserved when the bands are well separated in frequency.

When the on-signal and flanking noise bands are approximately equal in level, spectral proximity of the bands can lead to a relatively large masking release. Schooneveldt and Moore (1987) have hypothesized that this effect is partly due to a *within*-channel (non-CMR) process. They noted that an interaction of closely spaced comodulated noise bands results in a highly regular beating pattern. When a signal is presented, the regular beating pattern is disturbed, providing a possible cue for signal detection. There is also some evidence from dichotic conditions that spectral proximity may facilitate the across-channel effect (Hall et al., 1990). Considering this, along with the preceding discussion of independence of information across channels, the spectral proximity of bands would appear to have multiple and sometimes antagonistic roles: (1) proximity is facilitative, partly due to the availability of within-channel cues; (2) for monaural presentation, proximity is deleterious to across-channel processes because it reduces channel independence; (3) for dichotic presentation, where channel independence is always maintained, proximity may enhance across-channel processing.

B. CMR and Stimulus Complexity

Recent experiments on CMR have investigated effects related to increasing levels of stimulus complexity. For example, in a study by Hall and Grose (1990) it was found that the CMR resulting from a set of seven comodulated noise bands was greatly reduced when two additional flanking bands (termed *codeviant bands*) were interposed between the on-signal band and its two nearest comodulated neighbors (see Figure 4). The codeviant bands were comodulated with respect to each other but not with respect to the other comodulated bands. Although part of the reduction in CMR can be explained by a within-channel corruption of the envelope of the on-signal band, part of the reduction appears to be due to a failure of the auditory system to separate the on-signal band from the proximal codeviant bands. This interpretation is supported by the fact that CMR recovers substantially when the number of codeviant bands is increased. This causes the codeviant band set to be heard as a distinct auditory source, separate from that formed by the comodulated bands. The results of this study imply that comodulation can serve as a cue for identifying separate auditory sources (see also Darwin and Carlyon, this volume, for a discussion of amplitude modulation as a grouping factor). These results also are useful in understanding CMR

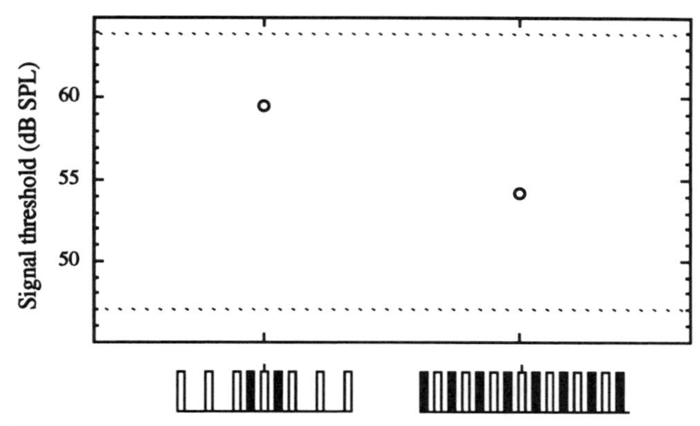

FIGURE 4 Signal threshold as a function of the number of codeviant bands present. The upper dotted line shows the threshold for the baseline condition (1000 Hz signal masked by a 20 Hz wide band centered on 1000 Hz); the lower dotted line shows the threshold in a condition where six comodulated bands were also present, centered on 400, 600, 800, 1200, 1400, and 1600 Hz. Two codeviant bands centered on 900 and 1100 Hz greatly reduced CMR (data point on left); when additional codeviant bands, centered on 300, 500, 700, 1300, 1500, and 1700 were added, CMR increased (data point on right).

under conditions of multiple, simultaneously present modulation patterns. Grose and Hall (1994) recently examined further conditions where multiple modulation patterns were simultaneously carried by two or more discrete sets of comodulated noise bands. The principal aim of this study was to determine whether simultaneous CMRs appear to occur for pure-tone signals distributed among each of the comodulated noise sets. Interestingly, concurrent CMRs do appear to occur, for conditions involving up to four tones presented in the context of four concurrent, independent modulation patterns (Grose and Hall, 1994). The results of these studies suggest that CMR can depend strongly upon stimulus features that affect auditory grouping. A further indication of this dependence is the fact that CMR can be greatly diminished if the temporal gating of the on-signal band is asynchronous with that of the comodulated flanking bands (Grose and Hall, 1993a).

Another example of CMR under a relatively high level of stimulus complexity has been provided recently by Eddins and Wright (1994). They investigated the effect of simultaneously employing two types of modulation carried by narrowband noise stimuli. One type of modulation was simply the random fluctuations inherent to the narrowband noise maskers. These modulations were either correlated across the noise bands (comodulated) or were random across the noise bands. The other type of modulation was a slower, sinusoidal fluctuation provided by amplitude modulating the narrow bands of noise. This type of modulation could be either correlated or random across the noise bands, depending upon the relative modulation phase across the noise bands. Eddins and Wright constructed conditions where one, both, or neither type of modulation was correlated across noise bands. They found that the masking release observed when both modulation types were correlated across frequency exceeded that observed when only one modulation type was correlated across frequency. This suggests that the auditory system can make use of two levels of envelope fluctuation simultaneously.

C. Monaural Envelope Correlation Perception and CMR

Several investigators have proposed general models to account for CMR. One general hypothesis is that CMR reflects a sensitivity of the auditory system to across-frequency differences in the pattern of modulation (Hall et al., 1984a). At that time, the literature was relatively sparse on the question of the sensitivity of the ear to correlations in temporal envelope of stimuli separated in frequency. Goldstein (1966) had demonstrated that listeners could discriminate across-frequency phase disparity of a low-frequency modulation of two carriers. Schubert and Nixon (1970) later found that, for relatively low-frequency carriers, the ear appeared to be insensitive to the across-frequency correlation of low-frequency, random modulation. Richards (1987, 1988, 1990), however, has found that, for higher frequency carriers, monaural envelope correlation discrimination for narrowband noise stimuli is relatively robust. In agreement with Schubert and Nixon, Richards found that sensitivity was poor for relatively low-frequency carriers, although later studies have shown that, performance can be relatively good even for low-frequency carriers (Moore & Emmerich, 1990). Sensitivity of the auditory system to across-frequency comodulation of envelopes has also been demonstrated recently using sinusuoidally amplitude modulated puretone carriers (Wakefield and Edwards, 1987; Yost and Sheft, 1989). These studies indicated that envelope phase disparities of carriers modulated at low rates could be discriminated for carriers separated by an octave or more. Although there appears to be a true across-channel component to such discriminations, it has been suggested that performance may also be influenced by within-channel cues (Strickland, Viemeister, Fantini, & Garrison, 1989) similar to those that may contribute to CMR (Schooneveldt & Moore, 1989a).

The results on monaural envelope correlation perception for narrowband noise stimuli raised the possibility that monaural envelope correlation perception and comodulation masking release may be two different aspects of the same basic auditory process (Fantini, 1991; Green, Richards, Ongan, 1990; Richards, 1987). CMR may be based upon the decorrelation of envelopes of comodulated bands caused by the presence of a signal. Some analyses of data have been broadly consistent with this hypothesis (Green et al., 1990; Richards, 1987). However, some data are difficult to reconcile with this hypothesis. For example, when CMR and envelope correlation perception are measured using the same listeners in both paradigms, the variation of CMR with phase and depth of modulation does not appear to be similar to that obtained in the envelope correlation perception task (Fantini, 1991). Also, whereas monaural envelope correlation perception becomes poorer as the noise bandwidth decreases, CMR improves with decreasing bandwidth (Moore & Emmerich, 1990). At this point, it appears unlikely that CMR and monaural envelope discrimination can be accounted for entirely by the same process(es).

D. The Role of Energy in Masker Dips

Buus (1985) proposed a model of CMR based upon the energy occurring in the masker dips. In this model, the comodulated flanking band essentially serves as a marker that indicates the optimal times to "listen" for the signal. When the flanking band envelope is high in amplitude, relatively little weight is given to the on-signal band, but when the flanking band envelope is low in amplitude, the on-signal band is weighted heavily. This rule would improve the signal-to-noise ratio at the signal frequency. In agreement with Buus's model, the results of several studies have indicated that the signal occurring in masker dip portions appears to contribute most to CMR. For example, when signals composed of three short (50 ms) tone bursts are added only in the dip regions of a masker envelope, a sizable masking release occurs. When the same signals are added in the peak regions of the masker, no masking release occurs (Grose and Hall, 1989; Moore, Glasberg, & Schooneveldt, 1990). Furthermore, signal energy placed in the masker peak regions does not result in a masking release, even when the signal–masker carrier phase relation is manipulated in such a way that the signal results in a material across-channel envelope pattern difference (Moore et al., 1990). Further studies by Hall and Grose (1991a) and by van den Brink, Houtgast, and Smoorenberg (1992) have highlighted the importance of energy in the dips of the masker. Such results lend support to the "dip-listening" model proposed by Buus (1985). Although most available data are in general agreement with Buus's model, it should be remembered that the situation where a signal affects only the peak portions of a masker is extremely unlikely to occur in the natural environment: Signals that alter the peak portions of a noise will alter the dip regions even more. It is possible, then, that information in the peak regions of the envelope will be utilized only when corresponding information also occurs in the dip regions (Hall and Grose, 1991a).

Another general model that has been proposed to account for CMR is based upon Durlach's (1963) equalization–cancellation (EC) model of the masking-level difference (Buus, 1985; Hall, 1986). Here, it is proposed that the envelopes at the output of different auditory filters are extracted and equalized in amplitude; one envelope is then subtracted from the other. A relatively large residual would indicate the presence of a signal. Although the results of some studies lend support to this model (Hall, Grose, & Haggard, 1988), a weakness is that the model does not easily account for the particular importance of information in the dip regions of the envelope. Furthermore, CMR has been demonstrated in conditions where the signal is the on-signal masking band added to itself (Hall and Grose, 1988). Under this circumstance, cancellation would not result in a residual, assuming prior envelope equalization.

E. Comodulation Detection Differences

Another phenomenon that may be related to CMR and envelope correlation perception and that depends critically upon the across-frequency correlation of stimulus envelopes has been termed *comodulation detection differences* (CDD). This refers to the finding that the detection threshold for a narrow band of noise (signal band) is lower when its envelope is uncorrelated with respect to a background of other noise bands than when all of the bands have correlated envelopes (Cohen and Schubert, 1987a; D. M. McFadden, 1987). As is true for CMR, CDD increases in magnitude with increasing number of noise bands (D. M. McFadden, 1987). CDD generally diminishes as the number of signal bands increases (fixed number of masker bands) (Wright, 1990). As with CMR, explanations have been proposed for CDD that are based upon envelope equalization–cancellation, dip listening, and envelope correlation detection (Cohen and Schubert, 1987a; D. M. McFadden, 1987; Wright, 1990). However, the explanation for CDD that has received the most attention is based upon auditory grouping principles. It has been argued that a signal band that is comodulated with the masking bands will perceptually fuse with the background (and therefore be hard to detect), whereas an uncorrelated signal band will "stand out" against a comodulated background (D. M. McFadden, 1987; Wright, 1990). Some of the data of Wright (1990) further indicate that the detection of a multiple noiseband signal embedded in multiple uncorrelated masking bands may be facilitated when the signal bands share common modulation. This finding may reflect processes that underlie grouping by common modulation.

A conceptually different view of CDD has been advanced by Fantini and Moore (1994), where part of the effect is attributed to within-channel processes. Fantini and Moore noted that the complexity of waveform fine structure at the output of the auditory filter centered on the signal is similar in both signal and nonsignal intervals in the case where all bands are comodulated. However, in the case where the signal band has an envelope that is different from that of a set of comodulated flanking bands, the waveform fine structure at the output of the auditory filter centered on the signal will contain a potential within-channel cue. When no signal is present, the fine structure will be relatively complex, but when a signal is present, the fine structure will alternate between relatively complex and relatively sinusoidal states. This cue may contribute to signal detection. One difficulty for this hypothesis, pointed out by Fantini and Moore (1994), is that a similar within-channel cue should be available for the case where all bands are uncorrelated (where detection performance is very poor). Fantini and Moore suggested that this poor performance may be based upon a different factor, perhaps due to a form of masking related to the spectral uncertainty associated with multiple, independent narrow bands of noise.

F. Comodulation and the Perception of Speech

One of the challenges presented by the phenomena considered previously is to determine what relevance they may have to the analysis of complex sounds in real environments. For example, it is possible that CMR reflects a noise reduction process that may allow environmental sounds to be detected and recognized at lower signal-to-noise ratios than would be possible by simply combining the outputs of the auditory filters containing signal information. It is not presently clear what the optimal paradigm is for evaluating CMR for speech signals. One paradigm has involved filtering the speech stimulus into a small number of frequency bands and using narrow comodulated noise bands to mask the speech. A CMR effect is then sought by adding further comodulated noise bands at frequencies having no speech energy: CMR would be indicated if the added noise bands resulted in an improvement in performance (Grose and Hall, 1992). Using this method, CMR effects have been found for the detection of speech in modulated noise, but effects related to speech *recognition* in comodulated noise have yet to be convincingly demonstrated. Grose and Hall found that recognition thresholds for a closed set of filtered vowel stimuli showed a CMR effect of about 6 dB, but noted that it was possible that vowel identification was based upon basic acoustic parameters of the highly filtered vowels rather than upon speech recognition. Measures involving recognition of speech signals that might be described as more difficult or less redundant (highly filtered monosyllabic words, differing only in terms of a single consonant, or highly filtered open set sentences) did not reveal a significant CMR. A general drawback of this speech CMR paradigm is that it involves relatively severe filtering of the speech signal, reducing the very "naturalness" that makes the signal of particular interest. Carrell and Opie (1992) have suggested a different way in which comodulation may be related to the analysis of speech sounds. Their basic speech stimuli were time-varying sinusoidal sentences (Remez, Rubin, Pisoni, & Carrell, 1981). These stimuli were composed of three tones, each of which followed the center frequency of one of the first three formants of naturally spoken sentences. The main result was that the intelligibility of these sentences was improved by amplitude comodulation of the component frequencies. The rates at which comodulation aids performance are roughly in line with those associated with CMR. Carrell and Opie suggest that the effect they found for speech may be related to the CMR phenomenon and that the natural comodulation of speech may be a factor that aids its segregation from background noise. Viewed in this way, the results of Carrell and Opie may be tied less specifically with the masking release of CMR than with the general idea that comodulation of particular frequencies favors the grouping of those frequencies into an auditory object (as is often assumed with regard to monaural envelope correlation perception and CDD). Indeed, the results of Carrell and Opie invite comparison with the CDD data of Wright (1990), which suggest that a signal composed of multiple noise bands may be better detected when the bands are comodulated.

V. CONCLUDING REMARKS

This chapter has surveyed three classes of phenomena that indicate that across-critical-band processes play a role in even relatively elementary aspects of auditory processing. The phenomenon of profile analysis demonstrates that, when the overall level of a stimulus is roving, the discrimination of a change in relative level of a component in one critical band can be facilitated by the presence of additional stationary components in surrounding critical bands. The level change at a discrete frequency region is detected as a perturbation in the spectral shape of an otherwise known spectral complex. The facilitation evident in profile analysis is maximized in conditions where the components making up the background complex are uniform in level, yielding a "flat" spectral profile.

The phenomenon of MDI is an example of the auditory system failing to monitor the output of just one critical band even when it would be clearly advantageous to do so. It appears that when modulation exists in multiple critical bands the auditory system has difficulty in processing the modulation in a single critical band independent of the output of other critical bands. While MDI occurs primarily when the flanking components are modulated, it can occur with unmodulated flanking components when they are synchronously gated. Synchronous gating may promote perceptual fusion of the components, thus hindering analysis of the individual components. Although not covered by this chapter, other examples exist of situations where the auditory system fails to monitor the output of a single critical band even when it would be advantageous to do (Neff & Green, 1987; Neff & Callaghan, 1987, 1988).

Finally, the chapter addressed effects that depend critically upon comodulation of envelopes in different frequency regions. The phenomena of CMR, CDD, and monaural envelope correlation perception indicate that the auditory system is sensitive to the across-frequency correlation of temporal envelopes. Under some circumstances, appreciable masking release can occur when comodulated flanking bands are present well away from the signal frequency. CMR appears to depend heavily upon an analysis of information in the dip regions of the envelope and is likely to be largest when several auditory channels simultaneously carry information about that envelope.

Perhaps the main reason that the phenomena examined in this chapter are of interest is that they may be instrumental in bridging the relatively wide gap that exists currently between what is known about very peripheral auditory processes (such as basic frequency analysis) and what is known about more complex central auditory processes (such as the perception of speech in noise). Results from profile analysis and CMR experiments were initially surprising, primarily because the paradigms used to measure the effects had previously been associated with highly successful single-channel explanations that were based upon peripheral auditory processes. In contrast, more central auditory processes probably contribute substantially to the effects found in profile analysis and CMR paradigms. If the initial results associated with these paradigms are viewed in a context of relatively central auditory analyses, they are no longer so unexpected. For example, a speech scientist would not be surprised at all by the notion that the auditory system combines information across several frequency channels; in fact, the surprise might be that a single-channel auditory filter explanation can account for so much auditory data so successfully. The experiments discussed in this chapter have used relatively well-controlled methods to explore the mechanisms that the auditory system may use to process and compare the output of different auditory filters. Perhaps the main challenge for future research is to determine how these mechanisms contribute to the formation of "auditory objects" or the construction of "the auditory scene" (see Chapters 11 and 12).

Acknowledgments

Preparation of this chapter was supported by the National Institute of Deafness and Other Communication Disorders and the Air Force Office of Scientific Research.

References

Bacon, S. P. & Grantham, D. W. (1989). Modulation masking: Effects of modulation frequency, depth and phase. *Journal of the Acoustical Society of America, 85,* 2575–2580.

Bacon, S. P., & Konrad, D. L. (1993). Modulation detection interference (MDI) with conditions favoring within or across-channel processing. *Journal of the Acoustical Society of America, 93,* 1012–1022.

Bacon, S. P. & Moore, B. C. J. (1993). Modulation detection interference: Some spectral effects. *Journal of the Acoustical Society of America, 93,* 3442–3453.

Bacon, S. P., & Opie, J. M. (1994). Monotic and dichotic modulation detection interference in practiced and unpracticed subjects. *Journal of the Acoustical Society of America, 95,* 2637–2641.

Berg, B. (1989). Analysis of weights in multiple observation tasks. *Journal of the Acoustical Society of America, 86,* 1743–1746.

Berg, B., & Green, D. M. (1990). Spectral weights in profile listening. *Journal of the Acoustical Society of America, 88,* 758–766.

Berg, B. G., & Green, D. M. (1992). Discrimination of complex spectra: Spectral weights and performance efficiency. In Y. Cazals, L. Demany & K. Horner (Eds.), *Auditory physiology and perception* (pp. 373–380). New York: Pergamon Press.

Bernstein, L., & Green, D. M. (1987a). Detection of simple and complex changes of spectral shape. *Journal of the Acoustical Society of America, 82,* 1587–1592.

Bernstein, L. R., & Green, D. M. (1987b). The profile-analysis bandwidth. *Journal of the Acoustical Society of America, 81,* 1888–1895.

Buus, S. (1985). Release from masking caused by envelope fluctuations. *Journal of the Acoustical Society of America, 78,* 1958–1965.

Carlyon, R. P. (1992). The psychophysics of concurrent sound segregation. *Philosophical Transactions of the Royal Society of London* (Series B), *336,* 347–355.

Carlyon, R. P., Buus, S., & Florentine, M. (1989). Comodulation masking release for three types of modulators as a function of modulation rate. *Hearing Research, 42,* 37–46.

Carrell, T. D., & Opie, J. M. (1992). The effect of amplitude comodulation on object formation in sentence perception. *Perception and Psychophysics, 52,* 437–445.

Cohen, M. F. (1991). Comodulation masking release over a three octave range. *Journal of the Acoustical Society of America, 90,* 1381–1384.

Cohen, M. F., & Schubert, E. D. (1987a). The effect of cross-spectrum correlation on the detectability of a noise band. *Journal of the Acoustical Society of America, 81,* 721–723.

Cohen, M. F., & Schubert, E. D. (1987b). Influence of place synchrony on detection of a sinusoid. *Journal of the Acoustical Society of America, 81,* 452–458.

Dai, H., & Green, D. M. (1991). Effect of amplitude modulation on profile detection. *Journal of the Acoustical Society of America, 90,* 836–845.

Durlach, N. I. (1963). Equalization and cancellation theory of binaural masking-level differences. *Journal of the Acoustical Society of America, 35,* 1206–1218.

Eddins, D. A., & Wright, B. A. (1994). Comodulation masking release for single and multiple rates of envelope fluctuation. *Journal of the Acoustical Society of America* (in press).

Fantini, D. A. (1991). The processing of envelope information in comodulation masking release (CMR) and envelope discrimination. *Journal of the Acoustical Society of America, 90,* 1876–1888.

Fantini, D. A., & Moore, B. C. J. (1994). Profile analysis and comodulation detection differences using narrow bands of noise and their relation to comodulation masking release. *Journal of the Acoustical Society of America, 95,* 2180–2191.

Fantini, D. A., Moore, B. C. J., & Schooneveldt, G. P. (1993). Comodulation masking release as a function of type of signal, gated or continuous masking, monaural or dichotic presentation of flanking bands, and center frequency. *Journal of the Acoustical Society of America, 93,* 2106–2115.

Feth, L. L., & Stover, L. J. (1987). Demodulation process in auditory perception. In W. A. Yost

& C. S. Watson (Eds.), *Auditory processing of complex sounds* (pp. 76–86). Hillsdale, NJ: Lawrence Erlbaum Associates.

Goldstein, J. L. (1966). An investigation of monaural phase perception. Doctoral dissertation, University of Rochester, Rochester, NY, University Microfilms, Ann Arbor, MI.

Green, D. M. (1988). *Profile analysis*. Oxford: Oxford University Press.

Green, D. M., & Dai, H. (1992). Temporal relations in profile comparisons. In Y. Cazals, L. Demany, & K. Horner (Eds.), *Auditory physiology and perception* (pp. 471–478). New York: Pergamon Press.

Green, D. M., & Kidd, G., Jr. (1983). Further studies of auditory profile analysis. *Journal of the Acoustical Society of America, 73,* 1260–1265.

Green, D. M., Kidd, G., Jr., & Picardi, M. C. (1983). Successive vs. simultaneous comparison in auditory intensity discrimination. *Journal of the Acoustical Society of America, 73,* 639–643.

Green, D. M., & Mason, C. R. (1985). Auditory profile analysis: Frequency, phase and Weber's law. *Journal of the Acoustical Society of America, 77,* 1155–1161.

Green, D. M., Mason, C. R., & Kidd, G., Jr. (1984). Profile analysis: Critical bands and duration. *Journal of the Acoustical Society of America, 75,* 1163–1167.

Green, D. M., & Nguyen, Q. (1988). Profile analysis: Detecting dynamic spectral changes. *Hearing Research, 32,* 147–163.

Green, D. M., Onsan, Z. A., & Forrest, T. G. (1987). Frequency effects in profile analysis and detecting complex spectral changes. *Journal of the Acoustical Society of America, 81,* 692–699.

Green, D. M., Richards, V. M., & Onsan, Z. A. (1990). Sensitivity to envelope coherence. *Journal of the Acoustical Society of America, 87,* 323–329.

Grose, J. H., & Hall, J. W. (1989). Comodulation masking release using SAM tonal complex maskers: Effects of modulation depth and signal position. *Journal of the Acoustical Society of America, 85,* 1276–1284.

Grose, J. H., & Hall, J. W. (1990). The effect of modulation coherence on signal threshold in frequency-modulated noise bands. *Journal of the Acoustical Society of America, 88,* 703–710.

Grose, J. H., & Hall, J. W. (1992). Comodulation masking release for speech stimuli. *Journal of the Acoustical Society of America, 91,* 1042–1050.

Grose, J. H., & Hall, J. W. (1993a). Comodulation masking release: Is comodulation sufficient? *Journal of the Acoustical Society of America, 93,* 2896–2902.

Grose, J. H., & Hall, J. W. (1993b). Gap detection in a narrow band of noise in the presence of a flanking band of noise. *Journal of the Acoustical Society of America, 93,* 1645–1648.

Grose, J. H., & Hall, J. W. (1994). Across-frequency processing of multiple modulation patterns. *Journal of the Acoustical Society of America* (submitted).

Haggard, M. P., Hall, J. W., & Grose, J. H. (1990). Comodulation masking release as a function of bandwidth and test frequency. *Journal of the Acoustical Society of America, 88,* 113–118.

Haggard, M. P., Harvey, A. D. G., & Carlyon, R. P. (1985). Peripheral and central components of comodulation masking release. *Journal of the Acoustical Society of America, 78,* S63.

Hall, J. W. (1986). The effect of across-frequency differences in masking level on spectro-temporal pattern analysis. *Journal of the Acoustical Society of America, 79,* 781–787.

Hall, J. W. (1987). Experiments on comodulation masking release. In W. A. Yost & C. S. Watson (Eds.), *Auditory Processing of Complex Sounds* (pp. 57–66). Hillsdale, NJ: Lawrence Erlbaum Associates.

Hall, J. W., Cokely, J., & Grose, J. H. (1988). Signal detection for combined monaural and binaural masking release. *Journal of the Acoustical Society of America, 83,* 1839–1845.

Hall, J. W., Davis, A. C., Haggard, M. P., & Pillsbury, H. C. (1988). Spectro-temporal analysis in normal-hearing and cochlear-impaired listeners. *Journal of the Acoustical Society of America, 84,* 1325–1331.

Hall, J. W., & Grose, J. H. (1988). Comodulation masking release: Evidence for multiple cues. *Journal of the Acoustical Society of America, 84,* 1669–1675.

Hall, J. W., & Grose, J. H. (1989). Spectro-temporal analysis and cochlear hearing impairment: Effects of frequency selectivity, temporal resolution, signal frequency and rate of modulation. *Journal of the Acoustical Society of America, 85,* 2550–2562.

Hall, J. W., & Grose, J. H. (1990). Comodulation masking release and auditory grouping. *Journal of the Acoustical Society of America, 88,* 119–125.

Hall, J. W., & Grose, J. H. (1991a). Relative contributions of envelope maxima and minima to comodulation masking release. *Quarterly Journal of Experimental Psychology, 43A,* 349–372.

Hall, J. W., & Grose, J. H. (1991b). Some effects of auditory grouping factors on modulation detection interference (MDI). *Journal of the Acoustical Society of America, 90,* 3028–3036.

Hall, J. W., Grose, J. H., & Haggard, M. P. (1988). Comodulation masking release for multi-component signals. *Journal of the Acoustical Society of America, 83,* 677–686.

Hall, J. W., Grose, J. H., & Haggard, M. P. (1990). Effects of flanking band proximity, number, and modulation pattern on comodulation masking release. *Journal of the Acoustical Society of America, 87,* 269–283.

Hall, J. W., Grose, J. H., & Moore, B. C. J. (1992). The influence of frequency selectivity on comodulation masking release in normal-hearing listeners. *Journal of Speech and Hearing Research, 36,* 410–423.

Hall, J. W., Haggard, M. P., & Fernandes, M. A. (1984). Detection in noise by spectro-temporal pattern analysis. *Journal of the Acoustical Society of America, 76,* 50–56.

Hall, J. W., Haggard, M. P., & Harvey, A. D. G. (1984). Release from masking through ipsilateral and contralateral comodulation of a flanking band. *Journal of the Acoustical Society of America, 76,* S76.

Kay, R. H. (1982). Hearing of modulation in sounds. *Physiological Reviews, 62,* 894–975.

Kidd, G., Jr. (1993). Individual differences in the improvement in spectral shape discrimination due to increasing number of nonsignal tones. *Journal of the Acoustical Society of America, 93,* 992–996.

Kidd, G., Jr. (1987). Auditory discrimination of complex sounds: The effects of amplitude perturbation on spectral shape discrimination. In W. A. Yost & C. S. Watson (Eds.), *Auditory processing of complex sounds* (pp. 16–25). Hillsdale, NJ: Lawrence Erlbaum Associates.

Kidd, G., Jr., Mason, C. R., & Green, D. M. (1986). Auditory profile analysis of irregular sound spectra. *Journal of the Acoustical Society of America, 79,* 1045–1053.

Kidd, G., Jr., Mason, C. R., Uchanski, R. M., Brantley, M. A., & Shah, P. (1991). Evaluation of simple models of auditory profile analysis using random reference spectra. *Journal of the Acoustical Society of America, 90,* 1340–1354.

Mason, C. R., Kidd, G., Jr., Hanna, T. E., & Green, D. M. (1984). Profile analysis and level variation. *Hearing Research, 13,* 269–275.

McFadden, D. (1986). Comodulation masking release: Effects of varying the level, duration, and time delay of the cue band. *Journal of the Acoustical Society of America, 80,* 1658–1667.

McFadden, D., & Wright, B. (1990). Temporal decline of masking and comodulation detection differences. *Journal of the Acoustical Society of America, 88,* 711–724.

McFadden, D. M. (1987). Comodulation detection differences using noiseband signals. *Journal of the Acoustical Society of America, 81,* 1519–1527.

Mendoza, L., Hall, J. W., & Grose, J. H. (1994a). Modulation detection interference (MDI) using random and sinusoidal amplitude modulation. *Journal of the Acoustical Society of America.* (in press)

Mendoza, L., Hall, J. W., & Grose, J. H. (1994b). Within- and across-channel processes in modulation detection interference. *Journal of the Acoustical Society of America.* (in press)
Moore, B. C. J., & Emmerich, D. S. (1990). Monaural envelope correlation perception, revisited: Effects of bandwidth, frequency separation, duration, and relative level of the noise bands. *Journal of the Acoustical Society of America, 87,* 2628–2633.
Moore, B. C. J., Glasberg, B. R., Gaunt, T., & Child, T. (1991). Across-channel masking of changes in modulation depth for amplitude- and frequency-modulated signals. *Quarterly Journal of Experimental Psychology, 43A,* 327–347.
Moore, B. C. J., Glasberg, B. R., & Schooneveldt, G. P. (1990). Across-channel masking and comodulation masking release. *Journal of the Acoustical Society of America, 87,* 1683–1694.
Moore, B. C. J., & Jorasz, U. (1992). Detection of changes in modulation depth of a target sound in the presence of other modulated sounds. *Journal of the Acoustical Society of America, 91,* 1051–1061.
Moore, B. C. J., & Schooneveldt, G. P. (1990). Comodulation masking release as a function of bandwidth and time delay between on-frequency and flanking maskers. *Journal of the Acoustical Society of America, 88,* 725–731.
Moore, B. C. J., Sek, A., & Shailer, M. J. (1994). Modulation discrimination interference for narrowband noise modulators. *Journal of the Acoustical Society of America* (in press).
Moore, B. C. J., & Shailer, M. J. (1991). Comodulation masking release as a function of level. *Journal of the Acoustical Society of America, 90,* 829–835.
Moore, B. C. J., & Shailer, M. J. (1992). Modulation discrimination interference and auditory grouping. *Philosophical Transactions of the Royal Society,* (Series B), *336,* 339–346.
Moore, B. C. J., Shailer, M. J., & Black, M. J. (1993a). Dichotic interference effects in gap detection. *Journal of the Acoustical Society of America, 93,* 2130–2133.
Moore, B. C. J., Shailer, M. J., Hall, J. W., & Schooneveldt, G. P. (1993b). Comodulation masking release in subjects with unilateral and bilateral hearing impairment. *Journal of the Acoustical Society of America, 93,* 435–451.
Neff, D. L., & Callaghan, B. P. (1987). Simultaneous masking by small numbers of sinusoids under conditions of uncertainty. In W. A. Yost & C. S. Watson (Eds.), *Auditory Processing of Complex Sounds* (pp. 37–46). NJ: Lawrence Erlbaum Associates.
Neff, D. L., & Callaghan, B. P. (1988). Effective properties of multicomponent simultaneous maskers under conditions of uncertainty. *Journal of the Acoustical Society of America, 83,* 1833–1838.
Neff, D. L., & Green, D. M. (1987). Masking produced by spectral uncertainty with multicomponent maskers. *Perception and Psychophysics, 41,* 409–415.
Remez, R. E., Rubin, P. E., Pisoni, D. B., & Carrell, T. D. (1981). Speech perception without traditional speech cues. *Science, 212,* 947–950.
Richards, V. M. (1987). Monaural envelope correlation perception. *Journal of the Acoustical Society of America, 82,* 1621–1630.
Richards, V. M. (1988). Components of monaural envelope correlation perception. *Hearing Research, 35,* 47–58.
Richards, V. M. (1990). The role of single channel cues in synchrony perception: The summed waveform. *Journal of the Acoustical Society of America, 88,* 786–795.
Richards, V. M., Onsan, Z., & Green, D. M. (1989). Auditory profile analysis: potential pitch cues. *Hearing Research, 39,* 27–36.
Schooneveldt, G. P., & Moore, B. C. J. (1987). Comodulation masking release (CMR): effects of signal frequency, flanking band frequency, masker bandwidth, flanking band level, and monotic vs. dichotic presentation flanking bands. *Journal of the Acoustical Society of America, 82,* 1944–1956.
Schooneveldt, G. P., & Moore, B. C. J. (1988). Failure to obtain comodulation masking re-

lease with frequency-modulated maskers. *Journal of the Acoustical Society of America, 83,* 2290–2292.

Schooneveldt, G. P., & Moore, B. C. J. (1989a). Comodulation masking release (CMR) as a function of masker bandwidth, modulator bandwidth and signal duration. *Journal of the Acoustical Society of America, 85,* 273–281.

Schooneveldt, G. P., & Moore, B. C. J. (1989b). Comodulation masking release (CMR) for various monaural and binaural combinations of the signal, on-frequency, and flanking bands. *Journal of the Acoustical Society of America, 85,* 262–272.

Schubert, E. D., & Nixon, J. C. (1970). On the relation between temporal envelope patterns at different points in the cochlea. *ONR Technical Report, 140,* 253.

Shailer, M. J., & Moore, B. C. J. (1983). Gap detection as a function of frequency, bandwidth and level. *Journal of the Acoustical Society of America, 74,* 467–473.

Spiegel, M. F., Picardi, M. C., & Green, D. M. (1981). Signal and masker uncertainty in intensity discrimination. *Journal of the Acoustical Society of America, 70,* 1015–1019.

Strickland, E. A., Viemeister, N. F., Fantini, D. A., & Garrison, M. A. (1989). Within- versus cross-channel mechanisms in detection of envelope phase disparity. *Journal of the Acoustical Society of America, 86,* 2160–2166.

Tansley, B. W., & Suffield, J. B. (1983). Time course of adaptation and recovery of channels selectively sensitive to frequency and amplitude modulation. *Journal of the Acoustical Society of America, 74,* 765–775.

van den Brink, W., Houtgast, T., & Smoorenburg, G. (1992). Signal detection in temporally modulated and spectrally shaped maskers. *Journal of the Acoustical Society of America, 91,* 267–278.

Versfeld, N. J., & Houtsma, A. J. M. (1991). Perception of spectral changes in multi-tone complexes. *Quarterly Journal of Experimental Psychology, 43A,* 459–480.

Wakefield, G. H., & Edwards, B. (1987). Discrimination of envelope phase disparity. *Journal of the Acoustical Society of America, 82,* S41.

Wilson, A. S., Hall, J. W., & Grose, J. H. (1990). Detection of frequency modulation (FM) in the presence of a second FM tone. *Journal of the Acoustical Society of America, 88,* 1333–1338.

Wright, B. A. (1990). Comodulation detection differences with multiple signal bands. *Journal of the Acoustical Society of America, 87,* 293–303.

Yost, W. A. (1992). Auditory image perception and amplitude modulation: Frequency and intensity discrimination of individual components for amplitude- modulated two-tone complexes. In Y. Cazals, L. Demany, & K. Horner (Eds.), *Auditory physiology and perception* (pp. 487–493). Oxford: Pergamon Press.

Yost, W. A., & Sheft, S. (1989). Across-critical-band processing of amplitude-modulated tones. *Journal of the Acoustical Society of America, 85,* 848–857.

Yost, W. A., & Sheft, S. (1990). A comparison among three measures of cross-spectral processing of amplitude modulation with tonal signals. *Journal of the Acoustical Society of America, 87,* 897–900.

Yost, W. A., & Sheft, S. (1992). Modulation detection interference: Long term practice effects and the role of modulation phase. 15th midwinter research meeting of the Association for Research in Otolaryngology, 68.

Yost, W. A., & Sheft, S. (1994). Modulation detection interference: Across-frequency processing and auditory grouping. *Hearing Research, 79,* 48–58.

Yost, W. A., Sheft, S., & Opie, J. (1989). Modulation interference in detection and discrimination of amplitude modulation. *Journal of the Acoustical Society of America, 86,* 2138–2147.

CHAPTER 8

Pitch Perception

Adrianus J. M. Houtsma

I. INTRODUCTION

Pitch is defined by the American National Standards Institute (1973) as "that attribute of auditory sensation in terms of which sounds may be ordered on a scale extending from high to low." Pitch is a particularly important attribute of sound. It is an essential element for features such as melody and harmony in music, and it conveys the bulk of the prosodic information in speech. Like loudness and timbre, it is a subjective attribute that cannot be expressed in physical units or measured by physical means. In the case of a *pure tone,* its primary objective correlate is the physical attribute frequency, but the tone's intensity, duration, and temporal envelope also have a well-established influence on its pitch. If a tone is *complex* and contains many sinusoids with different frequencies, which is usually the case with natural sounds, we may hear a single pitch as, for instance, in the case of a single note played by a clarinet. We may also hear a cluster of pitches as, for instance, a chord being played by a group of instruments. We may even hear individual partials as sinusoids, all having their own pitches. Even sounds that are not formed of well-defined discrete partials can evoke pitch sensations. This will be referred to as *nontonal* pitch.

Our auditory memory seems to be particularly good at storing and re-

Hearing

trieving pitch relationships, given that most people can easily recognize tunes or melodies and sing them more or less correctly. This ability to recognize and reproduce frequency ratios is often referred to as *perfect relative pitch.* Some people possess the ability to identify the pitch of sounds on an absolute, nominal scale without any explicit external reference. This relatively rare ability is referred to as *perfect absolute pitch.*

In this chapter we will first discuss the sensation of pitch evoked by pure tones, its dependence on various physical attributes of the signal, and our sensitivity to changes in frequency. We will consider complex tones and show how a single holistic pitch percept is determined by fundamental frequency as well as harmonic partials. A third class of sounds to be discussed consists of signals having continuous spectra, with a temporal or spectral regularity or with a spectral discontinuity. Such sounds can evoke pitch sensations corresponding with modulation frequency, spectral ripple, or edge frequency. We will discuss how pitch is internally represented, either as part of a musical scale or an intonation contour in speech, or in isolation, as in the case of absolute pitch. Finally, the multidimensional nature of pitch will be discussed in terms of the attributes pitch chroma, tone height, and harmonic proximity.

II. PURE TONES

A. The Mel Scale

There are various methods for measuring how the pitch of a pure tone depends on its frequency. One can obtain a pitch–frequency relation by magnitude estimation. One can also use a "doubling" or "halving" method in which subjects adjust the frequency of a comparison tone until it subjectively sounds twice or half as high as the pitch of a test tone with a frequency set by the experimenter. The classical result of such experiments is the *mel scale* measured by Stevens, Volkmann, and Newman (1937) and shown in Figure 1. The scale, obtained by the method of pitch halving, has an arbitrary pitch reference of 1000 mels at a frequency of 1000 Hz. A tone that sounds, on average, twice as high receives a value of 2000 mels, whereas a tone that sounds only half as high has a pitch of 500 mels. One can clearly see that pitch, expressed in mels (the unit is derived from *mel*ody), is not identical to frequency and not even linear in frequency. There is a direct and simple relationship between the mel scale of pitch and the critical-band scale (bark scale) for frequency resolution in the ear as discussed in Chapter 5. Zwicker and Feldtkeller (1967) pointed out that 1 bark is exactly 100 mels, which implies that the scales are essentially the same. This is because pitch, as measured by Stevens et al., is apparently determined by the center of excitation activity along the basilar membrane, which is also reflected in the

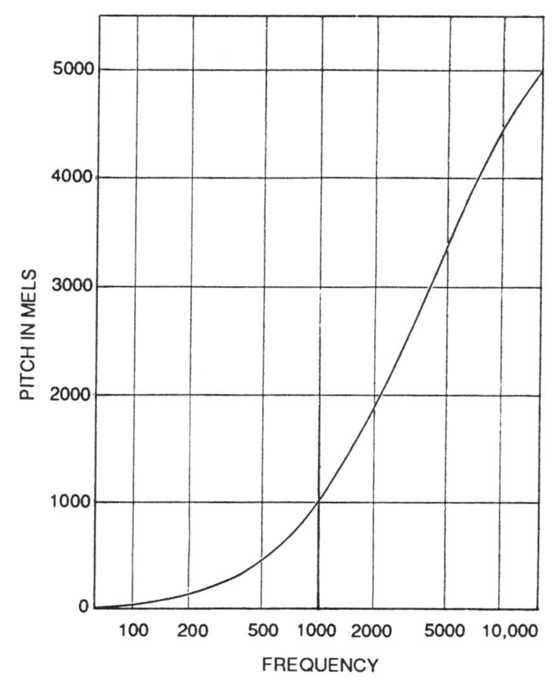

FIGURE 1 The relation of pitch (in mels) to the frequency of a pure tone. A 1000 Hz tone is arbitrarily assigned a value of 1000 mels. (From Stevens et al., 1937, reprinted with permission.)

critical-band or bark scale. Although the mel scale is based on empirical and scientific results, musicians may find it difficult to reconcile such a scale with the familiar subjective musical intervals of fifths, octaves, or semitones that they tend to use as relative scale units. It is probably for that reason that the mel scale never became quite as popular as the comparable *sone scale* for loudness.

B. Dependence on Intensity

Although the mel scale suggests that the pitch of a pure tone is simply determined by its frequency, the perceived pitch also depends on some other factors, one being intensity. If one measures for a group of subjects how, on average, the pitch of a pure tone changes with the tone's intensity, one typically finds that (1) for tones below 1000 Hz the pitch decreases with increasing intensity, (2) for tones between 1000 and 2000 Hz the pitch remains rather constant, and (3) for tones above 2000 Hz the pitch tends to rise with increasing intensity. Stevens (1935) reported the first data on this phenomenon, coming mostly from one listener, which are shown in Figure 2. Subsequent investigations have shown that for most people the magnitude of the pitch shift effect is smaller than was reported by Stevens (Verschuure & van Meeteren, 1975) and that the effect varies considerably between individual listeners (Morgan, Garner, & Galambos, 1951; Terhardt, 1974a). Interquartile ranges found by Morgan et al. (1951) have been superimposed on Stevens' data in Figure 2. For very short tone bursts, less than 40 ms, an increase in intensity always seems to lower the pitch, regardless of the tone's frequency (Rossing & Houtsma, 1986). This is probably also the reason why the pitch of such a very short tone burst depends on the shape of its temporal envelope, with the lowest pitch always being obtained with a constant-amplitude on–off gate function (Hartmann, 1978; Rossing, & Houtsma, 1986).

C. Influence of Partial Masking

The simultaneous presence of other tones or noise may also alter the perceived pitch of a pure tone. If the interfering tone or noise band is just below the frequency of a test tone, the pitch of this test tone is always increased, sometimes by as much as a semitone (Terhardt & Fastl, 1971), as is illustrated in Figure 3. Interfering sounds above the test tone frequency have a much less consistent effect.

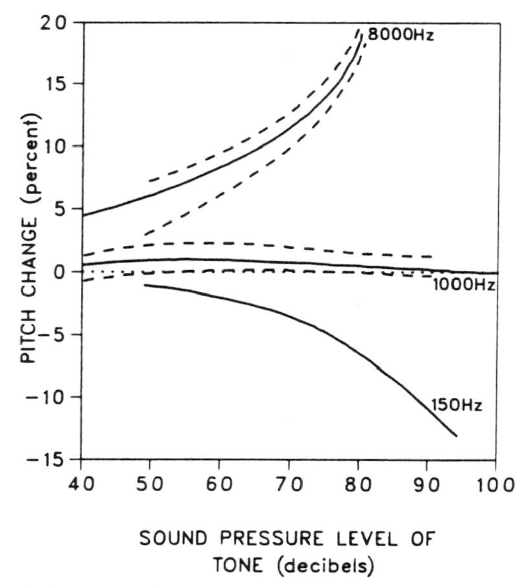

FIGURE 2 Pitch change as a function of sound pressure level of a pure tone. Solid curves: mean data from Stevens (1935). Dashed curves: 25th (lower) and 75th (upper) percentile of distribution of pitch changes in 18 ears, measured by Morgan et al. (1951).

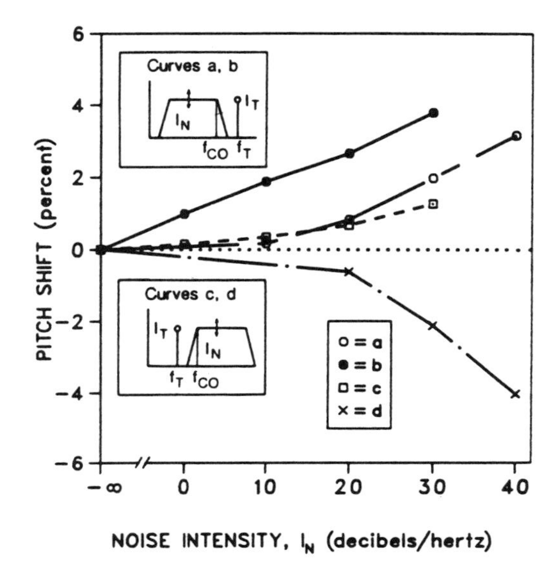

FIGURE 3 Pitch shift of a sinusoidal test tone induced by bandpass noise just below (a and b) and just above (c and d) the test tone frequency. Sound pressure levels of test tones were 50 dB, and frequencies were (a) 300 Hz, (b) 3800 Hz, (c) 3400 Hz, and (d) 100 Hz. (After Terhardt & Fastl, 1971.)

D. Binaural Diplacusis

Finally, the pitch sensation of a pure tone also typically depends somewhat on the ear to which it is presented. If a subject is asked to adjust the frequency of a comparison tone in one ear so that it matches the pitch of a test tone in the other ear, the frequencies will often come out slightly but consistently different. This effect, which is found to some extent in every listener, is known as *binaural diplacusis.* Interaural pitch differences are normally less than 2%, and the interaural frequency difference function may change slowly with time, as has been measured by van den Brink (1970) and is shown in Figure 4.

E. Frequency Discrimination

If two sinusoidal tones of different frequency are presented sequentially, there is some smallest frequency difference below which listeners can no longer tell consistently which of the tones is higher. A frequency difference resulting in 75% correct responses in a two-interval, two-alternative forced-choice paradigm is usually referred to as a *just noticeable difference,* or jnd. Classical frequency jnd data were measured by Shower and Biddulph (1931) with a method of frequency modulation detection. Modern data were provided by Moore (1973), who measured frequency jnds as a function of tone

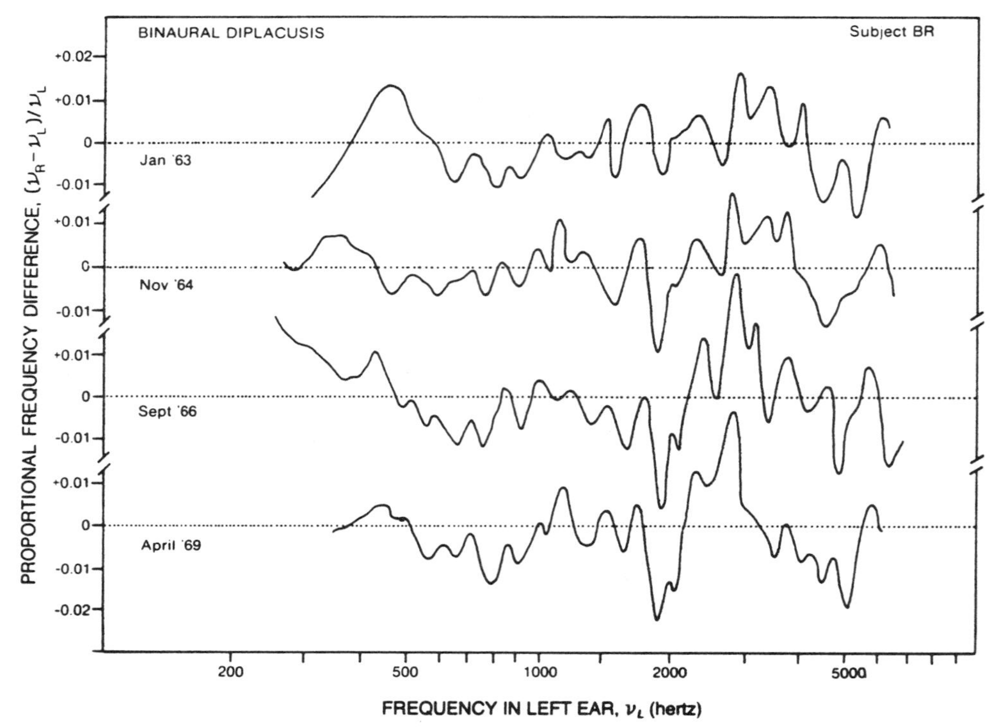

FIGURE 4 Binaural diplacusis patterns of one subject measured at intervals of several years. (From Brink, 1970, reprinted with permission.)

duration, and by Wier, Jesteadt, and Green (1977), who measured jnds as a function of tone intensity. A summary of Moore's data is shown in Figure 5, and of the data by Wier et al. in Figure 6.

The data of Figure 5 and 6 clearly show that the accuracy of our hearing system for distinguishing sequential tones of different frequency is much greater than the ability to resolve these tones (see Chapter 5). The large difference between the 0.1–0.2% frequency discrimination threshold and the approximately 10% frequency separation required to resolve simultaneous tones has sometimes been presented as a paradox and has been a reason for assuming the presence of "neural sharpening" mechanisms in the central auditory system. The reader should realize, however, that frequency discrimination behavior and frequency resolution in the auditory periphery have, in principle, very little to do with one another. Discrimination limits are imposed primarily by the amount of noise in the system. If there were no noise, one would be able to discriminate tones with an arbitrarily small frequency difference, no matter how steep or shallow were the slopes of peripheral auditory filters. Given that the frequency encoding process in the

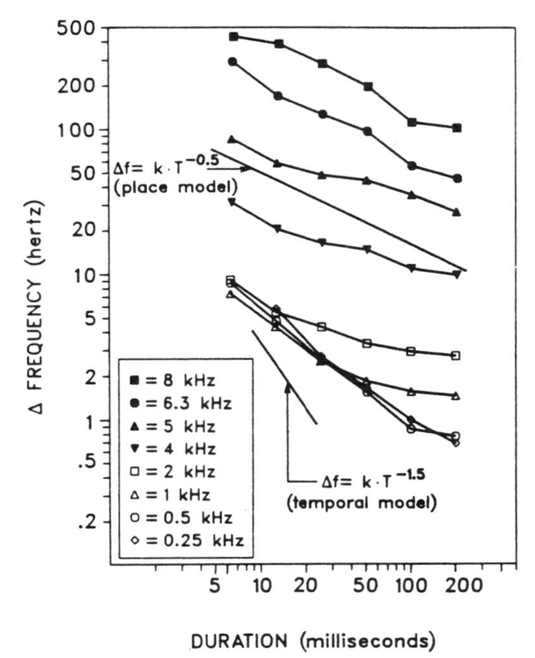

FIGURE 5 Just noticeable frequency differences as a function of stimulus duration. Sinusoidal tones had a constant loudness level of 60 phons. Line segments represent predictions of Siebert's (1970) place model (top, slope = −0.5) and temporal model (bottom, slope = −1.5). (Data from Moore, 1973.)

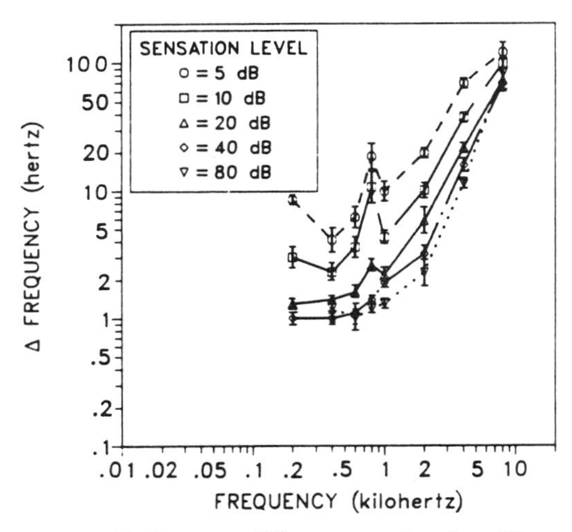

FIGURE 6 Just noticeable frequency differences as a function of frequency, with sensation level as parameter. Tone durations were 500 ms. (Data from Wier et al., 1977, reprinted with permission.)

auditory system is noisy, however, the resolution power at the periphery will show up as a model parameter in any stochastic frequency-coding model. For instance, Siebert (1970) has shown that optimal use of neural firing rate information across fibers of the auditory nerve, assuming cochlear filtering in accordance with the classical observations by von Békésy, predicts a frequency discrimination performance comparable to the data of Shower and Biddulph (1931). Performance is predicted to be proportional to the inverse square root of tone duration. Optimal use of all temporal information in firing patterns yields a predicted performance that is much better than is observed under any condition and is proportional to duration raised to the power -1.5. One can see from the time-dependence slopes shown in Figure 5 that high frequencies tend toward predictions of the place model, whereas the lowest frequencies (250 and 500 Hz) show a duration dependence that is more in accordance with a time model.

III. COMPLEX TONES

A. Historical Background

Between 1840 and 1850 an interesting discussion took place in the *Annalen für Physik und Chemie* between Ohm and Seebeck about the pitch of a complex tone. Such a tone is composed of several sinusoidal tones, the lowest in frequency being the *fundamental,* and the others (*harmonics*) having frequencies that are multiples of the frequency of the fundamental (see also Chapter 1). Seebeck (1841) presented observations on sounds made with a mechanical siren. These sounds were periodic, containing controllably suppressed odd harmonics. Seebeck described how the pitch he associated with the sound as a whole always seemed to follow the fundamental, even if this fundamental component was very weak. He concluded that the fundamental frequency is not the only determinant of pitch, but that the upper harmonics also contribute to the subjective pitch sensation. Ohm (1843) argued that our ears perform a real-time frequency analysis similar to the mathematical formulation of Fourier, where the frequency of the lowest spectral component determines the pitch of the complex, and the other frequencies determine the sound's timbre. The strong fundamental pitch sensation in the absence of acoustic power reported by Seebeck therefore had to be based on an illusion. Twenty years later Helmholtz (1863) chose the side of Ohm in this debate and thereby settled the issue for almost a century to follow.

Just before the Second World War Schouten (1938) rekindled the Ohm–Seebeck debate by demonstrating that Seebeck's conclusion was basically correct. With his optical equipment he could generate periodic complex tones devoid of any acoustical power at the fundamental frequency. Schouten was able to show that the pitch sensation associated with the *missing fundamental*, as it later became known, could not be explained as a nonlinear difference tone generated at the auditory periphery, as first Helmholtz (1863) and later Fletcher (1924) had argued. According to Schouten, the pitch sensation is caused by neural detection of periodic fluctuations in the envelope pattern of clusters of harmonics that the ear fails to resolve. If spectral resolution is insufficient, two or more summed harmonics will appear at the output of the cochlear filter. The periodicity of the envelope of such a summed signal is the same as the periodicity of the fundamental, even if the fundamental is physically absent. It can be picked up through phase locking by fibers of the auditory nerve and transmitted to central parts of the brain. Since insufficient cochlear resolution is an essential element of Schouten's pitch theory, this theory became known as the *residue theory of pitch* (Schouten, 1940).

Soon it became clear, however, that Schouten's residue theory also failed to provide an adequate explanation of new experimental findings. Ritsma (1962) found a clear upper limit to the harmonic order beyond which no tonal residue, that is, pitch, is heard. He also found that the existence region for the tonal residue extends to combinations of harmonics that the cochlea should be able to resolve, which is in contradiction with the essence of the residue theory. Some years later Ritsma (1967) and Plomp (1967) found that the best harmonics to convey a pitch sensation of a missing fundamental are on the order of 3, 4, and 5. In this so-called dominant region, harmonic frequencies differ by 25% or more and should, as has been discussed in Chapter 5, be well resolved in the periphery of the auditory system. Perhaps the most direct evidence against the residue theory was the finding by Houtsma and Goldstein (1972) that two successive simultaneous harmonics with frequencies nf_0 and $(n + 1) f_0$, presented to different ears, evoke an equally effective fundamental pitch percept as a monotic or diotic presentation of the same two harmonics. In the dichotic case, with each harmonic going to a different ear, there is no physical interference or cochlear residue. The experimental results force one to conclude that the pitch of complex tones is mediated primarily by a central mechanism that operates on neural signals derived from those stimulus harmonics spectrally resolved in the cochlea. Modern pitch theories therefore almost always contain elements of *central* pitch processing.

B. Template Theories of Pitch

In this class of theories it is assumed that, at some central processing stage in the brain, a spectral template is matched to frequencies or frequency transformations of those stimulus partials that are resolved in the cochlea. One of these pitch theories is the optimum processor theory of Goldstein (1973), schematically illustrated in Figure 7. The theory assumes that the frequen-

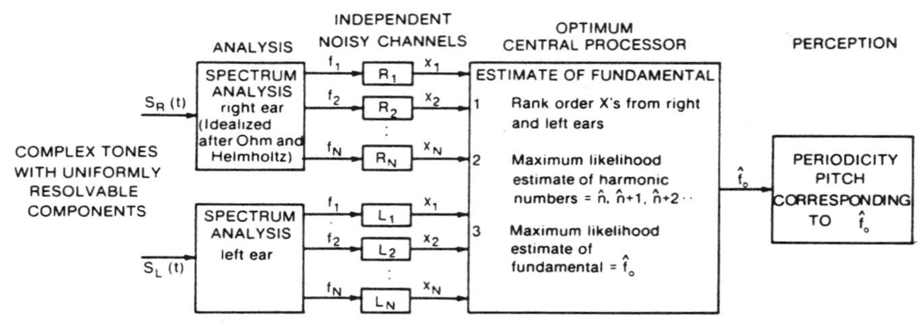

FIGURE 7 Schematic representation of the optimum processor model of Goldstein (1973). (Reprinted with permission.)

cies f_i of spectrally resolved stimulus partials are transformed into Gaussian random variables x_i, with means equal to f_i and variances that are functions of f_i only. All amplitude and phase information is ignored. A central processor assumes that the input numbers x_i are noisy representations of harmonic frequencies and makes an optimal estimate of the unknown harmonic numbers and fundamental frequency. The variance function is the only free parameter of the model. This function represents the noise in the frequency coding process in our auditory system. It causes the central processor to sometimes make an incorrect estimate of the harmonic order of a set of partials, the probabilities of which can be computed exactly with the theory.

The virtual pitch theory of Terhardt (1972, 1979) gives an alternative account of the central pitch percept. It is formulated in a deterministic manner, unlike the optimum processor theory, and is schematically illustrated in Figure 8. The theory assumes that spectral frequencies are transformed in the auditory periphery into spectral pitch cues according to cer-

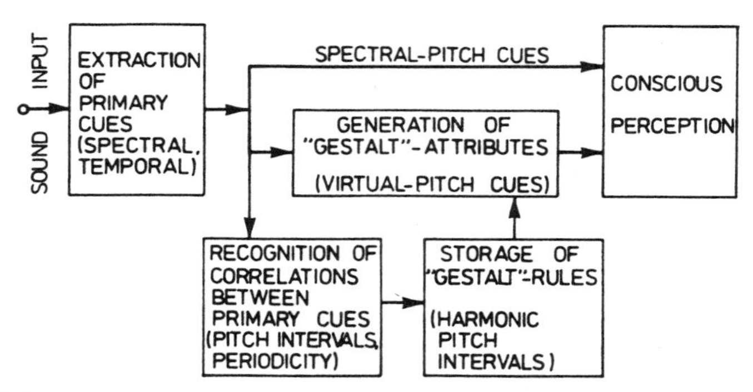

FIGURE 8 Schematic representation of principles underlying the virtual pitch theory of Terhardt (1974b).

tain empirical rules that reflect, for instance, pure-tone pitch shift phenomena discussed in Sections II.B–D. Virtual pitch cues are centrally derived from spectral pitch cues by finding common subharmonics. The model is similar to the optimum processor theory of Goldstein in the sense that both are spectral template matching models. The virtual pitch theory, however, also considers intensities of partials and masking effects. The output of the model is a list of virtual pitch candidates, each with an associated strength, that can be computed from details of the physical stimulus with an algorithm provided by Terhardt (1979).

C. The Role of Unresolved Harmonics

Despite the general development of our understanding that pitch perception is primarily a central process, the question still remains whether totally abandoning Schouten's residue theory is justified. Given the obvious shortcomings of the residue theory it remains true, for instance, that a periodic pulse train retains a certain pitch quality even if all low-order resolvable harmonics have been removed (Moore & Rosen, 1979). Hoekstra (1979) found that the jnd for the missing fundamental of an octave-band wide tone complex remains finite, at about 5 Hz, if the missing fundamental becomes very low and the octave band contains many closely spaced harmonics. Houtsma and Smurzynski (1990) studied pitch identification as well as pitch discrimination performance with complex tones composed of 11 successive harmonics. All complexes had missing fundamental frequencies between 200 and 300 Hz, and harmonic spectra starting between the 7th and the 25th harmonic. Phase relations were either zero (sine) phase, giving a waveform with very distinct peaks, or "negative Schröder" phase (Schröder, 1970) that minimizes the crest factor (peakedness) of the complex-tone signal at the cochlear output. The outcome of the experiments was that, if the number of resolvable harmonics in the complex was progressively reduced, identification performance dropped from near perfect to a low but clearly above-chance level. As shown in Figure 9, jnds increased from about 0.5 to about 5 Hz. The phase relation between harmonics seemed to matter very little. If, on the other hand, the tone complexes contained no resolved harmonics, that is, if the lowest harmonic was on the order 12 or higher, identification and discrimination performance levels remained constant and independent of the harmonic order of the stimulus. Phase, however, turned out to be of great influence on performance, with jnds being almost a factor of 2 larger with the Schröder-phase relation than with the sine-phase relation.

The conclusion from all these experimental results is that not only do low-order harmonics, resolved in the cochlea, contribute to the percept of pitch of a complex tone, but also high-order unresolved harmonics. Their degrees of contribution, however, are quite different. Resolved components

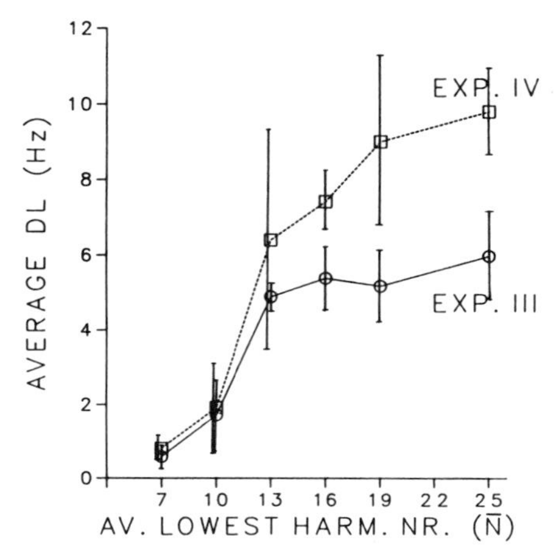

FIGURE 9 Just noticeable differences in fundamental frequency (around 200 Hz) of a complex tone with 11 successive harmonics. The abscissa designates the lowest harmonic number of the 11 harmonics, which are in sine phase (solid curve) or negative "Schröder" phase (dashed curve). Bars indicate standard deviations of mean jnds of four subjects. (From Houtsma and Smurzynski, 1990.)

evoke a stronger, more salient and sharply defined pitch image than unresolved components, as shown by much higher identification scores in, for instance, melodic interval identification tasks and by lower jnds in pitch discrimination tasks. If complex-tone stimuli are broadband, that is, if resolved as well as unresolved partials are present, the latter will generally dominate in determining the perceived pitch, except for complex tones with very low fundamentals (Moore & Peters, 1992).

D. Hybrid Models

In terms of models, one could conclude that two separate neural mechanisms lead to a pitch percept. One operates on neural signals derived from partials, which are resolved in the cochlea, is located centrally because it derives these signals from inputs in the left and right ear simultaneously, and is insensitive to phase relations between complex-tone harmonics. The other operates on temporal properties of cochlear output, similar to the residue mechanism proposed by Schouten (1940). One can also argue, however, that there is only one neural pitch mechanisms in the central auditory system, which yields different performance levels or parameter dependencies for different stimulus conditions. One such model, proposed by Srulovicz and Goldstein (1983), might be a fitting candidate, and it is illustrated in Figure 10. A central spectral magnitude is determined at each frequency by the response of the eighth nerve fiber with characteristic frequency f_c. The

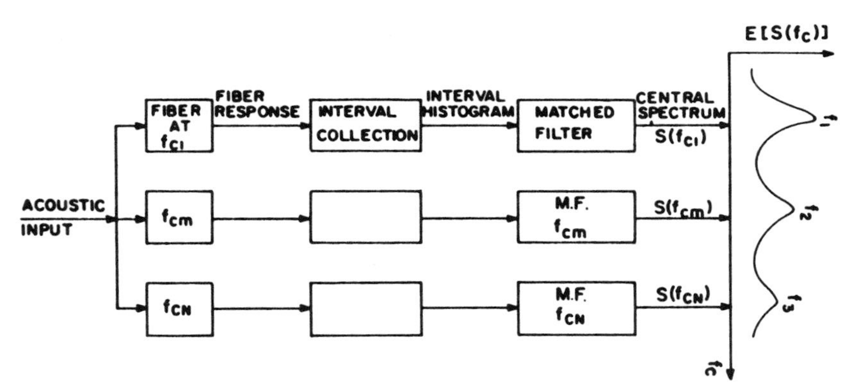

FIGURE 10 Schematic outline of pitch model by Srulovicz and Goldstein (1983).

interspike interval histogram (see Chapter 3) of each fiber is passed through a filter that is matched to its f_c, yielding a single-valued output as contribution to a central spectrum. Frequencies resolved in the cochlea will show up in the central spectrum as well-identified peaks, and a holistic fundamental pitch percept can be derived from this central spectrum in a way described by the optimum processor theory of Goldstein (1973). Degraded pitch performance for unresolved harmonics is predicted by this model because it makes very inefficient use of the abundant fundamental-period information present in the firing patterns of eighth nerve and cochlear nucleus fibers (Horst, Javel, & Farley, 1986; Kim, Rhode, & Greenberg, 1986) for stimuli containing many unresolvable harmonics. Instead of directly computing the inverse of the principal peak in interspike interval histograms, the model maps this phase-locking information at the level of the central spectrum into all the possible harmonics of the fundamental, after which a harmonic-template estimate is made on this central spectrum to find the missing fundamental. This coding and central recovery scheme for the missing fundamental of complex tones with high-order harmonics may seem unnecessarily complicated and very inefficient, but appears consistent with the relatively weak pitch image evoked by such high-order harmonics compared with low-order resolved harmonics.

Another computational model for pitch identification and phase sensitivity with complex-tone stimuli has recently been proposed by Meddis and Hewitt (1991a, 1991b). The model, which is illustrated in Figure 11, has many elements also seen in earlier models (Wightman, 1973; Terhardt, 1972; Srulovicz & Goldstein, 1983), but combines them in a rather unique way. It is composed of (1 and 2) a linear bandpass filter representing the outer and middle ear, (3) a bank of 128 overlapping critical-band (gammatone) filters representing basilar membrane action. It is then followed for

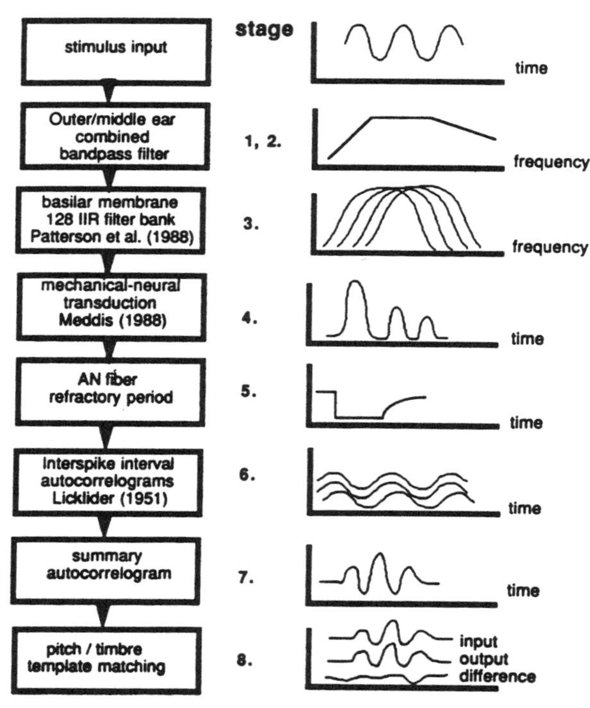

FIGURE 11 Schematic outline of pitch model by Meddis and Hewitt (1991a). Right column shows signal transformations at the various stages. (Reprinted with permission.)

each channel by (4) a hair-cell simulation model (Meddis, 1988), (5) a simple refractory-period model for nerve fibers, and (6) an interspike interval autocorrelation process (Licklider, 1951). Finally, (7) a summary autocorrelogram is formed by averaging the stage 5 output across channels. The pitch percept is represented in this summary autocorrelogram by peak locations, indicating pitch candidates, and peak height indicating relative strength or salience of these pitch candidates.

Although the model incorporates the stochastic nature of neural processes, it uses only their average statistics and is therefore in principle a deterministic model, similar to Wightman's or Terhardt's model. It can identify pitch candidates for any complex stimulus and make a prediction about their relative strengths, at least in an ordinal sense. It is not a discrimination model for making quantitative predictions about pitch confusions, because the output of the model is noiseless and no specific description of a decision model is included. Nevertheless, the model has provided a convincing qualitative account of many known properties of complex-tone pitch, such as the weakening of the sensation with increasing harmonic order, ambiguity, the inharmonic frequency shift effect (Schouten, Ritsma, & Cardozo, 1962), the existence and dominant regions, as well as the effects of repetition pitch and amplitude-modulated noise pitch, which are still to be discussed. To account for dichotic pitch phenomena (Houtsma & Goldstein, 1972; Bilsen & Goldstein, 1974), it seems that the model can be adapted in a fairly simple manner; for instance, by forming the summary autocorrelogram of the last stage from averages across all left and right channels.

E. Pitch of Simultaneous Complex Tones

When listening to music, we are often exposed to complex tones presented simultaneously. If the notes C, E, and G are played simultaneously on three different musical instruments, we can easily perceive the major triad C-E-G. This implies that each of the pitches C, E, and G must be perceived. From an acoustical point of view, the fundamentals of the three notes may be weak or totally absent, and the partials of all three notes are mixed together. Apparently our central auditory system is able to reconstruct groups of harmonically related partials from the total of all resolved partials it receives from the cochlear output.

Beerends and Houtsma (1986, 1989) investigated to what extent our auditory system is able to recognize the two (missing) fundamentals of two simultaneous two-tone harmonic complexes, as a function of the harmonic order of the partials and the manner in which partials were distributed between the ears. They found that deterioration of pitch identification performance with increase in harmonic order was about the same for all presentation conditions; that is, it did not matter very much whether partials of each complex tone went to different ears, all partials went to both ears, or partials of each tone were divided between ears. The conclusion was that frequency information about resolved partials must all end up in the same central pool and that grouping of partials for pitch processing is based on principles other than binaural spatial information (see Chapter 11 for further discussion of this topic). It can be shown that template models like the optimum processor or virtual pitch theories are, in principle, able to account for the observed phenomena (Beerends, 1989).

F. Pitch Ambiguity

Before we leave the topic of tonal pitch, we may wonder whether it is even correct to speak of "the pitch" of a complex tone. On music paper, the sound of a musical instrument is usually represented by a single note, which is thought of as having a certain pitch, duration, and timbre. Laboratory experiments on discrimination, identification, or matching of pitches typically show, however, that the pitch of complex tones can be ambiguous, especially if low-order harmonics are weak or missing or if only a few harmonics are present. Modern pitch theories can adequately account for this ambiguity. Furthermore, another source of ambiguity is, on the one hand, well known but, on the other hand, less well understood and considerably more difficult to model. As in the popular saying about the forest and the trees, the auditory system can perceive a sound complex holistically, where it usually evokes a sensation of a single pitch and some timbre, and also analytically, where it perceives many pitches of individual harmonics or partials. Some models explicitly recognize the existence of holistic and analytic pitch percepts. Terhardt's (1972) virtual pitch theory distinguishes spectral and virtual pitch cues, and Goldstein's (1973) theory distinguishes noisy transformations of resolved frequencies from central estimates of periodicity. None of the theories is able to explain, however, what conditions decide whether analytic or holistic pitch cues are used. Experimental attempts to control and measure conditions for analytic or holistic pitch perception (Smoorenburg, 1970; Houtsma & Fleuren, 1991) generally show that it is difficult to control the perception mode by experimental conditions in individual listeners. Some listeners have a strong inclination toward analytic perception behavior, others show a strong tendency toward holistic behavior, and still others show inconsistent behavior. Only group-averaged behavior—for instance, under a condition where holistic and analytic perception modes lead to opposite responses—shows some definite tendencies. The most important one is that, for complex tones with high-order harmonics, listeners' responses tend to divide about half-and-half into analytic and synthetic responses, whereas for tones with low-order harmonics analytic responses dominate. Lowering the harmonic order of a complex tone enhances both the holistic pitch percept, because of the dominant region effect (Ritsma, 1967), and the analytic pitch percept, because of the increased spectral resolution in the auditory periphery (see Chapter 5). Apparently, the effect of the latter is stronger than that of the former, at least with two-tone complexes. Much more systematic experimental evidence is required, however, before serious modeling attempts can be undertaken to describe the precise relationship between analytic and holistic pitch perception behavior.

IV. NONTONAL PITCH

A. Repetition Pitch

While visiting the French castle of Chantilly de la Cour, the Dutch physicist Christiaan Huygens noticed that the garden fountain, located in a vertical recess surrounded by marble steps, produced a noisy sound with a distinct musical pitch. Huygens described the pitch as corresponding with the sound of an open organ pipe of a length matching the depth of the stairs. Since that observation in 1693 it has been found that, in general, if an arbitrary sound $s(t)$ and its echo $s(t - T)$ are added together, a repetition pitch is heard that corresponds with a pure tone of frequency $1/T$. The sound $s(t)$ may be a simple click, a burst of white noise, a sample of speech, or just about any other broadband sound. The effect has been studied systematically for monotic and diotic conditions, where $s(t)$ and $s(t - T)$ go either to one ear or to both ears (Bilsen, 1968; Yost, Hill, & Perez-Falcon, 1978; Yost & Hill, 1978), and also for dichotic conditions, where the signals $s(t)$ and $s(t - T)$ go to different ears (Bilsen & Goldstein, 1974). Repetition pitch effects are typically found for delay times between 1 and 10 ms, yielding pitches varying from 100 to 1000 Hz. The effect is even stronger if there are many repeated echoes, for instance a signal $s(t) = x(t) + a_1x(t - T) + a_2x(t - 2T) + \ldots + a_nx(t - nT)$. Such signals are often referred to as *comb-filtered signals,* because the repeated echoes in the time domain cause more sharply defined maxima to occur in the spectrum at frequencies $f_n = n/T$ ($n = 0, 1, 2, 3, \ldots$). For this reason repetition pitch phenomena can, at least in principle, be accounted for by the same models used to describe the pitch of complex tones. Other models that are based on the interaural cross-correlation between similarly tuned channels have been developed by Blauert (1974) and Bilsen (1977).

The repetition pitch phenomenon can sometimes be used very creatively for special effects in electronic music. It can also be a nuisance. For instance, in a concert hall, a wrongly placed wall or other reflecting surface may cause at a particular seat an echo with a delay between 1 and 10 ms, producing an audible sound coloration.

B. Huggins Pitch and Edge Pitch

There are other conditions under which broadband noise can evoke a pitch sensation. One of these conditions is known in the literature as *Huggins pitch* (Cramer & Huggins, 1958). It arises if broadband noise signals are dichotically presented, identical in every respect except for an interaural phase shift over a small frequency region below 1500 Hz. A faint pitch is heard that appears to correspond to the center frequency of the phase shift region. The phenomenon is regarded as evidence of an interaural subtraction process such as, for instance, in the equalization and cancellation model of Durlach (1972) discussed in Chapter 10. The faint pitch is then attributed to a central narrow band of noise, which remains after the subtraction process.

If broadband noise is filtered, either highpass or lowpass, and if the filter's cutoff slope is sufficiently steep, a vague pitch is heard at or around the spectral edge (Small & Daniloff, 1967; Fastl, 1971). If the spectrum of a complex tone is abruptly terminated at some harmonic of high order, a much more pronounced pitch is evoked near the spectral edge frequency, which can even be comparable in salience to that of a pure tone if the phase spectrum is optimal (Kohlrausch & Houtsma, 1992). Although both pitch phenomena deal with sensations associated with some spectral discontinuity, they behave quantitatively in such different ways that they are probably based on entirely different auditory mechanisms (Kohlrausch & Houtsma, 1992).

The noise edge pitch described by Small and Daniloff and by Fastl can also be created dichotically. Klein and Hartmann (1981) described how, if both ears are stimulated with the same broadband noise signal, except for an interaural phase transition function that steps from 0 to 180° at a frequency f_c, two faint pitches are heard, one slightly below and the other slightly above the phase transition frequency f_c. Frijns, Raatgever, and Bilsen (1986) found a fairly unimodal distribution around f_c for this faint pitch. Binaural edge pitch is not very salient. Subjects can tell to some extent whether one sensation is higher or lower than another or match the binaural stimulus in pitch to a pure tone with some degree of accuracy. It has never been shown, however, that melodies or melodic pitch intervals can be recognized if the phase step frequency is given discrete values on a musical scale.

C. Pitch of Amplitude-Modulated Noise

If broadband noise is periodically gated or amplitude modulated by a sine wave, a pitchlike phenomenon associated with the envelope periodicity is observed. Miller and Taylor (1948) showed that subjects could discriminate noise-gating frequencies below 100 Hz with about the same precision as they could pure tones. This is shown in Figure 12. Beyond 100 Hz, the jnds for the interruption rate become much larger than pure-tone jnds, the whole interruption-pitch phenomenon fading away at about 300 Hz. Similar to binaural edge pitch, it is not entirely clear whether AM or interruption pitch is a true pitch phenomenon, especially in the musical sense. The fact that subjects can discriminate between high and low pitches, as in Miller and Taylor's experiment, suggests that the percept satisfies the official definition of pitch. Moreover, Burns and Viemeister (1976) have shown that subjects could identify seven known melodies played with AM noise in the modulation-frequency range of 100–200 Hz at a 90% correct level. The same authors found, however, just as Houtsma, Wicke, and Ordubadi (1980) did, that musically trained subjects were not able to score better than about 50% correct if asked to identify random melodic intervals that differed by semitone steps.

The fact that periodically modulated noise appears to evoke pitch sensations has traditionally been regarded as direct evidence that pitch is medi-

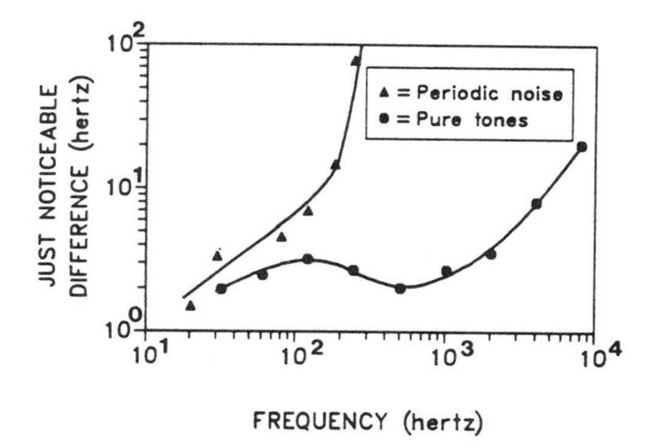

FIGURE 12 Just noticeable differences in the interruption rate of white noise (circles) and the frequency of a pure tone (triangles). (After Miller and Taylor, 1948.)

ated by temporal mechanisms in the auditory system, because the long-term average spectra of these signals are flat. Pierce, Lipes, and Cheetham (1975) have shown, however, that this argument does not necessarily hold, because short-term spectra do contain information about the modulation or interruption frequency. Houtsma et al. (1980) compared predictions by both temporal and short-term spectral models with measured pitch recognition scores and found that measured behavior generally supports the temporal view.

V. PITCH SCALES: RELATIVE AND ABSOLUTE

A. Relative Pitch

The role of pitch in music is based primarily on pitch relations and not on absolute pitch values. Sets of notes that people have used to make music throughout history, from the Greek tetrachords and Gregorian church modes to present-day major, minor, and chromatic scales, all have well-defined mutual relationships without the necessity for an absolute reference. When musicians play in an ensemble or sing together, it is usually sufficient that they tune their scales to one another, with the instruments that are most difficult to tune (such as piano or organ) being taken as the reference. The international convention to fix the fundamental frequency of the A_4 (the A in the fourth octave on a piano keyboard) at 440 Hz is only of rather recent origin, is typical only of Western music, and is still not endorsed by some of our major symphony orchestras.

Given the rather low priority obviously placed on absolute frequency standards in music, it will not be surprising that pitch perception is relation oriented rather than absolute. Many people, for instance, are able to sing a tune without knowing the key in which they are singing. Melodic steps, that is, sequential frequency ratios, can be produced with great accuracy without the necessity or even the awareness of any absolute reference. The first formal musical training children receive in school is the do-re-mi scale, which is relative: all tone steps relative to the note "do" are well defined, but "do" itself can be taken as any convenient frequency.

Formal musical pitch perception training, called *solfeggio,* consists mainly of strengthening and formalizing a natural ability to recognize, memorize, and reproduce certain sets of frequency ratio steps. Music students learn to associate names such as octave, fifth, or minor third with simultaneous or sequential sounds they hear, and they learn to sing melodic intervals from written music notation. In this way an *absolute* sense of *relative* pitch is developed, which is considered to be a standard skill of every professional musician.

Musical scales are in principle built from arbitrary frequency steps. Our modern diatonic and chromatic scales represent only a particular historical development in our Western culture. Scales used in other cultures often contain intervals quite different from those in Western tone scales. Some basic intervals, however, for instance the octave (ratio 2:1) and fifth (ratio 3:2), are found in many of the non-Western scales, probably because they occur between clearly audible elements (the first, second, and third harmonics) in natural periodic sounds. Especially if music is polyphonic or harmonic, which has been a characteristic of Western music since the twelfth century, the necessity of avoiding beats between partials forces one to choose melodic scale steps that are matched as well as possible to the frequency ratios of partials in instrumental or vocal sounds. The fact that a perfect match is mathematically impossible and therefore compromises must be made has led to the development of various tuning systems or *temperaments,* such as the Pythagorean, just or natural, mean-tone, and equal temperament. The introduction of alternative tone systems in this century, for instance, the quarter-tone system (Hába, 1927) or the 31-tone system (Fokker, 1949), have all met with limited success because of the unpleasant-sounding beats that occur between mistuned partials. Such beats do not occur with the tuning system proposed by Mathews and Pierce (1980), where the frequencies of partials of the sounds to be used are chosen to match the novel melodic tone steps obtained, for instance, by dividing an octave into 13 equal-ratio frequency steps. Such sounds may be difficult to find in the natural world, but can easily be synthesized with modern digital techniques.

A developed absolute sense of relative pitch can also be used for the psychoacoustical study of pitch perception. Requiring trained subjects to identify or reproduce aurally presented musical intervals or short tone sequences with experimental test sounds is a good alternative to the more conventional techniques of pitch matching or low–high discrimination, especially if it is not clear from the start that the sensation being studied is a real pitch phenomenon (Houtsma, 1984).

B. Pitch Contours in Speech

When we speak, our voice produces either periodic sounds for vowels and voiced consonants or noisy and aperiodic sounds for fricatives and unvoiced stop consonants. With voiced sounds our vocal cords vibrate at a rate f_0, which is therefore the fundamental of the vowel or voiced-consonant sound. During a spoken sentence the value of f_0 varies with time, forming a more or less continuous pattern. This so-called *intonation pattern* carries important prosodic information and follows very specific language-dependent rules (Hart, Collier, & Cohen, 1990).

One might wonder whether, from a perceptual viewpoint, pitch interval relationships are the same in running speech as they are in music. One might expect this to be the case because a spoken vowel is, in principle, no different from any other musical sound and the human voice is, after all, the most frequently used musical instrument. On the other hand, there is a clear categorical difference between f_0 contours in speech and in music. The former are always continuous, the latter almost always discrete and restricted to a limited number of f_0 values on the chosen musical scale.

Hermes and van Gestel (1991) have found evidence that pitch relations in speech and in music are perceptually different. They presented subjects with two "ma-*ma*-ma" utterances, each in a different octave range, with the middle syllable being accented by making an up–down f_0 sweep. One pitch accent was fixed by the experimenter, while the other could be adjusted by subjects to match the prominence of the accent. Analysis of the matched frequency excursions showed that equal accent prominence was not given by equal frequency or log-frequency excursions, but rather by equal excursions along an equivalent rectangular band (ERB) scale as discussed in Chapter 5. This implies that the prominence of a pitch accent in speech is determined by the number of critical bands the fundamental f_0 is swept through, a thought that was also the basis of the mel scale of Stevens and Zwicker discussed in Section II.A. An unresolved problem with this notion, however, is that the mel scale was intended for pure tones, whereas the pitch of most speech signals is mostly virtual, with very little energy in the fundamental frequency component.

C. Absolute Pitch

Absolute pitch refers to the ability of some people to identify musical sounds by their proper note name, or to name the key of a piece of music, without the use of any obvious external reference. Despite the rather large number of studies that have been devoted to this topic, information is still mostly empirical and sketchy, and our understanding of the phenomenon is still rather poor.

Perhaps the most systematic and comprehensive studies on the topic were done by Bachem (1937, 1940, 1954). Among people appearing to possess absolute pitch he distinguishes between genuine and acquired absolute pitch skills. Possessors of *genuine* absolute pitch typically make quick absolute identifications, accurate within a semitone, with octave confusions being the principal source of errors. *Acquired* skills are behaviorally characterized by slow judgments, as if subjects are trying to recall some learned reference like the A_4 for orchestra musicians or an extreme of the vocal range for singers. Given enough time, these subjects can make fairly accurate absolute pitch judgments, but if forced to respond quickly they will typically make large errors.

Bachem (1954) measured free-field pure-tone frequency jnds as a function of the temporal separation between tones which varied between one second and one week. Figure 13 shows jnds expressed as a percentage of the

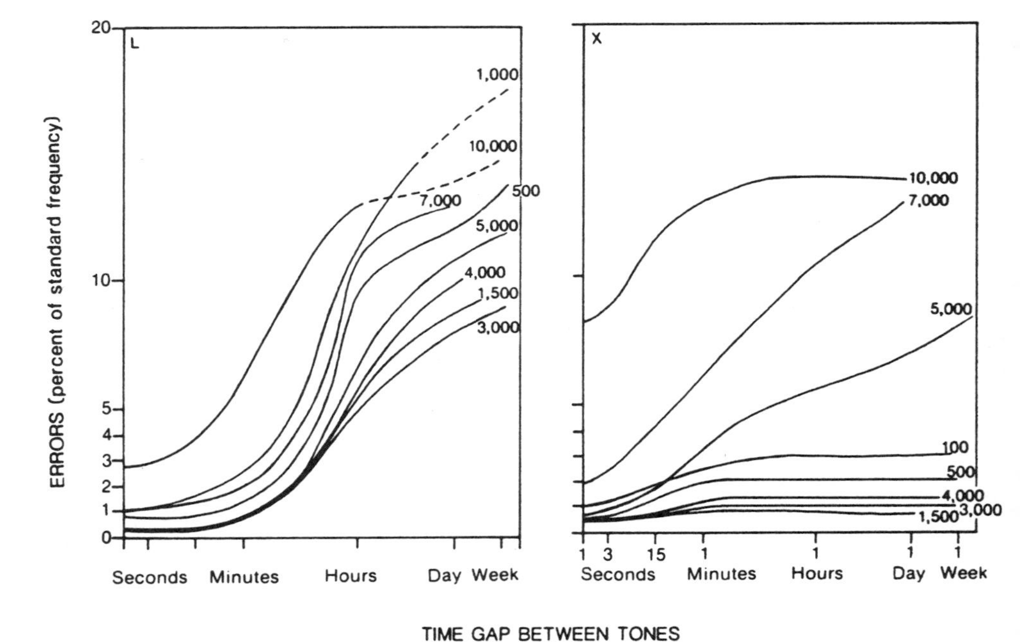

FIGURE 13 Just noticeable differences for pure-tone frequency as a function of the intertone time gap from two subjects. Duration of tones was 2 s and presentation was freefield. Subject *X* claimed absolute pitch, subject *L* did not. (Data from Bachem, 1954, reprinted with permission.)

tone's frequency of two subjects, one without (*L*) and the other with absolute pitch (*X*). Subject *L* shows a jnd that grows steadily with time, indicating a degrading memory trace for the pitch of the first note. Subject *X* shows a fairly constant jnd of about 3%, at least for frequencies below 5000 Hz, independent of how much time has lapsed between tones. This subject apparently labels the perceived pitches with some verbal code and ultimately compares the labels. Labeling is done by this subject with an accuracy of 3%.

It is still not known whether the observed behavioral differences between those who do and do not possess absolute pitch reflect actual physiological differences. There is some indication (Bachem, 1940) that absolute pitch requires an innate ability, combined with the right exposure during a critical development period at an early age.

VI. MULTIDIMENSIONAL ASPECTS OF PITCH

Up to this point the attribute pitch has been treated in this chapter as a one-dimensional entity. This seems to be in accordance with the ANSI (1973) definition, which describes pitch as a sensory attribute that enables ordering of sounds on a scale extending from low to high. It also appears to be consistent with the centuries-old practice of staff notation, where pitches of musical sounds are represented by the places of musical notes on a staff. Such a staff is actually nothing other than a visually convenient graphical representation of objects on a one-dimensional scale.

If one looks at musical practice and sees how pitch and pitch relationships are treated in music theory and composition analysis, however, one can hardly avoid drawing the conclusion that there must be more than a single dimension to the sensation of pitch. If, for instance, one takes the dimension that underlies conventional music notation as the only dimension of pitch, one has difficulty explaining why a C_5 sounds closer to a C_4 than an $F\sharp_4$. If one runs up or down a diatonic or chromatic scale, one clearly perceives a circularity, where in every octave pitches seem to repeat themselves in some sense. One thereafter might want a perceptual representation of pitch in which stimuli that are close together sound similar, and stimuli that are far apart sound dissimilar.

Octave similarity has formed the basis of the two-dimensional helix representation of pitch shown in Figure 14(a) (Revesz, 1954; Shepard, 1964). Going around one turn of the helix, one traverses the 12 chromatic pitches in an octave (C, C#, D, etc.). Completion of one turn brings one almost but not quite back to the place of origin. The dimension one varies by going around the helix is often called *pitch chroma,* and the axial distance one travels is referred to as *tone height*. The degree of octave similarity is represented by the amount of stretch in the helix along its axis. In the extreme case of a

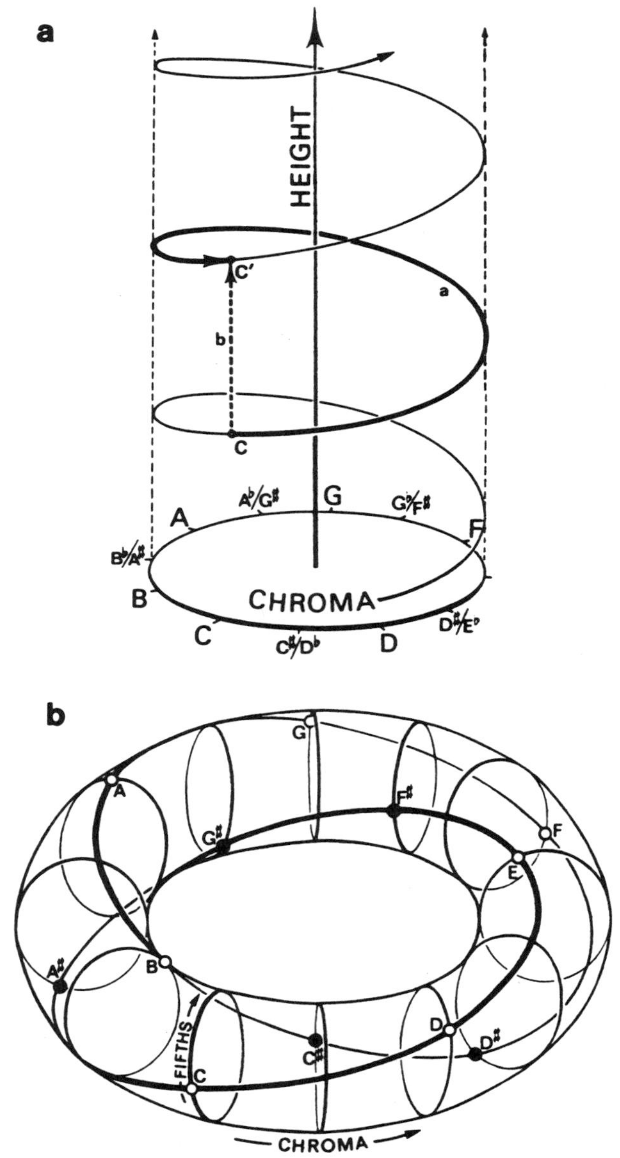

FIGURE 14 Multidimensional representations of pitch. (a) Simple helix representing pitch chroma and tone height. (b) Torus with double helix, representing the chroma circle and the circle of fifths. (From Shepard, 1982, reprinted with permission.)

completely stretched helix, we have no octave similarity and have, in fact, returned to the one-dimensional representation which was discussed earlier. The other extreme of the helix being compressed to a circle yields a pitch scale with octave identity. Shepard (1964) has shown how to make stimuli that have this sensory circular property. Stimuli moving around the circle evoke a sensation of an infinitely rising or falling pitch. These and similar stimuli have been used by composers as special sound effects and have occasionally been used as research tools in psychoacoustic experiments (Shepard, 1964; Allik, Dzhafarov, Houtsma, Ross, & Versfeld, 1989; Deutsch, 1991).

In addition to chromatic distance and octave similarity are other principles in music theory that determine proximity or distance between pitches. The circle of fifths, for instance, recognizes the close relationship between notes that are a fifth apart and have a tonic-dominant or tonic-subdominant relationship. Harmony in traditional Western music is based on this principle, and one can find it directly, for instance, in the physical layout of the bass keys of an accordion. If one combines the principles of the chroma circle and the circle of fifths, one obtains the pitch representation of a double helix wrapped around a torus, shown in Figure 14(b). These and other even more complex pitch representations can be found in a review chapter on this topic by Shepard (1982).

Many of the principles other than chromatic distance, particularly octave similarity and the circle of fifths, are of harmonic rather than melodic origin. They follow from frequency relationships that are found between overtones of natural periodic sounds such as the human voice, strings, or wind instruments. The relevance of such principles should therefore depend heavily on the spectral composition of sounds used to evoke pitch sensations. A multidimensional pitch space for harmonic sounds may be very different from a pitch space for sounds like church bells (Houtsma & Tholen, 1987) or sounds with stretched partials (Mathews & Pierce, 1980). There is therefore good reason to have doubts about the general validity of any abstract theory of multidimensional pitch space that does not deal with the issue of orchestration.

References

Allik, J., Dzhafarov, E. N., Houtsma, A. J. M., Ross, J., & Versfeld, N. J. (1989). Pitch motion with random chord sequences. *Perception and Psychophysics, 46,* 513–527.

American National Standards Institute. (1973). American national psychoacoustical terminology. S3.20. New York: Author.

Bachem, A. (1937). Various types of absolute pitch. *Journal of the Acoustical Society of America, 9,* 146–151.

Bachem, A. (1940). The genesis of absolute pitch. *Journal of the Acoustical Society of America, 11,* 434–439.

Bachem, A. (1954). Time factors in relative and absolute pitch determination. *Journal of the Acoustical Society of America, 26,* 751–753.
Beerends, J. G. (1989). Pitches of simultaneous complex tones. Ph.D. dissertation, Eindhoven University of Technology.
Beerends, J. G., & Houtsma, A. J. M. (1986). Pitch identification of simultaneous dichotic two-tone complexes. *Journal of the Acoustical Society of America, 80,* 1048–1056.
Beerends, J. G., & Houtsma, A. J. M. (1989). Pitch identification of simultaneous diotic and dichotic two-tone complexes. *Journal of the Acoustical Society of America, 85,* 813–819.
Bilsen, F. A. (1968). On the interaction of a sound with its repetitions. Ph.D. thesis, Delft, Waltman.
Bilsen, F. A. (1977). Pitch of noise signals: Evidence for a "central spectrum." *Journal of the Acoustical Society of America, 61,* 150–161.
Bilsen, F. A., & Goldstein, J. L. (1974). Pitch of dichotically delayed noise and its possible spectral basis. *Journal of the Acoustical Society of America, 55,* 292–296.
Blauert, J. (1974). *Räumliches Hören.* Stuttgart: Hirzel Verlag.
Burns, E. M., & Viemeister, N. F. (1976). Nonspectral pitch. *Journal of the Acoustical Society of America, 60,* 863–869.
Brink, G. van den (1970). Experiments on binaural diplacusis. In R. Plomp & G. F. Smoorenburg (Eds.), *Frequency analysis and periodicity detection in hearing* (pp. 362–374). Leiden: Sijthoff.
Cramer, E. M., & Huggins, W. H. (1958). Creation of pitch through binaural interaction. *Journal of the Acoustical Society of America, 30,* 413–417.
Deutsch, D. (1991). Pitch proximity in the grouping of simultaneous tones. *Music Perception, 9,* 185–198.
Durlach, N. I. (1972). Binaural signal detection: equalization and cancellation theory. In J. V. Tobias (Ed.), *Foundations of modern auditory theory* (Vol. 2, pp. 369–462). New York: Academic Press.
Fastl, H. (1971). Über Tonhöhenempfindungen bei Rauschen. *Acustica, 25,* 350–354.
Fletcher, H. (1924). The physical criterion for determining the pitch of musical tone. *Physical Review, 23,* 427–437.
Fokker, A. D. (1949). *Just intonation.* The Hague: M. Nijhoff.
Frijns, J. H. M., Raatgever, J., & Bilsen, F. A. (1986). A central spectrum theory of binaural processing. The binaural edge pitch revisited. *Journal of the Acoustical Society of America, 80,* 442–451.
Goldstein, J. L. (1973). An optimum processor theory for the central formation of the pitch of complex tones. *Journal of the Acoustical Society of America, 54,* 1496–1516.
Hába, A. (1927). *Neue Harmonielehre des diatonischen, chromatischen, Viertel-, Drittel-, Sechstel- und Zwölfteltonsystems.* Leipzig: Kirstner and Siegel.
Hart, J.'t, Collier, R., & Cohen, A. (1990). *A perceptual study of intonation.* Cambridge: Cambridge University Press.
Hartmann, W. M. (1978). The effect of amplitude envelope on the pitch of sinewave tones. *Journal of the Acoustical Society of America, 63,* 1105–1113.
Helmholtz, H. L. F. von (1863). *Die Lehre von den Tonempfindungen als physiologische Grundlage für die Theorie der Musik,* Braunschweig: F. Vieweg & Sohn. (English trans. A. J. Ellis; New York: Dover, 1954).
Hermes, D. J., & van Gestel, J. C. (1991). The frequency scale of speech intonation. *Journal of the Acoustical Society of America, 90,* 97–102.
Hoekstra, A. (1979). Frequency discrimination and frequency analysis in hearing. Ph.D. thesis, University of Groningen.
Horst, J. W., Javel, E., & Farley, G. R. (1986). Coding of spectral fine structure in the auditory

nerve. I. Fourier analysis of period and interspike interval histograms. *Journal of the Acoustical Society of America, 79,* 398–416.

Houtsma, A. J. M. (1984). Pitch salience of various complex sounds. *Music Perception, 1,* 296–307.

Houtsma, A. J. M., & Fleuren, J. F. M. (1991). Analytic and synthetic pitch of two-tone complexes. *Journal of the Acoustical Society of America, 90,* 1674–1676.

Houtsma, A. J. M., & Goldstein, J. L. (1972). The central origin of the pitch of complex tones: Evidence from musical interval recognition. *Journal of the Acoustical Society of America, 51,* 520–529.

Houtsma, A. J. M., & Smurzynski, J. (1990). Pitch identification and discrimination for complex tones with many harmonics. *Journal of the Acoustical Society of America, 87,* 304–310.

Houtsma, A. J. M., & Tholen, H. J. G. M. (1987). A carillon of major-third bells: II. A perceptual evaluation. *Music Perception, 4,* 255–266.

Houtsma, A. J. M., Wicke, R. W., & Ordubadi, A. (1980). Pitch of amplitude-modulated low-pass noise and predictions by temporal and spectral theories. *Journal of the Acoustical Society of America, 67,* 1312–1322.

Kim, D. O., Rhode, W. S., & Greenberg, S. R. (1986). Responses of cochlear nucleus neurons to speech signals: Neural encoding of pitch, intensity and other parameters. In B. C. J. Moore & R. D. Patterson (Eds.), *Auditory frequency selectivity* (pp. 281–288). London: Plenum Press.

Klein, M. A., & Hartmann, W. M. (1981). Binaural edge pitch. *Journal of the Acoustical Society of America, 70,* 51–61.

Kohlrausch, A., & Houtsma, A. J. M. (1992). Pitch related to spectral edges of broadband signals. *Philosophical Transactions of the Royal Society of London, 336* Part B, 375–382.

Licklider, J. C. R. (1951). A duplex theory of pitch perception. *Experientia, 7,* 128–133.

Mathews, M. V., & Pierce, J. R. (1980). Harmony and nonharmonic partials. *Journal of the Acoustical Society of America, 68,* 1252–1257.

Meddis, R. (1988). Simulation of auditory-neural transduction: Further studies. *Journal of the Acoustical Society of America, 83,* 1056–1063.

Meddis, R., & Hewitt, M. J. (1991a). Virtual pitch and phase sensitivity of a computer model of the auditory periphery. I. Pitch identification. *Journal of the Acoustical Society of America, 89,* 2866–2882.

Meddis, R., & Hewitt, M. J. (1991b). Virtual pitch and phase sensitivity of a computer model of the auditory periphery. II. Phase sensitivity. *Journal of the Acoustical Society of America, 89,* 2883–2894.

Miller, G. A., & Taylor, W. G. (1948). The perception of repeated bursts of noise. *Journal of the Acoustical Society of America, 20,* 171–182.

Moore, B. C. J. (1973). Frequency difference limens for short-duration tones. *Journal of the Acoustical Society of America, 54,* 610–619.

Moore, B. C. J., & Peters, R. W. (1992). Pitch discrimination and phase sensitivity in young and elderly subjects and its relationship to frequency selectivity. *Journal of the Acoustical Society of America, 91,* 2881–2893.

Moore, B. C. J., & Rosen, S. M. (1979). Tune recognition with reduced pitch and interval information. *Quarterly Journal of Experimental Psychology, 31,* 229–240.

Morgan, C. T., Garner, W. R., & Galambos, R. (1951). Pitch and intensity. *Journal of the Acoustical Society of America, 23,* 658–663.

Ohm, G. W. (1843). Über die Definition des Tones nebst daran geknupfter Theorie der Sirene und ähnlicher tonbildender Vorrichtungen. *Annalen für Physik und Chemie, 59,* 513–565.

Pierce, J. R., Lipes, R., & Cheetham, C. (1975). Uncertainty concerning the direct use of time

information in hearing: Place clues in white-spectra stimuli. *Journal of the Acoustical Society of America, 61,* 1609–1621.
Plomp, R. (1967). Pitch of complex tones. *Journal of the Acoustical Society of America, 41,* 1526–1533.
Revesz, G. (1954). *Introduction to the psychology of music.* Norman: University of Oklahoma Press.
Ritsma, R. J. (1962). The existence region of the tonal residue I. *Journal of the Acoustical Society of America, 34,* 1224–1229.
Ritsma, R. J. (1967). Frequencies dominant in the perception of the pitch of complex sounds. *Journal of the Acoustical Society of America, 42,* 191–198.
Rossing, T. D. and Houtsma, A. J. M. (1986). Effects of signal envelope on the pitch of short sinusoidal tones. *Journal of the Acoustical Society of America, 79,* 1926–1933.
Schouten, J. F. (1938). The perception of subjective tones. *Proceedings of the Koninklijke Akademie van Wetenschap, 41,* 1086–1093.
Schouten, J. F. (1940). The residue and the mechanism of hearing. *Proceedings of the Koninklijke Akademie van Wetenschap, 43,* 991–999.
Schouten, J. F., Ritsma, R. J., & Cardozo, B. L. (1962). Pitch of the residue. *Journal of the Acoustical Society of America, 34,* 1418–1424.
Schröder, M. R. (1970). Synthesis of low-peak-factor signals and binary sequences with low autocorrelation. *IEEE Transactions on Information Theory, IT-16,* 85–89.
Seebeck, A. (1841). Beobachtungen über einige Bedingungen der Entstehung von Tönen. *Annalen für Physik und Chemie, 53,* 417–436.
Shepard, R. N. (1964). Circularity in judgments of relative pitch. *Journal of the Acoustical Society of America, 36,* 2346–2353.
Shepard, R. N. (1982). Structural representations of musical pitch. In D. Deutsch (Ed.), *The psychology of music* (pp. 343–390). New York: Academic Press.
Shower, E. G., & Biddulph, R. (1931). Differential pitch sensitivity of the ear. *Journal of the Acoustical Society of America, 3,* 275–287.
Siebert, W. M. (1970). Frequency discrimination in the auditory system: Place or periodicity mechanisms. *Proceedings of the IEEE, 58,* 723–730.
Small, A. M., Jr., & Daniloff, R. G. (1967). Pitch of noise bands. *Journal of the Acoustical Society of America, 41,* 506–512.
Smoorenburg, G. F. (1970). Pitch perception of two-frequency stimuli. *Journal of the Acoustical Society of America, 48,* 924–941.
Srulovicz, P., & Goldstein, J. L. (1983). A central spectrum model: A synthesis of auditory-nerve timing and place cues in monaural communication of frequency spectrum. *Journal of the Acoustical Society of America, 73,* 1266–1276.
Stevens, S. S. (1935). The relation of pitch to intensity. *Journal of the Acoustical Society of America, 6,* 150–154.
Stevens, S. S., Volkmann, J., & Newman, E. B. (1937). A scale for the measurement of the psychological magnitude of pitch. *Journal of the Acoustical Society of America, 8,* 185–190.
Terhardt, E. (1972). Zur Tonhöhewahrnehmung von Klängen II: Ein Funktionsschema. *Acustica, 26,* 187–199.
Terhardt, E. (1974a). Pitch of pure tones: Its relation to intensity. In E. Zwicker and E. Terhardt (Eds.), *Facts and models in hearing* (pp. 353–360). Stuttgart: Springer Verlag.
Terhardt, E. (1974b). Pitch, consonance and harmony. *Journal of the Acoustical Society of America, 55,* 1061–1069.
Terhardt, E. (1979). Calculating virtual pitch. *Hearing Research, 1,* 155–182.
Terhardt, E., & Fastl, H. (1971). Zum Einfluss von Störtönen und Störgeräuschen auf die Tonhöhe von Sinustönen. *Acustica, 25,* 53–61.

Verschuure, J., & van Meeteren, A. A. (1975). The effect of intensity on pitch. *Acustica, 32,* 33–44.

Wier, C. C., Jesteadt, W., & Green, D. M. (1977). Frequency discrimination as a function of frequency and sensation level. *Journal of the Acoustical Society of America, 61,* 178–184.

Wightman, F. L. (1973). The pattern-transformation model of pitch. *Journal of the Acoustical Society of America, 54,* 407–416.

Yost, W. A., & Hill, R. (1978). Strength of the pitches associated with ripple noise. *Journal of the Acoustical Society of America, 64,* 485–492.

Yost, W. A., Hill, R., & Perez-Falcon, T. (1978). Pitch and pitch discrimination of broadband signals with rippled power spectra. *Journal of the Acoustical Society of America, 63,* 1166–1173.

Zwicker, E., & Feldtkeller, R. (1967). *Das Ohr als Nachrichtenempfänger.* Stuttgart: S. Hirzel Verlag.

CHAPTER 9

Spatial Hearing and Related Phenomena

D. Wesley Grantham

I. INTRODUCTION

A normal-hearing person has an immediate appreciation of auditory space in the sense that orientation toward acoustic events is natural, rapid, and in general, accurate. Although spatial acuity is poorer by up to two orders of magnitude in the auditory than in the visual domain, the auditory world has the advantage of extending in *all* directions around the observer, while the visual world is restricted to frontal regions. Perrott, Saberi, Brown, and Strybel (1990) have suggested that this differential perceptual geometry enables the auditory system to react first to environmental events and to "point" the observer's head toward events for more refined spatial analysis by the visual system. Of course, in addition to its role in directing the visual system, the auditory system is a sophisticated spatial processor in its own right, allowing organisms to detect and monitor the positions of auditory objects in three dimensions as well as facilitating the identification of these objects. This chapter reviews basic and new experimental findings related to human spatial hearing.

II. BINAURAL PROCESSING

Binaural processing refers in a broad sense to the functions underlying *any* human capabilities that are rendered possible or superior by the use of two

Hearing

ears rather than one. Such capabilities include sound source localization in three dimensions, speech identification in noisy environments, and loudness judgments, as well as performance in headphone tasks in which the signals to the two ears are independently manipulated. Here the term *binaural processing* is used in the more narrow sense to refer only to the last of these capabilities; that is, the processing underlying *headphone* tasks, in which optimum performance requires the use of both ears.

A. Lateralization

A stereo recording heard through headphones generally gives the impression that (1) the sound appears to come from inside the head, and (2) there is a definite sense of space, at least along the dimension stretching between the ears. Thus, a violin might appear to be at the left ear, while simultaneously a singer's voice might appear between the center of the head and the right ear. The within-the-head perception of such headphone-presented sound is refered to as *internalization* of the sound image(s) (Plenge, 1974); the differential placement of sound images along the imaginary line between the ears is referred to as *lateralization* of the images. By contrast, when a sound is presented from a loudspeaker, the perceived sound is generally described as being *externalized* (outside the head), and the description of its subjective position is a task of *localization*.

A sound presented identically to the two ears through headphones (a *diotic* stimulus) is usually lateralized in the center of the head. The image can be made to move toward the right ear in two different ways: (1) by introducing a time delay to the left ear's input, or (2) by making the right-ear signal more intense than the left-ear signal. For each of these two cues for lateralization, two questions have intrigued psychoacousticians. First, how does the lateral position of a sound image depend upon the magnitude of the interaural difference? Second, what is the *threshold* interaural difference? The first question involves identification: The subject must indicate *where* the sound image is lateralized. The second question involves *resolution:* the experimenter determines how *small* an interaural difference can be detected. Both questions are discussed here.

1. Identification Tasks

a. Lateral Position as a Function of Interaural Temporal Difference (ITD)

Yost (1981) had subjects judge the lateral position of tone bursts as a function of the interaural delay imposed. The subjects were asked to indicate the intracranial position of the image using a slide potentiometer for which the endpoints denoted the maximum lateral positions (the subject's ears). Selected results from Yost's study are shown in Figure 1, where image position is plotted for a 0.5 kHz tone as a function of interaural phase difference

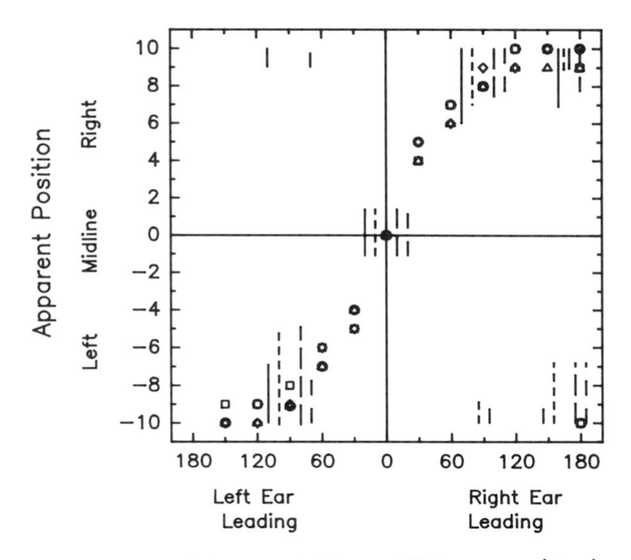

FIGURE 1 Lateral position of a 0.5 kHz tone as a function of interaural phase difference. The rating scale (ordinate) corresponds to the markings on a slide potentiometer that subjects used to indicate intracranial lateral position. The different symbols represent the median responses of four different subjects; vertical lines indicate the range of responses at selected values of IPD. For IPD = 180°, a bimodal distribution is observed, with indicated lateralizations on both sides of the subjects' heads. Note that, for a 0.5 kHz tone, IPD = 180° corresponds to an ITD of 1000 μs. (From Yost, 1981.)

(IPD) [for a pure tone, IPD (in degrees) = $360 \cdot f$(ITD), where f is the frequency of the tone, and ITD is in seconds]. As shown in the figure, the image position moved nearly linearly toward the leading ear as IPD was increased from 0° to 60°. As IPD was increased further (from 60°–120°), the image continued shifting toward the leading ear, but more gradually; finally, as the IPD approached the ambiguous value of 180°, the stimulus began to elicit responses toward the *opposite* (lagging) ear. Yost also found that the effect of IPD on lateral position decreased as stimulus frequency increased; for tones above 1.5 kHz there is little if any effect of ITD on perceived laterality (Schiano, Trahiotis, & Bernstein, 1986; see Yost & Hafter, 1987, for an extensive review of studies that have measured lateralization of sinusoids.)

b. Lateral Position as a Function of Interaural Level Difference (ILD)

Yost (1981) also measured lateralization as a function of ILD. Selected data are shown in Figure 2; about 75–80% of the maximum lateral position was reached for ILDs of 10–12 dB, more or less independent of stimulus frequency or intensity. As with the dependence on ITD, lateral position is a

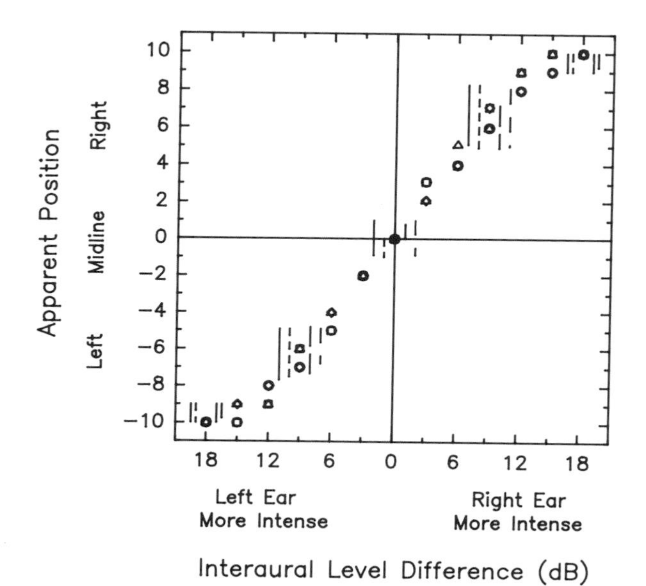

FIGURE 2 Lateral position of a 0.2 kHz tone as a function of interaural level difference. The different symbols represent the median responses of four different subjects; vertical lines indicate the range of responses at selected values of ILD. (From Yost, 1981.)

negatively accelerating function of ILD, although it appears that there is a larger linear segment in the case of ILD variation.

2. Resolution Tasks

In a typical task measuring *threshold* ITD or ILD, a subject is presented with two successive sounds: a stimulus with a small interaural difference favoring one ear, and after a pause, a stimulus with the same interaural difference favoring the other ear. The subject's task is to indicate whether the image moved left–right or right–left. *Threshold ILD* or *ITD* is defined as that magnitude of the interaural difference leading to a criterion level of performance in this type of task, such as 75% correct.

a. ITD Thresholds

Klumpp and Eady (1956) determined ITD thresholds for several types of stimuli. Some of their data are shown in Figure 3, which displays threshold ITD as a function of the (center) frequency of the stimulus. For sinusoids (open circles) with frequencies up to 1.0 kHz, threshold ITD decreases as a function of frequency, reaching a minimum of 11 μs at 1.0 kHz. Their data indicate that the binaural system is exquisitely sensitive to small time differences between the inputs to the two ears. For sinusoids above 1.0 kHz, ITD

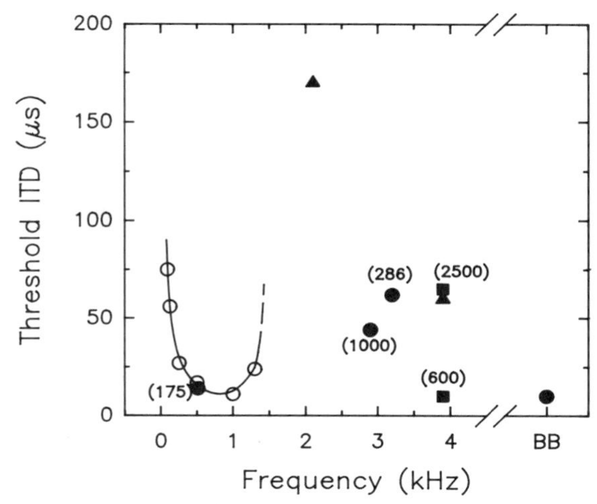

FIGURE 3 Threshold ITD as a function of center frequency for various stimuli: ○, pure tones (Klumpp & Eady, 1956); ●, noise bands (Klumpp & Eady, 1956); ■, noise bands (Henning, 1974b); ▲, sinusoidally amplitude-modulated tones, modulated at 300 Hz (Henning, 1974a). Noise bands were centered at the frequencies shown on the abscissa; numbers shown in parentheses indicate bandwidth. BB is broadband noise.

thresholds start to increase again; and for sinusoids above approximately 1.5 kHz, it is no longer possible to measure a threshold.

Although the binaural system is insensitive to ITDs in high-frequency *sinusoids,* ITD thresholds have been measured for complex tones and narrowband noises whose frequency content is restricted to high frequencies. As indicated by the filled symbols in Figure 3, ITD thresholds for these stimuli can in some cases be as low as those obtained with sinusoids. Thus, although the binaural system is evidently unable to use the cycle-by-cycle timing information in sinusoids whose frequency exceeds about 1.5 kHz (see Chapter 3), it can respond to the timing information carried by the relatively slowly fluctuating *envelopes* of high-frequency stimuli.

b. ILD Thresholds

The ILD threshold is generally between 0.5 and 1.0 dB for most types of stimuli (see Grantham, 1984b; and Yost & Hafter, 1987, for reviews of this work). This performance again indicates that the binaural system is very sensitive to small differences between the inputs to the two ears.

3. Interaction of ILDs and ITDs

Perhaps the prime motivation for the study of human sensitivity to ITDs and ILDs is that these interaural differences are generally considered to be the primary cues underlying our ability to localize sounds in the horizontal plane (see Section III.A). Of course, sounds in the free field do not produce ITDs and ILDs in isolation; therefore, it is of interest to study the interaction of these cues. This has been done either by setting the two cues in opposition to explore the degree of cancellation produced (the so-called trading-ratio experiments—e.g., G. G. Harris, 1960) or by measuring lateralization as a joint function of ILD and ITD (e.g., Domnitz & Colburn, 1977; Whitworth & Jeffress, 1961). These investigations have shown that ITD-based lateralization and ILD-based lateralization are tradable or additive to some extent. Based on this additivity, some early investigators (e.g., Jeffress, 1948) hypothesized that there is actually only *one* mechanism in the binaural system that responds to interaural differences—an interaural time-difference comparator—and that sensitivity to ILDs occurs because neural impulses are evoked with a shorter latency (i.e., earlier in time) as intensity increases.

Several lines of evidence argue against a strict interpretation of the latency hypothesis, not the least of which is that combinations of large ITDs and ILDs can lead to a perception of *two* images, which are differentially affected by ITD and ILD manipulations (e.g., Hafter & Jeffress, 1968). More recently, it has been shown that "unnatural" combinations of ILD and ITD (i.e., those that cannot normally occur in free field) lead to both greater variability in lateralization responses and to lower ratings of "naturalness" than stimuli with ITD-ILD combinations that do occur in nature (Gaik, 1993). Thus, a strict latency hypothesis can be rejected; it is generally believed that separate ILD and ITD mechanisms exist in the binaural system (e.g., Stern & Colburn, 1978; see Chapter 10).

On the other hand, it is clear that *some* degree of ILD-to-ITD conversion does occur in the nervous system and that lateralization mediated by the ITD-sensitive system is thus affected by both ITDs and ILDs in the stimulus. Furthermore, it has been suggested that the latency hypothesis may, in fact, apply in its more strict sense to low-frequency stimuli (Grantham, 1984b; Hafter, 1984). There is also evidence from a different type of experiment (Hafter, Dye, Wenzel, & Knecht, 1990) that ILDs and ITDs are coded in a common channel, at least for near-threshold magnitudes. Thus, it is possible that a separate ILD processor is invoked only for middle- to high-frequency stimuli with moderate- to large-magnitude interaural differences.

4. Interaural Correlation Discrimination

The stimuli described so far result in compact images in the head. Such compact images are generally produced when the stimuli to the two ears are *correlated* (here interaural *correlation* is loosely defined as the point-by-point correlation coefficient computed for a stimulus segment after an appropriate delay is imposed on one of the inputs to maximize the correlation). If an interaurally *un*correlated signal is presented (for example, by presenting the outputs of two independent noise generators to the two ears), the resulting image is no longer compact and lateralizable to a single position, but rather is diffuse and fills the entire head. Stimuli with varying degrees of interaural correlation can be created by appropriate mixtures of noise generators that are presented to the two ears either in common or independently (Pollack & Trittipoe, 1959).

Thresholds for interaural correlation discrimination have been determined by having subjects discriminate the diffuseness or compactness of images produced by presenting noise stimuli with varying degrees of interaural correlation. When the reference interaural correlation is 0.0, thresholds for a third-octave noise band centered at 0.5 kHz are 0.3–0.4 (Gabriel & Colburn, 1981; Grantham, 1982; Pollack & Trittipoe, 1959). That is, starting from a maximally diffuse image, an interaural correlation of 0.3–0.4 is required before the subject can detect that the diffuseness has decreased. For a reference correlation of 1.0, the threshold interaural correlation difference is about 0.02 (Gabriel & Colburn, 1981); in other words, a noise with an interaural correlation of 0.98 is discriminably less compact than the reference (diotic) noise. Thus, the binaural system is exquisitely sensitive to the decorrelation of a noise as well as to the ITD and ILD of low-frequency or wideband stimuli.

B. Binaural Detection

1. Historical Overview

The *detection* of signals under binaural conditions has been extensively studied. In the prototypical experiment, the subject is asked to detect a low-frequency (e.g., 0.5 kHz) tonal signal in the presence of a broadband masking noise. In a reference condition, the threshold is determined with signal and masker presented to only one ear. Of interest in the binaural detection experiments is how the threshold changes when the masker or signal is presented to *both* ears in various configurations.

When signal and masker are both identical at the two ears, there is *no* change in threshold relative to monaural presentation (e.g., Sever & Small, 1979). Similarly, if an interaural manipulation (say a 0.5 ms delay to the left ear) is applied to both the masker and the signal, the threshold does not change (Jeffress, Blodgett, & Deatherage, 1952); in these cases, detectability is the same when using two ears as when using one. However, if *different* interaural manipulations are applied to the masker and signal, there is often a reduction in signal threshold compared to the monaural case. For example, if the 0.5 kHz signal is interaurally delayed by 0.5 ms (IPD = 90°), while the noise is left identical in the two ears, about 10 dB less signal is required for detection than in the monaural case (Jeffress, Blodgett, & Deatherage, 1952); in other words, there is a "binaural release from masking" (also called a *masking-level difference,* or MLD) of 10 dB.

Hirsh (1948) investigated the effects on binaural detection of several variables, including signal frequency, masker level, and masker and signal interaural configuration. He found, as have many investigators since, that the maximum binaural release from masking occurred when the masker was identical in the two ears while the signal was phase-shifted by 180° (π radians) in one ear relative to the other (this configuration is known as *N*0-*S*π). Some of Hirsh's data are presented in Figure 4, which shows the signal level required for threshold in the presence of a broadband noise as a function of signal frequency. The upper curve shows that threshold in the reference diotic condition (*N*0-*S*0) does not change much as frequency is increased from 0.1 to 5.0 kHz. The threshold in the *N*0-*S*π configuration (lower curve) is always below that in the diotic condition, indicating the advantage realized when interaural differences are introduced. The maximum MLD—12–15 dB—occurs for signal frequencies of 0.2 and 0.5 kHz.

Since Hirsh's investigation, many aspects of binaural detection have been investigated. Variables that have been manipulated include the type of masker and type of signal (noise, tones, clicks, and speech have all been employed), the frequency content and bandwidth of the masker (Wightman, 1971), duration and relative temporal positions of masker and signal (Kohl-

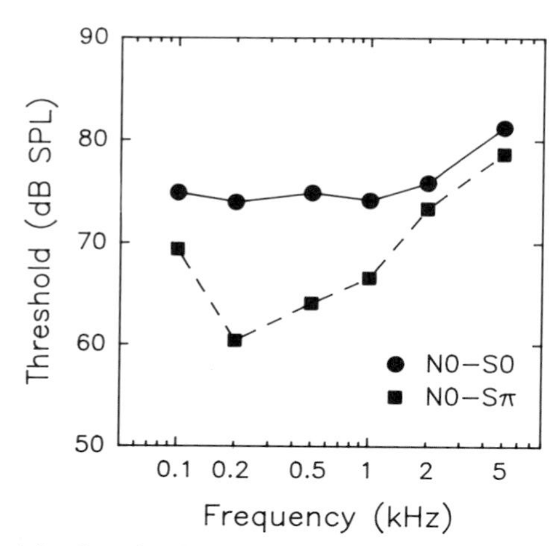

FIGURE 4 Signal level at threshold in the presence of wideband masking noise as a function of signal frequency. *N*0-*S*0, noise and signal are identical at the two ears, *N*0-*S*π, noise identical at the two ears, signal phase-inverted in one ear relative to the other. (Data from Hirsh, 1948.)

rausch, 1986), and masker interaural parameters, including ITD (Langford & Jeffress, 1964), ILD (McFadden, 1968), IPD (Jeffress et al., 1952), and interaural correlation (Robinson & Jeffress, 1963). In general, any interaural manipulation that is differentially applied to masker and signal may lead to an MLD. As shown in Figure 4, the maximum advantage is generally revealed at low signal frequencies; however, certain conditions have been found in which a large MLD is found at high frequencies as well (e.g., McFadden & Pasanen, 1974). Excellent reviews of the binaural detection literature are given in Durlach (1972) and Durlach and Colburn (1978).

2. The Relationship between Binaural Detection and Interaural Correlation Discrimination

As described in Subsection II.A.4, small decreases in interaural correlation can be detected: practiced observers can discriminate a diotic noise (interaural correlation = 1.0) from a noise that has been decorrelated to a value of 0.98. When an $S\pi$ signal is added to a $N0$ masker, the result is a slight decorrelation of the entire binaural stimulus. Several investigators have asked whether binaural detection in the tasks described previously might be based on a subject's detection of the change in interaural correlation when the signal is added to the noise.

Durlach, Gabriel, Colburn, and Trahiotis (1986) showed that the interaural correlation, ρ, for the stimulus produced with an $N0$–$S\pi$ configuration is

$$\rho = (1 - S/N)/(1 + S/N), \tag{1}$$

where S/N is the signal to noise ratio (power units). To investigate the relation between detection and discrimination, Koehnke, Colburn, and Durlach (1986) compared performance in two different experiments. In one, subjects were required to detect an $S\pi$ signal, either 0.5 kHz or 4.0 kHz, in the presence of an $N0$ noise, one-third octave wide and centered at the signal frequency. In the other, the same subjects had to discriminate a diotic noise (again one-third octave wide, centered at either 0.5 or 4.0 kHz) from the same noise with a given amount of decorrelation introduced. As shown in Figure 5, performance at both frequencies was almost identical in the two tasks when the stimulus measures were related according to Eq. (1). They concluded that binaural detection under these narrowband masker conditions was closely related to the binaural system's ability to discriminate interaural correlation.

More recent experiments by Bernstein and Trahiotis (1992b) and Jain, Gallagher, Koehnke, and Colburn (1991) confirm that, in many cases, binaural signal detection can be predicted by performance in interaural correlation discrimination tasks. The implication is that the salient and relevant cue

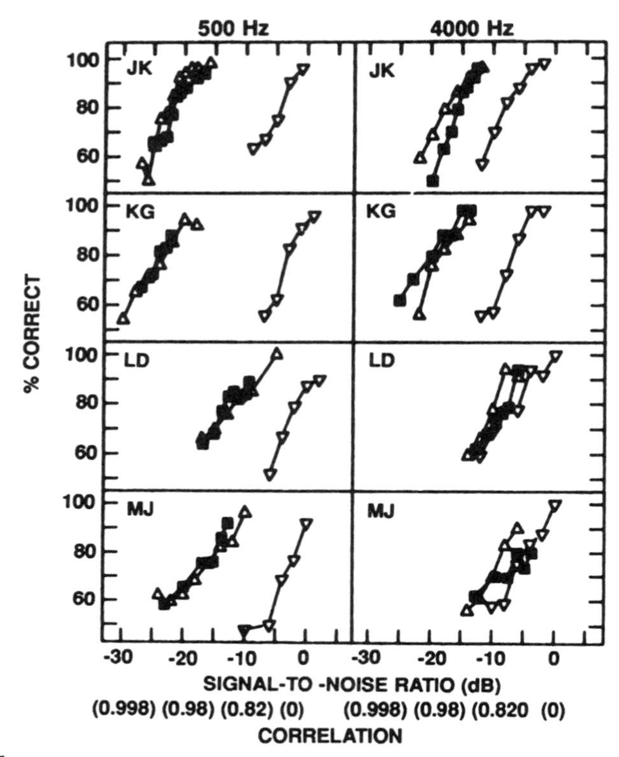

FIGURE 5 Comparison of interaural correlation discrimination and binaural signal detection. Data are shown for four subjects (the four rows) and for signal and masker frequencies centered at 0.5 kHz (left panels) and 4.0 kHz (right panels). ■, performance in interaural correlation (IAC) discrimination task as a function of the IAC of the comparison stimulus, indicated by the lower labels on the abscissa; △, performance in *N*0-*S*π signal detection task as a function of signal-to-noise (S/N) ratio, indicated by the upper labels on the abscissa. IAC and S/N are related according to Eq. (1) in text. (▽ is performance in the *N*0-*S*0 signal detection task.) (From Koehnke et al., 1986.)

in binaural detection tasks is the change in interaural correlation. However, not all binaural detection data can be explained in this way. Results obtained by Bernstein and Trahiotis (1992b), Durlach et al. (1986), and Jain et al. (1991) suggest a complex interplay between interaural correlation, spectral content, and the bandwidth of the stimulus. These interactions limit any blanket generalization one can make about the relationship between discrimination and detection.

C. Temporal Effects in Binaural Processing

Two separate aspects of binaural temporal processing have received attention in recent years. *Binaural adaptation* refers to the failure of the system, in a lateralization task, to optimally use portions of a binaural stimulus following its onset; that is, for certain ongoing stimuli, later portions of the stimulus contribute less than its onset to the lateralization threshold. *Binaural sluggishness* refers to the relative difficulty subjects have in detecting or discriminating dynamic *changes* in interaural differences. Both aspects will be discussed in this section.

1. Binaural Adaptation

If the binaural system were a perfect temporal integrator, threshold ITD should decrease as the square root of stimulus duration (e.g., Yost & Hafter, 1987; see also Chapter 6). In studies that have measured ITD thresholds as a function of stimulus duration it has generally been found that the improvement in threshold is *less* than predicted by a perfect integrator. For example, ITD thresholds from Tobias and Zerlin (1959), when plotted on a log–log plot, have a slope of −0.3 rather than the predicted −0.5. Yost and Hafter summarized the results of several other studies that measured ITD thresholds for both low-frequency and high-frequency stimuli as a function of stimulus duration; in most cases, integration was not optimal, indicating that information following the stimulus onset was contributing less than the onset.

To learn more about this failure of perfect temporal integration, Hafter and his colleagues have performed a number of experiments in which interaural difference thresholds were measured for trains of high-frequency clicks (typically, centered at 4.0 kHz). Hafter and Dye (1983) measured ITD thresholds as a function of the number of clicks (n) and the interclick interval (ICI). They found that the maximum rate of click presentation that the binaural system could follow without loss of information was about 100 clicks/s. For faster click rates (i.e., for ICI <10 ms) the slope of the threshold-vs.-n function was shallower than the predicted −0.5 (Figure 6). The process by which information from later-arriving portions becomes progressively less effective when information is presented at high rates has been called *binaural adaptation* (Hafter, Buell, & Richards, 1988b).

Hafter et al. (1988b) have summarized the following properties of binaural adaptation:

1. It occurs in monaural channels prior to the first site of binaural interaction. It occurs independently in different frequency (critical band) channels; however, it is not based on adaptation in eighth nerve fibers. A probable site for the mechanism is the cochlear nucleus.

2. Although information arriving at high input rates is lost as far as *lateralization* is concerned, other aspects of stimulus processing (i.e., pitch processing) are not subject to this adaptation.

3. The adapted system can be "restarted" by several types of trigger

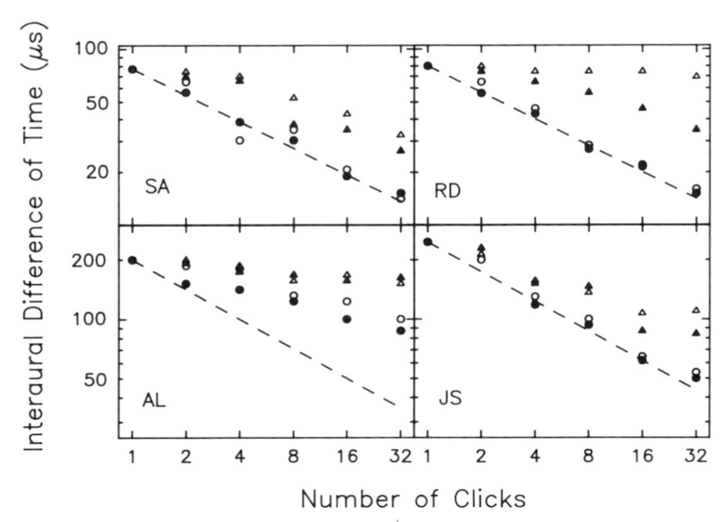

FIGURE 6 Threshold ITD as a function of the number of clicks (n) in the train. Data are shown for four subjects (in the four panels): △, ICI = 1 ms; ▲, ICI = 2 ms; ○, ICI = 5 ms; ●, ICI = 10 ms. Dashed lines have slopes of −0.5. (From Hafter & Dye, 1983.)

signal. The essential feature of this retriggering process may be a change in the spectrum, especially at a remote frequency region. For example, the presentation of a low-intensity 1 kHz tone after the onset of a click train is sufficient to restart the system, such that the relative effectiveness of the first click following the trigger is fully restored.

4. Binaural adaptation is probably *not* the primary mechanism underlying the phenomenon known as the *precedence effect,* which is thought to be mediated at the cortical level and is subject to cognitive influences (see Section IV).

Hafter and Buell (1985) and Hafter et al. (1988b) have drawn an analogy between binaural adaptation and the visual perceptual fading that occurs when an image is stabilized on the retina. In both cases persistent, nonchanging input results in loss of responsiveness, but an environmental change "restarts" the system, reinstating the initial level of sensitivity. According to this view, the auditory localizing system is normally quiescent and prompted to sample directional information only upon *changes* in the environment, such as the occurrence of a new sound. The system thus avoids overload from unnecessary (redundant) directional information.

2. Binaural Sluggishness

In contrast to binaural adaptation, it is assumed that binaural sluggishness occurs subsequent to the site of binaural interaction. Binaural sluggishness refers specifically to the response of the system when there is a change in the stimulus; in particular, when there is a change in its interaural configuration.

If a broadband stimulus is presented over headphones such that one input is delayed relative to the other (by, say, 500 μs), the subject perceives an intracranial image that is lateralized toward the leading ear (see Subsection II.A.1). If now the ITD is slowly changed from +500 μ to −500 μs and back to +500 μs, the subject can hear the position of the image move back and forth through the head. However, as the rate of ITD fluctuation increases, the subject has increasing difficulty following the movement: Even for rates as low as 2.4 Hz, the subject can no longer follow the complete excursions of the image (Blauert, 1972). As the rate of fluctuation increases further to 10–20 Hz, the subject ceases to hear any movement at all and perceives a steady "blur" (Grantham & Wightman, 1978). This relative insensitivity to dynamic variations in interaural differences has been called *binaural sluggishness* (Grantham & Wightman, 1979).

Several studies have measured the limits of the ability to follow ongoing variations in ITD or in interaural correlation (Blauert, 1972; Grantham & Wightman, 1978, 1979; Grantham, 1982). These studies provided a measure of the binaural system's response to periodic fluctuation. To observe its response to a *step change* in an interaural parameter, Kollmeier and Gilkey (1990) used a signal detection task in which a masker (a 750 ms burst of noise) was switched halfway through its presentation from $N\pi$ to $N0$ [see Figure 7(a)]. The signal, a 20 ms, 0.5 kHz tone burst in $S\pi$ configuration, was presented at one of several temporal locations on either side of the masker transition point. In a control condition that did not involve binaural processing [Figure 7(b)], the transition was a 15-dB decrease in the level of an $N\pi$ masker (signal still $S\pi$).

Thresholds for one of the four subjects are shown in Figure 8 as a func-

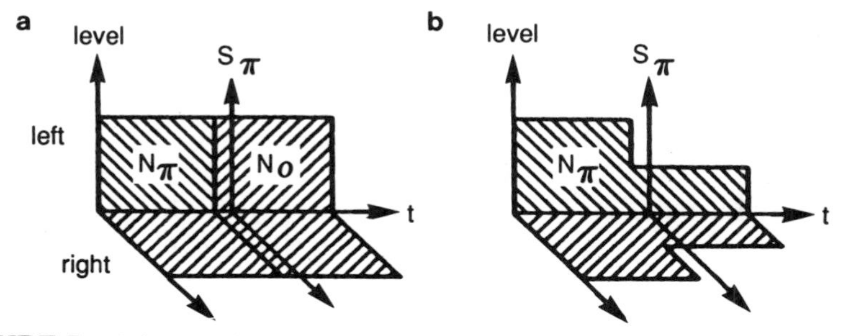

FIGURE 7 Schematic drawing showing the temporal pattern of the masker in two conditions in an experiment by Kollmeier and Gilkey (1990). (a) Masker switches from $N\pi$ to $N0$ halfway through its presentation. (b) Masker is attenuated by 15 dB halfway through its presentation. The vectors shown as $S\pi$ indicate the temporal position of the $S\pi$ probe signal. (From Kollmeier & Gilkey, 1990.)

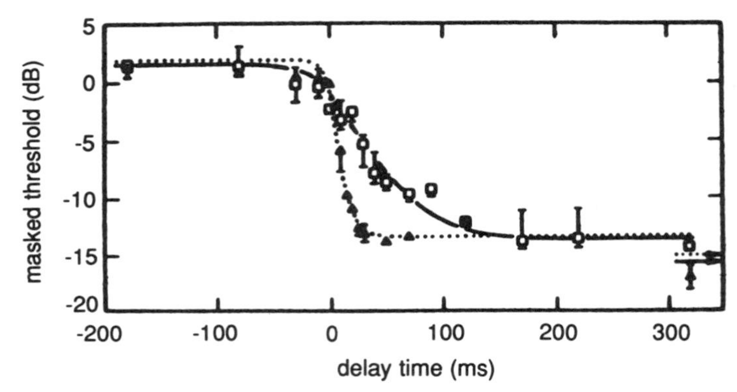

FIGURE 8 Masked threshold for one subject as a function of time between the masker transition and the offset of the probe signal. □, binaural condition; △, "monaural" control condition. (From Kollmeier & Gilkey, 1990.)

tion of the delay between the masker transition and the signal offset. The sluggish response of the binaural system is clearly seen by the relatively slow improvement signal detectability following the masker transition compared to that shown by the monaural system. From these data Kollmeier and Gilkey estimated that the binaural system's temporal response is two to three times slower than that of the monaural system.

3. Binaural Sensitivity to Fluctuating ILDs

Several studies have shown that the binaural system's response to fluctuating ILDs is not as sluggish as its response to fluctuating ITDs or fluctuating interaural correlation (Blauert, 1972; Grantham, 1984a; Bernstein & Trahiotis, 1992a). It is not clear why the temporal response to ILDs and ITDs should be different, nor is it clear what selective advantage might attach to the high degree of sluggishness found with respect to the processing of ITDs. This different sluggishness does provide further support for the notion that processing of ILDs and ITDs is independent.

III. LOCALIZATION AND SPATIAL RESOLUTION IN THE FREE FIELD

Headphone experiments have the advantage that they allow precise and independent control over interaural parameters such as ITD and ILD. However, headphones have typically been used to investigate the effects only of interaural differences and have not manipulated the spectral cues that are also known to be important in providing spatial information to human listeners (see Section III.B). Since the perception of most headphone-presented stimuli is of images inside the subject's head, the degree to which these results can be generalized to free-field situations is necessarily limited. To learn about *external* auditory spatial perception, it is necessary to study human performance in free-field experiments that employ remote sound sources (however, see Wightman & Kistler, 1989, for a description of procedures enabling subjects to externalize headphone-presented stimuli).

The distinction that was drawn in Section II between identification and discrimination experiments applies equally to free-field studies. *Identification* tasks in the free field are often referred to as *localization studies* or *source identification studies*. Typically a sound is presented from one of N potential positions, and the observer must respond by pointing or by stating which of the N sources emitted the sound. In *discrimination* tasks a subject typically hears two sounds presented successively and must say whether the second was to the right or left of (below or above, for vertical-plane study) the first. The derived measure is called the *minimum audible angle;* discrimination studies are sometimes referred to as *spatial resolution studies*.

A. Localization and Spatial Resolution in the Horizontal Plane

1. Localization Studies

Localization of sounds in the horizontal plane is mediated primarily by ILDs and ITDs. A source directly in front of a subject (0° azimuth) produces waveforms at the two ears that are very nearly identical (negligible ILDs and ITDs); but a source displaced from midline will arrive sooner and be more intense (depending on its frequency) at the near than the far ear. Figures 9 and 10 illustrate how the interaural differences vary as a function of source azimuth. The ITD varies systematically as a sound source is moved from 0° to 90° (opposite one ear), where it is at its maximum (Figure 9). As the source is moved from 90° to 180°, the ITD decreases again until it is near zero for positions directly behind the listener. In the case of the ILD (produced by the "head shadow"), the functions depend on the frequency of the stimulus (Figure 10); in general, these functions are more irregular than the ITD function, sometimes exhibiting maxima at azimuths other than 90° (Shaw, 1974). Both ILDs and ITDs vary over ranges within which humans have good sensitivity to these differences (see Section I.A and Figures 1–3).

Stevens and Newman (1936) were among the first investigators to measure localization in an anechoic environment. They presented various sounds to a subject seated on top of a 9 ft ventilator on the top of a building (thereby minimizing reflections from nearby surfaces). The sounds were presented from a boom-mounted loudspeaker that was 12 ft from the subject's position and could be rotated to any position around him at ear level. For each presentation the subject had to state the apparent direction of the

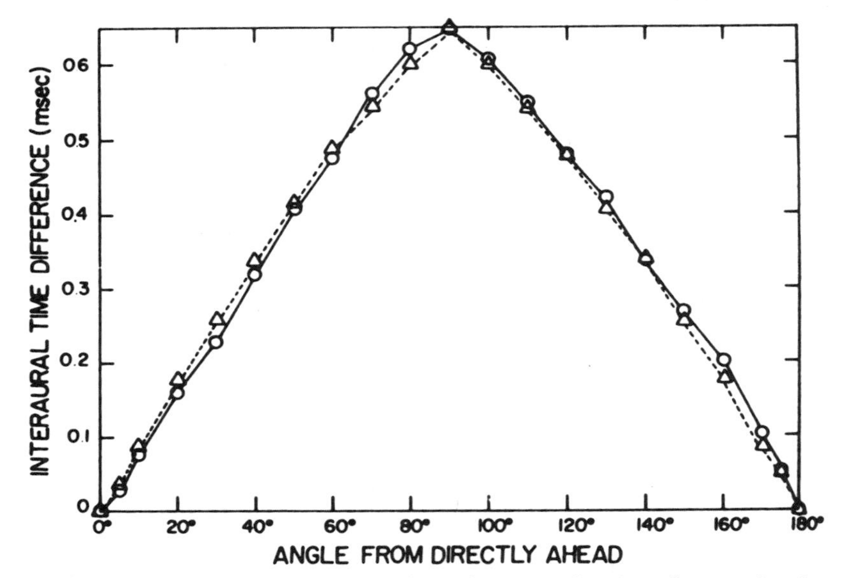

FIGURE 9 Measured and computed values of ITD as a function of a source's azimuth angle; 0° is directly ahead. (Adapted from Fedderson, Sandel, Teas, and Jeffress, 1957, after Green, 1976, Fig. 8.2, p. 204).

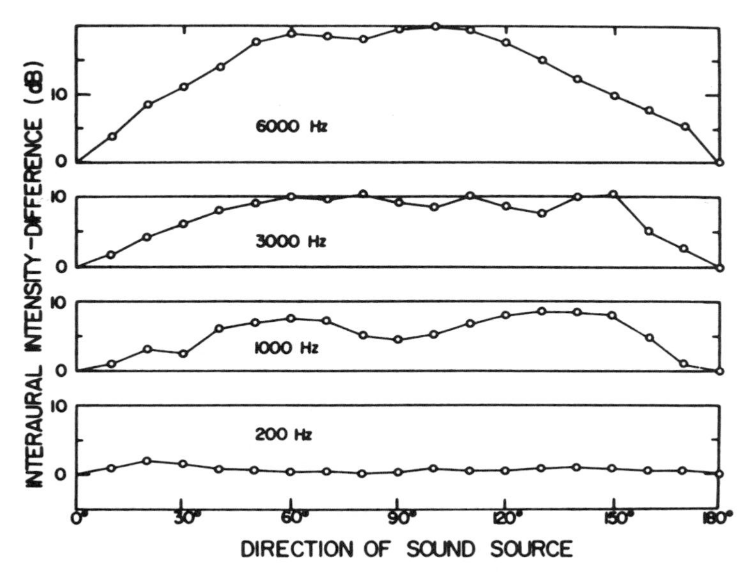

FIGURE 10 Measured values of ILD as a function of a source's azimuth angle, shown for frequencies from 200 to 6000 Hz; 0° is directly ahead. (Adapted from Fedderson et al., 1957, after Green, 1976, Fig. 8.3, p. 205).

source. The dependent variable was the "average absolute error," that is, the average discrepancy between the actual angular position of the source and the subject's response.

Figure 11 plots average error (collapsed over source position) as a function of stimulus frequency. There is a mid-frequency region (2.0–4.0 kHz) where performance was worse than at either lower or higher frequencies. Stevens and Newman explained the relatively poor performance in the mid-frequency region in terms of the duplex theory (Rayleigh, 1907). According to this theory, ITD is a salient cue for localization of low-frequency tones, but becomes less salient (and eventually useless) as frequency increases (see Figure 3). On the other hand, ILD provides a useful localization cue for high-frequency stimuli, but due to the frequency-dependent head-shadow effect (Figure 10), ILD information becomes unavailable as stimulus frequency decreases. In the mid-frequency range, neither ITD nor ILD provides optimal information for localization, and error scores thus increase. Today it is generally thought that a strict version of the duplex theory is untenable, because subjects *are* sensitive to ITDs in high-frequency *complex* stimuli (see Figure 3); however, as will be discussed later, recent findings have indicated that the duplex theory may be applicable to free-field localization performance even for nontonal stimuli (see Hafter, 1984, for an historical review of the duplex theory).

Another important finding from the Stevens and Newman study concerned the occurrence of front–back reversals. A front–back reversal is the error of responding that a source is in a rear quadrant when it is actually in the corresponding front quadrant (or vice versa). As can be seen from Figures 9 and 10, a particular interaural difference is generally associated with

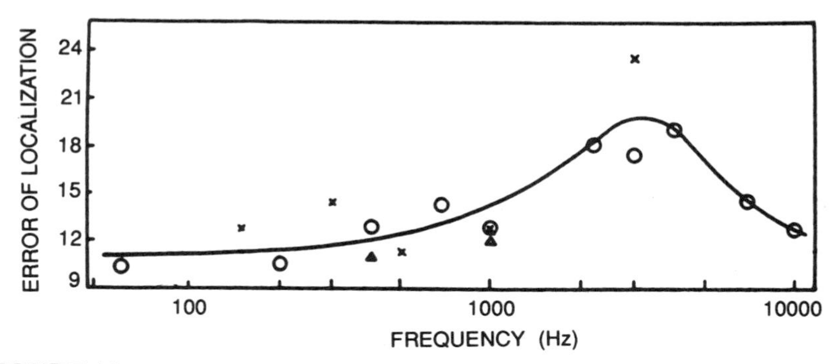

FIGURE 11 Average error of localization as a function of signal frequency. Data are averaged over source azimuth. The different symbols represent data collected from different experiments. (From Stevens and Newman, 1936, *American Journal of Psychology*. Copyright 1936 by the University of Illinois Press. Used by permission of the University of Illinois Press.)

(at least) two azimuths in the horizontal plane; for example, an ITD of 250 μs favoring the right ear can be produced by a source at 30° (in the front right quadrant) or by a source at 150° (in the rear right quadrant). Thus, a subject with only interaural difference information available will not be able to distinguish front from rear source positions.

Stevens and Newman found that front–back reversals were quite frequent for signal frequencies below 2.0 kHz, but decreased significantly for higher frequencies. The ability to discriminate front from rear for the high frequencies was attributed to the shadowing effect of the pinna, which attenuated high-frequency sounds from the rear, thus providing a loudness cue. Pinna-based attenuation did not occur for the lower frequencies because the wavelength of the sound was large compared to the dimensions of the ear (i.e., the sound passed around the pinna's outer flange without loss of energy).

a. Response Bias versus Random Variability

The errors in Figure 11 may reflect response bias, random variability in responding, or both. Recent studies of localization have provided more extensive error analyses that have allowed for independent evaluation of bias and variability. One such study (Rakerd & Hartmann, 1985) employed an eight-loudspeaker array to investigate localization of 0.5 kHz tones [Figure 12(a)]. In one condition, the tones were presented as 50 ms pulses in an anechoic chamber, and the subject had to state on each trial which of the eight loudspeakers produced the sound. Results are shown in Figure 12(b), which plots the average response (in terms of speaker number) as a function of the source number.

Rakerd and Hartmann derived three measures of error performance that have general utility in any source identification task. (1) Overall error (D) is the rms average of the discrepancy between the azimuth of a given source and the subject's responses to that source; this measure is analogous to the mean absolute error computed by Stevens and Newman and does not distinguish bias from random variability. (2) Variable error (s) is the standard deviation of the subject's responses; it differs from D in that it is a measure of the rms discrepancy between the subject's individual responses to a particular source and his or her *mean response* to that source (not the actual source position). In Figure 12(b) the variable error is represented by the standard deviation bars shown around the mean responses. (3) Constant error (C) is the average deviation of the mean response from the source; for a given source, it is represented by the vertical distances between the data points and the diagonal shown in Figure 12(b). [Note that a fourth measure, the signed error (E), is related to C, but takes into account the *sign* of the deviation when averaging across subjects or positions.] The values of these

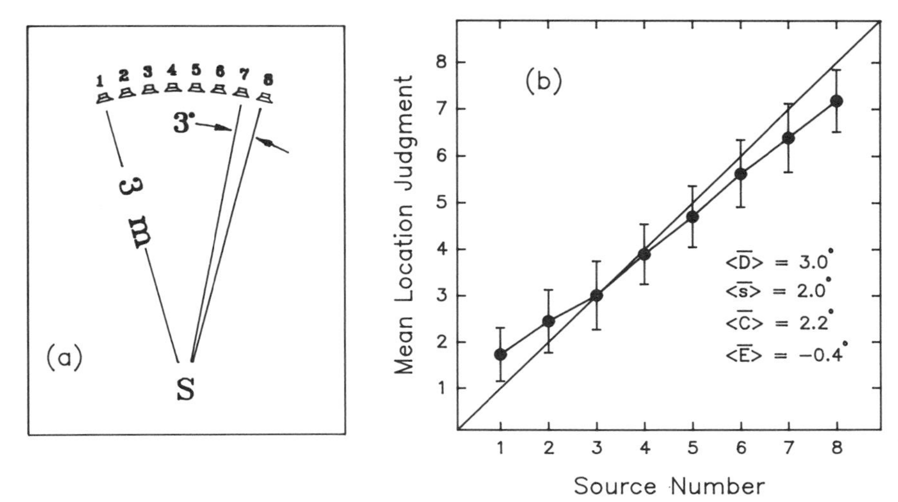

FIGURE 12 The arrangement of an eight-loudspeaker array (top view), similar to that employed by Rakerd and Hartmann (1985). (b) Mean localization judgment for each of the eight loudspeakers in the array (adjacent loudspeakers were separated by 3°). Data points are averaged over 10 subjects and 20 presentations. Error bars show standard deviation across replications, averaged over the 10 subjects. See text for a description of the error measures. (Adapted from Rakerd and Hartmann, 1985.)

statistics (and of E) for the data shown in Figure 12(b) are indicated within the figure; the relationship among them is

$$D^2 = C^2 + s^2. \tag{2}$$

In any complete model of localization both constant errors and variability should be accounted for.

2. Resolution Studies

The measure s (random variability) quantifies a subject's consistency or resolution in localizing sound sources. However, it is generally obtained within the context of an identification experiment and may well be affected by contextual variables such as the span of the loudspeaker array (e.g., Hartmann & Rakerd, 1989) or the type of response required (e.g., Makous & Middlebrooks, 1990). A more direct measure of spatial resolution is the minimum audible angle (MAA), which was first measured by Mills (1958). On each trial, his subjects were presented with two successive sinusoids, the first of which came from a reference azimuth (0, 30, 45, 60, or 75°) and the second from a position slightly to the left or right of the reference. The subject had to report whether the second tone was to the left or right of the first; the minimum audible angle was determined as that angular separation at which the subject responded at a particular criterion level above chance (see Hartmann & Rakerd, 1989, for a discussion of the implications of methodology for the interpretation of the MAA).

Some of Mills data are shown in Figure 13, where the MAA is plotted as a function of frequency. Under the best circumstances (low-frequency tone presented from directly in front of a subject) the MAA is about 1° of arc. Mills (1960) demonstrated, by consideration of the physical cues arriving to a subject's ears and of ILD and ITD thresholds obtained under headphones, that the basis for MAA task performance was, in fact, ITD discrimination at low frequencies and ILD discrimination at high frequencies, thus providing strong support for the duplex theory as it applies to pure tones.

Figure 13 also shows that spatial resolution is better (at all frequencies) for sound sources directly in front of the subject than for sources at 30° off midline. This general finding—that spatial resolution diminishes as sources move to the periphery—has been replicated many times, as illustrated in Figure 14 for low-frequency tones (left panel) and for high-frequency tones and wideband stimuli (right panel). Part of the degradation in resolution shown in Figure 14 can be attributed to the reduction in magnitude of the slopes of the ILD and ITD functions as a source is moved to the periphery (Figures 9 and 10); for example, a 10° azimuth change results in a smaller change in ITD and ILD at a reference azimuth of 75° than at midline. Another factor contributing to the loss of spatial resolution as a source moves to the periphery is the reduction of ITD and ILD sensitivity as

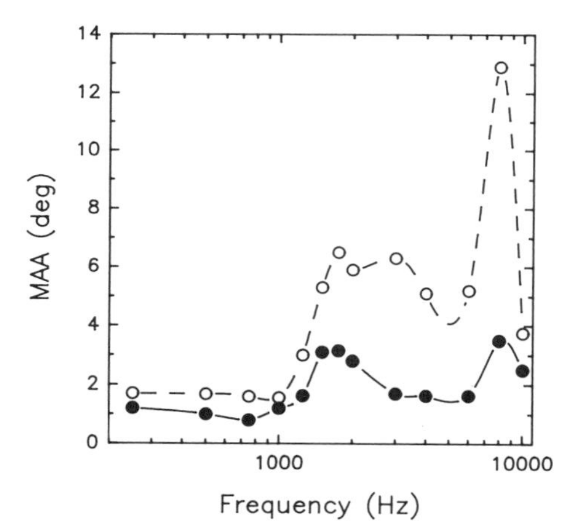

FIGURE 13 Minimum audible angle as a function of tonal frequency. ●, sources presented from 0° azimuth (directly in front of subject); ○, sources presented from 30° to the right of subject's midline. (Adapted from Mills, 1958.)

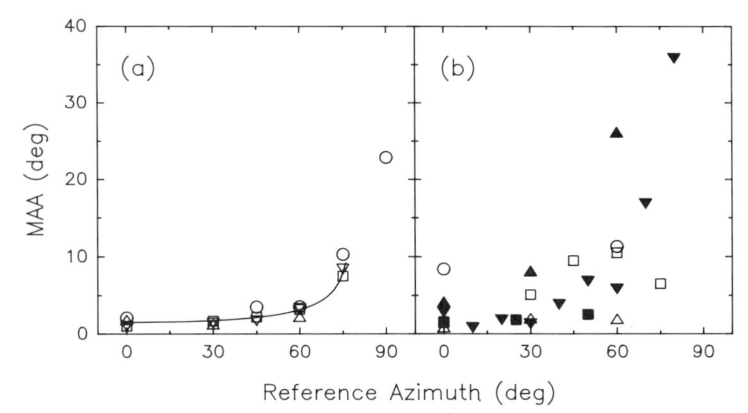

FIGURE 14 MAA in the horizontal plane as a function of a source's reference azimuth. Data are shown from several studies. (a) MAAs for low-frequency tones. □, Mills (1958), 500 Hz tone; ▽, Mills (1958), 1000 Hz tone; △, J. D. Harris (1972), 1250 Hz tone; ○, Grantham (1986), 500 Hz tone. (b) MAAs for high-frequency stimuli. □, Mills (1958), 3000 Hz tone; △, J. D. Harris (1972), 3200 Hz tone; ○, Chandler and Grantham (1992), 3000 Hz tone; ▲, Hafter, Buell, Basiji, and Shriberg (1988a), 4000 Hz click; ▼, Hafter, Saberi, Jensen, and Briolle (1991), 4000 Hz click; ■, Litovsky and Macmillan (1994), wideband noise.

measured under headphones when the reference ITD or ILD is varied from 0 (Hafter & DeMaio, 1975; Hafter, Dye, Nuetzel, & Aronow, 1977).

3. The Concurrent MAA

The relevance of the MAA as a measure of auditory spatial resolution is restricted to the case in which the two sound sources do not overlap temporally. The real auditory environment more typically consists of temporally overlapping sounds from multiple sources, so it is also of interest to measure spatial resolution for concurrently active sounds.

Perrott (1984a) presented two simultaneous tones of different frequencies to two different loudspeakers in the horizontal plane. The lower frequency tone (0.5 kHz) was presented from the left loudspeaker on half of the trials and from the right loudspeaker on the other half; in a single-interval, two-alternative forced-choice paradigm, the subject had to say whether the higher frequency tone appeared on the left or the right. Perrott thus defined the *concurrent minimum audible angle* (concurrent MAA) as that angular separation of the loudspeakers at which performance in this task reached 75% correct.

Concurrent MAAs from this experiment are plotted in Figure 15 as a function of the reference azimuth; the parameter is the frequency separation of the two tones. Performance at midline (0° on the abscissa) was quite good

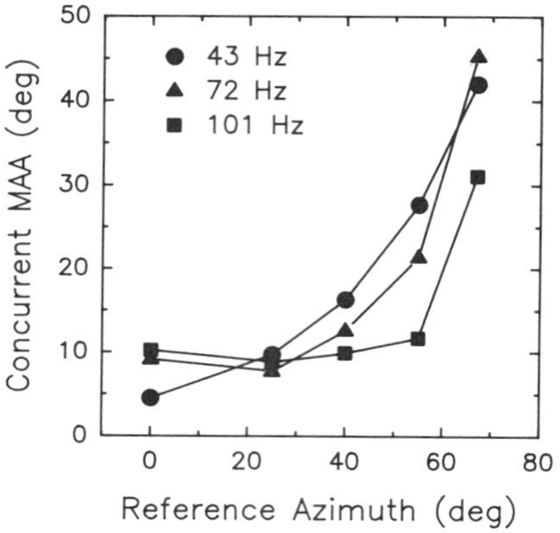

FIGURE 15 Concurrent MAA as a function of reference azimuth. Parameter is the frequency difference (in Hz) between the two sinusoidal signals. (Adapted from Perrott, 1984a.)

for the three frequency separations shown, even approaching the normal MAA of 1–2° (concurrent MAA = 5–10°); however, for a smaller frequency separation (15 Hz; data not shown), performance did not reach 75% correct, even for the widest angular separation employed (46°). Therefore, as the concurrent sounds became less spectrally resolvable (falling within a single critical band; see Chapter 5), subjects had increasing difficulty sorting out their respective sources. On the other hand, with any degree of temporal asychrony, resolution performance improved, even for very small frequency separations (see Chapters 11 and 12). Auditory spatial resolution is apparently a joint function of the spectral and temporal relationships among the sounds being discriminated.

Further experiments on the perception of concurrently active sound sources have been conducted by Perrott (1984b) and Divenyi and Oliver (1989).

4. The Role of Pinna Cues in Horizontal-Plane Localization

By placing a small microphone in a subject's ear canal, it has been shown that the pinna filters sound in a way that depends on its direction of incidence (e.g., Kuhn, 1987). This direction-dependent filtering by the pinna provides a potential cue to sound localization. Apart from its role in resolving front–back confusions for high-frequency stimuli (see Subsection III.A.1), it has generally been thought that the pinna was useful only for vertical-plane localization (or for monaural localization) (see Section III.B). However, several studies have demonstrated that the direction-dependent filtering effected by the pinnae can influence azimuthal localization as well, at least under certain circumstances. For example, Musicant and Butler

(1984a) found that localization of high-frequency sounds in the horizontal plane was significantly better when subjects' pinna cavities were open than when they were occluded, indicating that the pinna indeed contributed to localization. There is some debate over the relative weights given to pinna cues and interaural difference cues [Musicant and Butler (1985) found conditions in which the pinna cues actually were dominant over interaural differences cues, but Middlebrooks (1992) found that the interaural difference cues always dominated performance.] In any case, given the known importance of spectrally based pinna cues for *monaural* localization in the horizontal plane (e.g., Musicant & Butler, 1984b), it seems reasonable that stimulus situations can occur in which the pinna cues contribute to horizontal-plane localization.

5. Relative Weight of ITD and ILD Cues in Horizontal-Plane Localization

Searle, Braida, Davis, and Colburn (1976) concluded, based on a theoretical analysis of available data, that ITD and ILD cues provided the greatest contribution to horizontal-plane localization. They also concluded that pinna cues played some role. Unfortunately, these investigators were not able to derive differential weights for the ITD and ILD cues. However, Wightman and Kistler (1992) were able to independently assess the weight of these two cues. Their technique involved the measurement of head-related transfer functions (HRTFs; see Subsection III.B.2 for details) of individual subjects for a number of spatial locations, and the subsequent presentation of virtual free-field stimuli to these subjects through headphones. By digital manipulation of the signals, stimuli were created that had ITD cues associated with one location in space and ILD and pinna cues associated with a different position; subjects were required to indicate the location of these stimuli. The results were straightforward: For broadband noise, subjects' responses were governed completely by the ITD cue. For high-pass filtered noises, on the other hand, responses were governed by the ILD and pinna cues. Wightman and Kistler concluded that, in the horizontal plane, low-frequency ITDs are the dominant cues for localization of broadband sounds, while ILDs are the dominant cues for localizing high-frequency sounds. Apparently the ITDs in the envelopes of high-frequency signals do not contribute significantly to performance (see also Middlebrooks & Green, 1990).

The results of Wightman and Kistler (1992) provide compelling evidence that the duplex theory is, in fact, applicable to horizontal-plane sound localization after all: Stimuli with low-frequency content are localized based primarily on ITDs; stimuli without low-frequency content are localized based primarily on ILDs (including pinna cues). The caveat stipulating that the duplex theory applies only to pure tones apparently is no longer required.

B. Localization and Resolution in the Vertical Plane

For a subject using both ears, spatial resolution for sounds in the frontal region is not as good in the vertical plane as in the horizontal plane. For broadband signals the vertical-plane MAA for sources located in the median sagittal plane is on the order of 4° of arc (Blauert, 1983; Perrott & Saberi, 1990). This compares to a horizontal-plane MAA of about 1°, as described in the previous section. However, Makous and Middlebrooks (1990) found that the difference between horizontal-plane and vertical-plane localization performance decreased with increasing azimuth: For sounds near 90° azimuth the horizontal-plane error actually exceeded the vertical-plane error.

1. Cues for Vertical-Plane Localization

The primary cues for localization in the vertical plane are pinna-based spectral cues (e.g., Batteau, 1967; Kuhn, 1987). That the convolutions of the pinnae are crucial for making localization judgments was demonstrated by Gardner and Gardner (1973), who showed that error rate increased systematically as more and more of the pinnae's cavities were filled up (leaving the entrance to the meatus open). Furthermore, localization performance declines as high frequencies are removed from the stimulus, falling to near chance levels for stimulus containing no energy above about 7 kHz (Roffler & Butler, 1968; Hebrank & Wright, 1974). Because of their small dimensions, the folds of the pinna provide effective filtering only for frequencies above this value; hence, when these frequencies are removed, the primary basis for vertical-plane localization is eliminated.

Two other cues have been suggested for vertical-plane localization. Reflections from the torso contribute some information if the stimulus contains energy up to 2–3 kHz; generally this contribution is small relative to that of the other cues (Kuhn, 1987; Searle et al., 1976). Another potential cues is the *interaural* pinna disparity cue, which (presumably due to asymmetries in a person's two pinna shapes) can provide useful information if the stimulus contains frequencies above 8–10 kHz (Butler, 1969; Butler, Humanski, & Musicant, 1990; Middlebrooks, Makous, & Green, 1989; Searle, Braida, Cuddy, & Davis, 1975).

For stimuli located off the median sagittal plane (e.g., for an array of sources in the lateral vertical plane), subjects can and do use the ITDs and ILDs that arise from the source's change in elevation, in addition to the pinna-based spectral cues just mentioned (Butler and Humanski, 1992). Middlebrooks and Green (1991) have reviewed the recent literature dealing with localization of sources varying in both horizontal and vertical dimensions.

2. Spectral Peaks and Notches as Cues to Elevation

What specific spectral information does sound source elevation provide to subjects that they are able to use in a localization task? It is clear that the stimulus arriving at a subject's eardrum is filtered by the head, pinna, and ear canal, and that the total resultant filtering is dependent upon the direction from which the sound comes. For the remainder of this discussion it will be useful to distinguish between a signal's "distal" spectrum (that measured at the source) and the "proximal" spectrum (that measured near the subject's eardrum). The proximal spectrum divided by the distal spectrum yields the head-related transfer function (HRTF). Here, *HRTF* will be used synonymously with *directional transfer function* (DTF), although technically they are not identical functions.[1]

Hebrank and Wright (1974) and Butler and Belendiuk (1977), using small probe microphones, demonstrated that for a flat, broadband source, the proximal spectrum displayed certain characteristics that changed systematically with source elevation. For example, a notch in the frequency spectrum was found to migrate from lower to higher frequencies as the elevation of the broadband source was varied from −30° to +30° in front of the subject. Furthermore, when an explicit spectral feature was applied to the *source* (e.g., a notched noise was presented from the loudspeaker), subjects tended to localize the stimulus based on the frequency of that spectral feature and *not* on the actual location of the source. These effects were observed with spectral peaks as well as notches. For example, a stimulus with a spectral peak at 8.0 kHz tended to be localized overhead (independent of the actual loudspeaker position); this is the precise spatial position that, when activated with a broadband source, produced a peak in the proximal stimulus at 8.0 kHz (Blauert, 1969/1970; Hebrank & Wright, 1974).

The vertical-plane localizing system apparently responds as though the distal stimulus (at least if it is unfamiliar) is broadband and has a smooth spectrum. Distinct spectral features in the proximal stimulus are thus interpreted as arising from pinna-based filtering. This can explain the finding that the apparent elevation of narrowband signals is determined by the center frequency of the stimulus, not by the actual location of the source (Blauert, 1969/1970; Butler & Helwig, 1983); in this case, the system picks the elevation that would produce a relative peak at the stimulus frequency if

[1] The HRTF can be separated into two components: (1) filtering that is common to all source locations, which includes the contribution of the recording microphone and effects of ear canal resonance; and (2) filtering that depends on source position. The latter is referred to as the *DTF* (Middlebrooks & Green, 1990; Middlebrooks, 1992).

the distal stimulus had been broadband. Only when the bandwidth of the stimulus is increased sufficiently to include a relatively large flat region do the responses begin to correlate with the source positions.

Efforts to *quantify* the relationship between the spectrum of the proximal stimulus and perceived elevation have been hampered by the large degree of variability in subjects' pinna sizes and shapes (Butler & Belendiuk, 1977; Kuhn, 1987; Middlebrooks et al., 1989; Wightman & Kistler, 1989) and by the multiplicity of potential cues that are available in the proximal spectrum. One approach to this complex problem by has been taken by Middlebrooks (1992), who demonstrated that the overall spectral *pattern* of the proximal stimulus for a particular individual could predict the vertical component of that individual's localization responses quite well (at least for high-frequency, narrowband stimuli). In particular, given a complete set of DTFs for a particular subject (i.e., one for each of many source positions around him), the vertical localization response was well predicted as that elevation for which the associated DTF was most highly correlated (compared to all other DTFs) with the proximal stimulus. This correlational technique enabled quantitative description of the results without specification of which spectral features (peaks, notches, or other characteristics) were mediating performance.

An alternative approach to relating vertical-plane localization performance to pinna-based filtering has been the application of neural network functions to describe the transformation of proximal stimuli to sound source directions for the cat (Neti, Young, & Schneider, 1992). The latter technique has the advantage of allowing a determination of the relative contributions of different spectral regions of the proximal stimulus to the final solution. Neti et al. (1992) found, based on cat HRTFs, that the midfrequency region (5–18 kHz), which contained a prominent direction-dependent spectral notch, provided the best localization cues in their neural network model.

3. Covert Spectral Cues to Elevation

The spectral cues discussed in the previous section are *overt* spectral features; that is, they potentially provide localization information directly from the spectrum received at the eardrum. Other cues may be *covert;* these are not based on peaks and notches in the proximal spectrum. Rather, a covert cue is uncovered when the amplitude response of the pinna is plotted for a single frequency as a function of source *direction*. It represents, for a given frequency, that direction at which there is the largest boost (or attenuation) in amplitude compared to all other directions. The use of such a cue implies, therefore, that the observer has some template or memory of how all frequencies are affected at all directions. Measurements of sound pressure as a function of source elevation shown in Figure 16 (from Kuhn, 1987) illustrate how covert notches can vary systematically with stimulus elevation.

Covert cues were originally proposed to account for *monaural* localization performance in the horizontal plane (Musicant & Butler, 1984b). More recently, Butler and his colleagues have proposed that covert cues contribute to binaural, vertical-plane localization as well (Butler et al., 1990; Humanski & Butler, 1988). These investigators demonstrated that there was a positive correlation between the spectral location of covert features and subjects' elevation responses in cases in which there was no apparent relation between the *overt* peaks and troughs and subjects' responses. However, it is still not clear whether the auditory system is using these covert cues or whether it is using overt, but as yet unidentified features of the proximal stimulus.

C. Auditory Distance Perception

Compared to the study of localization in the horizontal and vertical planes, relatively little attention has been devoted to the study of auditory distance perception. Clearly, the ability of organisms to judge distances is important from an evolutionary point of view, and the auditory system's function as an "early warning system" suggests that auditory distance perception should be a well-developed capability. Coleman (1963) has reviewed some of the early research on auditory distance perception, and more recently Blauert (1983, pp. 116–137, 175–177) has provided a thorough discussion of the factors underlying humans' judgment and discrimination of auditory distance.

These reviews have identified four cues for auditory distance judgment:

1. Sound pressure level (the greater the SPL, the shorter is the judged distance);
2. The amount of reverberation (the greater the ratio of direct to reverberent energy in the received signal, the shorter is the judged distance);
3. Spectral shape of the received signal (the greater the high-frequency content of the stimulus, the shorter is the judged distance, at least for targets farther away than about 1 m);
4. Binaural cues (for sources off midline and closer than about 1 m, the greater the ITD or ILD, the shorter is the judged distance), however, the evidence on this last point is inconclusive (Blauert, 1983, pp. 176–177).

Discussion of auditory distance cues has emphasized the distinction between those cues that are *absolute* and those that are *relative* (Mershon &

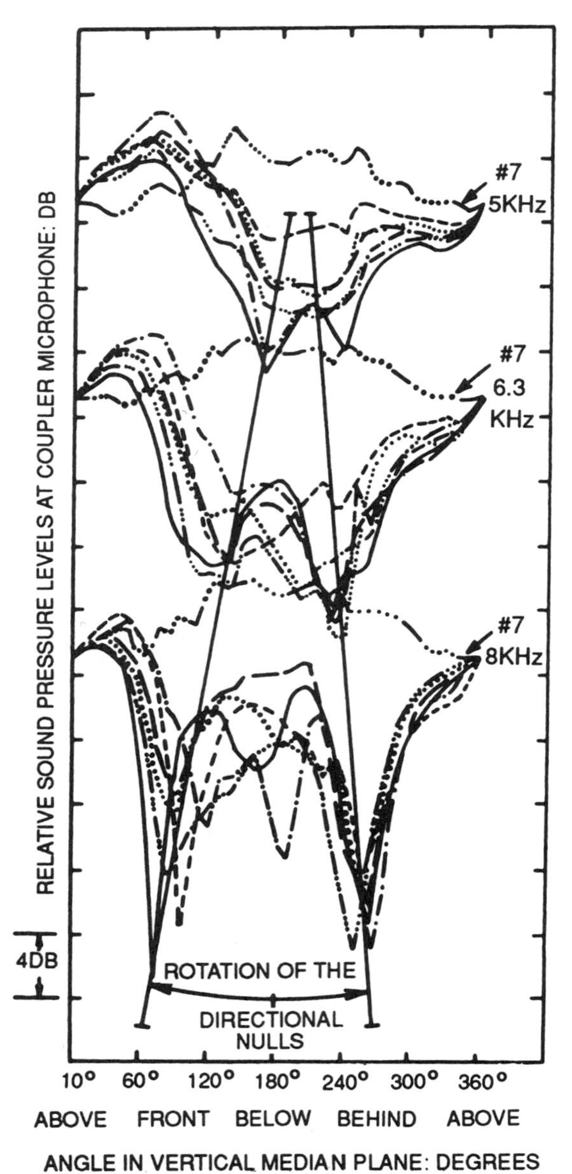

FIGURE 16 Relative sound pressure level measured in a manikin's ear canal as a function of source elevation in the midsagittal plane. Each group of curves represents measurements made at a different frequency. The different curves within a group were obtained with different pinnae (curve 7 was obtained with the pinna replaced by a flat plate). (From Kuhn, 1987.)

King, 1975; Mershon & Bowers, 1979). The former type of cue enables a judgment of a target's actual distance, whereas the latter enables a judgment of which of two targets is the nearer, without providing information about the actual distance of either. The distinction often hinges on the issue of *familiarity* of the target; a cue that generally provides only relative information about a target's distance (e.g., SPL) can become an absolute cue to distance if the subject has a priori information about its level or spectrum (i.e., if he or she is familiar with the sound). Other cues (to be described later) can be considered absolute even in the absence of such a priori information.

As before, the distinction should be borne in mind between identification and discrimination tasks. In the case of distance perception, an identification task requires subjects to indicate the actual distance of a target; a discrimination task typically involves successive presentation of two sounds, with the subject indicating which of the two was nearer. The latter task yields a measure that should be independent of a subject's familiarity with the sound (if the two sounds are identical as measured at their sources); however, performance in the former task can depend strongly on familiarity.

1. The Use of SPL as a Cue to Distance

According to the inverse square law, sound level decreases by about 6 dB for each doubling of distance from a point sound source in an anechoic environment. It has been shown that subjects can use the change in sound level to make distance discriminations. Strybel and Perrott (1984), employing the method of limits, measured the difference limen (DL) for distance of wideband noise stimuli. For distances greater than 6 m, they found that the DL for distance was about 5% of the reference distance, which is the predicted value, given the human intensity DL of about 0.5 dB (see Chapter 4). However, for distances less than 6 m, Strybel and Perrott found that the DL for distance increased to about 20% of the reference distance, thus greatly exceeding the predicted value based on intensity discrimination.

Ashmead, LeRoy, and Odom (1990) also measured distance discrimination for a wideband noise stimulus. Using a two-interval forced-choice procedure, these investigators found that the DL was as predicted from the use of intensity cues, even at short distances (1–2 m). These authors attributed the earlier investigators' finding of an elevated DL at short distances to their use of a method of limits, combined with the adoption of a conservative response criterion by their subjects. Using a criterion-free measure of performance, subjects can and do use the expected intensity change optimally to judge relative distances of auditory targets.

When the intensity cue was removed (by adjusting the intensity on each presentation such that the level arriving at the subject's ears was always the same) Ashmead et al. (1990) found that distance discrimination deteriorated significantly, indicating that it was indeed the intensity cue that mediated performance. Interestingly, Litovsky and Clifton (1992) found that, although adults were misled by randomization of intensity in a distance judgment task, 6-month-old infants were *not* misled by randomization of the intensity cue; evidently the infants relied primarily on other cues to distance.

In a more recent series of experiments, Ashmead, Davis, and Northington (1995) have found that the *change* in intensity that occurs when a subject is walking toward a sound source can afford an *absolute* cue to distance. In an anechoic environment the sound pressure produced by a source is

$$P = kr^{-1}, \tag{3}$$

where P is the sound pressure measured at the receiver, r is the distance between the receiver and the source, and k is a constant. Ashmead et al. pointed out that the rate of change of pressure with distance, relative to the pressure, is

$$(\mathrm{d}P/\mathrm{d}r)/P = -(kr^{-2})/(kr^{-1}) = -r^{-1}. \tag{4}$$

This measure is independent of the intensity of the source and thus can provide an absolute cue to distance (see Lee, 1980, for discussion of how this line of reasoning has been applied to other sensory systems).

Ashmead et al. (1995) found that subjects do, in fact, use this cue. In their experiment, subjects had to indicate the distance of sound sources by walking (blindfolded) to the point at which they judged the source to be. In one condition the 1.5 s noise stimulus was presented while the subject was stationary, and he or she had to begin walking after the source had terminated; in the other condition, the subject began walking prior to the initiation of the stimulus and thus had available the dynamic cue represented in Eq. (4). Intensity of the sources was randomly varied from trial to trial to make absolute intensity an unreliable cue. The results of this experiment are shown in Figure 17; performance in the moving condition was clearly superior to that in the stationary condition. The authors demonstrated in further experiments and analysis that the basis for the superior performance was indeed the dynamic change in pressure that occurred during the subject's motion.

2. Reverberation as a Cue to Auditory Distance

Mershon and King (1975) and Mershon and Bowers (1979) found that subjects' first distance judgments of unfamiliar stimuli were correlated with actual distance when the stimuli were presented in a reverberant environment. They concluded that the ratio of direct to reflected sound, which

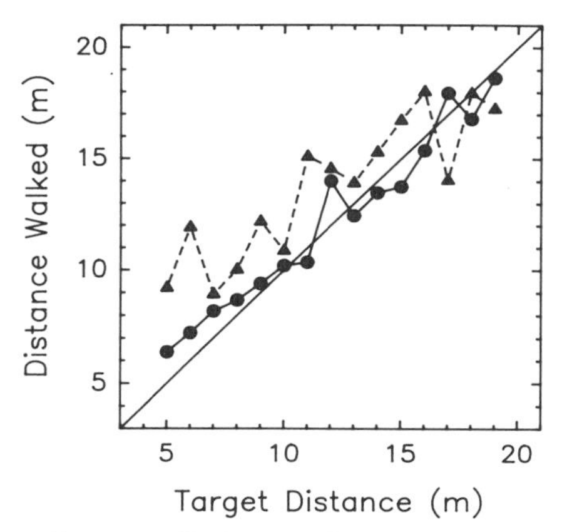

FIGURE 17 Average distance walked (nine subjects) as a function of target distance: ●, listener moving; ▲, listener stationary. (From Ashmead et al., 1995. Copyright 1995 by the American Psychological Association. Adapted by permission.)

would increase as a sound source was moved closer to a listener, provided an *absolute* cue to distance.

Results from a more recent experiment that compared distance judgments in anechoic vs. echoic conditions (Mershon, Ballenger, Little, & McMurtry, 1989) are displayed in Figure 18. Subjects (192) were assigned to one of eight conditions (two types of room × four loudspeaker distances) and were each given five trials. The first and fifth trials involved a distance

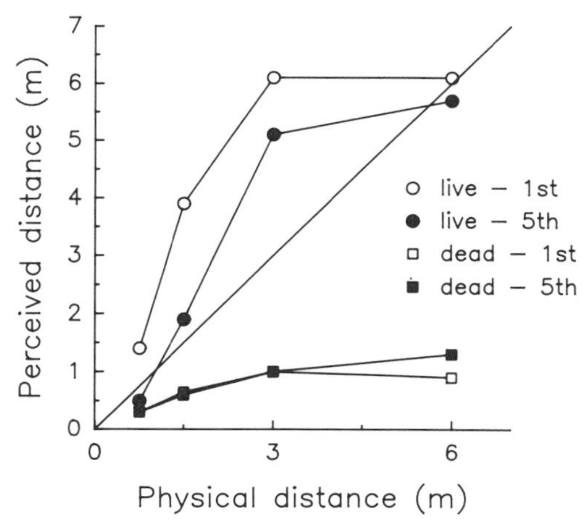

FIGURE 18 Mean judged distance as a function of physical distance for subjects' first and fifth distance judgments in either "live" or "dead" room. (Adapted from Mershon et al., 1989.)

judgment of the designated loudspeaker (trials 2–4 consisted of distance judgments of other loudspeakers, not reported here). Judgments in the "dead" room were almost independent of actual distance, whereas judgments in the "live" room were correlated much more with actual distance. Thus, reverberation does appear to provide an absolute cue to distance, although the accuracy of response is not impressive (there were large overestimates of distance in the middle ranges employed).

3. Spectral Content as a Cue to Auditory Distance

Sound traveling through air is attenuated more at high frequencies than at low frequencies; thus, the balance of high- and low-frequency energy is a potential cue to distance (Coleman, 1968).

By specifically manipulating the high-frequency cutoff of a broadband noise in a manner that mimicked the influence of the differential attenuation of air on sound transmission, Little, Mershon, and Cox (1992) investigated the salience of the spectral content cue (in one set of conditions overall level of the stimuli was held constant). They found, as expected, that the spectral content did not provide an *absolute* cue to distance (subjects did not respond differentially on their *first* judgments to stimuli with different cutoffs), but the cue did provide significant *relative* information (analysis of responses within subjects indicated that judged distance increased with decreasing high-frequency content).

4. Concluding Remarks

Two issues should be considered with regard to the study of human auditory distance perception. First, there is some evidence that "action" tasks (such as walking to the perceived source during or after it has emitted a signal) yield more accurate responses than the traditional magnitude estimation tasks (in which the observer remains seated and verbally specifies the distance) (e.g., Loomis, Fujita, Da Silva, & Fukusima, 1992). Thus, the mode of response should be included in any general model of distance processing. The second issue concerns the importance of *dynamic* cues in distance perception. The recent data of Ashmead et al. (1995) suggest that *changes* in sound pressure as an organism moves (or as the target moves) might provide information beyond that available when the target and observer are static. It is an interesting and important empirical question whether dynamic changes in spectral content or reverberation can provide as useful information as changes in sound pressure. Future research will discover the importance of these "new" cues and allow scientists to determine the relative weights of both dynamic and static cues to auditory distance.

D. Detection and Discrimination of the Motion of Auditory Targets

1. The Minimum Audible Movement Angle

Several experiments in recent years have extended the question of auditory spatial resolution to moving targets. In a typical experiment of this kind, the subject has to decide whether an auditory signal is moving to the left or right. The standard threshold measure is the minimum audible movement angle (MAMA); that is, the smallest angular distance a moving sound must traverse to be just discriminable either from a stationary source or from a source moving in the opposite direction. J. D. Harris and Sergeant (1971) were the first to report that, for very slowly moving sound sources in the horizontal plane, the MAMA was only slightly larger than the MAA: for a source moving in the horizontal plane at an angular velocity of 2.8°/s, the MAMA was between 2° and 4° of arc for tonal signals of 0.8 to 6.4 kHz. By comparison, the MAA for these signals varies from 1° to 3° (see Figure 13).

The MAMA depends on the frequency and velocity of the stimulus, as illustrated by the data of Perrott and Tucker (1988) shown in Figure 19. MAMAs show the same general dependence on signal frequency as the MAA (indicated by the lowest function): best performance is at low and high frequencies, and worst performance is in the midfrequency range from 1.5–3.0 kHz. Moreover, the faster the target travels, the greater the angular extent it must traverse for its direction to be detected. In no case was the MAMA for any stimulus *smaller* than the MAA determined at that frequency.

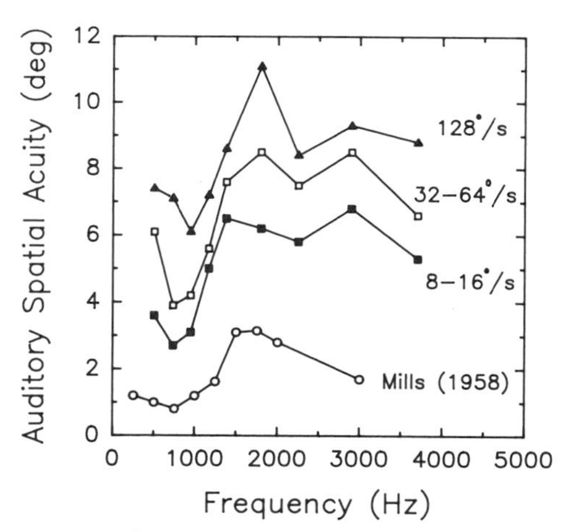

FIGURE 19 MAMAs as a function of tonal frequency for three different velocity ranges. MAAs from Mills (1958) are also shown. (Adapted from Perrott and Tucker, 1988.)

In a motion detection task the variables of velocity, duration, and angular extent traversed are interrelated; therefore, if velocity is held constant and angular extent is manipulated, stimulus duration also varies. This relationship is illustrated in Figure 20, which displays some data from Chandler and Grantham (1992). The left panel plots MAMA (for a 1 kHz tone) as a function of velocity; the increase in threshold is well fitted by a straight line. The right panel plots the same MAMAs as a function of stimulus *duration,* where duration = MAMA/velocity. The pattern in the right panel has been interpreted as evidence that a minimum integration time is required for subjects to perform optimally in a spatial resolution task (Grantham, 1986; Chandler & Grantham, 1992). Thus, for very brief durations (fast velocities) the MAMA is elevated because the system lacks sufficient time to capture and process movements whose extents are on the order of the MAA. However, as duration increases above some critical value (300–500 ms), performance begins to asymptote, approaching (although perhaps not reaching) the maximum resolution, as represented by the MAA.

2. "Snapshot" versus Motion-Sensitive Mechanism for Motion Discrimination

In his investigation of the detection and discrimination of motion of 500 Hz tonal stimuli, Grantham (1986) concluded that subjects based their decisions

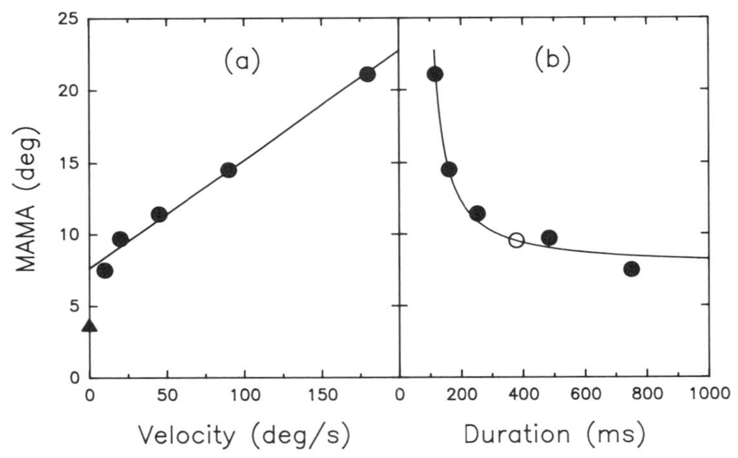

FIGURE 20 (a) MAMA for a 1.0 kHz tone for one observer as a function of source velocity. The MAA for this subject and stimulus is shown on the left ordinate by the filled triangle. (b) The same MAMAs are replotted as a function of duration (signal on-time), where duration = MAMA/velocity. The linear regression from panel (a) is also replotted as a function of duration. The open circle is the point on the fitted function that is 25% greater than the asymptote; the corresponding duration (379 ms here) is taken as an estimate of the subject's minimum integration time. (From Chandler and Grantham, 1992.)

about whether or how much a target moved on the spatial changes involved. For example, in a motion discrimination task (which of two targets moved faster?), Grantham found that for a given extent of movement of a reference target, about 4°–10° *additional* extent was required for discrimination, independent of stimulus duration and reference velocity. These and other results from that study suggested that subjects were not responding to velocity per se when they made judgments about the motion of targets, but were discriminating the distances traversed by the targets.

This view of auditory motion perception has been called the *snapshot* hypothesis (e.g., Dooley, 1987; Middlebrooks & Green, 1991); the general notion is that humans process motion, not via a direct appreciation of velocity, but by taking snapshots of the target's position at its onset and offset and comparing the spatial positions. If the positions are discriminably different, and if the time between the snapshots is "sufficient," the system concludes that motion occurred; velocity, if it need be known, is computed as distance divided by time.

Perrott and Marlborough (1989) conducted an experiment aimed at directly testing the snapshot hypothesis: Using broadband noise, they determined MAMAs under two conditions. In one, MAMAs were determined in the standard way: The stimulus was presented from a moving loudspeaker (velocity: 20°/s) for a duration that was varied adaptively to track the MAMA threshold. In the second condition, designed to mimic a snapshot mechanism, two brief (10 ms) pulses were presented in succession from the moving loudspeaker, and the duration between the pulses was varied adaptively to track a "marked endpoint" MAMA. According to the authors' reasoning, a snapshot hypothesis would predict no difference in performance between the two conditions, because a snapshot mechanism for discriminating motion would receive the same information in both cases. Perrott and Marlborough (1989) found the average MAMA for the standard condition (0.9°) to be significantly lower than for the marked condition (1.6°). They concluded that the snapshot hypothesis could be rejected; the implication was that MAMAs might be determined by special motion-sensitive mechanisms that benefit from stimulation through the entire trajectory spanned by a moving target.

Perrott, Constantino, and Ball (1993) presented further evidence supporting motion-sensitive mechanisms that are inconsistent with a literal snapshot view. They found that subjects could discriminate two sounds that traversed the same angular trajectory in the same temporal interval, but whose velocities varied in different ways (e.g., one accelerated, one decelerated). A strict snapshot mechanism should not be able to discriminate between these two stimuli.

On the other hand, no study has convincingly demonstrated that subjects respond directly to the motion of an auditory target in the same way that they *can* respond directly to the motion of a visual target (e.g., Lappin, Bell, Harm, & Kottas, 1975). All experiments that have compared the MAA to the MAMA (except that of Perrott & Marlborough, 1989, if their "marked MAMA" condition can be considered to be a MAA task), have found the MAA to be smaller than any MAMA obtained under similar conditions (Chandler & Grantham, 1992; Grantham, 1986; Harris & Sergeant, 1971; Perrott & Tucker, 1989). The snapshot hypothesis is a more parsimonious view to the extent that existing data can be explained in terms of static discrimination ability without the need to postulate a specialized motion-sensitive mechanism. The data obtained by Perrott and Marlborough (1989) and by Perrott et al. (1993), although inconsistent with a simplistic snapshot mechanism that literally samples only at the instants of onset and offset, are *not* inconsistent with a more sophisticated version of the snapshot model that assumes that the process of "taking" a snapshot requires time and that the ensuing snapshot is actually the result of some spatial and temporal integration.

Some indirect evidence might be interpreted as support for the existence of motion-sensitive mechanisms in the auditory system: (1) the existence of single units specifically responsive to moving auditory targets at several levels of the cat's auditory system (Altman, Bechterev, Radionova, Shmigidina, & Syka, 1976; Sovijärvi & Hyvärinen, 1974); and (2) the demonstration of an auditory motion aftereffect (Grantham, 1992), albeit weak and highly variable across subjects. However, even if such motion-sensitive mechanisms *do* exist in the human auditory system, it is not clear if and under what circumstances they are invoked in the everyday perception of auditory events. It is possible, of course, that motion is analyzed in parallel by two systems: one that responds directly to motion under the proper circumstances, and one that infers motion based on temporal and spatial resolution of snapshots. To date, available data do not permit a clear choice among these possibilities (motion mechanism, snapshot mechanism, or both), nor if there *is* a dual system, do current data suggest under what circumstances each mechanism is invoked.

IV. THE PRECEDENCE EFFECT

A. Background

An important issue in the study of sound localization concerns the effect of reflections (echoes) on the ability to localize or discriminate the positions of sounds. Early studies of sound localization in the horizontal plane (summarized in Wallach, Newman, & Rosenzweig, 1949) found that subjects responded accurately even in reverberant rooms. The fact such sounds were localizable with good precision, despite the presence of multiple reflections with conflicting directional information, has been attributed to the auditory system's ability to "snatch" and give extra weight to the leading sound (which contains appropriate interaural information) and less to the later sound (reflections). This ability to snatch and emphasize the leading sound in making a localization decision has been called the *precedence effect* (see Gardner, 1968; Zurek, 1987).

Experimental study of the precedence effect has typically employed a two-loudspeaker arrangement in an anechoic chamber, as illustrated in Figure 21 (Blauert, 1983). One loudspeaker (S_0 in the figure) delivers the first sound, and the second loudspeaker (S_T) delivers a simulated reflection from a different direction (a replica of the first sound after a variable delay). The diagram to the right of the figure illustrates the perception of events as a function of the echo delay when the signal is continuous speech. For simultaneous presentation of the two signals (echo delay = 0), a single fused "phantom" is perceived midway between the two loudspeakers. As echo delay is increased from 0 to 1 ms, the phantom migrates toward the leading loudspeaker (at +40° in the figure). The effect over this very short range of echo delays has been referred to as *summing localization* (Blauert, 1983, pp. 204ff); in this range the perceived location is presumably determined by the interaural differences in the resultant waveforms arriving at the two ears from the two sources.

For echo delays between 1 and 30 ms, the source is localized at the lead loudspeaker's position with no apparent contribution from the lag loudspeaker; in this region, the leading sound takes "precedence," and directional information from the lag loudspeaker is largely discounted. Other features of the precedence effect that have been discussed elsewhere (Blauert, 1983; Zurek, 1987) can be summarized as follows: (1) although the presence of a stimulus from the lag source does not appreciably affect the apparent

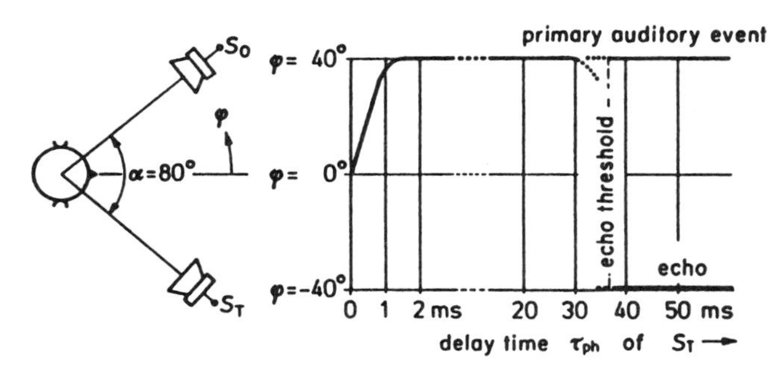

FIGURE 21 The left portion illustrates the loudspeaker configuration employed to mimic a direct sound from one direction (S_0, 40° to left of midline) and a delayed reflection from another direction (S_T, 40° to right of midline). The right panel shows the perceived location of the event (running speech) as a function of S_T delay. (From Blauert, 1983.)

position of the source, its presence is quite detectable—turning it off results in loudness, spaciousness, or timbre changes in the perceived image; (2) changes in the relative intensities of the primary source and echo affect the time course of events shown in Figure 21, but do not change the sequence of perceptual events over a wide range of intensity variation; and (3) the precedence effect is not perfect, that is, there is *some* influence of the lag source on apparent position. This issue will be discussed further.

When the echo delay exceeds 30–35 ms, the image "breaks up" into two parts, and the subject can separately localize the two events at the positions of the respective loudspeakers. The echo delay at which the image breaks from one into two parts is called the *echo threshold* (Blauert, 1983). The echo threshold has been shown to depend strongly on the type of stimulus employed; for connected speech, threshold is about 35 ms (Figure 21); for a single click or brief noise burst, the threshold decreases to 5–10 ms (Freyman, Clifton, & Litovsky, 1991); for continuous impulsive music, echo threshold *increases* to 50–75 ms (unpublished observation; see also Gardner, 1969; Zurek, 1987). Thus, echo threshold appears to be a complex function of the transient content in a stimulus and of duration.

B. The Localization Aspect: Localization of Sounds in Rooms

Rakerd and Hartmann (1985) measured the localization of 0.5 kHz sinusoids in the horizontal plane. In the conditions of interest here, the tones were either pulsed on for 50 ms with no rise–decay time (impulsive tones) or were turned on with a 7 s rise time and remained on until the subject responded (slow-onset tones). Additionally, the anechoic room used was either (1) empty or (2) contained a single reflecting surface (left wall), positioned 3 ft to the subject's left. Localization responses are shown in Figure 22. Error scores (D) are indicated within the figure (note that the empty-room, impulsive-tone condition is replotted from Figure 12).

For impulsive tones, there is confirmation of the existence of the precedence effect: Error scores with echo present (left-wall condition) are only slightly higher than scores with no echo (total error scores, 4.4° and 3.0°, respectively). That the precedence effect is imperfect is indicated by the fact that this small difference is statistically significant. For slow-onset tones (right panel), precedence fails completely: Although they can be localized well in the empty room ($D = 3.5°$), slow-onset tones are localized poorly when the wall is present ($D = 9.7°$). Rakerd and Hartmann showed that the large systematic response bias that occurred in the left-wall, slow-onset condition was due to the subject's apparent attempt to compute a compromise based on the often conflicting and unrealistic ILDs and ITDs created by the addition of the direct and reflected sound. When an abrupt onset was available to provide initial realistic location information, Rakerd and Hart-

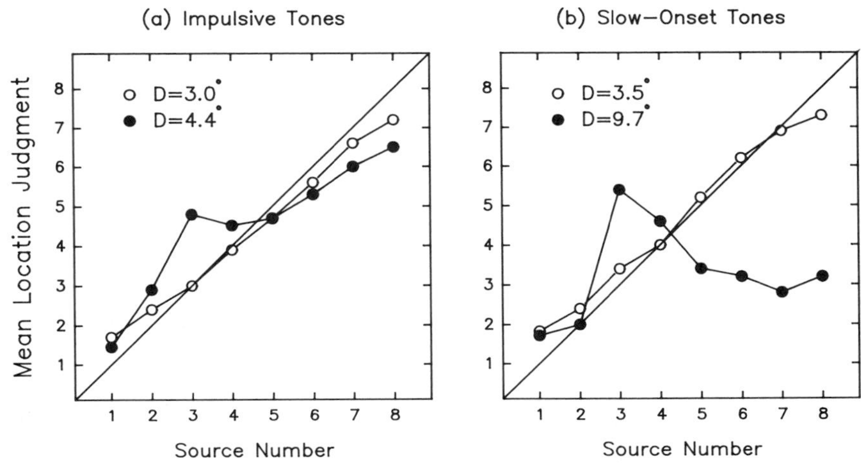

FIGURE 22 Mean localization judgments (10 subjects) for 500 Hz sinusoids as a function of loudspeaker number [adjacent loudspeakers separated by 3°; see Figure 12(a)]: ○, in anechoic room; ●, in anechoic room with a single reflecting surface located 3 ft to the subject's left. *D* represents overall error score. (Adapted from Rakerd and Hartmann, 1985.)

mann speculated that the later occurring "implausible" cues would be discounted.

The precedence effect apparently works less well with noise stimuli than with pure tones. Giguère and Abel (1993) found a 10–15% decrease in percent correct performance when subjects localized octave-band noise signals in a reverberent room compared to their performance in an absorbant room, even when brief (5 ms) rise times were employed. If the ongoing random envelope fluctuations in the noise stimuli provided a continuous reactivation of the precedence mechanism, it may have been more difficult for subjects to discount misleading cues contained in the ongoing portion of the stimulus. Still, localization performance remained well above chance in the reverberent-room conditions.

It has been suggested that the relative lack of contribution from an echo to the perceived location of a target may be based on temporary insensitivity to interaural differences that occurs following an impulsive event (e.g., Shinn-Cunningham, Zurek, & Durlach, 1993). For example, in a headphone experiment Zurek (1980) showed that the ILD and ITD thresholds for a brief noise burst were elevated significantly when the burst followed a leading noise burst by a time interval of 0.5–10 ms. Such temporary insensitivity has also been demonstrated in free-field studies in which it has been shown that the MAA for the *second* sound is elevated when the delay between the leading and lagging sounds is less than 10 ms (Perrott, Marlborough, Merrill, & Strybel, 1989; Litovsky & Macmillan, 1994). At this point, it is not clear how this 10 ms period of interaural insensitivity can account for a precedence effect that can extend 30–75 ms (Figure 21).

C. Buildup of Echo Suppression

Some recent experiments on the long-term effects of precedence have focused on echo suppression; that is, the degree to which the second sound is heard as a separate event. Freyman et al. (1991) investigated an effect that they called the *buildup of echo suppression*. They found that the degree of echo suppression depends upon the immediately preceding history of impulsive information experienced by a subject. For example, if a single click with a simulated 8-ms delayed reflection is presented in isolation, the subject clearly hears both events (the lagging event is above the echo threshold). However, if this click pair is presented repeatedly, say a rate of 4/s, the subject hears the echo event only for the first 4–8 occurrences; thereafter, the echo becomes inaudible as a separate event. In other words, echo threshold increases during the course of a sequence of identically presented pulses. This buildup of echo suppression indicates that the precedence effect has a dynamic component that extends over relatively long periods of time (measured in seconds).

To quantify the buildup of echo suppression, Freyman et al. (1991) measured echo threshold for a click that was either presented in isolation or preceded by a train of identical clicks; the number of clicks in the train and the click rate were independent variables. For a click presented in isolation the mean echo threshold for four subjects was 5.2 ms. Threshold elevations for conditions in which there was a preceding train are shown in Figure 23. The critical feature of the preceding train is the *number* of clicks in the train; buildup of echo suppression appears to reach its maximum amount after nine preceding clicks, independent of click rate (and train duration).

The buildup of echo suppression is dependent on the locations of the leading and lagging sounds. Once suppression has built up for a particular source in a particular position, a change in the source position (in particular, a swap between the positions of leading and lagging sounds) results in a "restarting" of the system (Clifton, 1987). Thus, immediately after such a switch (even if the rhythm of the train is not disrupted), the echo threshold is restored to its baseline (or *below* its baseline) level; for the example of the 8-ms delayed reflection, the echo event becomes audible again immediately after a switch. Clearly, high-level complex processing must underlie these long-term effects; Clifton, Freyman, Litovsky, and McCall (1994) have suggested that the buildup effect may be an adaptive phenomenon that inhibits redundant information in a reflective environment as it becomes familiar. So far, the critical variables underlying the buildup of echo sup-

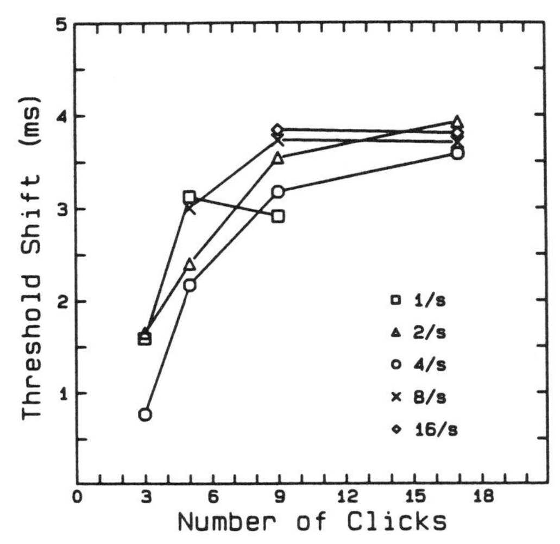

FIGURE 23 Elevation in echo threshold for a test click as a function of the number of identical clicks in a preceding train. The parameter is the rate of click presentation. (From Freyman et al., 1991.)

pression and its breakdown following a stimulus change have not been completely investigated.

D. Cortical Basis for the Precedence Effect

Many lines of evidence suggest that the precedence effect is mediated at a very high level in the auditory system, presumably at the cortex. Zurek (1987) has reviewed some of this evidence, including ablation-behavior studies with animals. Behavioral studies with special human populations (Clifton, Morrongiello, & Dowd, 1984, with infants; Cornelisse & Kelly, 1987, with patients with cerebrovascular accidents) have indicated that subjects with an underdeveloped or a damaged auditory cortex do not respond normally to precedence-effect stimuli, though they do respond normally to single auditory events. Finally, Saberi and Perrott (1990), employing a paradigm similar to that of Zurek (1980) for measuring the ITD threshold for a second event, found that the elevation in threshold for the second sound was practically eliminated when subjects were given sufficient practice. This result implies that the responsible mechanism is probably not located in the peripheral auditory system.

Although the discussion to this point has focused on headphone studies and on the precedence effect as it is observed in the horizontal plane, the precedence effect is not exclusively a binaural phenomenon. Blauert (1971) and Rakerd and Hartmann (1992) have shown that the precedence effect occurs (although not as strongly as in the horizontal plane) in the median plane, thus indicating that the effect is more generally an effect of apparent locations of events than an effect related to binaural processing.

V. CONCLUSIONS

This review of topics related to human spatial hearing has highlighted several new experimental findings and trends in the areas of binaural hearing, auditory localization, and echo suppression. Some of the newer and more intriguing experimental results that are likely to guide future research in these areas may be summarized as follows:

1. The binaural system rapidly adapts to unchanging periodic inputs with high repetition rates (Hafter et al., 1988b). This peripherally mediated process may serve to prevent possible overload by blocking the relay of redundant information to higher centers for binaural analysis. The adapted binaural system can be readily retriggered by spectral "surprises."

2. To a first approximation, the perceived location of an auditory object in three dimensions is governed independently by subsystems that determine its horizontal position based on interaural difference cues, its vertical position based on pinna-based spectral cues, and its distance by a constellation of cues including intensity, reverberation, and spectral content. Accuracy in responding to the distance of an object is influenced by subject's familiarity with the sound; accuracy in responding to elevation depends upon the spectral composition of the stimulus (the spectrum must be wide, including frequencies above about 7.0 kHz, and must be relatively smooth for optimum performance).

Familiarity with the stimulus is not required for accurate localization of sounds in the horizontal plane. In this case, sounds containing low frequencies are localized using ITDs whereas sounds that contain no low frequencies are localized using ILDs; subjects appear *not* to localize sounds based on ITDs in the envelopes of high-frequency stimuli (Wightman & Kistler, 1992; Middlebrooks et al., 1989). The duplex theory of sound localization first proposed by Rayleigh (1907) appears to provide an accurate description of horizontal-plane localization for complex stimuli as well as for tones.

3. Debate continues over whether motion of auditory targets in the horizontal plane is analyzed by special motion detectors in the auditory system or by a snapshot mechanism that infers motion via prior analysis of spatial and temporal changes (Grantham, 1986; Perrott & Marlborough, 1989). The possibility exists, of course, that auditory motion is analyzed in parallel by two systems: one that makes decisions based on spatial and temporal

changes associated with the endpoints (or with other, yet unknown trajectory points of the stimulus) and one that is directly sensitive to motion itself. Binaural sluggishness, as measured under headphones, would imply that if a motion-sensitive system exists, it would be sensitive only to long-duration, slow-velocity movements. However, to date, definitive experiments have not been conducted.

4. Through the operation of a precedence mechanism, the apparent positions of transient sound sources are not much affected by room reflections (Wallach et al., 1949). However, the precedence effect is not absolute, because reflections can exert a small influence on localization and resolution performance (Hafter et al., 1991; Rakerd & Hartmann, 1985). The echo-suppression aspect of the precedence effect operates within a long-term temporal context (measured in seconds), in which its effectiveness builds up or is "released" depending on the type and position of preceding acoustic transient information (Freyman et al., 1991).

5. A variety of factors can influence humans' sense of auditory space. In addition to the proximal acoustic cues that have been emphasized in this chapter, many "higher level" factors have been shown to contribute to auditory spatial perception. Such factors include prior and cotemporal auditory experience (Steiger & Bregman, 1982), visual influences, (Shelton & Searle, 1980; Warren, Welch, & McCarthy, 1981), and proprioceptive input (Lackner & Shenker, 1985).

Any complete description of human spatial hearing must ultimately account for these higher level influences that are typically operative in real-world settings. However, at the current stage of our understanding of spatial hearing, it seems appropriate to focus specific modeling efforts on the influences of the more immediate acoustic environment. Such modeling efforts are the topic of the next chapter.

Acknowledgments

The author would like to thank Daniel H. Ashmead, Alan D. Musicant, and the editor, Brian C. J. Moore, for many helpful suggestions and comments on earlier versions of this manuscript. He also thanks Don Riggs and Xuefeng Yang for assistance in the preparation of the figures.

References

Altman, J. A., Bechterev, N. N., Radionova, E. A., Shmigidina, G. N., & Syka, J. (1976). Electrical responses of the auditory area of the cerebellar cortext to acoustic stimulation. *Experimental Brain Research, 26,* 285–298.

Ashmead, D. H., Davis, D., and Northington, A. (1995). The contribution of listeners' approaching motion to auditory distance perception. *Journal of Experimental Psychology: Human Perception and Performance, 21,* 239–256.

Ashmead, D. H., LeRoy, D., & Odom, R. D. (1990). Perception of the relative distances of nearby sound sources. *Perception and Psychophysics, 47,* 326–331.

Batteau, D. W. (1967). The role of the pinna in human localization. *Proceedings of the Royal Society of London* Series B, *168,* 158–180.

Bernstein, L. R., & Trahiotis, C. (1992a). Detection of antiphasic sinusoids added to the envelopes of high-frequency bands of noise. *Hearing Research, 62,* 157–165.

Bernstein, L. R., & Trahiotis, C. (1992b). Discrimination of interaural envelope correlation and its relation to binaural unmasking at high frequencies. *Journal of the Acoustical Society of America, 91,* 306–316.

Blauert, J. (1969/1970). Sound localization in the median plane. *Acustica, 22,* 205–213.

Blauert, J. (1971). Localization and the law of the first wavefront in the median plane. *Journal of the Acoustical Society of America, 50,* 466–470.

Blauert, J. (1972). On the lag of lateralization caused by interaural time and intensity differences. *Audiology, 11,* 265–270.

Blauert, J. (1983). *Spatial hearing: The psychophysics of human sound localization,* trans. J. S. Allen. Cambridge, MA: MIT.

Butler, R. A. (1969). Monaural and binaural localization of noise bursts vertically in the median sagittal plane. *Journal of the Auditory Research, 3,* 230–235.

Butler, R. A., & Belendiuk, K. (1977). Spectral cues utilized in the localization of sound in the median sagittal plane. *Journal of the Acoustical Society of America, 61,* 1264–1269.

Butler, R. A., & Helwig, C. C. (1983). The spatial attributes of stimulus frequency in the median sagittal plane and their role in sound localization. *American Journal of Otolaryngology, 4,* 165–173.

Butler, R. A., & Humanski, R. A. (1992). Localization of sound in the vertical plane with and without high-frequency spectral cues. *Perception and Psychophysics, 51,* 182–186.

Butler, R. A., Humanski, R. A., & Musicant, A. D. (1990). Binaural and monaural localization of sound in two-dimensional space. *Perception, 19,* 241–256.

Chandler, D. W., & Grantham, D. W. (1992). Minimum audible movement angle in the horizontal plane as a function of stimulus frequency and bandwidth, source azimuth, and velocity. *Journal of the Acoustical Society of America, 91,* 1624–1636.

Clifton, R. K. (1987). Breakdown of echo suppression in the precedence effect. *Journal of the Acoustical Society of America, 82,* 1834–1835(L).

Clifton, R. K., Freyman, R. L., Litovsky, R. Y., and McCall, D. (1994). Listeners' expectations about echoes can raise or lower echo threshold. *Journal of the Acoustical Society of America, 95,* 1525–1533.

Clifton, R. K., Morrongiello, B. A., & Dowd, J. M. (1984). A developmental look at an auditory illusion: The precedence effect. *Developmental Psychobiology, 17,* 519–536.

Coleman, P. D. (1963). An analysis of cues to auditory depth perception in free space. *Psychological Bulletin, 60,* 302–315.

Coleman, P. D. (1968). Dual role of frequency spectrum in determination of auditory distance. *Journal of the Acoustical Society of America, 44,* 631–632(L).

Cornelisse, L. E., & Kelly, J. B. (1987). The effect of cerebrovascular accident on the ability to localize sounds under conditions of the precedence effect. *Neuropsychologia, 25,* 449–452.

Divenyi, P. L., & Oliver, S. K. (1989). Resolution of steady-state sounds in simulated auditory space. *Journal of the Acoustical Society of America, 85,* 2042–2052.

Domnitz, R. H., & Colburn, H. S. (1977). Lateral position and interaural discrimination. *Journal of the Acoustical Society of America, 61,* 1586–1598.

Dooley, G. J. (1987). The perception of auditory dynamic stimuli. *Ph.D. dissertation, University of Cambridge.*

Durlach, N. I. (1972). Binaural signal detection: Equalization and cancellation theory. In J. V. Tobias (Ed.), *Foundations of modern auditory theory* (Vol. 2, pp. 371–462). New York: Academic Press.

Durlach, N. I., & Colburn, H. S. (1978). Binaural phenomena. In E. C. Carterette & M. P. Friedman (Eds.), *Handbook of perception* (Vol. 4, pp. 365–466). New York: Academic Press.

Durlach, N. I., Gabriel, K. J., Colburn, H. S., & Trahiotis, C. (1986). Interaural correlation discrimination. II. Relation to binaural unmasking. *Journal of the Acoustical Society of America, 79,* 1548–1557.

Fedderson, W. E., Sandel, T. T., Teas, D. C., & Jeffress, L. A. (1957). Localization of high-frequency tones. *Journal of the Acoustical Society of America, 29,* 988–991.

Freyman, R. L., Clifton, R. K., & Litovsky, R. Y. (1991). Dynamic processes in the precedence effect. *Journal of the Acoustic Society of America, 90,* 874–884.

Gabriel, K. J., & Colburn, H. S. (1981). Interaural correlation discrimination. I. Bandwidth and level dependence. *Journal of the Acoustical Society of America, 69,* 1394–1401.

Gaik, W. (1993). Combined evaluation of interaural time and intensity differences: Psychoacoustic results and computer modeling. *Journal of the Acoustical Society of America, 94,* 98–110.

Gardner, M. B. (1968). Historical background of the Haas and/or precedence effect. *Journal of the Acoustical Society of America, 43,* 1243–1248.

Gardner, M. B., & Gardner, R. S. (1973). Problem of localization in the median plane: Effect of pinnae cavity occlusion. *Journal of the Acoustical Society of America, 53,* 400–408.

Giguère, C., & Abel, S. M. (1993). Sound localization: Effects of reverberation time, speaker array, stimulus frequency, and stimulus rise/decay. *Journal of the Acoustical Society of America, 94,* 769–776.

Grantham, D. W. (1982). Detectability of time-varying interaural correlation in narrow-band noise stimuli. *Journal of the Acoustical Society of America, 72,* 1178–1184.

Grantham, D. W. (1984a). Discrimination of dynamic interaural intensity differences. *Journal of the Acoustical Society of America, 76,* 71–76.

Grantham, D. W. (1984b). Interaural intensity discrimination: Insensitivity at 1000 Hz. *Journal of the Acoustical Society of America, 75,* 1191–1194.

Grantham, D. W. (1986). Detection and discrimination of simulated motion of auditory targets in the horizontal plane. *Journal of the Acoustical Society of America, 79,* 1939–1949.

Grantham, D. W. (1992). Adaptation to auditory motion in the horizontal plane: Effect of prior exposure to motion on motion detectability. *Perception and Psychophysics, 52,* 144–150.

Grantham, D. W., & Wightman, F. L. (1978). Detectability of varying interaural temporal differences. *Journal of the Acoustical Society of America, 63,* 511–523.

Grantham, D. W., & Wightman, F. L. (1979). Detectability of a pulsed tone in the presence of a masker with time-varying interaural correlation. *Journal of the Acoustical Society of America, 65,* 1509–1517.

Green, D. M. (1976). *An introduction to hearing.* Hillsdale, NJ: Lawrence Erlbaum.

Hafter, E. R. (1984). Spatial hearing and the duplex theory: How viable is the model? In G. M. Edelman, W. E. Gall, & W. M. Cowan (Eds.), *Dynamics aspects of neocortical function* (pp. 425–448). New York: Wiley.

Hafter, E. R., & Buell, T. N. (1985). The importance of transients for maintaining the separation of signals in auditory space. In M. Posner & C. Marin (Eds.), *Attention and performance XI* (pp. 337–354). Hillsdale, NJ: Lawrence Erlbaum.

Hafter, E. R., Buell, T. N., Basiji, D. A., & Shriberg, E. E. (1988a). Discrimination of direction for complex sounds presented in the free-field. In H. Duifhuis, J. W. Horst, & H. P. Wit (Eds.), *Basic issues in hearing: Proceedings of the 8th international symposium on hearing* (pp. 394–401). London: Academic Press.

Hafter, E. R., Buell, T. N., & Richards, V. M. (1988b). Onset-coding in lateralization: Its form, site, and function. In G. M. Edelman, W. E. Gall, & W. M. Cowan (Eds.), *Auditory function: Neurobiological bases of hearing* (pp. 647–676). New York: Wiley.

Hafter, E. R., & De Maio, J. (1975). Difference thresholds for interaural delay. *Journal of the Acoustical Society of America, 57,* 181–187.

Hafter, E. R., & Dye, R. H., Jr. (1983). Detection of interaural differences of time in trains of high-frequency clicks as a function of interclick interval and number. *Journal of the Acoustical Society of America, 73,* 644–651.

Hafter, E. R., Dye, R. H., Jr., Nuetzel, J. M., & Aronow, H. (1977). Difference thresholds for interaural intensity. *Journal of the Acoustical Society of America, 61,* 829–834.

Hafter, E. R., Dye, R. H., Jr., Wenzel, E. M., & Knecht, K. (1990). The combination of interaural time and intensity in the lateralization of high-frequency complex signals. *Journal of the Acoustical Society of America, 87,* 1702–1708.

Hafter, E. R., & Jeffress, L. A. (1968). Two-image lateralization of tones and clicks. *Journal of the Acoustical Society of America, 44,* 563–569.

Hafter, E. R., Saberi, K., Jensen, E. R., & Briolle, F. (1991). Localization in an echoic environment. In Y. Cazals, K. Horner & L. Demany (Eds.), *Auditory physiology and perception* (pp. 555–561). Oxford: Pergamon Press.

Harris, G. G. (1960). Binaural interactions of impulsive stimuli and pure tones. *Journal of the Acoustical Society of America, 32,* 685–692.

Harris, J. D. (1972). A florilegium of experiments on directional hearing. *Acta Oto-Laryngologica,* Suppl. 298, 1–26.

Harris, J. D., & Sergeant, R. L. (1971). Monaural/binaural minimum audible angles for a moving sound source. *Journal of Speech and Hearing Research, 14,* 618–629.

Hartmann, W. M., & Rakerd, B. (1989). On the minimum audible angle—A decision theory approach. *Journal of the Acoustical Society of America, 85,* 2031–2041.

Hebrank, J., & Wright, D. (1974). Spectral cues used in the location of sound sources on the median plane. *Journal of the Acoustical Society of America, 56,* 1829–1834.

Henning, G. B. (1974a). Detectability of interaural delay in high-frequency complex waveforms. *Journal of the Acoustical Society of America, 55,* 84–90.

Henning, G. B. (1974b). Lateralization and the binaural masking-level difference. *Journal of the Acoustical Society of America, 55,* 1259–1265.

Hirsh, I. J. (1948). The influence of interaural phase on interaural summation and inhibition. *Journal of the Acoustical Society of America, 20,* 536–544.

Humanski, R. A., & Butler, R. A. (1988). The contribution of the near and far ear toward localization of sound in the sagittal plane. *Journal of the Acoustical Society of America, 83,* 2300–2310.

Jain, M., Gallagher, D. T., Koehnke, J., & Colburn, H. S. (1991). Fringed correlation discrimination and binaural detection. *Journal of the Acoustical Society of America, 90,* 1918–1926.

Jeffress, L. A. (1948). A place theory of sound localization. *Journal of Comparative and Physiological Psychology, 41,* 35–39.

Jeffress, L. A., Blodgett, H. C., & Deatherage, B. H. (1952). The masking of tones by white noise as function of the interaural phases of both components. I. 500 cycles. *Journal of the Acoustical Society of America, 24,* 523–527.

Klumpp, R. G., & Eady, H. R. (1956). Some measurements of interaural time difference thresholds. *Journal of the Acoustical Society of America, 28,* 859–860.

Koehnke, J., Colburn, H. S., & Durlach, N. I. (1986). Performance in several binaural-interaction experiments. *Journal of the Acoustical Society of America, 79,* 1558–1562.

Kohlrausch, A. (1986). The influence of signal duration, signal frequency and masker duration on binaural masking level differences. *Hearing Research, 23,* 267–273.

Kollmeier, B., & Gilkey, R. H. (1990). Binaural forward and backward masking: Evidence for sluggishness in binaural detection. *Journal of the Acoustical Society of America, 87,* 1709–1719.

Kuhn, G. F. (1987). Physical acoustics and measurements pertaining to directional hearing. In W. A. Yost & G. Gourevitch (Eds.), *Directional hearing* (pp. 3–25). New York: Springer-Verlag.

Lackner, J. R., & Shenker, B. (1985). Proprioceptive influences on auditory and visual spatial localization. *Journal of Neuroscience, 5,* 579–583.

Langford, T. L., & Jeffress, L. A. (1964). Effect of noise crosscorrelation on binaural signal detection. *Journal of the Acoustical Society of America, 36,* 1455–1458.

Lappin, J. S., Bell, H. H., Harm, O. J., & Kottas, B. (1975). On the relation between time and space in the visual discrimination of velocity. *Journal of Experimental Psychology: Human Perception and Performance, 1,* 383–394.

Lee, D. N. (1980). The optic flow field. *Transactions of the Royal Society of London,* Series B, *290,* 169–179.

Litovsky, R. Y., & Clifton, R. K. (1992). Use of sound-pressure level in auditory distance discrimination by six-month-old infants and adults. *Journal of the Acoustical Society of America, 92,* 794–802.

Litovsky, R. Y., & Macmillan, N. A. (1994). Sound localization precision under conditions of the precedence effect: Effects of azimuth and standard stimulus. *Journal of the Acoustical Society of America, 96,* 752–758.

Little, A. D., Mershon, D. H., & Cox, P. H. (1992). Spectral content as a cue to perceived auditory distance. *Perception, 21,* 405–416.

Loomis, J. M., Fujita, N., Da Silva, J. A., & Fukusima, S. S. (1992). Visual space perception and visually directed action. *Journal of Experimental Psychology: Human Perception and Performance, 18,* 906–921.

Makous, J. C., & Middlebrooks, J. C. (1990). Two-dimensional sound localization by human listeners. *Journal of the Acoustical Society of America, 87,* 2188–2200.

McFadden, D. (1968). Masking-level differences determined with and without interaural disparities in masker intensity. *Journal of the Acoustical Society of America, 44,* 212–223.

McFadden, D., & Pasanen, E. G. (1974). High-frequency masking-level differences with narrow-band noise signals. *Journal of the Acoustical Society of America, 56,* 1226–1230.

Mershon, D. H., Ballenger, W. L., Little, A. D., & McMurtry, P. L. (1989). Effects of room reflectance and background noise on perceived auditory distance. *Perception, 18,* 403–416.

Mershon, D. H., & Bowers, J. N. (1979). Absolute and relative cues for the auditory perception of egocentric distance. *Perception, 8,* 311–322.

Mershon, D. H., & King, E. (1975). Intensity and reverberation as factors in the auditory perception of egocentric distance. *Perception and Psychophysics, 18,* 409–415.

Middlebrooks, J. C. (1992). Narrow-band sound localization related to external ear acoustics. *Journal of the Acoustical Society of America, 92,* 2607–2624.

Middlebrooks, J. C., & Green, D. M. (1990). Directional dependence of interaural envelope delays. *Journal of the Acoustical Society of America, 87,* 2149–2162.

Middlebrooks, J. C., & Green, D. M. (1991). Sound localization by human listeners. *Annual Review of Psychology, 42,* 135–159.

Middlebrooks, J. C., Makous, J. C., & Green, D. M. (1989). Directional sensitivity of sound-pressure levels in the human ear canal. *Journal of the Acoustical Society of America, 86,* 89–108.

Mills, A. W. (1958). On the minimum audible angle. *Journal of the Acoustical Society of America, 30,* 237–246.

Mills, A. W. (1960). Lateralization of high-frequency tones. *Journal of the Acoustical Society of America, 32,* 132–134.

Musicant, A. D., & Butler, R. A. (1984a). The influence of pinnae-based spectral cues on sound localizations. *Journal of the Acoustical Society of America, 75,* 1195–1200.

Musicant, A. D., & Butler, R. A. (1984b). The psychophysical basis of monaural localization. *Hearing Research, 14,* 185–190.

Musicant, A. D., & Butler, R. A. (1985). Influence of monaural spectral cues on binaural localization. *Journal of the Acoustical Society of America, 77,* 202–208.

Neti, C., Young, E. D., & Schneider, M. H. (1992). Neural network models of sound localization based on directional filtering by the pinnna. *Journal of the Acoustical Society of America, 92,* 3140–3156.

Perrott, D. R. (1984a). Concurrent minimum audible angle: A re-examination of the concept of auditory spatial acuity. *Journal of the Acoustical Society of America, 75,* 1201–1206.

Perrott, D. R. (1984b). Discrimination of the spatial distribution of concurrently active sound sources: Some experiments with stereophonic arrays. *Journal of the Acoustical Society of America, 76,* 1704–1712.

Perrott, D. R., Constantino, B., & Ball, J. (1993). Discrimination of moving events with accelerate or decelerate over the listening interval. *Journal of the Acoustical Society of America, 93,* 1053–1057.

Perrott, D. R., & Marlborough, K. (1989). Minimum audible movement angle: Marking the end points of the path traveled by a moving sound source. *Journal of the Acoustical Society of America, 85,* 1773–1775.

Perrott, D. R., Marlborough, K., Merrill, P., & Strybel, T. Z. (1989). Minimum audible angle thresholds obtained under conditions in which the precedence effect is assumed to operate. *Journal of the Acoustical Society of America, 85,* 282–288.

Perrott, D. R., & Saberi, K. (1990). Minimum audible angle thresholds for sources varying in both elevation and azimuth. *Journal of the Acoustical Society of America, 87,* 1728–1731.

Perrott, D. R., Saberi, K., Brown, K., & Strybel, T. Z. (1990). Auditory psychomotor coordination and visual search performance. *Perception and Psychophysics, 48,* 214–226.

Perrott, D. R., & Tucker, J. (1988). Minimum audible movement angle as a function of signal frequency and the velocity of the source. *Journal of the Acoustical Society of America, 83,* 1522–1527.

Plenge, G. (1974). On the differences between localization and lateralization. *Journal of the Acoustical Society of America, 56,* 944–951.

Pollack, I., & Trittipoe, W. J. (1959). Binaural listening and interaural noise cross correlation. *Journal of the Acoustical Society of America, 31,* 1250–1252.

Rakerd, B., & Hartmann, W. M. (1985). Localization of sound in rooms. II. The effects of a single reflecting surface. *Journal of the Acoustical Society of America, 78,* 524–533.

Rakerd, B., & Hartmann, W. M. (1992). Precedence effect with and without interaural difference—Sound localization in three planes. *Journal of the Acoustical Society of America, 92,* 2296(A).

Rayleigh, L. (1907). On our perception of sound direction. *Philosophical Magazine, 13,* 214–232.

Robinson, D. E., & Jeffress, L. A. (1963). Effect of varying the interaural noise correlation on the detectability of tonal signals. *Journal of the Acoustical Society of America, 35,* 1947–1952.

Roffler, S. K., & Butler, R. A. (1968). Factors that influence the localization of sound in the vertical plane. *Journal of the Acoustical Society of America, 43,* 1255–1259.

Saberi, K., & Perrott, D. R. (1990). Lateralization thresholds obtained under conditions in which the precedence effect is assumed to operate. *Journal of the Acoustical Society of America, 87,* 1732–1737.

Schiano, J. L., Trahiotis, C., & Bernstein, L. R. (1986). Lateralization of low-frequency tones and narrow bands of noise. *Journal of the Acoustical Society of America, 79,* 1563–1570.

Searle, C. L., Braida, L. D., Cuddy, D. R., & Davis, M. F. (1975). Binaural pinna disparity: Another auditory localization cue. *Journal of the Acoustical Society of America, 57,* 448–455.

Searle, C. L., Braida, L. D., Davis, M. F., & Colburn, H. S. (1976). Model for auditory localization. *Journal of the Acoustical Society of America, 60,* 1164–1175.

Sever, J. C., Jr., & Small, A. M., Jr. (1979). Binaural critical masking bands. *Journal of the Acoustical Society of America, 66,* 1343–1350.

Shaw, E. A. G. (1974). Transformation of sound pressure level from the free field to the eardrum in the horizontal plane. *Journal of the Acoustical Society of America, 56,* 1848–1861.

Shelton, B. R., & Searle, C. L. (1980). The influence of vision on the absolute identification of sound-source position. *Perception and Psychophysics, 28,* 589–596.

Shinn-Cunningham, B. G., Zurek, P. M., & Durlach, N. I. (1993). Adjustment and discrimination measurements of the precedence effect. *Journal of the Acoustical Society of America, 93,* 2923–2932.

Sovijärvi, A. R. A., & Hyvärinen, J. (1974). Auditory cortical neurons in the cat sensitive to the direction of sound source movement. *Brain Research, 73,* 455–471.

Steiger, H., & Bregman, A. S. (1982). Negating the effects of binaural cues: Competition between auditory streaming and contralateral induction. *Journal of the Experimental Psychology: Human Perception and Performance, 8,* 602–613.

Stern, R. M., & Colburn, H. S. (1978). Theory of binaural interaction based on auditory-nerve data. IV. A model for subjective lateral position. *Journal of the Acoustical Society of America, 64,* 127–140.

Stevens, S. S., & Newman, E. B. (1936). The localization of actual sources of sound. *American Journal of Psychology, 48,* 297–306.

Strybel, T. Z., & Perrott, D. R. (1984). Discrimination of relative distance in the auditory modality: The success and failure of the loudness discrimination hypothesis. *Journal of the Acoustical Society of America, 76,* 318–320(L).

Tobias, J. V., & Zerlin, S. (1959). Lateralization threshold as a function of stimulus duration. *Journal of the Acoustical Society of America, 31,* 1591–1594.

Wallach, H., Newman, E. B., & Rosenzweig, M. R. (1949). The precedence effect in sound localization. *American Journal of Psychology, 62,* 315–336.

Warren, D. H., Welch, R. B., & McCarthy, T. J. (1981). The role of visual-auditory "compellingness" in the ventriloquism effect: Implications for transitivity among the spatial senses. *Perception and Psychophysics, 30,* 557–564.

Whitworth, R. H., & Jeffress, L. A. (1961). Time vs. intensity in the localization of tones. *Journal of the Acoustical Society of America, 33,* 925–929.

Wightman, F. L. (1971). Detection of binaural tones as a function of masker bandwidth. *Journal of the Acoustical Society of America, 50,* 623–636.

Wightman, F. L., & Kistler, D. J. (1989). Headphone simulation of free-field listening. I. Stimulus synthesis. *Journal of the Acoustical Society of America, 85,* 858–867.

Wightman, F. L., & Kistler, D. J. (1992). The dominant role of low-frequency interaural time differences in sound localization. *Journal of the Acoustical Society of America, 91,* 1648–1661.

Yost, W. A. (1981). Lateralization position of sinusoids presented with interaural intensive and temporal differences. *Journal of the Acoustical Society of America, 70,* 397–409.

Yost, W. A., & Hafter, E. R. (1987). Lateralization. In W. A. Yost & G. Gourevitch (Eds.), *Directional hearing* (pp. 49–84). New York: Springer-Verlag.

Zurek, P. M. (1980). The precedence effect and its possible role in the avoidance of interaural ambiguities. *Journal of the Acoustical Society of America, 67,* 952–964.

Zurek, P. M. (1987). The precedence effect. In W. A. Yost & G. Gourevitch (Eds.), *Directional hearing* (pp. 85–105). New York: Springer-Verlag.

CHAPTER 10

Models of Binaural Interaction

Richard M. Stern
Constantine Trahiotis

I. INTRODUCTION: CROSS-CORRELATION MODELS OF BINAURAL PERCEPTION

The human binaural system has attracted the attention of auditory theorists since Lord Rayleigh formulated the duplex theory in 1907. The "modern era" of binaural modeling can be said to have begun in 1948 with Jeffress's prescient paper suggesting a neural coincidence mechanism to detect interaural time differences and, coincidentally, the original descriptions of the binaural masking level difference provided independently by Hirsch (1948) and Licklider (1948). The next 25 years witnessed an explosion in experimental studies in subjective lateralization, binaural detection, and interaural discrimination (as discussed in Chapter 9 of this volume, and in the complementary review by Hafter and Trahiotis, 1994). A number of significant efforts were made to describe these data in terms of quantitative models. For example, Sayers and Cherry (1957) formulated the first explicit cross-correlation model that was directly compared to experimental data. Webster (1951) and Jeffress, Blodgett, Sandel, and Wood (1956) called attention to the importance of stimulus variability through their seminal papers that highlighted the interaural time delay (ITD) produced by a vectorial combination of the target and masker components of stimuli used in binaural

Hearing

detection experiments. This "vector" model was later elaborated on by Hafter and others (e.g., Hafter, 1971; Yost, 1972). Other important models developed during that period include the equalization–cancellation model of Durlach (1963, 1972) and the model based on auditory-nerve activity of Colburn (1973, 1977), among several others. Two review chapters by Durlach and Colburn (1978) and Colburn and Durlach (1978), respectively, elegantly summarize most of the major experimental results and theoretical models up to 1972. Colburn (1995) has recently written a comprehensive review of binaural models that contains a thorough discussion of computational models of physiological processing.

Colburn and Durlach (1978) argued that all of the then-current binaural models could be thought of as a particular realization of the generic model of binaural interaction shown in Figure 1. This generic structure includes a series of peripheral processing steps consisting of bandpass filtering, rectification, stochastic neural representation of the signals, comparison of interaural timing information over a limited range of internal delays using a correlation or coincidence mechanism, consideration of interaural intensity differences of the outputs of the monaural processors, and a subsequent decision-making mechanism. Since 1978 there has been a general acceptance of this basic structure, and especially the cross-correlation mechanism used for the extraction of interaural timing information. The basic cross-correlation mechanism has been extended in a number of ways within more recent models. Other significant recent trends in binaural modeling include an increased reliance on computational (as opposed to analytical) approaches to predict the phenomena, as well as initial efforts to make use of head-related transfer functions to understand and reproduce out-of-head localization phenomena. In recent years our understanding of the binaural system has begun to be applied to the simulation of room acoustics for sound

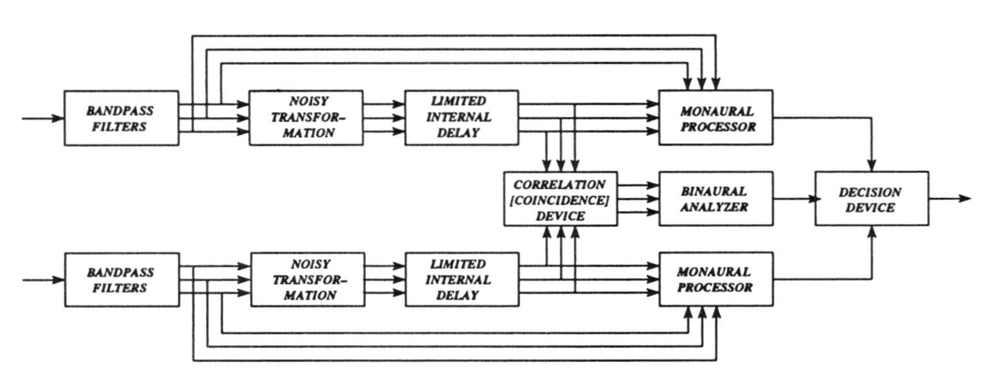

FIGURE 1 Generic model of binaural processing proposed by Colburn and Durlach (1978). The parallel sets of arrows indicate multiple parallel channels of information in the model.

presented through headphones (e.g., Bodden, 1993), to the reduction of error rates of speech recognition systems (e.g., DeSimio and Anderson, 1993; Sullivan and Stern, 1993), and in the simulation of out-of-head images for virtual environments and displays (e.g., Wenzel, 1992).

The goal of this chapter is to provide the reader with an intuitive understanding of how cross-correlation-based binaural models work and an appreciation of their capabilities and limitations in describing a variety of binaural phenomena. This chapter is more tutorial but less comprehensive than the one recently written by Colburn (1994), and our expectation is that both types of discussion of this work will be useful, especially to new researches in the field. We review the initial formulations of the cross-correlation model in Section II and describe how this structure has been recently modified in Section III. In Section IV we describe how the general cross-correlation models have been applied to psychophysical data.

II. STRUCTURE OF BINAURAL CROSS-CORRELATION-BASED MODELS

A. The Original Forms of the Cross-Correlation Model

Modern binaural models are all based on Jeffress's (1948) original conception of a neural "place" mechanism that would enable the extraction of interaural timing information. Jeffress suggested that external interaural delays could be inferred by central units that record coincidences of neural impulses from pairs of more peripheral nerve fibers. Each central unit was presumed to compare information from the two ears after a series of internal time delays, as shown in the block diagram of Figure 2. The delay mechanism is commonly conceptualized in the form of a ladder-type delay line as in Figure 2, but such a structure is not the only possible realization. A key parameter in the analysis of the outputs of such a mechanism is the interaural difference of the total delay incurred by the two monaural signals arriving at a given coincidence detector. This variable will be referred to as the *net internal delay* τ for that particular unit. The short-term average of the set of coincidence outputs plotted as a function of their internal delay τ is an approximation to the short-term cross-correlation function of the neural signals arriving at the coincidence detectors. Licklider (1959) proposed that such a mechanism could also be used to achieve an autocorrelation of neural signals for use in models of pitch perception.

A different, analytical, model based on cross-correlation was developed by Sayers and Cherry (1957) to describe their early measurements of fusion and laterality. We briefly review the major features of this model because it was the first quantitative application of cross-correlation, and because it contains many of the elements of modern models of binaural perception. In

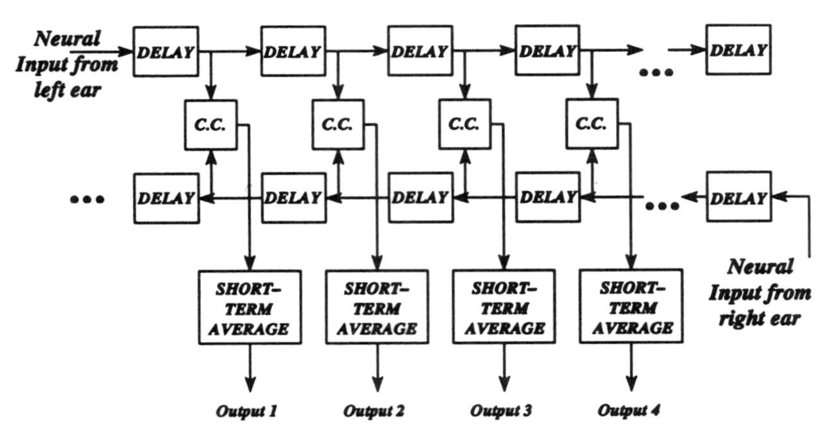

FIGURE 2 Schematic representation of the Jeffress place mechanism. The blocks labeled *C.C.* record coincidences of neural activity from the two ears (after the delays are incurred).

Sayers and Cherry's formulation the *short-term running cross-correlation function* $R(\tau, t)$ is formed from the signals to the two ears, according to the equation

$$R(\tau,t) = \int_{-\infty}^{t} x_L(\alpha)x_R(\alpha - \tau)w(t - \alpha)p(\tau)\, d\alpha$$

where $x_L(t)$ and $x_R(t)$ are the signals to the left and right ears. The function $w(t)$ represents the temporal weighting of the short-term correlation operation, and typically took on an exponential form in most of Sayers and Cherry's calculations. The function $p(\tau)$ was typically of the form $e^{-k|\tau|}$ and served to emphasize the contributions of internal delays of small magnitude. We later refer to this type of emphasis as *centrality*. As is the case with all cross-correlation-based models, an additional mechanism is needed to account for the *effects of interaural intensity difference* (IID). Sayers and Cherry added a constant proportional to the intensity of the left-ear signal to values of $R(\tau,t)$ for which τ was less than 0 and a (generally different) constant proportional to the intensity of the right-ear signal to values of $R(\tau,t)$ for which τ was greater than 0. A *judgment mechanism* then extracted subjective lateral position using the statistic

$$\hat{P} = \frac{I_L - I_R}{I_L + I_R}$$

where I_L and I_R are the integrals of the intensity-weighted short-term cross-correlation function over negative and positive values of τ, respectively.

It is important to note that the Sayers and Cherry model is based on processing the cross-correlation of the *original* signals to the two ears, rather than the cross-correlation of the neural representation of these signals after filtering by the auditory periphery. Figure 3 shows examples of such cross-correlation function, for a 500-Hz tone and for an ideal bandpass noise with center frequency 500 Hz and bandwidth 200 Hz. Each signal contains an ITD of 500 μs (0.5 ms). Both functions exhibit positive peaks spaced at the reciprocal of the center frequency, and the amplitudes of the peaks for the bandpass noise decrease at a rate that is proportional to the signal bandwidth. In each case the maximum value of the cross-correlation function occurs at the internal delay equal to the ITD of the stimulus, but for the sine wave, this maximum value also occurs at several other locations along the τ axis.

B. Colburn's Auditory-Nerve-Based Model

An influential quantification of the Jeffress hypothesis was formulated by Colburn (1973, 1977) who compared the information that could be extrac-

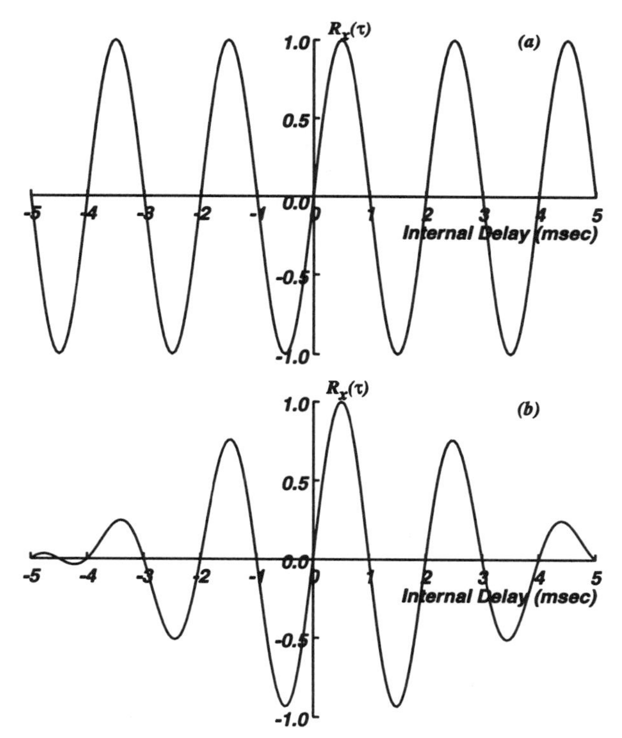

FIGURE 3 Examples of cross-correlation functions for (a) a 500-Hz pure tone and (b) bandpass noise with a center frequency of 500 Hz and bandwidth of 200 Hz. In each case the signals are presented with an ITD of 0.5 ms.

ted from the responses of populations of auditory-nerve fibers to performance in binaural detection and interaural discrimination experiments. Colburn's model consisted of two parts: a model of auditory-nerve activity and a central processor that analyzes and displays comparisons of firing times from ear to ear.

1. The Model of Auditory-Nerve Activity

The model of auditory-nerve activity used in the original Colburn model was adapted from an earlier formulation by Siebert (1970). It consists of a bandpass filter, a lowpass filter, and an exponential rectifier, followed by a mechanism that generates firing times of a nonhomogeneous Poisson process at a rate that is proportional to the output of the rectifier. Stern and Shear (1995) later modified this model by changing the shape of the nonlinear rectifier and interchanging the order of the rectifier and the low-pass filter. Functional models of similar form have been used in the work of several other researchers including Duifhuis (1973), Blauert and Cobben (1978), and Lindemann (1986a).

Colburn used the nonhomogeneous Poisson process to characterize the response of auditory-nerve fibers to sound because it is the simplest stochastic process that can realistically be applied to model the neural firing times. Each fiber is characterized by a *rate function*, $r(t)$, that describes the instantaneous rate of firing assumed to be produced by that fiber. This time-varying function depends on both the characteristic frequency (CF) of the nerve fiber and the spectral–temporal characteristics of the stimulus. Using an explicit analytical model like the Poisson process, one can calculate means and variances of the predicted outputs of the coincidence counters. These statistics can then be used to predict discrimination and detection thresholds either by application of the Cramer–Rao bound (cf. Van Trees, 1968), or by direct prediction of performance obtained by assuming that the decision variable is normally distributed. The general success of this approach notwithstanding, constructing and evaluating quantitative neurally based models is inevitably a compromise between analytical tractability and faithfulness to the known physiological results. For example, it is well known that the peripheral auditory system is both time varying (due to the refractory nature of the auditory response) and nonlinear. The Poisson-process model ignores the refractoriness in the response and analytical predictions can be developed easily only for a limited set of stimuli with quasi-static interaural differences (including pure tones, tones in noise, and bandpass noise). Furthermore, predictions are easily developed only for a limited set of assumed peripheral nonlinearities (including exponential and half-wave power-law rectifiers).

Because of the difficulty in developing analytical predictions for many interesting stimuli, including signals with significant transients and stimuli that give rise to the precedence effect, more computationally oriented (and more physiologically accurate) models of the peripheral auditory response to sound are now becoming increasingly popular (e.g., Carney, 1993; Payton, 1988; Meddis, Hewitt, & Shackleton, 1990). For example, the Meddis et al. model of auditory nerve activity has been incorporated into the binaural processing model of Shackleton, Meddis, & Hewitt (1992).

2. The Model of Central Processing

Colburn first considered for the central processing model a binaural analyzer that used general comparisons of timing information from the responses of the fibers from the two ears, but he found that more information was available to such a model than appeared to be used by humans. He found that predictions of a model that compares interaural timing information emanating only from fibers with the same CF, and with only a single internal interaural delay, were consistent with performance observed in several interaural time and intensity discrimination tasks (Colburn, 1973). The more restricted model described the processing that would be provided by the ensemble of coincidence-counting units originally proposed by Jeffress. Colburn subsequently compared the predictions of the restricted model with available data on binaural detection thresholds and found that the model could predict virtually all of those data as well (Colburn, 1977).

The response of each coincidence counter of the Colburn model is characterized by two parameters: the internal delay and CF. Colburn (1977) assumed that a coincidence is achieved only when the firings from the two input fibers are nearly simultaneous. This enabled him to assume that the output of the coincidence counter is also a Poisson process, with mean

$$E[L(\tau,f)] = T_W \int_0^T r_L(t) r_R(t - \tau)\, dt$$

where $L(\tau,f)$ represents the number of coincidences recorded by a unit with internal delay and CF equal to τ and f, respectively. The functions $r_L(t)$ and $r_R(t)$ are the Poisson rate functions of the two input fibers, T_W represents the duration of the coincidence window, and T represents the duration of the stimulus. For a given CF, the expected number of coincidences plotted as a function of the internal delay parameter (τ) describes the *cross-correlation of the neural representation* of the binaural signal, as determined by the rate functions of the fibers from the two ears at that CF. The function $E[L(\tau,f)]$ can also be thought of as a special case of the corresponding running cross-correlation function of the Poisson rate functions $r_L(t)$ and $r_R(t)$,

$$E[L(\tau,f)] = E[L(\tau,t,f)] = \int_{-\infty}^{t} r_L(\alpha) r_R(\alpha - \tau) w_c(t - \alpha) p(\tau)\, d\alpha$$

The function $w_c(t)$ in this equation is a temporal weighting function and emphasizes the most recent values of the cross-correlation of $r_L(t)$ and $r_R(t)$, just as the function $w(t)$ used by Sayers and Cherry provided temporal weighting for the cross-correlation of the original stimulus. For most of the research of Colburn, Stern, and colleagues, $w_c(t)$ is assumed to be a constant for $0 \leq t \leq T$ and 0 otherwise. Stern and Bachorski (1983) have developed some predictions for the statistics of the coincidence-counter outputs using an exponentially shaped $w_c(t)$, similar to the exponential function $w(t)$ proposed by Sayers and Cherry (1957).

Colburn and Durlach (1978) have noted that Colburn's auditory-nerve-based model can also be regarded as a generalization of the equalization–cancellation (EC) model of Durlach (1963). The EC model has been most successful in predicting the results of binaural detection experiments. Predictions are obtained by applying a combination of ITD and IID that produce the best "equalization" of the masker components of the stimuli presented to each of the two ears and allow "cancellation" of the resulting signals by subtraction of one from the other. Compensation for ITD has a greater impact on predictions than compensation for IID in the equalization stage of the EC model. The internal interaural delays of the fiber pairs of the Jeffress–Colburn model perform the same function as the ITD-equalizing operation of the EC model. As a result, many detection-threshold predictions for the two models are similar in form. Minor differences in predictions occur because the EC model assumes that only a single best delay is available in the equalization operation, whereas the Jeffress–Colburn structure implies that many delays are simultaneously available for processing the signals.

C. Physiological Support for Cross-Correlation in the Binaural System

A number of physiological studies have described cells that are likely to be relevant to binaural processing, as summarized by the discussion in Chapter 3, along with the review chapters of Kuwada and Yin (1987) and Colburn (1995). For example, cells that appear to record IIDs have been reported in the superior olivary complex, inferior colliculus, and other sites (e.g., Boudreau & Tsuchitani, 1968; Goldberg & Brown, 1969). Of particular interest to the developers of models based on the Jeffress–Colburn coincidence mechanism are cells first reported by Rose, Geisler, and Hind (1966) in the inferior colliculus that appear to be maximally sensitive to signals presented with a specific interaural delay, independent of frequency. This delay is referred to as a *characteristic delay*. Cells with a similar response have been reported by others in the medial superior olive (e.g., Goldberg & Brown, 1969; Crow, Rupert, & Moushegian, 1978; Yin & Chan, 1990), and the dorsal nucleus of the lateral lemniscus (e.g., Brugge, Anderson, & Aitkin, 1970). A series of measurements has been performed that characterizes the distribution of ITD-sensitive cells in the inferior colliculus (Yin & Kuwada, 1984; Kuwada, Stanford, & Batra, 1987), and the medial geniculate body (Stanford, Kuwada, & Batra, 1992). Although most cells exhibit characteristic delays that fall within the maximum delay possible for a point source in a free field for a particular animal, a substantial number of ITD-sensitive cells have characteristic delays that fall outside the "physically plausible" range.

The anatomical origin of the characteristic delays has been the source of some speculation. Many physiologists believe that the delays are of neural origin, caused either by slowed conduction velocity or synaptic delays (e.g., Smith, Joris, & Yin, 1993; Carr & Konishi, 1990; Young & Rubel, 1983). Schroeder (1977) proposed an alternative hypothesis, suggesting that the characteristic delays could also be obtained if higher processing centers compare timing information derived from auditory-nerve fibers with different CFs. This hypothesis is also a part of the model proposed more recently by Shamma, Shen, and Gopalaswamy (1989), which they call the *stereausis model*. Shamma's model for central processing is very similar to the general structure proposed by Jeffress and quantified by Colburn, and it has been implemented as an integrated circuit by Lazzaro, Mead, and colleagues (Lazzaro, 1991; Mead, Arreguit, & Lazzaro, 1991). The stereausis model is not currently specified in sufficient detail to enable its predictions to be compared critically either to predictions of other models or to the corresponding experimental data. In general, the predictions of binaural models are unaffected by whether the internal delays are assumed to be caused by neural or mechanical phenomena.

D. Temporal Integration of the Coincidence Display

Although the binaural system can be shown to resolve static ITDs as small as tens of microseconds (e.g., Klumpp & Eady, 1956; Zwislocki & Feldman, 1956), experiments measuring responses to *time-varying* ITDs (e.g., Licklider, Webster, & Hedlun, 1950; Grantham & Wightman, 1978) indicate a much more "sluggish" response, with limits on the order of tens of milliseconds, as discussed in Chapter 9 of this volume. To understand these apparently diverging sets of results, one must recall that results in static ITD-discrimination experiments reflect changes in place of activity of the coincidence-counting units along the internal-delay axis. Hence, thresholds would reflect the density of fiber pairs with respect to internal delay at each

CF. Resolution of time-varying interaural differences, on the other hand, reflects temporal integration or the averaging of instantaneous responses over *running time* (as opposed to interaural time delay).

To understand binaural "sluggishness," it is helpful to think of the temporal averaging of the matrix of coincidence-counting units as the output of a linear filter that has an impulse response equal to the temporal weighting function $w(t)$ and an input equal to the instantaneous interaural cross-correlation of the neural response to the signals, $r_L(t)r_R(t - \tau)$. Figure 4 demonstrates how $E[L(t,\tau,f)]$, the expected value of the instantaneous number of coincidences, varies as a simultaneous function of internal delay (τ) and running time (t) with and without temporal integration. The stimulus in each case is a 500-Hz tone, and responses are depicted for fibers with that frequency as the CF. In the upper panel the instantaneous value of $r_L(t)r_R(t - \tau)$ is plotted without averaging over running time. Note that peaks appearing in this panel are limited to particular intervals of the running time, t (as well as occurring at particular values of τ). The lower panel of Figure 4 shows the same function, but *after* integration by convolution

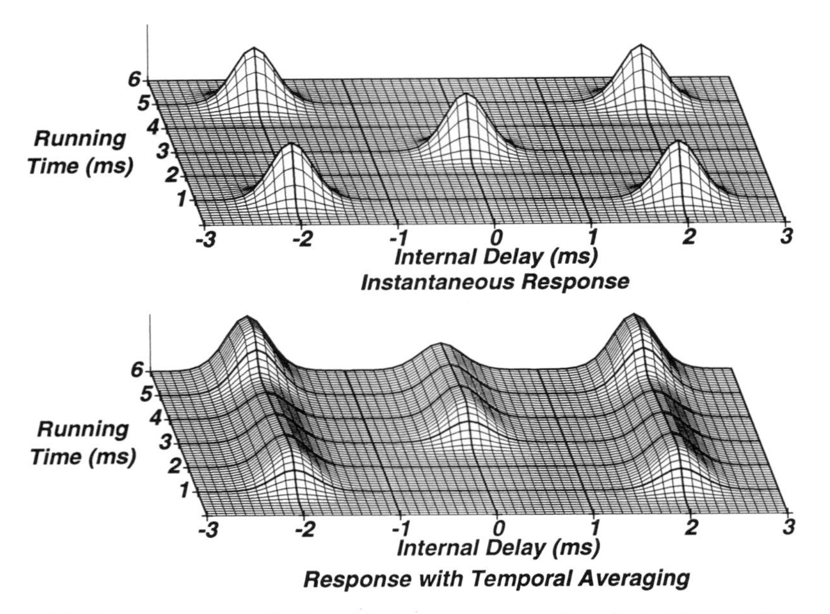

FIGURE 4 The expected value of the instantaneous number of coincidences, $E[L(t, \tau, f)]$, as a simultaneous function of running time t and internal delay τ to a 500-Hz tone with zero ITD. The response is shown using no temporal integration (upper panel) and using temporal integration by an exponentially shaped temporal weighting function with an effective cutoff frequency of 5 Hz in the frequency domain.

with an exponential time window of the form $w(t) = e^{-10\pi t}$ for positive values of t. This is the impulse response of a single-pole low-pass filter with cutoff frequency 5 Hz, which is typical of the types of integrating filters considered by Grantham and Wightman (1978) and others for accounting for sensitivity to temporal modulation of ITDs. Note that temporal integration causes the isolated peaks in the instantaneous cross-correlation shown in the upper panel to be transformed to smoother ridges that are parallel to the t-axis. We believe that this enables the binaural system to provide a stable spatial representation of the acoustic world.

III. EXTENSIONS OF THE CROSS-CORRELATION APPROACH

Since the time of Colburn's original formulation, several research groups have extended the structure and application of the frequency-dependent cross-correlation analyzer in a number of different ways. We summarize a number of these extensions to the Jeffress–Colburn model in this section.

A. Extensions by Stern, Colburn, and Trahiotis

The goals of the work of Stern and his colleagues (Stern & Colburn, 1978, 1985; Stern & Bachorski, 1983; Stern, Shear, & Zeppenfeld, 1988a; Stern & Trahiotis, 1992; Stern & Shear, 1995) have been to determine ways in which the subjective lateral position of binaural stimuli can be related to the activity of the coincidence-counting units and in turn to examine the extent to which objective interaural discrimination and binaural detection results can be related to changes in predicted lateral position. Colburn's coincidence-counting mechanism was extended by adding explicit assumptions concerning time–intensity interaction as well as a mechanism for extracting subjective lateral position from the modified display (Stern & Colburn, 1978). This extension of the Jeffress–Colburn model is referred to as the *position-variable model* by Stern and his colleagues. To predict quantitatively various data concerning lateral position, Stern and colleagues (Stern et al., 1988a; Stern & Shear, 1995) also slightly modified the characterization of auditory-nerve activity and the description of the function that specifies the distribution of internal delays. In addition they proposed a second coincidence-based mechanism that emphasizes the impact of ITDs that are consistent over a range of frequencies (Stern and Trahiotis, 1992).

The function specifying the distribution of internal delays plays an important (but frequently unrecognized) role in developing predictions of subjective lateral position. Colburn (1969, 1977) originally assumed that the density function for internal delays, called $p(\tau)$, was independent of frequency, and he fitted the shape of $p(\tau)$ to predict the relative masking level differences for stimuli in the N_0S_π *vs.* $N_\pi S_0$ conditions, as defined in Chapter 9 of this volume. More recently, Stern and Shear (1995) made this function weakly dependent on frequency and changed its shape slightly to describe the observed dependence of the lateralization of tonal stimuli with a fixed ITD on stimulus frequency (Schiano, Trahiotis, & Bernstein, 1986).

The effect of the frequency-dependent density function for internal delay, $p(\tau,f)$, on the representation of a 500-Hz pure tone with an ITD of +0.5 ms is demonstrated in Figure 5. This figure shows the average total number of coincidences recorded by the coincidence-counting units as a joint function of internal delay (along the horizontal axis) and CF (along the oblique axis). Note that even though the stimulus is tonal, the spread of excitation result-

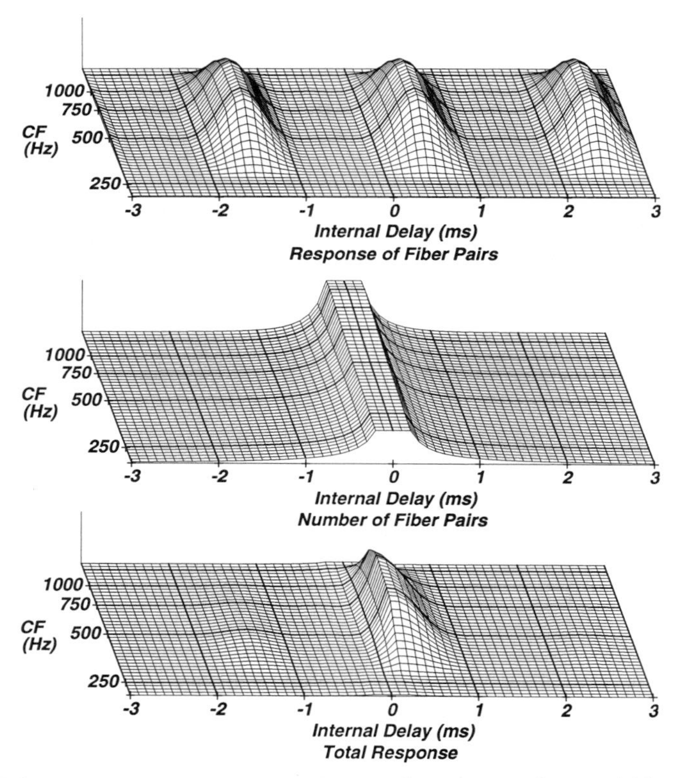

FIGURE 5 Cross-correlation patterns showing the response of an ensemble of binaural fiber pairs to a 500-Hz pure tone with a 0.5-ms ITD. Upper panel: the relative number of coincidences per fiber pair as a function of internal delay τ (in ms) and CF of the auditory-nerve fibers (in Hz). Central panel: the function $p(\tau,f)$, which describes the assumed distribution of internal delays as a function of CF. Lower panel: $E[L_T(\tau,f)\ ]$, the expected total number of coincidences as a function of internal delay and CF, which is the product of the upper and central panels.

ing from finite-bandwidth peripheral filtering produces a synchronized response over a fairly wide range of CFs (cf. Pfeiffer and Kim, 1975). The upper panel shows $E[L(\tau,f)]$, the average number of coincidences per fiber pair. The center panel shows the function $p(\tau,f)$ that describes the distribution of fiber pairs as a function of internal delay and CF. The lower panel displays $E[L_T(\tau,f)]$, the average total number of coincidences at each internal delay and CF, which is the product of the number of counts per fiber pair [$L(\tau,f)$, upper panel] and the number of fiber pairs [$p(\tau,f)$, central panel]. At each CF there is a distinct maximum in the cross-correlation function at a value of internal delay that is close to that of the original interaural delay of the stimulus.

The form of the function $p(\tau,f)$ implies that there are relatively more coincidence-counting units with internal interaural delays of smaller magnitude, which has been confirmed by physiological measurements (e.g., Kuwada et al., 1987). Nevertheless, a substantial fraction of the coincidence counters is assumed to have internal delays that are much greater in magnitude than the largest delays that are physically attainable with free-field stimuli. The existence of very long internal delays is in accord with psychoacoustical as well as physiological data.

As mentioned earlier in connection with the model of Sayers and Cherry (1957), any correlation-based binaural model must include an additional mechanism to incorporate the effects of IIDs on lateralization. The approach taken by Stern and Colburn (1978) was to multiply the function $L_T(\tau,f)$ by a Gaussian-shaped weighting function, referred to as $L_I(\tau,f)$, with location along the τ axis dependent on IID:

$$L_p(\tau,f) = L_T(\tau,f)L_I(\tau,f)$$

Stern and Colburn (1978) proposed that the predicted lateral position of a stimulus, $\hat{P}$, can be obtained by computing the centroid along the τ axis of the position function $L_p(\tau,f)$ while averaging over frequency:

$$\hat{P} = \frac{\int_{-\infty}^{\infty}\int_{-\infty}^{\infty} \tau L_P(\tau,f)\, d\tau df}{\int_{-\infty}^{\infty}\int_{-\infty}^{\infty} L_P(\tau,f)\, d\tau df}$$

This definition of predicted lateral position was originally adopted by Stern and Colburn for reasons of computational simplicity, and it has been employed by Blauert and his colleagues (e.g., Lindemann, 1986a) as well. It should be noted, however, that models like the position-variable model that predict the intracranial location of only a single image are unable to explain experimental results that suggest the existence of multiple images, such as the studies by Moushegian and Jeffress (1959), Whitworth and Jeffress (1961), and Hafter and Jeffress (1968).

Recently, Stern and Trahiotis (1992) have incorporated an additional modification to the model that is designed to emphasize the modes of the function $E[L(\tau,f)]$ that appear at the same internal delay over a range of CFs. These modes are referred to as the *straight* modes of $E[L(\tau,f)]$, and this weighting mechanism will be discussed in detail in Subsection IV.A.2.

B. Extensions by Blauert, Cobben, Lindemann, and Gaik

Blauert and his colleagues have made important contributions to correlation-based models of binaural hearing over an extended period of time. Their efforts have been directed primarily toward understanding how the binaural system processes more complex sounds in real rooms and have tended to be computationally oriented. This approach is complementary to that of Colburn and his colleagues, who have focused on explaining "classical" psychoacoustical phenomena using stimuli presented through earphones. In recent years Blauert and his colleagues have been applying knowledge gleaned from fundamental research in binaural hearing to help develop a "cocktail party processor" that can identify, separate, and enhance individual sources of sound in the presence of other, interfering sounds.

In the first English-language description of their modeling efforts, Blauert and Cobben (1978) combined the running cross-correlator of Sayers and Cherry (1957) with the model of the auditory periphery suggested by Duifhuis (1973). This model of peripheral processing is functionally similar to the model proposed by Siebert and adopted by Colburn. Blauert and Cobben described the response of the model to single clicks and to pairs of clicks, presented from spatially separated loudspeakers in an anechoic chamber. They found that a characterization of the average response to a sound in a particular frequency-specific auditory-nerve channel was adequate for their purposes (and that a characterization of the individual stochastic neural firing times was unnecessary).

Blauert and his colleagues subsequently developed a series of mechanisms that explicitly introduced the effects of stimulus IIDs into the modeling process. One of the most interesting and best known of these mechanisms is the one proposed by Lindemann (1986a), which may be regarded as an extension and elaboration of an earlier hypothesis of Blauert (1980). Lindemann extended the original Jeffress coincidence-counter model in two ways, adding (1) a mechanism that inhibits outputs of the coincidence counters when activity is produced by coincidence counters at adjacent internal delays, and (2) monaural-processing mechanisms at the "edges" of the display of coincidence-counter output that become active when the intensity of the signal to one of the two ears is extremely small.

The inhibitory mechanisms of the Lindemann model produce a "sharpening" of the peaks of the coincidence-counter outputs along the internal-delay axis. This is illustrated in Figure 6, which compares the temporal evolution of the response of the outputs of coincidence counters with a CF of 500 Hz to a pure tone with a sinusoidally varying ITD with and without the types of inhibition proposed by Lindemann (Palm, 1989). As in Figure 4, the oblique axis represents *running* time, and the horizontal axis represents internal delay. In calculating these responses we included both the static-inhibition and dynamic-inhibition components defined by Lindemann (1986a).

A very interesting property of the Lindemann model is that the interaction of the inhibition mechanism and the monaural processing mechanism causes the locations of peaks of the coincidence-counter outputs along the

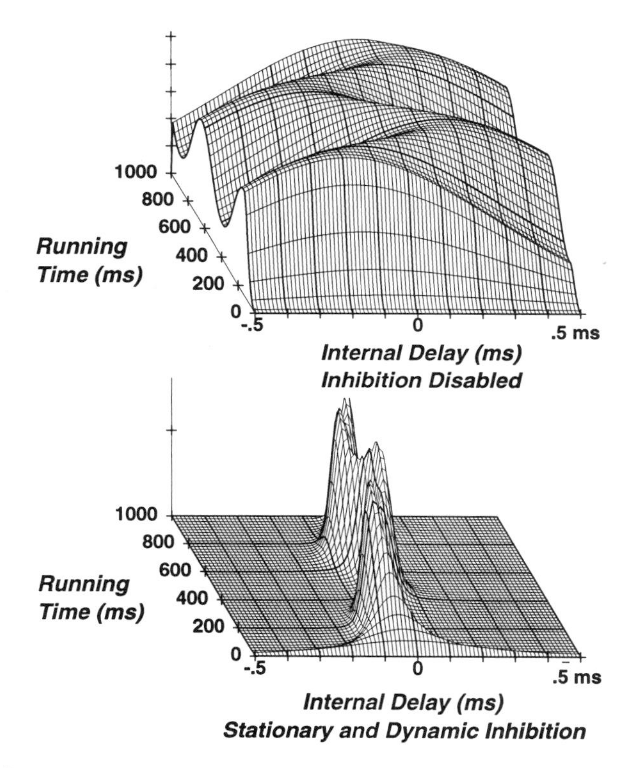

FIGURE 6 Plots of $E[L(t, \tau, f)]$, the expected value of the instantaneous number of coincidences as a simultaneous function of running time t and internal delay τ, to a 500-Hz tone with an ITD that is sinusoidally modulated at a frequency of 2 Hz. Response is shown without the stationary and dynamic inhibition mechanisms of the model of Lindemann (1986a) (upper panel) and with these inhibition mechanisms (lower panel).

internal-delay axis to shift with changes in IID. In other words, this model produces a time–intensity trading mechanism at the level of the coincidence-counter outputs. Although the net effect of IIDs on the patterns of coincidence-counter outputs in the Lindemann model is not unlike the effect of the intensity-weighting function $L_I(\tau,f)$ in the model of Stern and Colburn (1978), the time–intensity interaction of the Lindemann model is more esthetically satisfying to many because it arises naturally from the fundamental assumptions of the model rather than as the result of the imposition of an arbitrary weighting function. It has not yet been possible to determine the physiological plausibility of the Lindemann inhibition mechanism, but this issue is the currently the object of current investigation in the laboratories of Tom C. T. Yin and Shigeyuki Kuwada.

Gaik (1993) extended the Lindemann mechanism by adding a further weighting to the coincidence-counter outputs that reinforces naturally occurring combinations of ITD and IID. This has the effect of causing physically plausible stimuli to produce coincidence outputs with a single prominent peak that is compact along the internal-delay axis and consistent over frequency. Conversely, very unnatural combinations of ITD and IID (which tend to give rise to multiple spatial images) produce response patterns with more than one prominent peak along the internal-delay axis.

C. Mechanisms for Time–Intensity Interaction and Image Formation

1. Time–Intensity Interaction

As noted earlier, all correlation-based models must include some mechanism to describe the effects of IID on subjective lateral position. At one time, it was felt that the effects of IIDs in binaural lateralization could be accounted for by the decrease in latency of the auditory nerve response that occurs as the intensity of the signals is increased. This *peripheral* time–intensity trading mechanism, known as the *latency hypothesis,* was discussed by Jeffress in 1948 and later elaborated by David, Guttman, and van Bergeijk (1958) and Deatherage and Hirsh (1959). This hypothesis was at least qualitatively supported by early lateralization studies that utilized small ITDs and IIDs, but it cannot describe either lateralization data over a wider range of stimulus conditions such as the stimuli presented by Sayers (1964), Domnitz and Colburn (1977), and Bernstein and Trahiotis (1985), or the inability to trade time and intensity differences completely, as first shown by Hafter and Carrier (1972).

We have described three explicit *more central* mechanisms that have been proposed to account for the effects of IID: the multiplicative intensity-weighting pulse of Stern and Colburn (1978), the mechanism involving lateral inhibition of the coincidence-counting response along adjacent delays proposed by Lindemann (1986a), and the separate weighting of the left and right halves of the cross-correlation function of the data as used in the model of Sayers and Cherry (1957). The Stern–Colburn and Lindemann models provide similar predictions for the lateralization of 500-Hz pure tones as a joint function of ITD and IID, which are not as well described by the original Sayers and Cherry model.

In general, no focused attempt has been made to evaluate critically these or other intensity-weighting mechanisms on the basis of their ability to describe the lateralization of stimuli other than 500-Hz pure tones. Nevertheless, the general consensus among contemporary theoreticians is that the time–intensity interaction takes place at the level at which timing information from the signals to the two ears is first compared, if not more centrally, rather than at the level of the auditory nerve.

2. Image Formation

Thus far we have described two specific ways of predicting lateral position from the display of interaural coincidence-counting units: computation of the centroid along the internal-delay axis, as is done by Stern and Colburn (1978) and Lindemann (1986a), and the comparison of activity along the right and left halves of the internal-delay axis, as is done by Sayers and Cherry (1957). Other ways of predicting lateral position exist as well. One plausible alternative is to assume that position can be related to the location of the peaks of the cross-correlation function (as opposed to the centroid). The locations of the peaks of the cross-correlation function allow one to account for the multiple images that can occur for tonal stimuli presented interaurally out of phase (e.g., Sayers, 1964; Yost, 1981), as well as for the secondary "time image" observed for some stimuli presented with conflicting ITDs and IIDs (e.g., Whitworth & Jeffress, 1961; Hafter & Jeffress, 1968). Lateralization mechanisms based on the peaks of functions characterizing the response of the coincidence counters have been discussed by Lindemann (1986a) in conjunction with some of the predictions of his model. More recently, Shackleton et al. (1992) described a model that assumes that the listener computes either the centroid along the τ-axis or the locations of the peaks of the responses of coincidence-counting units, choosing the statistic that more accurately describes the results for a given experiment. Although definitely not parsimonious, this type of approach may be necessary to account for the data in all their complexity.

Few focussed efforts have been made to assess the relative merits of the various ways of generating an estimate of lateral position from the pattern of activity of the coincidence-counting units. For the most part, the specific lateralization mechanism adopted by a given researcher appears to have been selected more for convenience than on the basis of strongly held principles.

IV. ABILITY OF CROSS-CORRELATION MODELS TO DESCRIBE PSYCHOACOUSTICAL DATA

In this section we describe how the patterns of activity of the matrix of coincidence-counting units of the Jeffress–Colburn model and its extensions can describe some of the phenomena that have been important for researchers in binaural perception. In each case, we provide intuitive examples of some of the various ways in which interaural timing information may be utilized in making observations and forming decisions. We then summarize the characteristics and limitations of the ability of the models to describe the phenomena.

A. Subjective Lateral Position

1. Lateralization of Pure Tones

The comparison of theoretical predictions to experimental data concerning lateralization of pure tones is complicated by the fact that the experimental data differ across studies. For example, Domnitz and Colburn (1977) describe a subjective image that returns to the center of the head as the interaural phase difference (IPD) of 500-Hz pure tones approaches ± 180°, while other researchers such as Sayers (1964) and Yost (1981) describe multiple images appearing at each of the two ears at these IPDs. By all accounts, the perceived image (or images) tends to be diffuse and labile under these conditions.

The response of the ensemble of coincidence-counting units to a pure tone has already been discussed in Section III.A, and it is shown in Figure 5. As discussed in Section III.A, models that base their predictions of lateralization on the location of a single centroid along the internal-delay axis cannot describe the perception of multiple images. This shortcoming notwithstanding, models that use the centroid to compute a single image, such as the position-variable model of Stern and Colburn (1978), provide reasonably accurate predictions for a number of fundamental aspects of the lateralization of pure tones based on *ongoing* ITDs and IIDs. These aspects include (1) the periodicity of lateral position with respect to ITD; (2) the joint dependence of the lateralization of low-frequency pure tones on ITD and IID as seen in lateralization studies that describe time–intensity trading, the cue-reversal phenomenon (in which the direction of apparent motion of the image reverses as ITDs approach half the period of the tone) and, as IID increases, the inability to describe the dependence of lateralization on a peripheral conversion of IIDs into equivalent ITDs (e.g., Sayers, 1964; Domnitz & Colburn, 1977); (3) the approximately constant lateral position of a pure tone with a fixed ITD over a range of frequencies (Schiano et al., 1986); and (4) the trajectories of images produced by low-frequency stimuli presented with small interaural frequency differences (i.e., the so-called binaural beats) (e.g., Licklider et al., 1950).

2. Lateralization of Low-Frequency Bandpass Noise

In recent years greater attention has been focused on the lateralization of spectrally and temporally more complex stimuli, such as bandpass noise and amplitude-modulated tones. We discuss the lateralization of bandpass noise and amplitude-modulated tones separately, as different issues arise in understanding the processes by which they are lateralized.

Figure 7 shows the response of the coincidence-counting units to narrowband noise presented with a center frequency of 500 Hz and two different bandwidths, 50 Hz (upper panel) and 800 Hz (lower panel). In both cases the stimuli have an ITD of −1.5 ms. The response pattern for the noise with the bandwidth of 50 Hz looks very similar to the pattern that is observed for 500-Hz tones presented with the same ITD. [For 500 Hz tones, an ITD of −1.5 ms is equivalent to an ITD of +0.5 ms, which is shown in Figure 5 (upper panel).] The dichotic stimulus with the 50-Hz bandwidth is, in fact, lateralized on the "wrong" side of the head (i.e., the right side), whereas for larger bandwidths the sound becomes lateralized toward the left side of the head (Stern, Zeiberg, & Trahiotis, 1988b; Trahiotis and Stern, 1989). This occurs because the response to noise with a bandwidth of 50 Hz exhibits parallel maxima that appear alike (as in the upper panel of Figure 7), and it is not obvious that the true stimulus delay is −1.5 ms in this case (rather than, for example, +0.5 ms). With greater stimulus bandwidths, however, the cross-correlation function exhibits modes at internal delays of −1.5 ms over a broad range of frequencies, and it becomes obvious that this is the true ITD. This effect is illustrated for a noise with an 800-Hz bandwidth in the lower panel of Figure 7.

We refer to the consistency over frequency of the maxima of the coincidence-count response that indicates the true ITD as *straightness*. By independently manipulating ITD, IPD, and bandwidth, we have found that the binaural system appears to apply special emphasis to the straight components of the response to bandpass-noise stimuli such as those shown in Figure 7 (Stern et al., 1988; Trahiotis and Stern, 1989). We have developed two extensions of the general cross-correlation model to describe this concept in a quantitative fashion. The first model was a black-box formulation called the *weighted-image model* (Stern et al., 1988b). In this model, each ridge of maxima of the two-dimensional function $L_T(\tau,f)$ is weighted by an *ad hoc* function that is approximately proportional to the reciprocal of the variance of the internal delay of the mode over frequency. This function serves as an empirical estimate of the straightness of the mode. The weighted-image model successfully described the phenomena to which it

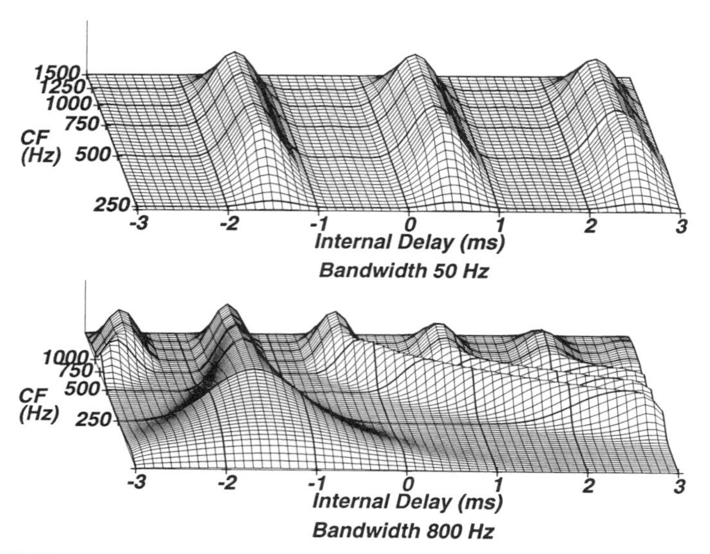

FIGURE 7 The response of an ensemble of coincidence-counting units to low-frequency bandpass noise with a center frequency of 500 Hz and an ITD of −1.5 ms. Upper panel: response to bandpass noise with a bandwidth of 50 Hz. Lower panel: response to bandpass noise with a bandwidth of 800 Hz.

has been applied. Nevertheless, it has always been considered to be an interim formulation because it cannot easily be generalized to enable predictions for many interesting stimuli that do not produce isolated ridges of maxima of $L_T(\tau,f)$.

A more satisfying explanation for the straightness-weighting phenomenon was more recently proposed by Stern and Trahiotis (1992). This model, referred to as the *extended position-variable model,* assumes that the outputs of the coincidence-counting units are passed through a *second* level of coincidence-counting units. Each set of inputs to this second layer of temporal processing is assumed to come from first-level coincidence counters representing a range of CFs, but with a common internal delay. The effect of this type of processing is illustrated in Figure 8, which compares the response of the original model (without any additional straightness weighting) and the response of the extended model. The stimulus in this figure is bandpass noise centered at 500 Hz with an ITD of −1.5 ms and a bandwidth of 400 Hz. The sets of points denoted by the filled circles in the upper panel of Figure 8 are examples of combinations of CF and internal delay that would constitute inputs to the second-level coincidence counters. The center panel of Figure 8 shows the effect of weighting by the relative number of fiber pairs, which suppresses the effects of the responses at the true ITD of −1.5 ms. The lower panel of Figure 8 shows the dramatic effects of applying the second level of coincidences, which provides much greater emphasis to the straight ridge at −1.5 ms. This occurs because, for that ridge, all of the first-level coincidence counters are firing at rates at or near their maximum output. In contrast, the ridge closer to the midline (i.e., at an ITD of approximately 0) is attenuated because of the minimal response at characteristic frequencies below approximately 600 Hz at that ITD. In addition, this manner of weighting straightness also sharpens the ridges of the two-dimensional cross-correlation function along the internal-delay axis. It is

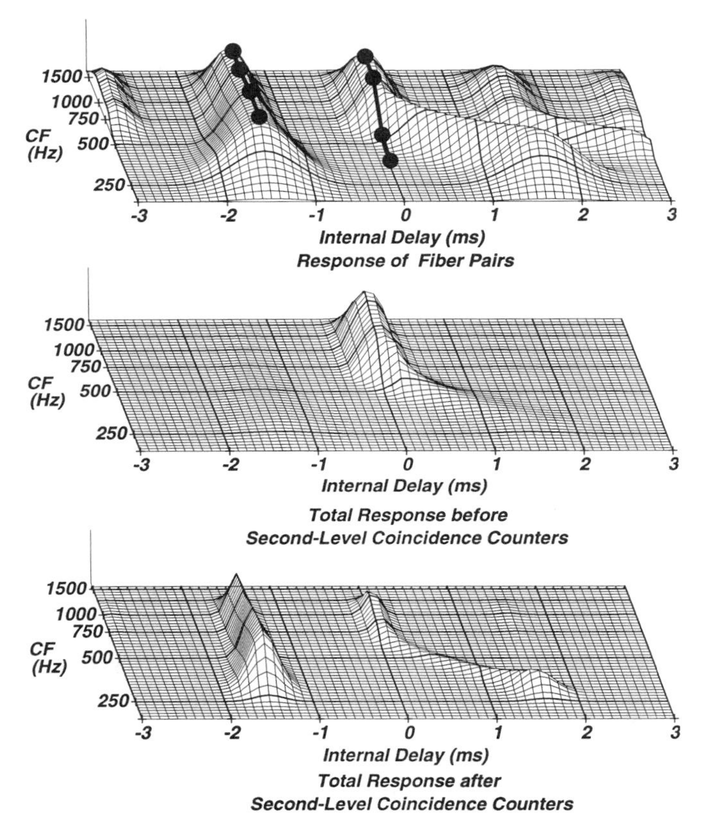

FIGURE 8 The effect of the putative secondary level of coincidence-counting units that produce "straightness weighting." Upper panel: cross-correlation patterns showing the response of an ensemble of binaural fiber pairs to noise with an ITD of −1.5 ms, a center frequency of 500 Hz, and a bandwidth of 400 Hz. Locations of constant internal delay but different CFs are identified by filled circles joined by lines. Central panel: same as upper panel, but incorporating the effects of the relative number of fiber pairs, as specified by the function $p(\tau,f)$. Lower panel: same as central panel, but after further processing by the second-level units that compute coincidences over frequency of the outputs of the original coincidence counters with the same internal delay.

important to note that "sharpening" along the internal-delay axis can occur without the explicit lateral-inhibition network proposed by Lindemann (1986a).

Shackleton et al. (1992) provide a different point of view, arguing that these data can be predicted by simply averaging the response of the coincidence-counting units over frequency, *without* any explicit mechanism that weights more heavily the straighter modes of the two-dimensional cross-correlation functions. There are at least two possible reasons why Shackleton et al. and the present authors differ in their conclusions concerning the modeling of these data. First, the model of Shackleton et al. lacks an explicit function like $p(\tau,f)$ to specify the distributions of fiber pairs with respect to internal delay and CF. It is not obvious that it will predict the much wider range of phenomena addressed by the position-variable model. Second, Shackleton et al. make use of the more detailed computational description of the auditory-nerve response to the stimuli based on the work of Meddis et al. (1990), rather than the analytical characterization used by Colburn, Stern, and their colleagues, which is simpler but not as descriptive of the physiological data. Setting aside these distinctions, we believe that the experimental data of Stern et al. (1988b) are more accurately described by the predictions of the extended position-variable model that explicitly includes straightness weighting (Stern and Trahiotis, 1992, Fig. 6) than by the predictions of the model of Shackleton et al. (1992, Figs. 2a and 2c). In our opinion, the predictions obtained by Stern and Trahiotis without straightness weighting (1992, Fig. 6) are not unlike those generated by the model of Shackleton et al.

The position-variable model as extended by Stern and Trahiotis (1992) appears to be able to describe most experimental results on the lateralization of low-frequency bandpass noise as a joint function of ITD, IPD, and bandwidth, when the signals are presented with equal amplitude to the two ears. Nevertheless, it does not describe the some of the results of Buell, Trahiotis, and Bernstein (1994), which describe the joint dependence of the lateral position of bandpass noise on ITD, IID, IPD, and bandwidth. (For example, the lateralization of stimuli with an ITD of 0 ms and an IPD of 270° is not affected by bandwidth for any IID, whereas the lateralization of stimuli with an ITD of 1.5 ms and an IPD of 0° is greatly affected by bandwidth for many IIDs.) Despite a concerted effort by Tao (Tao, 1992; Tao & Stern, 1992), which included the development of an alternative additive combination of interaural timing and intensity information, these data of Buell et al. (1994) appear to pose a continuing challenge for all models of binaural interaction.

3. Lateralization of Low-Frequency Amplitude-Modulated Tones

It has been known since the mid-1970s that the binaural system can lateralize high-frequency stimuli on the basis of the ITDs of their low-frequency envelopes (e.g., Henning, 1974; McFadden & Pasanen, 1976). In contrast, it had been believed until more recently that the lateralization of low-frequency stimuli was based solely on the ITD of their fine structure and that the ITD of the envelope had no impact on subjective lateralization. Using amplitude-modulated (AM) 500-Hz tones, Bernstein and Trahiotis (1985) demonstrated that the lateral position of low-frequency AM stimuli was affected, albeit by a small amount, by the ITD of the envelope of the stimulus, as well as by the ITD of its fine structure.

Figure 9 shows the response of the coincidence-detecting units to a 500-Hz low-frequency tone presented without amplitude modulation (upper panel), and with 100% sinusoidal amplitude modulation at a rate of 50 Hz (lower panel). The ongoing interaural delay is −1.5 ms in both cases. Bernstein and Trahiotis (1985) have shown that, although both stimuli are lateralized toward the right side of the head, the perceived location of the signal with the amplitude modulation (producing the response in the lower panel) is slightly to the left of that of the pure tone that produces the response curves in the upper panel. There are several possible causes for this. For example, Bernstein and Trahiotis suggested that the AM tone is perceived to the left of the pure tone because of the salience of low-frequency envelope cues. Stern et al. (1988b) suggested that the AM tone is

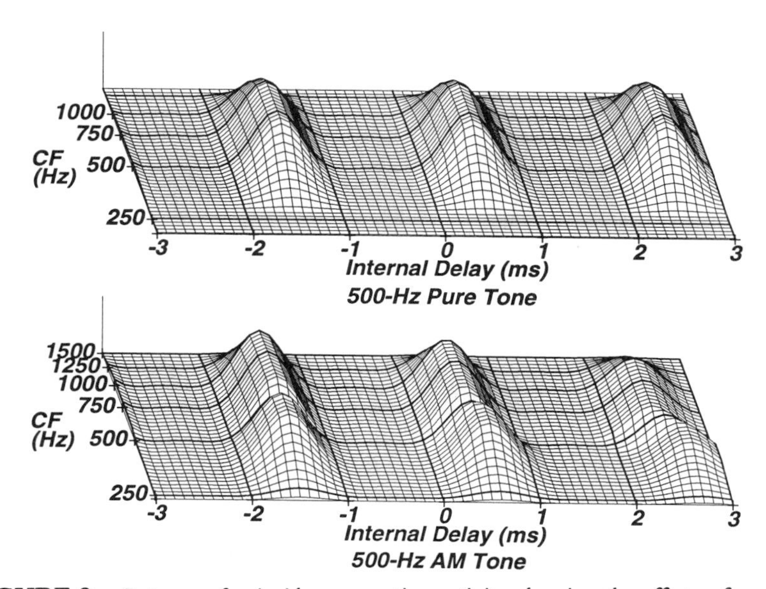

FIGURE 9 Patterns of coincidence-counting activity showing the effects of amplitude modulation on low-frequency tones. Upper panel: the response to a 500-Hz tone with a waveform ITD of −1.5 ms. Lower panel: the response to a 500-Hz tone with the same waveform delay and amplitude modulated with a modulation frequency of 50 Hz.

perceived to the left of the pure tone because the ridge at +0.5 ms is not as straight as the ridge at −1.5 ms. Nevertheless, we now believe that the most likely reason for the AM tone to be perceived to the left of the pure tone is simply because the peaks in the response to the AM stimulus are unequal in amplitude, with the peak (in the lower panel) at the true ITD (−1.5 ms) being greatest in size. In contrast, the peaks of the response to the pure tone (depicted in the upper panel) are all of equal amplitude. This would cause the centroid along the internal-delay axis of the response to the AM tone to be "pulled" farther toward the left side.

Stern et al. (1988a) and Stern and Shear (1995) have reported that the extended position-variable model correctly predicts the dependence of lateral position for 500-Hz AM tones on ongoing ITD and modulation frequency measured by Bernstein and Trahiotis (1985). The extended model also describes other aspects of the lateralization of low-frequency stimuli with complex envelopes, including the dependence of lateral position on pure modulator delay measured by Bernstein and Trahiotis (1985) (Stern, Zeppenfeld, & Shear, unpublished).

4. Lateralization of High-Frequency Amplitude-Modulated Tones and Bandpass Noise

As noted in the preceding section, the lateral position of high-frequency binaural stimuli with low-frequency envelopes such as AM tones and bandpass noise can be affected by the ITD of the envelope. Figure 10 illustrates how such stimuli are represented by the ensemble of coincidence-counting units. These plots were produced without the use of any additional explicit envelope extraction mechanism other than the low-pass filtering in the model for auditory-nerve activity. The low-pass filter has a frequency response that decreases linearly from 1200 Hz to 5200 Hz, as suggested by the physiological data of Johnson (1980), and the minor ripples in the plots show the effects of the residual energy at the relatively high carrier frequency after processing by the low-pass filter. The upper panel of Figure 10 shows the relative number of coincidences observed in response to a pure tone of frequency 3900 Hz. The central panel depicts the response to an AM tone with a carrier frequency of 3900 Hz and a modulation frequency of 300 Hz. The lower panel of the same figure shows the response to a bandpass noise with a center frequency of 3900 Hz and a bandwidth of 600 Hz. Each stimulus has an ITD of −1.5 ms. Lateralization of the AM tones and bandpass noise is dominated by the mode of the *envelope* of the cross-correlation functions, which in each of these two examples occurs at an internal delay of approximately −1.5 ms.

These observations are in accord with the conclusions of Colburn and Esquissaud (1976), who first suggested that cross-correlation-based models could be used to predict high-frequency binaural processing based on only

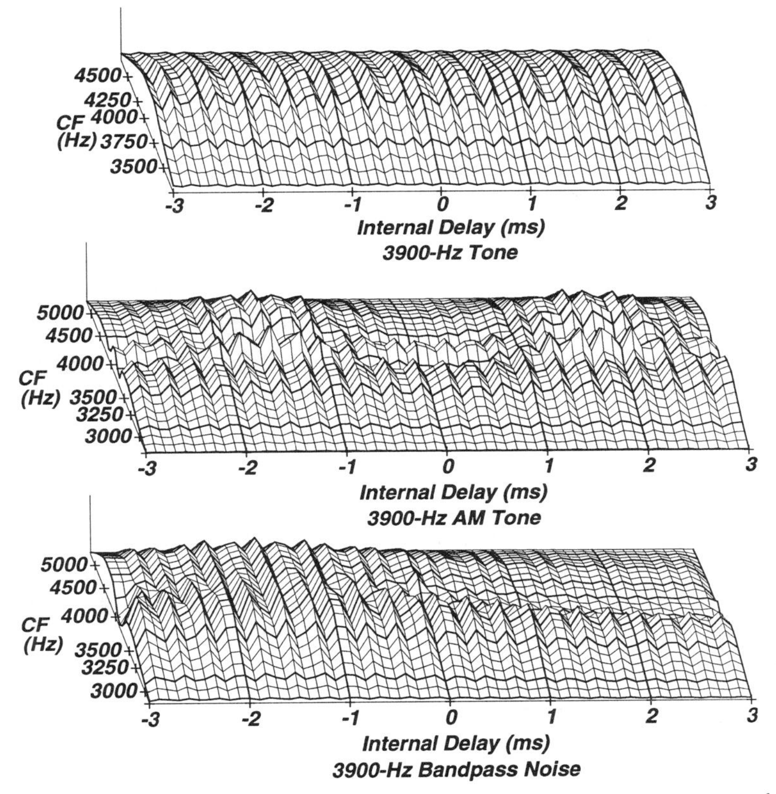

FIGURE 10 The response of an ensemble of coincidence-counting units to several types of high-frequency stimuli with an ITD of −1.5 ms. Upper panel: response to pure tones with a frequency of 3900 Hz. Central panel: response to AM tones with a modulation frequency of 300 Hz. Lower panel: response to bandpass noise with a center frequency of 3900 Hz and a bandwidth of 600 Hz.

the implicit envelope-extraction properties of the peripheral auditory system. For example, the response at the carrier frequency (3900 Hz) in Figure 10 is greatly attenuated by the implicit envelope-detecting effects of the cascaded combination of bandpass filter, nonlinear rectifier, and lowpass filter in the model that characterizes the processing of the peripheral auditory system. Although models such as the extended position-variable model should in principle be able to describe most high-frequency lateralization data based on envelope delays, few detailed attempts have been made to compare the predictions of these models to the experimental data. Stern et al. (1988a) did observe that the extended position-variable model in its present form fails to predict the unexpected observation by Trahiotis and Bernstein (1986) that bandpass noise with a given ITD tends to be lateralized further from the center of the head than AM tones of similar carrier frequency, modulation frequency, and effective bandwidth.

5. Other Lateralization Phenomena

The discussions of lateralization mechanisms in the preceding sections have all concerned simple stimuli that have been used in "classical" psychoacoustical experiments. Several recent studies have shown that direct application of the cross-correlation-based binaural processing models described in this chapter can describe more complex phenomena as well. For example, Hafter and Shelton (Hafter, Shelton, & Green, 1980; Hafter & Shelton, 1991) described the lateralization of diotic bandpass noise gated by brief rectangular pulses that themselves had an ITD. Surprisingly, these stimuli are frequently lateralized toward the ear receiving the gating signal that is *lagging* in time. It was later shown (Stern et al., 1991) that these counterintuitive lateralization effects are predicted quite elegantly by the models described in Section III of this chapter. Similarly, Bilsen and Raatgever (1973) have described a "dominant region" effect in which frequency components in the neighborhood of about 700 Hz appeared to be more salient than higher-frequency and lower-frequency components in the lateralization of broadband noise. Again, it was later shown (Stern et al., 1988a; Stern & Shear, 1995) that this phenomenon can be naturally accounted for by the shape of the density function for internal delays of the fiber pairs, $p(\tau,f)$. Finally, Lindemann (1986b) has shown that his extensions to the Jeffress–Colburn model can describe (at least qualitatively) the minimum temporal separation between the onsets of pairs of bandpass-filtered binaural clicks that is needed for echo perception, as well as the laterality of the fused image of binaural click pairs with short temporal separations, and the laterality of the echoes produced by binaural click pairs with longer temporal separations.

Until now, the application of binaural models to more complex stimuli has been limited by the difficulty in developing analytical expressions to characterize the response to these stimuli at the levels of the auditory nerve and the ensemble of coincidence-counting units. It is expected that, as the use of realistic computational models of the peripheral auditory response to sound becomes more widespread, and as the cost of computational resources decreases, the breadth of phenomena that are successfully predicted by the cross-correlation-based models will continue to increase.

B. Interaural Discrimination Phenomena Related to Subjective Lateral Position

The perceptual cue used by subjects in many interaural discrimination experiments is a change in subjective lateral position of the stimuli. We summarize in this section some of the ways in which several of the models of binaural lateralization have used lateral position to predict results of interaural discrimination experiments.

Models that describe the lateral position of binaural stimuli can be directly applied to discrimination experiments by computing or estimating the variance as well as the mean values of the predicted lateral positions of the stimuli, using optimal decision theory to estimate the best possible discrimination performance (cf. Van Trees, 1968). Most of the early black-box binaural models (e.g., Jeffress et al., 1956; Hafter, 1971) implicitly assumed that the variance of the position estimate is constant, which is likely to be valid if the changes in the ITD and IID of the stimuli are of sufficiently small magnitude. For example, models that assume that position is a linear combination of ITD and IID and that position variance is constant can predict the results of many tone-on-tone and noise-on-noise experiments typified by the data of Jeffress and McFadden (1971) and Yost, Nielsen, Tanis, and Bergert (1974).

Colburn (1973) and Stern and Colburn (1985) have provided predictions for interaural discrimination experiments using expressions for the variance of predicted position that were derived from the Poisson variability inherent in the auditory-nerve model used to describe the response to the stimuli. Colburn (1973) based his predictions on the amount of information in the ensemble of coincidence-counting units (without making any assumptions about the perceptual cue used by the subjects) and predicted the dependence of just-noticeable differences (jnds) in ITD and IID on baseline ITD, IID, and overall level (Hershkowitz & Durlach, 1969). Stern and Colburn (1985) derived an analytical expression for the variance of the predicted position variable $\hat{P}$. Calculating predictions on the basis of the mean and variance of $\hat{P}$ for the stimuli of each experiment, Stern and Colburn (1985) found that the original position-variable model correctly predicted many of the trends of interaural time and amplitude jnds (e.g., Domnitz & Colburn, 1977), and studies of masking using correlated targets and maskers (e.g., Yost et al., 1974; Jeffress & McFadden, 1971) at 500 Hz. To date, no attempts have been made to generate predictions for similar data at other frequencies. As noted in Subsection III.C.2, the model is unable to account for the results of certain other discrimination experiments concerning time–intensity tradability (e.g., Hafter & Carrier, 1972; Gilliom & Sorkin, 1972) because the data imply the use of multiple perceptual images and the theoretical predictions are based only on the dominant time–intensity traded image of the stimuli.

Stern et al. (1988) have developed a small number of predictions for the extended position-variable model involving the discrimination of high-frequency stimuli with low-frequency envelopes. They observed that this model correctly describes the general dependence of discrimination performance on modulation frequency for high-frequency AM tones (Henning,

1974), and it describes the dependence of the sensitivity to ITD on the interaural frequency difference of the carrier frequency (Nuetzel & Hafter, 1981). Because it is more difficult to calculate the variance of the position estimate for the extended position-variable model than for the original position-variable model, these predictions were obtained by assuming constant position variance.

C. Binaural Masking-Level Differences

The binaural masking-level difference (or MLD) is an extremely well known and robust binaural phenomenon, discussed in some detail in Chapter 9. Figure 11 illustrates how the ensemble of coincidence-counting units accounts for this phenomenon. The figure shows the cross-correlation patterns that result when a 500-Hz tonal target and a broadband masking noise are presented in the N_0S_π (masker interaurally in phase, target interaurally out of phase) and N_0S_0 (masker and target both interaurally in phase) configuration. The plots in Figure 11 include the effects of the relative number of fiber pairs, as specified by the function $p(\tau,f)$. Note that when the N_0 masker is presented alone (Figure 11, lower panel), the ridge of maxima at 0 internal delay has approximately constant amplitude over a broad range of frequencies. The addition of an in-phase (S_0) target to the masker at a target-to-masker intensity ratio of -20 dB has virtually no effect on the pattern of coincidence-counting activity, because the interaural time differences of the combined target and masker are unchanged (Figure 11, central panel). On the other hand, the addition of the 500-Hz out-of-phase (S_π) target to the in-phase masker cancels masker components at that frequency, causing a "dimple" to appear in the central ridge for CFs near the target frequency (Figure 11, upper panel). The target in the N_0S_π configuration is easily detected at -20 dB SNR because the pattern of responses in the upper panel of Figure 11 is easily discriminated from that in the lower panel. The N_0S_0 stimulus is not detected because the response of the binaural system is unaffected by whether the target is present or absent (central and lower panels of Figure 11).

Colburn (1977) was able to describe virtually all of the "classical" data obtained in experiments measuring binaural masking-level differences on the basis of the predicted outputs of the coincidence counters. His predictions were developed using the simplifying assumption that experimental performance is limited by the variability of the auditory-nerve response to the signals, as opposed to the intrinsic variability of the masker components. This assumption has since been shown to be invalid for some stimuli by Siegel and Colburn (1983). More recently, Gilkey and his colleagues (e.g., Gilkey, Robinson, & Hanna, 1985; Hanna & Robinson, 1985; Gilkey & Robinson, 1986) have presented a number of results using "frozen-noise"

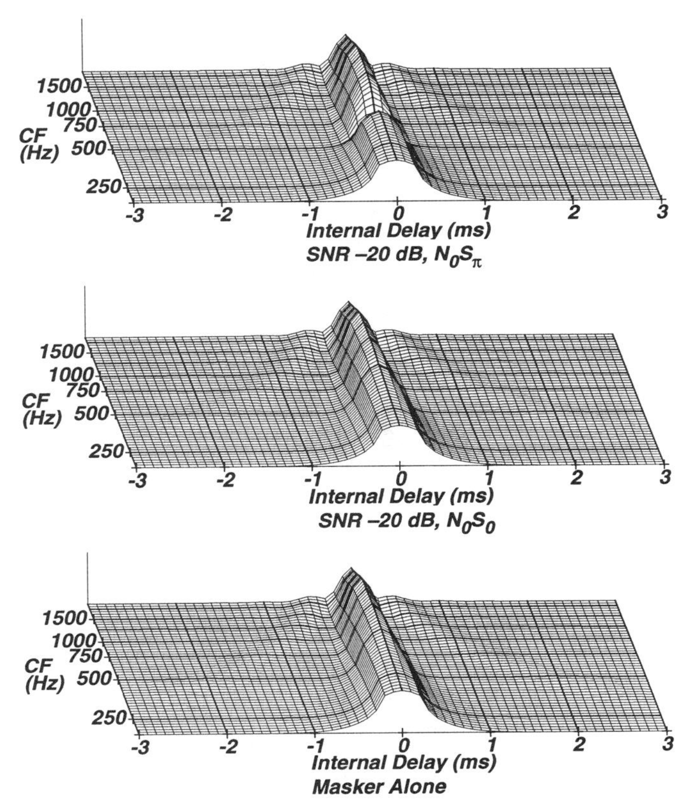

FIGURE 11 Patterns of coincidence-counting activity showing the response to stimuli used in N_0S_π and N_0S_0 binaural masking-level difference experiments. The target is presented at 500 Hz, either interaurally in phase or out of phase, as indicated, and the masker is broadband diotic noise. These plots include the effects of the relative number of fiber pairs, as specified by the function $p(\tau,f)$.

maskers in which the actual variability of the masker component of the stimulus can be experimentally controlled. To date no binaural model has been able to account for differences of detectability associated with the individual masker waveforms used in these studies.

The outputs of the coincidence-counting units are used to obtain predictions for all experiments. Nevertheless, we believe that binaural detection phenomena are mediated by a reading of the information from the display of coincidence-counting units that is different from that used for subjective lateral position and interaural discrimination. Specifically, the subjective lateral position of binaural stimuli and the ability to perform certain interaural discrimination tasks based on changes in lateral position both appear to depend on the *locations* of the ridges of the cross-correlation function along the τ axis. In contrast, successful predictions for binaural detection tasks can be obtained by quantifying the *decrease in amplitude* of these ridges at the target frequency produced by the addition of the target to the masker.

D. Dichotic Pitch Phenomena

Many dichotic broadband stimuli can produce a clear sensation of pitch when presented simultaneously to the two ears even though no pitch is perceived when the respective signals to the two ears are presented monaurally (e.g., Cramer & Huggins, 1958; Bilsen & Goldstein, 1974; Bilsen, 1976). This phenomenon, referred to as *dichotic pitch,* can be created by IPDs that change as a function of frequency, as discussed in Chapter 8.

Bilsen and his colleagues have argued convincingly that such pitch phenomena can also be explained in terms of the outputs of the coincidence counters (e.g., Bilsen, 1977; Raatgever & Bilsen, 1986; Frijns, Raatgever, & Bilsen, 1986). To illustrate, the upper panel of Figure 12 shows the pattern of coincidence-counter outputs produced by a "Huggins-pitch" stimulus, which has a rapid transition in IPD from $-\pi$ to π radii in a narrow range of

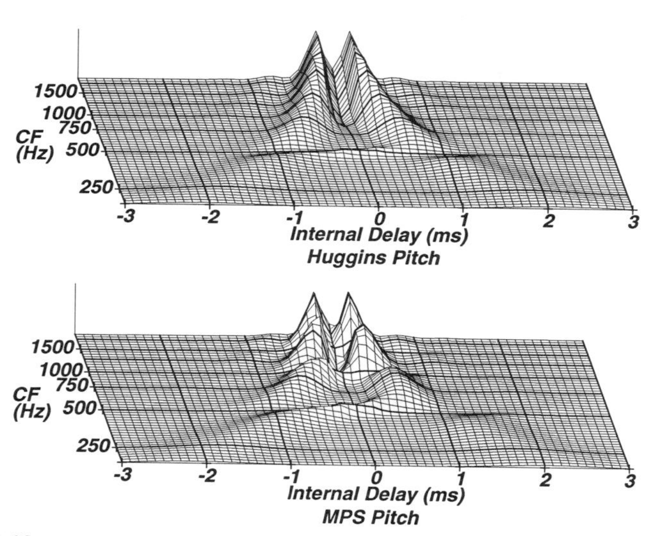

FIGURE 12 Upper panel: response of an ensemble of coincidence-counting units to a "Huggins-pitch" stimulus with an interaural phase transition from $-\pi$ to π rad, occurring in a narrow range of frequencies about 500 Hz. Lower panel: response of an ensemble of coincidence-counting units to a multiple-phase-shift stimulus producing dichotic pitch perceptions at 500 Hz. These plots include the effects of the relative number of fiber pairs, as specified by the function $p(\tau,f)$.

frequencies about 500 Hz. The lower panel of the same figure illustrates responses for a typical "multiple-phase shift" (MPS) stimulus, which contain an IPD that undergoes rapid transitions at integer multiples of 500 Hz. The plots in Figure 12 also include the effects of the relative number of fiber pairs, as specified by the function $p(\tau,f)$. The MPS stimuli produce a particularly strong pitch (Bilsen, 1976). The pitch sensation is presumably caused by the peaks of activity observed along the f-axis (at 0 internal delay) that appear at 500 Hz for the Huggins-pitch stimulus and at integer multiples of 500 Hz for the MPS-pitch stimulus. These figures closely resemble similar plots produced by the "central spectrum model" of Bilsen, Raatgever, and their colleagues.

In previous sections of this chapter we described two ways in which the auditory system appeared to use the activity from the ensemble of outputs of coincidence-counting units: The locations of peaks along the internal-delay axis appear to provide information needed to estimate auditory lateralization, and the decreases in activity can signal the presence of the target in a binaural MLD experiment. Bilsen, Raatgever, and their colleagues suggest that the information from the coincidence counters can also be used to estimate the pitch of dichotic-noise stimuli, by considering the patterns of activity of coincidence counters with 0 internal delay.

Raatgever and Bilsen (1986) have also measured the lateral position of the dichotic-pitch image in stimuli such as Huggins pitch and MPS pitch and found that the lateral position of this image is relatively unaffected by IID. This result reinforces the hypothesis that the interaction between ITD and IID in the lateralization process takes place centrally, combining information from the coincidence-counting units in a fashion that is different from that used in producing dichotic pitch. Raatgever and Bilsen (1986) and Frijns et al. (1986) have also estimated the pattern of activity of the ensemble of coincidence counters in response to dichotic-pitch stimuli by measuring the binaural MLDs that they produce for tonal targets. A good correspondence was observed between their experimental data and theoretical predictions.

The development of a theoretical framework for interpreting dichotic-pitch stimuli in the context of cross-correlation-type models represents a major step forward in developing a unified view of the way we process complex signals to the two ears.

E. Temporal Effects

In the previous sections we have considered primarily stimuli containing interaurally static binaural cues. In recent years, there has been increasing interest in the perception of stimuli with dynamically changing ITDs, IIDs, and interaural correlation. Many of the primary data are summarized in

Chapter 9. In this section we will summarize the theoretical formulations that have been used to explain the data within the context of the binaural display of coincidence-counting units.

1. Binaural "Sluggishness"

As noted in Section II.A, the binaural system is somewhat "sluggish" in its response to stimuli with time-varying interaural differences. For example, subjects are unable to track the instantaneous values of ITD or interaural correlation if they are varied with a frequency of more than a few Hz.

For the general model described in Section II.D, the instantaneous outputs of the coincidence-counting units undergo temporal integration using the temporal weighting function $w(t)$. This type of temporal integration inevitably causes temporal "sluggishness" because the duration of the integration window limits the resolution with which one can observe time-varying interaural differences of complex stimuli.

On the basis of some initial studies (Bachorski, 1983; Stern & Bachorski, 1983), we believe that many, if not all, of the sluggishness phenomena can be explained in terms of simple temporal integration of the coincidence-counter outputs, provided that the time constants for processing ITDs and for IIDs are allowed to differ (Grantham & Wightman, 1978; Grantham, 1984). Gabriel (1983) developed a black-box model that incorporated separate time constants for processing ITD and IID. The type of temporal averaging that is likely to mediate binaural sluggishness also provides at least a qualitative explanation for the disappearance of binaural beats at high beat frequencies (Licklider et al., 1950).

2. The Precedence Effect

The precedence effect refers to the dominant role that early-arriving, direct components of a sound (as opposed to later, reflected wavefronts) play in determining the location of that sound (e.g., Wallach, Newman, & Rosenzweig, 1949; Haas, 1951; Gardner, 1968). This phenomenon has motivated a large number of contemporary experimental studies, many of which are summarized in Zurek (1987) and in Chapter 9 of this volume. Although this is also a "temporal" effect, it is likely to be mediated by a different aspect of the binaural system from the mechanisms producing binaural sluggishness.

Lindemann's extension of the cross-correlation model is able to describe at least qualitatively several observations related to the precedence effect and other dynamic binaural phenomena including summing localization, the law of the first wavefront, and echo suppression (Lindemann, 1986b). McFadden (1973) also provided an insightful discussion of localization of sound and suppression of echoes, including a consideration of how each may be accomplished via inhibitory interactions of the outputs of the coincidence counters.

A number of contemporary researchers have conducted headphone studies to measure sensitivity to binaural cues using transient stimuli that are believed to give rise to the precedence effect (e.g., Yost and Soderquist, 1984; Zurek, 1987; Clifton, 1987). Nevertheless, there have been relatively few efforts to date to develop quantitative correlation-based models that can predict the results of such experiments. Progress has been hampered by the difficulty in extending analytical approaches such as those used by Colburn (1973, 1977) and Stern and Colburn (1978) to include time-varying interaural differences and by the lack of an explicit source of internal noise in computational models such as those of Blauert and colleagues (e.g., Lindemann, 1986a; Gaik, 1993). These limitations may be overcome in the future by combining a computational model that simulates the stochastic response of the auditory nerve to arbitrary stimuli (e.g., Payton, 1988; Meddis et al., 1990; Carney, 1993) with central processing as formulated either along the lines suggested by Lindemann model or according to one of the computational models of binaural processing in the brainstem (as reviewed by Colburn, 1994).

Setting these issues aside, it should be mentioned that Litovsky and Yin (1993) have noted that neural responses measured in the inferior colliculus to the second of a pair of binaural clicks may be moderately or severely suppressed depending on the temporal relations of the stimuli. Identifying these neural responses with the precedence effect would be premature at this time, and it is always possible that some of the "precedence-effect" phenomena are mediated by more central processes. Nevertheless, the temporal relations of the physiological stimuli used by Yin and his colleagues parallel those of the psychophysical studies, and it seems probable that some type of neural inhibition of the type postulated by Lindemann is actually occurring within the nervous system.

V. SUMMARY AND CONCLUSIONS

In this chapter we have reviewed the evolution and development of current theories of binaural interaction. We have noted that most current models now include as an intermediate display the interaural cross-correlation as a function of CF after processing by the peripheral auditory system. We believe that this display of information can be used as a powerful tool for understanding the perception of several quite different types of binaural phenomena, and we have attempted to facilitate this understanding by providing examples of these responses to a number of classical stimuli used in binaural experiments.

It is likely that different types of psychophysical tasks are mediated by different ways of interpreting the information contained in the binaural display. Lateralization phenomena can be predicted by considering the locations of the modes of the cross-correlation function along the internal-delay axis. Binaural detection thresholds can be predicted by consideration of the depth of notches of the ridges of the cross-correlation function at or near the target frequency. Many dichotic-pitch phenomena can be described at least qualitatively by examining the locations of modes of the cross-correlation function along the frequency axis at 0 internal delay. Consideration of the implicit temporal integration of the cross-correlation function of the stimuli goes a long way toward describing many binaural "sluggishness" phenomena, and the cross-correlation display provides a jumping-off point for more detailed efforts to characterize mechanisms underlying the perception of binaural stimuli with dynamically changing interaural time and intensity differences.

Over the next several years we expect to see cross-correlation-based models being applied to a wider variety of stimuli, with widespread use of computational simulations of the auditory-nerve response to stimuli and of neural interactions at more central sites. Similarly, we expect to see increasing use of signal processing schemes motivated by our knowledge of how the human binaural system functions in many diverse areas, including sound reproduction, environment simulation, and speech enhancement and recognition.

Acknowledgments

Preparation of this manuscript has been supported by NSF Grant IBN 90-22080 to Richard Stern and by NIH Grant DC-00234 and AFOSR Grant 89-0030 to Constantine Trahiotis. We also thank Les Bernstein and Steve Colburn for their comments on previous drafts of this manuscript and other valuable contributions to this work. Preparation of the figures has been facilitated by the efforts of Carl Block, Wonseok Lee, Steve Palm, Glenn Shear, Sammy Tao, Xaohong Xu, Andreas Yankopolus, and Torsten Zeppenfeld.

References

Bachorski, S. J. (1983). Dynamic cues in binaural perception. M.S. thesis, Carnegie Mellon University.

Bernstein, L. R., & Trahiotis, C. (1985). Lateralization of low-frequency complex waveforms: The use of envelope-based temporal disparities. *Journal of the Acoustical Society of America, 77,* 1868–1880.

Bilsen, F. A. (1976). Pronounced binaural pitch phenomenon. *Journal of the Acoustical Society of America, 59,* 467–468.

Bilsen, F. A. (1977). Pitch of noise signals: Evidence for a "central spectrum." *Journal of the Acoustical Society of America, 61,* 150–161.

Bilsen, F. A., & Goldstein, J. L. (1974). Pitch of dichotically delayed noise and its possible spectral basis. *Journal of the Acoustical Society of America, 55,* 292–296.

Bilsen, F. A., & Raatgever, J. (1973). Spectral dominance in binaural lateralization. *Acustica, 28,* 131–132.

Blauert, J. (1980). Modelling of interaural time and intensity difference discrimination. In G. van den Brink & F. A. Bilsen (Eds.), *Psychophysical, physiological, and behavioural studies in hearing* (pp. 421–424). Delft: Delft University Press.

Blauert, J. & Cobben, W. (1978). Some consideration of binaural cross-correlation analysis. *Acustica, 39,* 96–103.

Bodden, M. (1993). Modeling human sound source localization and the cocktail-party-effect. *Acta Acustica, 1,* 43–55.

Boudreau, J. C., & Tsuchitani, C. (1968). Binaural interaction in the cat superior olive S segment. *Journal of Neurophysiology, 31,* 442–454.

Brugge, J. F., Anderson, D. J., & Aitkin, L. M. (1970). Response of neurons in the dorsal nucleus of the lateral lemniscus of the cat to binaural stimuli. *Journal of Neurophysiology, 33,* 441–458.

Buell, T. N., Trahiotis, C., & Bernstein, L. R. (1994). Lateralization of bands of noise as a function of combinations of interaural intensitive differences, interaural temporal differences, and bandwidth. *Journal of the Acoustical Society of America, 95,* 1482–1489.

Carney, L. H. (1993). A model for the responses of low-frequency auditory nerve fibers in cat. *Journal of the Acoustical Society of America, 93,* 401–417.

Carr, C. E., & Konishi, M. (1990). A circuit for detection of interaural time differences in the brainstem of the barn owl. *Journal of Neuroscience, 10,* 3227–3246.

Clifton, R. K. (1987). Breakdown of echo suppression in the precedence effect. *Journal of the Acoustical Society of America, 82,* 1834–1835 (L).

Colburn, H. S. (1969). Some physiological limitations on binaural performance. Ph.D. dissertation, MIT.

Colburn, H. S. (1973). Theory of binaural interaction based on auditory-nerve data. I. General strategy and preliminary results on interaural discrimination. *Journal of the Acoustical Society of America, 54,* 1458–1470.

Colburn, H. S. (1977). Theory of binaural interaction based on auditory-nerve data. II. Detection of tones in noise. *Journal of the Acoustical Society of America, 61,* 525–533.

Colburn, H. S. (1995). Computational models of binaural processing. In H. Hawkins & T. McMullin (Eds.), *Auditory computation.* New York: Springer-Verlag.

Colburn, H. S., & Durlach, N. I. (1978). Models of binaural interaction. In E. C. Carterette & M. P. Friedman, (Eds.) *Handbook of perception* (pp. 467–518). New York: Academic Press.

Colburn, H. S., & Esquissaud, P. (1976). An auditory-nerve model for interaural time discrimination of high-frequency complex stimuli. *Journal of the Acoustical Society of America, 59,* S23(A).

Cramer, E. M., & Huggins, W. H. (1958). Creation of pitch through binaural interaction. *Journal of the Acoustical Society of America, 30,* 413–417.

Crow, G., Rupert, A. L., Moushegian, G. (1978). Phase-locking in monaural and binaural medulary neurons: Implications for binaural phenomena. *Journal of the Acoustical Society of America, 64,* 493–501.

David, E. E., Guttman, N., & van Bergeijk, W. A. (1958). On the mechanism of binaural fusion. *Journal of the Acoustical Society of America, 30,* 801–802.

Deatherage, B. H., & Hirsh, I. J. (1959). Auditory localization of clicks. *Journal of the Acoustical Society of America, 31,* 486–492.

DeSimio, M. P., & Anderson, T. R. (1993). Phoneme recognition with binaural cochlear models and the stereoausis representation. Proceedings of the IEEE international conference on acoustics, speech, and signal processing, *1,* 521–524.

Domnitz, R. H., and Colburn, H. S. (1977). Lateral position and interaural discrimination. *Journal of the Acoustical Society of America, 61,* 1586–1598.

Duifhuis, H. (1973). Consequences of peripheral frequency selectivity for nonsimultaneous masking. *Journal of the Acoustical Society of America, 54,* 1471–1488.

Durlach, N. I. (1963). Equalization and cancellation theory of binaural masking-level differences. *Journal of the Acoustical Society of America, 35,* 1206–1218.

Durlach, N. I. (1972). Binaural signal detection: Equalization and cancellation theory. In J. V. Tobias (Ed.), *Foundations of modern auditory theory* (pp. 369–462). New York: Academic Press.

Durlach, N. I. & Colburn, H. S. (1978). Binaural phenomena. In E. C. Carterette & M. P. Friedman, (Eds.) *Handbook of perception* (pp. 365–466). New York: Academic Press.

Frijns, H. M., Raatgever, J., & Bilsen, F. A. (1986). A central spectrum theory of binaural processing: The binaural edge pitch revisited. *Journal of the Acoustical Society of America, 80,* 442–451.

Gabriel, K. J. (1983). Binaural interaction in hearing impaired listeners. Ph.D. dissertation, MIT.

Gaik, W. (1993). Combined evaluation of interaural time and intensity differences: Psychoacoustic results and computer modeling. *Journal of the Acoustical Society of America, 94,* 98–110.

Gardner, M. B. (1968). Historical background of the Haas and/or precedence effect. *Journal of the Acoustical Society of America, 43,* 1243–1248.

Gilkey, R. H., & Robinson, D. E. (1986). Models of auditory masking: A molecular psychophysical approach. *Journal of the Acoustical Society of America, 79,* 1499–1510.

Gilkey, R. H., Robinson, D. E., & Hanna, T. E. (1985). Effects of masker waveform and signal-to-masker phase relation on diotic and dichotic masking by reproducible noise. *Journal of the Acoustical Society of America, 78,* 1207–1219.

Gilliom, J. D., & Sorkin, R. D. (1972). Discrimination of interaural time and intensity. *Journal of the Acoustical Society of America, 52,* 1635–1644.

Goldberg, J. M., & Brown, P. B. (1969). Response of binaural neurons of dog superior olivary complex to dichotic tone stimuli: Some physiological mechanisms of sound localization. *Journal of Neurophysiology, 32,* 613–636.

Grantham, D. W. (1984). Discrimination of dynamic interaural intensity differences. *Journal of the Acoustical Society of America, 76,* 71–76.

Grantham, D. W., & Wightman, F. L. (1978). Detectability of varying interaural temporal differences. *Journal of the Acoustical Society of America, 63,* 511–523.

Haas, H. (1951). Uber den Einfluss eines Einfachechos auf die Horsamkeit von Sprache. *Acustica, 1,* 49–58. English translation in H. Haas (1972), The influence of a single echo on the audibility of speech, *Journal of the Audio Engineering Society, 20,* 146–159.

Hafter, E. R. (1971). Quantitative evaluation of a lateralization model of masking-level differences. *Journal of the Acoustical Society of America, 50,* 1116–1122.

Hafter, E. R., & Carrier, S. C. (1972). Binaural interaction in low-frequency stimuli: The inability to trade time and intensity completely. *Journal of the Acoustical Society of America, 51,* 1852–1862.

Hafter, E. R., & Jeffress, L. A. (1968). Two-image lateralization of tones and clicks. *Journal of the Acoustical Society of America, 44,* 563–569.

Hafter, E. R., Shelton, B. R., & Green, D. M. (1980). A reversal in lateralization, with images appearing on the side of the delay. *Journal of the Acoustical Society of America, 68,* S16 (A).

Hafter, E. R. & Shelton, B. R. (1991). Counterintuitive reversals in lateralization using rectangularly modulated noise. *Journal of the Acoustical Society of America, 90,* 1901–1907.

Hafter, E. R., & Trahiotis, C. (1994). Functions of the binaural system. In M. J. Crocker (Ed.), *Handbook of acoustics.* New York: Wiley.

Hanna, T. E., & Robinson, D. E. (1985). Phase effects for a sine wave masked by reproducible noise. *Journal of the Acoustical Society of America, 77,* 1129–1140.
Henning, G. B. (1974). Detectability of interaural delay in high-frequency complex waveforms. *Journal of the Acoustical Society of America, 55,* 84–90.
Hershkowitz, R. M., & Durlach, N. I. (1969). Interaural time and amplitude jnds for a 500-Hz tone. *Journal of the Acoustical Society of America, 46,* 1464–1467.
Hirsch, I. J. (1948). The influence of interaural phase on interaural summation and inhibition. *Journal of the Acoustical Society of America, 29,* 536–544.
Jeffress, L. A. (1948). A place theory of sound localization. *Journal of Comparative and Physiological Psychology, 41,* 35–39.
Jeffress, L. A., Blodgett, H. C., Sandel, T. T., & Wood, C. L., III. (1956). Masking of tonal signals. *Journal of the Acoustical Society of America, 28,* 416–426.
Jeffress, L. A., & McFadden, D. (1971). Differences of interaural phase and level in detection and lateralization. *Journal of the Acoustical Society of America, 49,* 1169–1179.
Johnson, D. H. (1980). The relationship between spike rate and synchrony in responses of auditory-nerve fibers to single tones. *Journal of the Acoustical Society of America, 68,* 1115–1122.
Klumpp, R. G., & Eady, H. R. (1956). Some measurements of interaural time difference thresholds. *Journal of the Acoustical Society of America, 28,* 859–860.
Kuwada, S., Stanford, T. R., & Batra, R. (1987). Interaural phase-sensitive units in the inferior colliculus of the unanesthetized rabbit: effects of changing frequency. *Journal of Neurophysiology, 57,* 1338–1360.
Kuwada, S., & Yin, T. C. T. (1987). Physiological studies of directional hearing. In W. A. Yost, & G. Gourevitch (Eds.), *Directional hearing* (pp. 146–176). New York: Springer-Verlag.
Lazzaro, J. (1991). A silicon model of an auditory neural representation of spectral shape. *IEEE Journal of Solid-State Circuits, 26,* 772–777.
Licklider, J. C. R. (1948). The influence of interaural phase relations upon the masking of speech by white noise. *Journal of the Acoustical Society of America, 20,* 150–159.
Licklider, J. C. R. (1959). Three auditory theories. In S. Koch (Ed.), *Psychology: A study of a science* (pp. 41–144). New York: McGraw-Hill.
Licklider, J. C. R., Webster, J. C., & Hedlun, J. M. (1950). On the frequency limits of binaural beats. *Journal of the Acoustical Society of America, 22,* 468–473.
Lindemann, W. (1986a). Extension of a binaural cross-correlation model by contralateral inhibition. I. Simulation of lateralization for stationary signals. *Journal of the Acoustical Society of America, 80,* 1608–1622.
Lindemann, W. (1986b). Extension of a binaural cross-correlation model by contralateral inhibition. II. The law of the first wavefront. *Journal of the Acoustical Society of America, 80,* 1623–1630.
Litovsky, R. Y., & Yin, T. C. T. (1993). Single-unit responses to stimuli that mimic the precedence effect in the inferior colliculus of the cat. Abstracts of the 16th midwinter research meeting of the Association for Research in Otolaryngology, St. Petersburg, FL, p. 128(A).
Lyon, R. F. (1983). A computational model of binaural localization & separation. Proceedings of the IEEE international conference on acoustics, speech, and signal processing, pp. 1148–1151.
McFadden, D. (1973). Precedence effects and auditory cells with long characteristic delays. *Journal of the Acoustical Society of America, 54,* 538 (L).
McFadden, D., & Pasanen, E. G. (1976). Lateralization at high frequencies based on interaural time differences. *Journal of the Acoustical Society of America, 59,* 634–639.
Mead, C. A., Arreguit, X., & Lazzaro, J. (1991). Analog VLSI model of binaural hearing. *IEEE Transactions on Neural Networks,* NN-*2,* 230–236.

Meddis, R., Hewitt, M. J., & Shackleton, T. M. (1990). Implementation details of a computational model of the inner hair-cell/auditory-nerve synapse. *Journal of the Acoustical Society of America, 87,* 2866–2882.
Moushegian, G., & Jeffress, L. A. (1959). Role of interaural time and intensity differences in the lateralization of low-frequency tones. *Journal of the Acoustical Society of America, 31,* 1441–1445.
Nuetzel, J. M. & Hafter, E. R. (1981). Discrimination of interaural delays in complex waveforms: spectral effects. *Journal of the Acoustical Society of America,* 69, 1112–1118.
Palm, S. (1989). Enhancement of reverberated speech using models of the human binaural system. M.S. thesis, Carnegie Mellon University.
Payton, K. L. (1988). Vowel processing by a model of the auditory periphery: A comparison to eighth-nerve responses. *Journal of the Acoustical Society of America, 83,* 145–162.
Pfeiffer, R. R., & Kim, D. O. (1975). Cochlear nerve fiber responses: Distribution along the cochlear partition. *Journal of the Acoustical Society of America, 58,* 867–869.
Raatgever, J. & Bilsen, F. A. (1986). A central spectrum theory of binaural processing. Evidence from dichotic pitch. *Journal of the Acoustical Society of America, 80,* 429–441.
Rose, J. E., Geisler, C. D., & Hind, J. E. (1966). Some neural mechanisms in the inferior colliculus of the cat which may be relevant to localization of a sound source. *Journal of Neurophysiology, 29,* 288–314.
Sayers, B. M. (1964). Acoustic-image lateralization judgments with binaural tones. *Journal of the Acoustical Society of America, 36,* 923–926.
Sayers, B. M., & Cherry, E. C. (1957). Mechanism of binaural fusion in the hearing of speech. *Journal of the Acoustical Society of America, 29,* 973–987.
Schiano, J. L., Trahiotis, C., & Bernstein, L. R. (1986). Lateralization of low-frequency tones and narrow bands of noise. *Journal of the Acoustical Society of America, 79,* 1563–1570.
Schroeder, M. R. (1977). New viewpoints in binaural interactions. In E. F. Evans and J. P. Wilson (Eds.), *Psychophysics and physiology of hearing* (pp. 455–467). London: Academic Press.
Shackleton, T. M., Meddis, R., & Hewitt, M. J. (1992). Across frequency integration in a model of lateralization. *Journal of the Acoustical Society of America, 91,* 2276–2279 (L).
Shamma, S. A., Shen, N., & Gopalaswamy, P. (1989). Binaural processing without neural delays. *Journal of the Acoustical Society of America, 86,* 987–1006.
Siebert, W. M. (1970). Frequency discrimination in the auditory system: place or periodicity mechanisms. *Proceedings of the IEEE, 58,* 723–730.
Siegel, R. A., & Colburn, H. S. (1983). Internal and external noise in binaural detection. *Hearing Research, 11,* 117–123.
Smith, P. H., Joris, P. X., & Yin, T. C. T. (1993). Projections of physiologically characterized spherical bushy cell axons from the cochlear nucleus of the cat: Evidence for delay lines to the medial superior olive. *Journal of Computational Neurology, 331,* 245–260.
Stanford, T. R., Kuwada, S., & Batra, R. A. (1992). A Comparison of the interaural time sensitivity of neurons in the inferior colliculus and thalamus of the unanesthetized rabbit. *Journal of Neuroscience, 12,* 3200–3216.
Stern, R. M., Jr., & Bachorski, S. J. (1983). Dynamic cues in binaural perception. In R. Klinke & R. Hartmann (Eds.), *Hearing—Physiological bases and psychophysics* (pp. 209–215). Berlin: Springer-Verlag.
Stern, R. M., Jr., & Colburn, H. S. (1978). Theory of binaural interaction based on auditory-nerve data. IV. A model for subjective lateral position. *Journal of the Acoustical Society of America, 64,* 127–140.
Stern, R. M., & Colburn, H. S. (1985). Lateral-position-based models of interaural discrimination. *Journal of the Acoustical Societv of America,* 77, 753–755.

Stern, R. M., & Shear, G. D. (1995). Lateralization and detection of low-frequency binaural stimuli: Effects of distribution of internal delay. *Journal of the Acoustical Society of America* (in revision).

Stern, R. M., Shear, G. D., & Zeppenfeld, T. (1988a). High-frequency predictions of the position-variable model. *Journal of the Acoustical Society of America, 84,* S60 (A).

Stern, R. M., & Trahiotis, C. (1992). The role of consistency of interaural timing over frequency in binaural lateralization. In Y. Cazals, K. Horner, & L. Demany (Eds.), *Auditory physiology and perception* (pp. 547–554). Oxford: Pergamon Press.

Stern, R. M., Zeiberg, A. S., & Trahiotis, C. (1988b). Lateralization of complex binaural stimuli: A weighted image model. *Journal of the Acoustical Society of America, 84,* 156–165.

Stern, R. M., Zeppenfeld, T., & Shear, G. D. (1991). Lateralization of rectangularly modulated noise: Explanations for counterintuitive reversals. *Journal of the Acoustical Society of America, 90,* 1901–1907.

Sullivan, T. M., & Stern, R. M. (1993). Multi-microphone correlation-based processing for robust speech recognition. Proceedings of the IEEE international conference on acoustics, speech, and signal processing, 2, 91–94.

Tao, S. H. (1992). Additive versus multiplicative combination of differences of interaural time and intensity. M.S. thesis, Carnegie Mellon University.

Tao, S. H., & Stern, R. M. (1992). Additive versus multiplicative combination of differences of interaural time and intensity. *Journal of the Acoustical Society of America, 91,* 2414(A).

Trahiotis, C., & Bernstein, L. R. (1986). Lateralization of bands of noise and sinusoidally amplitude-modulated tones: Effects of spectral locus and bandwidth. *Journal of the Aoustical Society of America, 79,* 1950–1957.

Trahiotis, C., & Stern, R. M. (1989). Lateralization of bands of noise: Effects of bandwidth and differences of interaural time and intensity. *Journal of the Acoustical Society of America, 86,* 1285–1293.

Van Trees, H. L. (1968). *Detection, estimation, and modulation theory, Part I.* New York: Wiley.

Wallach, H., Newman, E. B., & Rosenzweig, M. R. (1949). The precedence effect in sound localization. *Journal of Psychology, 52,* 315–336.

Webster, F. A. (1951). The influence of interaural phase on masked thresholds. I. The role of interaural time-deviation. *Journal of the Acoustical Society of America,* 23, 452–462.

Wenzel, E. M. (1992). Localization in virtual acoustic displays. *Presence: Teleoperators and Virtual Environments, 1,* 80–107.

Whitworth, R. H., & Jeffress, L. A. (1961). Time versus intensity in the localization of tones. *Journal of the Acoustical Society of America, 33,* 925–929.

Wightman, F. L., & Kistler, D. J. (1989). Headphone simulation of free-field listening. I. stimulus synthesis. *Journal of the Acoustical Society of America,* 85, 858–867.

Yin, T. C. T., & Chan, J. C. L. (1990). Interaural time sensitivity in medial superior olive of cat. *Journal of Neurophysiology, 65,* 465–488.

Yin, T. C. T., & Kuwada, S. (1984). Neuronal mechanisms of binaural interation. In G. M. Edelman, W. C. Cowan, & W. E. Gall (Eds.), *Dynamic Aspects of Neocortical Function* (pp. 263–313). New York: Wiley.

Yost, W. A. (1972). Tone-on-tone masking for three listening conditions. *Journal of the Acoustical Society of America, 52,* 1234–1237.

Yost, W. A. (1981). Lateral position of sinusoids presented with intensive and temporal differences. *Journal of the Acoustical Society of America, 70,* 397–409.

Yost, W. A., Nielsen, D. W., Tanis, D. C., & Bergert, B. (1974). Tone-on-tone binaural masking with an antiphasic masker. *Perception and Psychophysics,* 15, 233–237.

Yost, W. A., & Soderquist, D. R. (1984). The precedence effect revisited. *Journal of the Acoustical Society of America, 76,* 1377–1383.

Young, S. R., and Rubel, E. W. (1983). Frequency-specific projections of individual neurons in chick brainstem auditory nuclei. *Journal of Neuroscience, 3,* 1373–1378.

Zurek, P. (1987). The precedence effect. In W. A. Yost & G. Gourevitch (Eds.), *Directional hearing* (pp. 85–105), New York: Springer-Verlag.

Zwislocki, J. & Feldman, R. S. (1956). Just noticeable differences in dichotic phase. *Journal of the Acoustical Society of America, 28,* 860–864.

CHAPTER 11

Auditory Grouping

C. J. Darwin
R. P. Carlyon

I. INTRODUCTION

Imagine that you are walking along one of the enclosing arms of a harbor on a calm day. Could you, by looking at the waves entering the harbor, describe the events happening out at sea? Clearly not, yet a comparable feat is continually performed by your auditory system, which, in many everyday situations, is presented with an acoustic waveform made up from a mixture of sounds originating from a variety of sources. In the space of a few seconds these sounds might include a number of different people speaking, a car passing, music from next door's radio, a door slamming, and the wind whistling through a crack in the window. The computational problem facing the auditory system is to interpret this complex waveform as sound-producing events. Each event must be assigned the appropriate instantaneous properties, such as location, timbre, and pitch, and their variation over time tracked to obtain such properties as melodic line, speech articulation, or spatial trajectory. The purpose of this chapter is to review the growing body of experiments that will help us determine how our auditory systems perform this exacting task.

The problem of interpreting sound in terms of separate events is closely related to the visual problem of interpreting, in terms of three-dimensional objects, the pair of two-dimensional images formed on the two retinas. An important difference is that, unlike vision, where adjacent retinal regions tend to be stimulated by light from the same object, the output of adjacent auditory frequency channels is not so constrained; frequency components from two separate sounds may be interleaved across the whole frequency spectrum.

Partly because of the technical difficulty of synthesizing controlled auditory stimuli, work on auditory grouping has lagged substantially behind that in vision. However, there is now an expanding body of experimental evidence on the heuristic principles used by the brain to organize auditory input, accompanied by a growing number of attempts to implement these principles in computer programs. Albert Bregman's book *Auditory Scene Analysis* (1990) thoroughly reviews the now extensive body of experimental evidence on auditory grouping. This chapter aims to complement rather than repeat its coverage. In particular, we will consider studies carried out since Bregman's book appeared and dwell more on the psychoacoustic and physiological ramifications of work on auditory grouping.

II. PERIPHERAL CONSIDERATIONS

Before we consider the central processing mechanisms involved in auditory grouping, we must take due account of how the peripheral auditory system analyses incoming sound. This analysis has been extensively reviewed in chapters 2 to 5, but it is worth describing here three aspects of peripheral processing which affect auditory grouping:

- *Frequency Resolution*. In the absence of any other sound, the normal human auditory system can resolve the first eight or so harmonics of a periodic complex tone whose harmonics are all equal in amplitude; these low-numbered harmonics can thus be explicitly identified as discrete frequency components, and we can sensibly speak of particular frequency components being grouped together. Individual higher numbered harmonics cannot be resolved unless they are substantially more intense than adjacent harmonics; the raw material on which auditory grouping can operate is now mixed within the output of an auditory channel. In addition, when other sounds are present at appropriate amplitudes, their frequency components can mask otherwise audible components or reduce the ability of the peripheral system to resolve even the low-numbered harmonics.

- *Phase Locking*. It is well known that auditory nerve fibers can phase lock to low-frequency tones. This ability decreases with increasing frequency and has disappeared by 4–5 kHz in the cat (Johnson, 1980). Psychophysical estimates in humans also put the limit at around 4–5 kHz (Moore, 1973). The auditory nerve will therefore phase lock to the low-frequency, resolved components of a complex tone, whereas it will do so less effectively to individual high-frequency components. However, auditory nerve fibers with high center frequencies will usually respond to several high-frequency components and phase lock to the envelope resulting from their interaction. For a periodic complex tone this envelope repeats at a rate equal to the fundamental frequency.

• *Adaptation*. Immediately after the onset of a sound, the firing rate of the auditory nerve increases then decreases rapidly (Kiang, Watanabe, Thomas, & Clark, 1965; Smith, 1979). This adaptation may highlight change and the onset of new sounds.

III. MECHANISMS OF AUDITORY GROUPING

A basic assumption underlying research on auditory grouping is that, before properties of an auditory event can be established, some decision must have been made concerning which components of the input are relevant to that event. Bregman (1990, p. 408) distinguishes two types of mechanism that can be used to decide which components belong to a particular sound source: *Primitive grouping mechanisms* partition the input on the basis of simple stimulus properties, whereas *schema governed mechanisms* select an array of data that meet criteria defined by a particular sound (or class of sounds such as speech). The primitive grouping mechanisms use simple general properties of a single sound source (such as a common onset time or a harmonic relationship among components) to which the listener may be innately attuned, and so do not necessarily depend on specific experience. The schemata that govern the second mechanism, on the other hand, are generally learned and so depend on the listener's specific experience.

The attraction of primitive grouping mechanisms is that they potentially exploit general properties of sound sources—the listener does not need to know what he or she is going to hear to be able segregate it from the background. They also allow an evolving auditory system to develop incrementally. For example, expertise in speech perception can take advantage of any sound separation ability available to the prelingual mammalian auditory system in order to help it to segregate speech from background sounds (Darwin, 1991). The "context independence" of these mechanisms makes them particularly well suited to psychoacoustically based research, where it may be desirable to use arbitrary stimuli, and where one may seek to develop relatively straightforward models of the processes underlying auditory grouping. For these reasons, we will concentrate our discussion on mechanisms underlying primitive, rather than schema-based, mechanisms.

In this chapter we will first look at the experimental evidence for listeners being able to detect simple properties (or cues) such as onset asynchrony, inharmonicity, and differences in the patterns of frequency or amplitude modulation among components. Next, we will examine how well listeners can use these cues to group together either particular frequency components or the output of particular auditory filters for the purpose of determining different properties of a complex sound. Perhaps surprisingly, we will find that different cues are apparently more effective in controlling grouping for the calculation of one attribute (such as vowel quality) than for another (such as pitch): Grouping is graded, not all-or-none.

Many early experiments on auditory grouping required subjects to make judgments which were directly related to the phenomenal experience of the number of sound sources that they heard. This work is extensively reviewed by Bregman (1990). Although it has been extremely valuable in indicating which cues are potentially used by primitive grouping mechanisms and in establishing a theoretical framework for research, it does not address the issue of whether a particular cue is actually used, or how much weight is given to it, for the purpose of determining a particular attribute of a complex sound. It is logically possible, and indeed experimentally demonstrable, that subjects may hear a complex as consisting of two sound sources while treating the whole sound as a single complex for the purpose of calculating a particular attribute such as pitch. Consequently, this chapter concentrates on work that has considered the effect of auditory grouping on the perception of particular attributes of sound such as pitch, timbre, and lateral position. A second focus of interest will be in determining not only which cues are used by the auditory system, but also in elucidating the mechanisms used to exploit them.

IV. HARMONICITY

All periodic sounds, such as voiced speech and those produced by most musical instruments, have frequency spectra that contain energy only at the fundamental frequency ($F0$), which corresponds to the repetition rate of the waveform, and at *harmonics* of that fundamental (see Chapters 1 and 8). This harmonicity gives rise to a highly integrated percept: To hear out even a low-numbered harmonic, it is usually necessary either to increase its level or to direct the listener's attention to it by, for example, immediately preceding the complex with that harmonic presented in isolation (Plomp, 1964; Plomp & Mimpen, 1968). However, as we shall see, mistuning that harmonic by even a small amount can cause the listener to perceptually segregate it from the remainder of the complex. In the first part of the next section, we will compare experiments that required listeners simply to detect that a harmonic has been mistuned to those requiring them to "hear out" that component from the rest of the complex. Although this class of experiment often uses sounds that are not very "life like," the very arbitrariness of the stimuli has allowed experimenters some degree of control over which auditory mechanisms are being used to perform the task and has permitted an evaluation of precisely which parameters determine whether or not mistuning can be detected. Having obtained this basic information, we will go on to consider other types of experiment, which provide evidence on the influence of mistuning on the integration of the mistuned harmonic into the pitch and phonetic percept of the rest of the complex.

A. Detecting Inharmonicity and "Hearing Out" Mistuned Components

1. Mistuning a Single Harmonic

One of the most useful distinctions to make when considering the processing of mistuning and, indeed, of other grouping cues is between processing which requires a comparison between the outputs of different auditory filters and that based on changes within a single auditory filter's output. That the two types of processing can lead to quite different patterns of results is illustrated by a study performed by Moore and his colleagues (Moore, Peters, & Glasberg, 1985b). They presented listeners with a harmonic complex and, on one half of each trial, mistuned one component slightly, either up or down in frequency. They found that thresholds, defined as the percentage mistuning necessary for listeners to identify which interval contained the mistuned harmonic, decreased progressively with increasing harmonic number. However, listeners reported that the nature of the task differed markedly between the low and the high harmonics: When a low-numbered harmonic was mistuned, they heard it "stand out" from the complex as a whole, but mistuning of high-numbered harmonics was identified by the perception of "a kind of beat." Moore et al. attributed this beating to an interaction, within a single auditory filter, between the mistuned harmonic and its neighbors and reasoned that the usefulness of this cue could be reduced by decreasing the signal duration, so that less than a full cycle of beating could occur within the presentation time of the stimulus. When they did so, by reducing the signal duration from 400 ms down to 50 ms, they found that thresholds for the high-numbered harmonics increased markedly to about 2–3%, whereas those for the low-numbered harmonics were less affected. At this short duration, thresholds were roughly independent of harmonic number.

Moore et al.'s experiment makes the important point that simple stimulus manipulations can affect the way in which listeners process a cue—in this case by reducing their ability to detect beats—and can markedly alter the pattern of results obtained. However, their listeners were required only to detect the mistuning, and as their informal reports suggested, this does not necessarily mean that the mistuned harmonic was perceived as a separate tone from the rest of the complex. This was demonstrated in a further experiment performed by Moore, Glasberg, and Peters (1986), who used similar stimuli but required a rather different type of response. Instead of encouraging listeners (by the use of feedback lights) to use any cue or percept necessary to detect mistuning, they presented a 400-ms complex sound that might, or might not, have one of its harmonics mistuned. Listeners were asked to judge whether they heard "a single sound with one pitch, or two sounds—a complex tone and a component with a pure-tone quality not belonging to the complex." In this new experiment, in which no feedback was given, the results differed from those obtained previously: For harmonics up to the sixth, threshold was quite low—about 2%, but for higher harmonic numbers, listeners frequently could not perform reliably. Thus it seems that listeners can hear out a low-numbered harmonic as soon as they can detect that it has been mistuned, but cannot do so for high harmonics, even when mistuned by an amount several times that necessary for detection. Moore et al. attributed this poor performance to the fact that a mistuned high harmonic is insufficiently resolved from its neighbors to be heard out as a separate tone. For these higher harmonics, the mistuning is detectable as roughness rather than as a separate tonal sound (Moore et al., 1985b).

As well as being influenced by harmonic number, there is evidence that the ability to hear out a mistuned harmonic depends on its absolute frequency. Hartman, McAdams, and Smith (1990) asked listeners to match the pitch of a pure tone presented in isolation to that of a component that had been mistuned from a harmonic complex and varied both the *F*0 of the complex and the harmonic number of the mistuned component. They argued that the observed decrease in performance at high harmonic numbers was related to the absolute component frequency, rather than to harmonic number per se, with performance deteriorating for harmonics with frequencies above about 2000 Hz. In a more recent experiment, Moore and Ohgushi (1993) presented listeners with a complex whose components were equally spaced in terms of auditory-filter bandwidths (i.e., on an ERB scale; see Chapter 5) and asked them to compare the frequency of a pure tone presented in isolation to that of one of the components of the complex. Their results suggested that listeners' ability to hear out a component depended *both* on the extent to which it was resolved from its neighbors *and* on its absolute frequency, with performance deteriorating for component frequencies above about 1000 Hz.

Both Harmann et al. and Moore and Ohgushi attributed the deterioration in performance due to a component having a high frequency per se to a breakdown in phase locking at the level of the auditory nerve. However, there is some evidence that the limitation is more central. Demany, Semal, and Carlyon (1991) required listeners to discriminate between a two-component complex composed of the fundamental and second harmonic of a given *F*0 and one whose components were mistuned from this harmonic relationship. They reported that thresholds increased with increases in *F*0, from about 2% at an *F*0 of 300 Hz to more than 10% when the *F*0 was 1000 Hz. Thus, like the studies of Hartmann et al. and of Moore and Ohgushi, there was a deterioration in performance at high frequencies independent of harmonic number. However, Demany et al. argued against an explanation in terms of phase locking on the basis of a second experiment, in which listeners had to perform a (sequential) frequency discrimination of the second harmonic, which was mistuned in opposite directions on the two halves of each trial. They found that in this new task, in which listeners used the absolute frequency of the second harmonic, rather than its frequency relative to the fundamental, thresholds did not increase with increases in *F*0. They argued that information on the frequency of the second harmonic must have been available at the level of the auditory nerve, concluded that, for their high-frequency conditions, "the detection of inharmonicity was limited by factors other than those responsible for fine pitch discriminations," and suggested that this was a central, rather than a peripheral, limitation. An alternative explanation is that the sequential task of their second experiment did not require such accurate phase locking as does the detection of mistuning (as in their first experiment) or the segregation of a component from a complex tone (Hartmann et al., 1990; Moore & Ohgushi, 1993).

Finally, it is worth describing the influence of two further factors on listeners' ability to detect mistuning and to "hear out" a component of a complex tone. First, a study by Demany and Semal (1988) in which listeners had to detect mistuning between two tones presented to opposite ears found thresholds that were similar to those obtained in later experiments with monaural presentation (Demany et al., 1991). As we shall see, this observation that listeners can process mistuning regardless of differences in ear of presentation between components extends to many other paradigms. Second, Roberts, and Bregman (1991) presented evidence that an individual harmonic may become more perceptually prominent by virtue of it deviating from a particular pattern of spectral spacing. They used sounds consisting of the odd-numbered harmonics of a 100, 200, or 400 Hz fundamental, to which had been added a single even-numbered harmonic. At the two lower fundamentals, listeners rated the perceived clarity of the even-numbered harmonic higher than that of its odd-numbered neighbors. This result is striking because, if perceived clarity were determined by how well a particular component was resolved, the even-numbered harmonic should have been *less* clear than its odd-numbered neighbors.

2. $F0$ Differences between Groups of Harmonics

Unlike the research on the mistuning of a single harmonic, experiments on the detection of differences in $F0$ ($\Delta F0$s) between groups of harmonics have used rather different stimuli from those requiring listeners to judge the number of sources present. Indeed, most studies of the detection of $\Delta F0$s between groups of harmonics have required listeners to detect differences between two *sequentially* presented complex tones, rather than to perform a simultaneous across-frequency comparison between the $F0$s of two different groups. For example, two studies (Hoekstra, 1979; Shackleton & Carlyon, 1994) have shown that, for harmonic complexes passed through a bandpass filter, frequency difference limens (DLs) are substantially lower when the combination of $F0$ and filter setting results in the harmonics being resolvable by the peripheral auditory system than when the harmonics are unresolved. A broadly similar conclusion was reached by Carlyon and Shackleton (1994), who *did* require listeners to perform an across-frequency comparison between the $F0$s of two simultaneously presented groups of components. They reported that performance was much better when both groups were resolved by the peripheral auditory system than when one group was unresolved. On average, thresholds for two groups of resolved components were about 2%, compared to about 10% when one group was unresolved.

Experiments on the number of sources heard have generally used speech-like stimuli. In speech, a voiced vowel consists of a large number of adjacent harmonics, some of which have been amplified by the resonances (or *formants*) of the vocal tract. These resonances are spaced on average about 1 kHz apart, and when more than one person is talking at once, the formant frequencies of one speaker will often be interleaved with those of another. A classic experiment by Broadbent and Ladefoged (1957) was concerned with the question of how, under such circumstances, listeners can group together appropriate formant frequencies. They synthesized sounds composed of the first two formants of a phrase and played one formant to each ear. When the two formants had the same $F0$, listeners generally reported hearing only a single voice, even though the formants were being led to different ears, but when the two formants had $F0$s that differed by 20%, listeners reported hearing two different speakers. This finding provided early evidence that $F0$ differences can be a more powerful cue to perceptual sound segregation than the "ear of presentation" (see the work of Demany and his colleagues, described earlier). Subsequent work using simple syllable-length sounds has shown that an $F0$ difference of only 2% between the first two formants can be sufficient to give an impression of two sound sources, whether they are led to the same ear or to different ears (Cutting, 1976; Gardner, Gaskill, & Darwin, 1989).

The distinction between resolved and unresolved harmonics is also relevant here. Darwin (1992) played the second formant of a four-formant syllable on a different *F*0 to that of the remaining formants and measured the percentage $\Delta F0$ necessary for listeners to hear two sound sources. When the syllable as a whole had a low *F*0, so that the second formant was defined by unresolved harmonics, this $\Delta F0$ was about 10%, but it dropped to only about 2.5% at high *F*0s, where the harmonics making up the second formant were resolved. This finding is similar to that described by Carlyon and Shackleton (1994) for the detection of *F*0 differences between two groups of components.

3. Summary

In an experiment where listeners are presented with a complex tone in which one of the harmonics has been mistuned, the results depend on the harmonic number of the mistuned component, its frequency, and the nature of the task required. For low harmonic numbers, a mistuning of approximately 2% is usually detectable and will lead to the mistuned harmonic being perceived as a separate tone. The exception is if the complex has a high *F*0, so that the frequency of the mistuned harmonic is above about 1000–2000 Hz, in which case much larger mistunings are required. For high-numbered harmonics, very small mistunings (<1%) can sometimes be detected, but this detection is based on a sensation of beating between the mistuned harmonic and its neighbors; the harmonic is not "heard out" as a separate tone, even when much larger mistunings are applied. When the usefulness of beating cues is reduced (but not necessarily eliminated) detection thresholds also increase, but still correspond to mistunings smaller than those necessary for the component to be heard out.

When two groups of harmonics (or two formants) are present simultaneously, an *F*0 difference between them can make the listener hear two sound sources, and this occurs almost as soon as the *F*0 difference is detectable. Such a difference is easier to detect when all the harmonics are resolved than when the harmonics in one group are unresolved. Unlike the case where an individual harmonic is mistuned, there is no evidence that the center frequencies of the groups of harmonics affect either the detection of the *F*0 difference or the number of sound sources heard.

B. Integration into Pitch Percepts

A comprehensive review of the literature on pitch perception appears in Chapter 8. Here, we restrict our discussion to one technique for determining the conditions affecting the integration of a single frequency component into the pitch of a complex tone. A plausible heuristic that the auditory system could use to determine which components are relevant to a particular *F*0 would be to consider only those that lie sufficiently close to a harmonic frequency. This principle underlies the "harmonic sieve" (Duifhuis, Willems, & Sluyter, 1982), which effectively excludes from the calculation of pitch any component whose frequency lies more than some fixed percentage from a harmonic of *F*0. The value of that percentage has been addressed experimentally by Moore and his colleagues (Moore, Glasberg, & Peters, 1985a). They mistuned one harmonic of a 12-harmonic complex and measured the consequent shift in the pitch of the complex. For small mistunings (less than about 3%) the pitch shift was a roughly linear function of the mistuning, but for larger mistunings the pitch shift of the complex *decreased,* approaching zero by about 8% mistuning. Therefore it seems that the harmonic sieve does not work on an "all or none" basis but, rather, that a harmonic makes progressively less contribution to the pitch of the complex as its mistuning increases from 3% to beyond 8%. Note that a component mistuned by up to 8% still makes some contribution to the pitch of the complex, even though a mistuning of only 2% is sufficiently large for it to be heard out as a separate tone from the rest of the complex (Moore et al., 1986).

Moore's results have been replicated by one of us (Darwin, 1992; Darwin & Ciocca, 1992), using a larger number of values of mistuning. These data are shown by the open symbols in Figure 1; they can be well fitted by assuming that the contribution that a particular harmonic makes to the pitch of a complex sound varies according to a Gaussian function of the amount of mistuning.

Figure 1 also shows that, when the mistuned harmonic arrives at the ear opposite from the remainder of the complex (filled symbols), it makes almost as large a contribution to its pitch as when it is presented to the same ear (open symbols). This surprisingly small effect of ear of presentation (or more generally of lateralization) to grouping for pitch perception echoes the similar findings already described for the detection of mistuning (Demany & Semal, 1988) and the identification of the number of sources heard (Broadbent & Ladefoged, 1957).

C. Identification of Speech Sounds

1. Mistuning a Single Harmonic

So far, we have reviewed evidence that mistuning a single harmonic allows the listener to hear it as a separate tone and match its pitch to that of a pure tone presented in isolation and that the mistuning reduces the contribution that the harmonic makes to the pitch of the complex. There is also evidence that mistuning a component can reduce the contribution that it makes to the

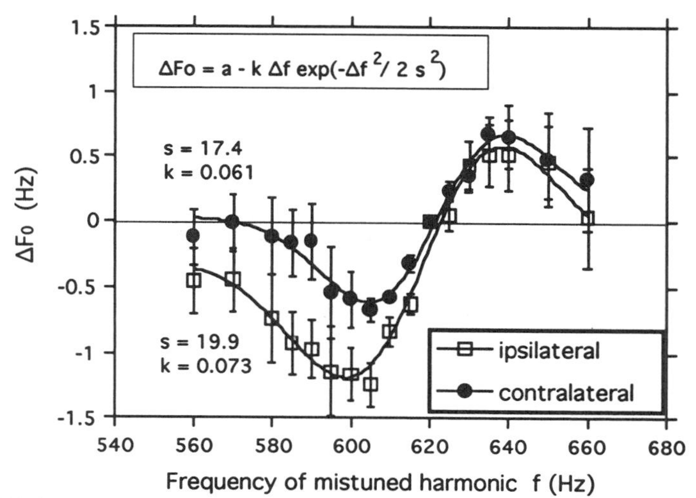

FIGURE 1 Matched pitch shifts (from 155 Hz) produced by mistuning the fourth harmonic of a 12-harmonic complex with a fundamental of 155 Hz. The mistuned component could be presented either to the same ear as the remaining harmonics or to the opposite ear. The fitted curves assume that the contribution that a harmonic makes to the perceived pitch varies according to a Gaussian function as it is progressively mistuned. Contralateral presentation results in a slightly smaller overall contribution (*k*) and a slightly narrower Gaussian (*s*).

phonetic categorization of a speech sound. Darwin and Gardner (1986) exploited the fact that an /I/ vowel can be turned into an /e/ simply by increasing the frequency of the first formant (*F*1). The first formant frequency itself is specified by the relative amplitudes of the harmonics close to that frequency—the harmonic amplitudes define a spectral envelope whose peak is the formant frequency. Consequently, it is possible to change a vowel from an /I/ to an /e/ simply by changing the relative amplitudes of the harmonics close to the first formant frequency (around 450 Hz). Physically removing, say, the fourth harmonic of a 125 Hz fundamental (i.e., 500 Hz) will shift *F*1 down in frequency and so produce a more /I/-like vowel. Darwin and Gardner showed that mistuning the fourth harmonic by about 8% gave a change in vowel quality similar to that produced by physically removing it. This change of 8% is substantially larger than that needed to hear the harmonic out as a mistuned component, but is similar to the mistuning needed to remove a harmonic from the calculation of pitch.

2. Mistuning a Group of Harmonics

Cutting (1976, Exp. 4) played the first formant of a /ba/,/da/, or /ga/ syllable to one ear, and the second formant to the other, either on the same F0 (100 Hz) or a different *F*0 (180 Hz). Subjects identified the appropriate syllable around 80% of the time, regardless of whether or not the two formants shared the same *F*0. Using similar stimuli, he showed that an *F*0 difference of only 2 Hz was sufficient to give the impression of two sound sources. However, although his experiments strongly suggest that his listeners were indeed phonetically combining information across sounds that were heard as different sound sources, the experiments suffered from a logical weakness. His subjects may have identified the syllables on the basis of the second formant alone (the first formant was constant across the three stop consonant categories). This possibility was explicitly controlled for by Darwin (1981), who required subjects to identify diphthongs that were uniquely specified by combinations of two different first and second formant transitions. The results confirmed Cutting's conclusions: Differences in *F*0 that were sufficient to give a clear percept of two sound sources did not impair classification of the diphthongs. Moreover, control conditions in which each formant was presented in isolation showed that subjects were not simply doing the task on the basis of information from a single formant.

Although these experiments found no evidence that formants are perceptually grouped together on the basis of *F*0, sufficiently large Δ*F*0s *can* play a role in perceptual grouping for phonetic categorization. Darwin (1981) presented listeners with a four-formant sound that was a combination of the syllables /li/ and /ru/. All four formants (*F*1, *F*2, *F*3, *F*4) played together on the same fundamental sounded predominantly like /ru/, but when the second formant was physically removed, the three remaining formants (*F*1, *F*3, *F*4) sounded like /li/. Playing *F*2 on a different *F*0 (174 Hz) to the other formants (110 Hz) caused listeners to hear two sounds, consisting predominantly of /li/ (*F*1, *F*3, *F*4), plus a "buzz" corresponding to the perceptually segregated *F*2. So, a difference in *F*0 was sufficient both to increase the number of sound sources heard, and to alter the phonetic category of the vowel. The experiment was later replicated by Gardner et al. (1989) who showed that a much smaller (4 Hz) Δ*F*0 was sufficient for subjects to hear the second formant as a separate sound source, but did not cause a change in phonetic percept. Thus, the impression of more than one sound source clearly requires a smaller Δ*F*0 than does phonetic segregation.

3. Separating Two Voices by *F*0 Differences

If a difference in F0 *can* produce phonetically relevant segregation of different formant regions, we would expect that the interference caused by mixing together two voices would be reduced when the voices were on different fundamentals. There are two lines of evidence that this is indeed so. First, Brokx and Nooteboom (1982) presented Dutch nonsense sentences against a background of speech. The intelligibility of the content words of the sentences increased from 45% to 65% correct as the *F*0 difference between them and the interfering speech increased from 0 to 3 semitones (approximately 0–18%), but dropped back to near 50% at the octave (open squares in Figure 2). A number of different perceptual mechanisms may be contributing to this overall effect.

Further evidence that F0 differences aid the perceptual segregation of concurrent sounds comes from the class of "double vowel" experiments first performed by Scheffers (1983) and more recently replicated and extended by Assmann and Summerfield (1990). They presented listeners with pairs of simultaneous vowels, each with a duration of 200 ms, and measured the percentage of trials on which both vowels were identified correctly as a function of the *F*0 difference between them. They observed a marked increase in performance as the *F*0 difference increased from 0 to 0.5 semitones (about 3%), with a more gradual increase as the *F*0 difference increased further; the results of a replication of their experiment by Culling and Darwin (1993) are shown by the filled circles in Figure 2. As the figure shows, the results of these "double-vowel" experiments differ from those of the experiments that we have described so far in that they show a phonetic advantage for very small *F*0 differences, with little if any further improvement for larger *F*0 differences.

Some further experiments performed by Culling and Darwin (1993) go some way to resolving this discrepancy. They presented evidence that the large improvement in performance between 0 and 0.5 semitones was due almost entirely to the Δ*F*0 in the region of the first formant (*F*1): A similar improvement occurred even when the higher formants of the two vowels

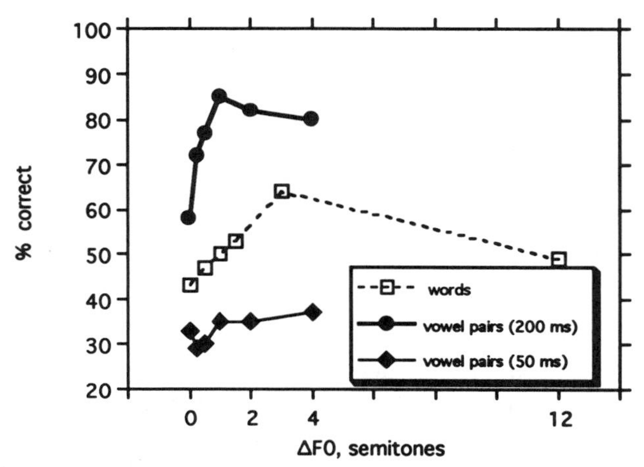

FIGURE 2 Improvement in speech identification with changes in Δ*F*0 for words in sentences masked by continuous speech (Brokx & Nooteboom, 1982) and for pairs of 200 ms or 50 ms steady-state vowels (Culling & Darwin, 1993).

were played on the same *F*0 as each other, provided that there was an *F*0 difference in the *F*1 region. Playing the first formants of the two vowels on the same *F*0 and imposing the Δ*F*0 only on the upper formants removed this abrupt improvement in performance, although a more gradual improvement, over a range of larger Δ*F*0s, remained.

Culling and Darwin's finding that larger *F*0 differences are needed for perceptual segregation in the higher formants than in the *F*1 region echoes the larger frequency difference limens for complex tones consisting of only unresolved harmonics, compared to those containing at least some resolved harmonics (Hoekstra, 1979; Houtsma & Smurzynski, 1990). At first sight, it is tempting to conclude that the dominance of the *F*1 region in the exploitation of small *F*0 differences is due to the resolvability of the low-frequency components, with the pitch of each vowel being extracted via a "pattern recognition" type pitch analyzer (Duifhuis et al., 1982; Goldstein, 1973; see Chapter 8). However, although such a mechanism could operate in frequency regions where the harmonics of one vowel's fundamental are substantially more intense than those of the other vowel, this is unlikely to be the whole story, particularly at frequencies where the harmonics from the two vowels are similar in amplitude.

As Figure 3 shows, although the low-frequency components of a vowel may be resolved *from each other,* this does not prevent them from interacting with adjacent components from the competing vowel. Culling and Darwin (1994) have argued that the beating between adjacent low-frequency components produces much of the improvement at small *F*0 differences. Adjacent components beat together at a rate equal to their frequency separation so that, for two vowels with *F*0s of 100 and 102 Hz, their fundamentals will beat at a rate of 2 Hz, the second harmonics at a rate of 4 Hz, the third at 6 Hz, and so on. Culling and Darwin showed that, during each cycle of beating between vowel pairs on different fundamentals, there were times when each vowel became more identifiable than the other. They presented a computational model of how the auditory system might exploit these beats by identifying each vowel in turn during the portion of the stimulus when it was dominant. Vowels lasting about 200 ms (as used by Scheffers, by Assmann and Summerfield, and by Culling and Darwin) give at least one cycle of beating between harmonics above the second, and both the model and experimental subjects could exploit the changing identifiability of the constituent vowels of a pair. However, when the vowel duration is reduced to 50 ms, there is insufficient time for adjacent harmonics to beat through an entire cycle, and accordingly, neither subjects (Assmann & Summerfield, 1990, see the filled diamonds in Figure 2; Culling & Darwin, 1993) nor the model show an abrupt improvement at small Δ*F*0s. Additional experimental data in support of Culling and Darwin's model has been obtained by Assmann and Summerfield (1994). They showed that while merely repeat-

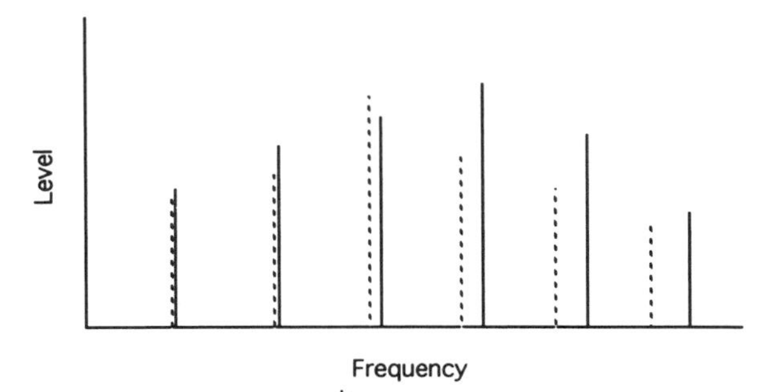

FIGURE 3 Schematic spectrum of the first formant frequency region of two vowels with different $F0$s. The components belonging to the vowel with the lower $F0$ are indicated by the dashed lines.

ing the same 50 ms segment of a double vowel four times did not improve identification over a single presentation, presenting four different successive 50 ms segments did improve performance.

Culling and Darwin's work goes some way to resolving the difference in the pattern of improvement observed between Brokx and Nooteboom's (1982) "running speech" experiment and those performed by Scheffers and by Assmann and Summerfield using double vowels (Figure 2). Because most vowels in running speech have durations much shorter than 200 ms, listeners would not be able to use the "beating" cue present in Scheffers's study, and so Brokx and Nooteboom's results may reflect the operation of a genuine across-frequency mechanism that separates concurrent sounds on the basis of differences between their $F0$s. Certainly, their finding of a gradual improvement with increases in $\Delta F0$ up to at least three semitones (19%, see Figure 3) provides the data most relevant to the perception of "everyday" speech.

D. Lateralization

Individual sound sources have a consistent subjective position; different frequency components of the sound do not appear to be coming from different directions. Yet in a reverberant or noisy environment, the cues to the direction of individual components may be unreliable or even indicate that different components should be localized in different directions. To maintain perceptual coherence, there is therefore a prima facie case that the perceived location of a sound source is determined *after* grouping of the individual frequency components has occurred. Although the basic calculations involved in comparing the phase in individual auditory channels occur at a very early stage of neural processing (Goldberg & Brown, 1969; Yin & Chan, 1990), there is both neurophysiological and psychophysical evidence for subsequent integration across frequency of these original frequency-specific calculations. Much of this evidence is described in detail in Chapters 3, 9, and 10, but it is worthwhile mentioning here two experiments demonstrating that harmonicity can affect the binaural processing of complex tones. Buell and Hafter (1991) reported that the threshold for detecting an interaural phase disparity of a "target" sinusoid could be elevated when a second "distracter" tone was presented in phase to the two ears, but only when the target and distracter were harmonically related. They commented on their findings: "The results . . . suggest that tones in a complex that bear a simple harmonic relation are combined into a single auditory object, regardless of their spatial coherence. Thus it would seem that harmonicity is a more compelling factor for object recognition than common spatial location" (p. 1899). Note that this is the same conclusion reached by Broadbent and Ladefoged in their 1957 study of the effects of *F*0 and ear-of-presentation differences on the number of sound sources reported by listeners.

The second finding, reported by Hill and Darwin (1993), was based on the phase ambiguity that occurs when a sinusoid is presented with a time delay between the two ears. For example, when a 500 Hz tone is presented binaurally and with a 1.5 ms delay to the right ear, listeners report hearing it on the right: in effect, they interpret the 1.5 ms phase lag as a 0.5 ms phase *lead* (recall that the period of a 500 ms tone is 2 ms, and see also Chapter 10 for a further discussion of this phenomenon). They expanded on the earlier work of Trahiotis and Stern (1989) and of Jeffress (1948) to show that this ambiguity could be eliminated if some additional tones, all harmonics of 500 Hz, were also presented with a 1.5 ms delay to the right ear. Under these conditions, all the components of the harmonic complex, including the 500 Hz tone, were heard on the left: This is the "correct" lateralization, in the sense that all components reached the left ear first, as they would if they had originated from a "real life" source on the subject's left. However, mistuning the 500 Hz tone caused it to be heard out as a separate sound source, and more important, as the mistuning increased above 1%, the 500 Hz tone was heard progressively more to the right, with its subjective position reaching an asymptote by about 6% mistuning.

E. Summary

A wide body of evidence has shown that listeners can detect and exploit differences in *F*0 between sounds to segregate them perceptually. When frequency components are well resolved by the peripheral auditory system, listeners can detect differences as small as 1–2% and use these differences both to judge how many sound sources are present and to match the pitch of a mistuned component to that of a pure tone presented in isolation. When one group of components is unresolved, differences of up to 10% are required. In both cases, there is a range of $\Delta F0$s where listeners will integrate a formant into their judgment of the category of a vowel, even though its *F*0 is sufficiently different from the others to be perceived as coming from a separate source. However, larger mistunings will cause its exclusion from the phonetic categorization.

Finally, where there are two or more sound sources with overlapping spectra, not even the low-numbered harmonics will always be resolved from adjacent components of a competing source. Whether or not there are sufficient "truly resolved" harmonics to identify even the first formant of a target vowel will depend not only on the frequency of that formant and the *F*0 at which it is spoken, but also on the intensity and frequency of the components of competing speech. Therefore, the 1–2% differences referred to previously are not relevant to all situations, even in the low-frequency region of the spectrum. Instead, it seems that differences in *F*0 of 6–12% are needed for effective separation of sounds with overlapping spectra (Brokx & Nooteboom, 1982).

V. ONSET AND OFFSET ASYNCHRONY

Components that start or stop at roughly the same time are more likely to have originated from the same sound source than components that start or stop at different times. There is evidence from a wide variety of paradigms that onset asynchrony in particular is a powerful cue for segregating a component.

A. Detecting and Discriminating Onset and Offset Asynchronies

As discussed in Chapter 6, the ear can detect differences in the time of arrival of energy in different frequency regions of as little as 1–2 ms. For example, Green (1971) showed that listeners could discriminate between pairs of "Huffmann sequences" (click-like sounds that have identical power spectra but can differ in the time of arrival of energy across the frequency spectrum) when the duration of each sequence was as short as 2 ms. However, this does not necessarily demonstrate that listeners use such small timing differences for concurrent sound segregation: Even though the stimuli were discriminable, Green's listeners still reported hearing each click as a single sound: clicks differed from each other only in quality (e.g., "tick" vs. "tock"). To get an idea of the range of onset asynchronies that affect auditory grouping, then, we have to turn to experiments requiring listeners to make slightly more complex judgments.

B. Order Discrimination

Onset-time differences about an order of magnitude larger are needed for subjects to tell without feedback which of two overlapping tones started earlier. For example, with 300-ms tones listeners need around 15–20 ms onset asynchrony to be able to say whether or not the higher tone came first. This threshold is reduced to about 10 ms when the overall duration of the tones decreases from 300 to 10 ms (Pastore, Harris, & Kaplan, 1982).

C. Timbre Judgments

When a single frequency component starts earlier than the remainder of a complex, an onset asynchrony of about 30 ms is needed to reduce its influence on the timbre of the complex. Bregman and Pinker (1978) played subjects an alternating sequence of two sounds, the first a pure tone (A), the second a two-tone complex ($B + C$). Typical frequencies for the A, B, and C tones were 559 Hz, 527 Hz, and 300 Hz, respectively. When the onsets of the two tones of the complex ($B + C$) differed by 29 or by 58 ms, subjects were more likely to hear the complex as decomposed into two separate tones—judging the timbre of the $B + C$ complex as purer and also being more likely to hear the upper note (B) join A to form a separate auditory stream. Subsequent experiments (Dannenbring & Bregman, 1978) refined this conclusion: Asynchronies of 35 ms will help to "remove" a tone from a complex only if that tone either starts before the remainder or finishes after them—conditions that also make the tone more audible (Rasch, 1981) and aid discrimination of onset asynchronies (Zera & Green, 1993).

Further evidence that onset asynchronies can affect the way in which listeners make across-frequency comparisons comes from two different types of discrimination experiment. In the paradigm known as *profile analysis,* listeners have to detect a difference between the level of one component of a complex sound and that of the other components (see Chapter 7). Green and Dai (1992) have shown that subjects need much larger increments to perform the task if there is an onset asynchrony of about 50 ms between the target component and the remainder, particularly if the target starts first, compared to the case where all components are turned on together. Analogous findings for the detection of mistuning, where thresholds can be greatly elevated by playing a "precursor" sound in the frequency region of the target component, have recently been reported by Carlyon (1994a).

D. Identifying Speech Sounds

Cutting (1976) performed some important early experiments on the influence of onset asynchrony on the fusion of complex sounds. In one study, he presented listeners with the syllable /da/ to both ears and found that interaural delays of less than 5 ms caused listeners to report the presence of two, rather than one, sound source. However, larger onset asynchronies were required to interfere with the fusion of formants into a single phonetic percept. For example, when the first and second formants of /da/ were played to opposite ears, listeners were able to fuse the two formants, thereby identifying the syllable correctly, for interaural asynchronies as long as 20–40 ms.

This latter finding is not restricted to integration of information across the two ears. Darwin (1984; Darwin & Sutherland, 1984) showed that augmenting the level of one harmonic of a vowel could change its perceived identity (/I/ vs. /e/), but that starting the augmented component before, and ending it after, the rest of the vowel could partially reverse the effect. An asynchrony of 40 ms was sufficient to reverse the effect somewhat, but not completely. He argued that the effect of onset asynchronies was due to grouping processes rather than to the effective level of the leading component being reduced by adaptation, for two reasons: (1) offset asynchronies alone also had some influence, albeit less than that of onset asynchronies, and (2) the effect of onset asynchrony could be partly removed by an additional grouping manipulation. The grouping manipulation is illustrated in Figure 4. In the left-hand panels the leading part of the added tone is assumed to continue into the vowel and so the vowel is heard as having a fourth harmonic that is correspondingly less intense. In the right-hand panels, when an additional captor tone is presented simultaneously with the leading part of the augmented harmonic (and an octave higher than it in frequency) the leading portion of the tone is segregated from its continuation and is heard as part of a new complex with the captor. Its continuation is heard as part of the vowel, whose fourth harmonic is then perceived as correspondingly more intense (Darwin & Sutherland, 1984).

The effect of onset asynchrony on speech identification is less marked when whole formants, rather than individual components, are made asynchronous. For example, Darwin (1981) found that staggering the onsets and/or offsets of the first three formants of each of 10 different vowels by up to 100 ms did not impair their identification, although the number of sound sources heard increased markedly with staggering. However, using the /ru/–/li/ paradigm described earlier, he was able to show that a 300 ms forward extension of the second formant of a composite syllable produced a significant increase in the number of /li/ responses, indicating that the second formant had been perceptually removed from the composite syllable. Therefore, we can tentatively place the onset asynchrony necessary for a formant to be removed from the phonetic identity of a syllable somewhere in the range between 100 and 300 ms.

It is not clear why larger asynchronies are required between formants than between individual frequency components for perceptual segregation to take place. Although the formants are more separated in frequency than are the individual frequency components, detection of onset asynchrony does not become more difficult as the components become more separated

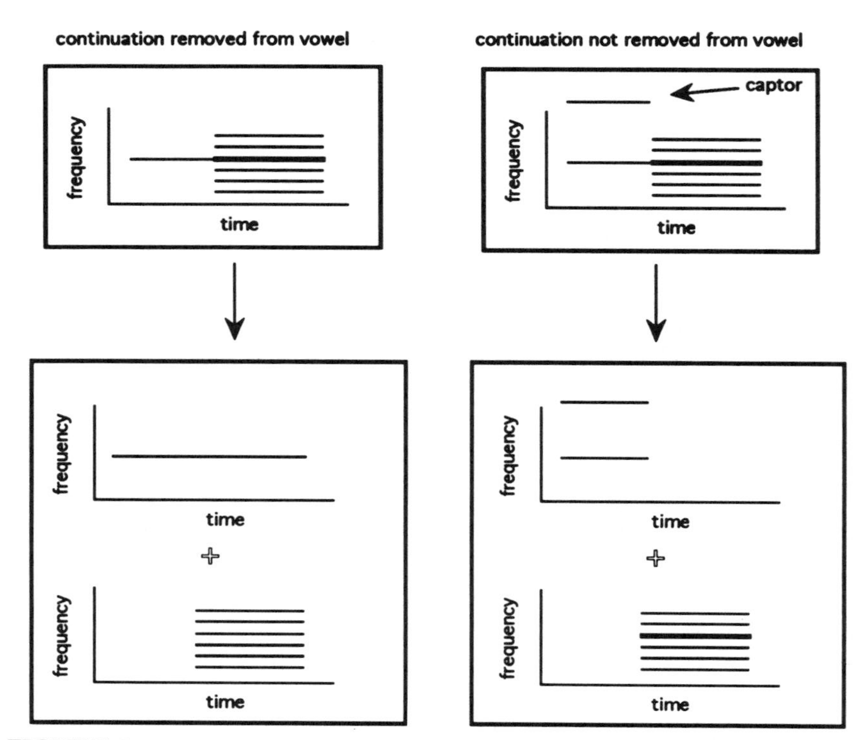

FIGURE 4 Stimulus manipulation used by Darwin and Sutherland (1984) to demonstrate that the effect of onset asynchrony on perceptual grouping was not entirely attributable to peripheral adaptation. In the left-hand panels a vowel has a tone added to it, which increases the intensity of the fourth harmonic. The tone starts earlier than the vowel. Listeners hear the original vowel quality, indicating that the additional tone is being perceptually segregated from the vowel. In the right-hand panels a captor tone causes the leading part of the added tone to perceptually segregate from its continuation into a vowel. The vowel is then heard as having a more intense fourth harmonic, which results in a change of vowel quality.

in frequency (Hirsh, 1959; Summerfield, 1982). The difference may reflect a distinction between perceptual mechanisms responsible for extracting formant frequencies from the spectral envelope of resolved harmonics and those responsible for assembling formant frequency tracks into phonetic percepts. Because the different components of a formant are less likely to start asynchronously than are different formants (which may start many tens of ms apart, as in aspirated stop consonants), the former mechanism may be less tolerant of asynchrony than the latter.

E. Pitch Perception

The paradigm developed by Moore et al. (1985a) for measuring the pitch of a complex tone that has one of its harmonics mistuned has been used to demonstrate that onset asynchrony is also important in pitch perception. If the mistuned harmonic starts more than about 80 ms before the rest of the complex, it exerts a smaller influence on the pitch of the complex than when it is synchronous. The mistuned harmonic has no effect on the pitch of the complex when it leads by about 300 ms (Darwin & Ciocca, 1992).

Using a different paradigm, Vos (1993) has found that an onset-time difference of 25 ms is not sufficient for pairs of complex tones to be segregated for the purpose of establishing their pitch. The failure of both Vos and of Darwin and Ciocca to find any effect of short onset asynchronies on perceptual grouping for pitch perception contrasts with the marked effect that onset asynchrony has on vowel quality at similarly short asynchronies. The size of onset asynchrony needed to remove a harmonic from the calculation of pitch is considerably larger than that required to remove a harmonic from the calculation of vowel quality in stimuli that are otherwise broadly similar (Hukin & Darwin, 1995).

F. Lateralization

Onset asynchrony can also produce segregation in lateralization experiments. As described earlier, the thresholds for interaural time differences of energy in one frequency region can be raised by the presence of energy in other frequency regions (McFadden & Pasanen, 1976; Zurek, 1985). This effect, though, can be substantially reduced if the interfering sound is presented continuously, rather than being gated simultaneously with the target (Trahiotis & Bernstein, 1990). However, there is evidence that less dramatic onset asynchronies, even as long as 200–250 ms, do not reduce the deleterious effect of interfering sounds on interaural time difference thresholds (Woods & Colburn, 1992; Stellmack & Dye, 1993). Note that in these latter experiments, a deterioration was observed even though the onset asynchrony was easily large enough for the interfering and target sounds to be perceived as separate sources.

Although modest onset asynchrony differences may have only a small effect in the threshold experiments referred to previously, there is evidence that they can affect suprathreshold lateralization judgments. As we mentioned in Section IV.D, phase ambiguity causes a 500-Hz tone presented in isolation and with an interaural time delay of 1.5 ms in the right ear to be heard as coming from that side. If the tone is one of a group of harmonics, all of which are delayed by 1.5 ms in the right ear, then listeners hear the whole complex on the left; the localization of the 500-Hz tone has been integrated with that of the other components. However, if the 500-Hz component is made to start 40-ms later than the other components, then it is localized separately from the others—it is heard toward the right, although not as far to the right as it would have been when heard in isolation (Hill & Darwin, 1993). The 40-ms asynchrony necessary to produce this effect is comparable with that required to segregate a harmonic from a vowel for the purpose of calculating vowel quality.

VI. AMPLITUDE-MODULATION PHASE DIFFERENCES

Speech contains two sources of amplitude modulation (AM). The way in which listeners exploit differences in the rate of vocal fold vibration between groups of components has already been dealt with in the section on Δ*F*0s. In this section we will briefly discuss work on the processing of across-frequency differences in the phase of these rapid amplitude fluctuations ("pitch pulses asynchronies," or PPAs; see Figure 5). However, most recent experiments on AM phase disparities have been concerned with the role of the slower fluctuations, which arise from dynamic variations in the characteristics of the vocal tract and which, like the faster variations caused by vocal fold vibration, are correlated across frequency. For example, when the lips close to produce a stop consonant, as in the middle of the word *about,* all the formants of the voice decrease in energy, only to swell again when the lips reopen. In contrast, the formants of other voices will show a pattern of AM that differs both in the rate and phase of modulation. Here we review the evidence that listeners can first detect, and second use in grouping, phase differences in amplitude modulation across frequency: Curiously, there is relatively little evidence on the role of across-frequency differences in AM rate over the subpitch range.[1] Chapter 7 reviews some related experiments on the release from masking produced by amplitude modulation in wide-band signals.

A. Detection of AM Phase Differences

A number of recent experiments have presented listeners with two amplitude-modulated tones and measured the smallest detectable phase disparity between the (usually sinusoidal) low-frequency modulators (Wakefield & Edwards, 1987; Strickland, Viemeister, Fantini, & Garrison, 1989; Yost & Sheft, 1989). In such detection experiments, care must be (and was) taken to prevent listeners from detecting the phase disparity using a within-channel cue, arising from the interaction between two components in auditory filters tuned between them. A typical precaution is to place a narrow band of noise between the two carriers, thereby masking any such interaction. When such precautions are taken, the results indicate that for sinusoidal modulators with frequencies below 64 Hz, threshold is constant at about 30–60°, with the exact value varying across studies. At higher modulation

[1] An exception is the study by Q. Summerfield and Culling (1992).

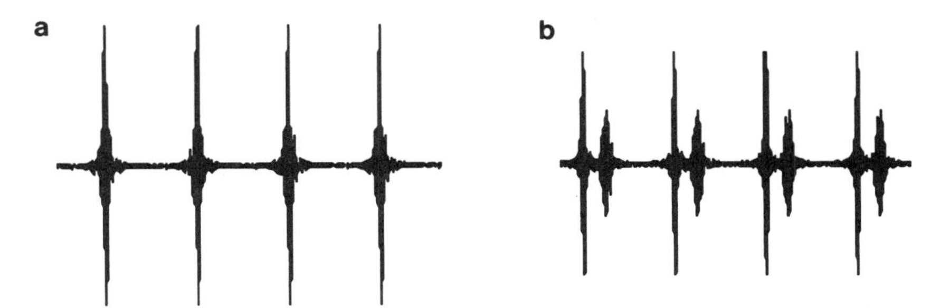

FIGURE 5 Composite waveforms of two synthetic formants, with pitch pulses (a) in synchrony and (b) out of synchrony.

frequencies, such as 128 Hz (Strickland et al., 1989), across-channel sensitivity becomes degraded and performance is highly dependent on within-channel cues. The results of these studies suggest that the usefulness of AM phase differences as a cue to sound segregation should decrease as modulation rate is increased from below about 50 Hz up into the pitch range. This conclusion also holds for asynchronies between pitch pulses in different formant regions, although for such PPAs threshold corresponds to a constant value of about 2.5 ms, rather than the constant 30–60° seen for sinusoidal AM (Carlyon, 1994b).

B. Grouping by AM Phase Differences

The role of amplitude modulation in fusing two sounds was reported as early as 1963, by von Békésy, who presented listeners with a 750-Hz tone to one ear and an 800-Hz tone to the other. Listeners reported hearing two separate tones when the stimuli were played at a steady level, but a single, fused sound when a common modulation, with a frequency of between 5 and 50 Hz, was applied to the two components. Implicit in von Békésy's report is the observation that AM at rates above 50 Hz did not produce fusion.

The highest modulation rate at which a sinusoidal AM phase difference *has* been shown to be effective in segregation is 100 Hz. Bregman, Abramson, Doehring, and Darwin (1985) presented listeners with a two-tone complex with component frequencies of 500 Hz and about 1500 Hz. The task was to judge the frequency of the upper component relative to that of a 1500 Hz tone presented in isolation. They reasoned that the task would be easier when the two synchronous components were perceptually separated from each other and included conditions where the tones were modulated at a rate of 100 Hz, either in phase or with an 180° phase disparity. Performance was better in the out-of-phase condition, and they interpreted their results as evidence for a difference in AM phase acting as a cue for segregation. Given that detection of *across-frequency* differences in AM phase breaks down at high rates, Bregman et al.'s data with a 100 Hz modulation rate either reflect the upper limits on the use of the cue or were mediated by the use of within-channel cues. The latter possibility exists because their stimuli were presented in quiet and auditory filters with characteristic frequencies (CFs) between the two carrier frequencies would have shown a pattern of beating that indicated the presence of two sound sources. Indeed, as Figure 6 shows, differences in AM phase could have facilitated the frequency discrimination task by allowing listeners to attend to those half-cycles of modulation when the response of a filter with a CF between the two tones was dominated by the 1500 Hz target component.

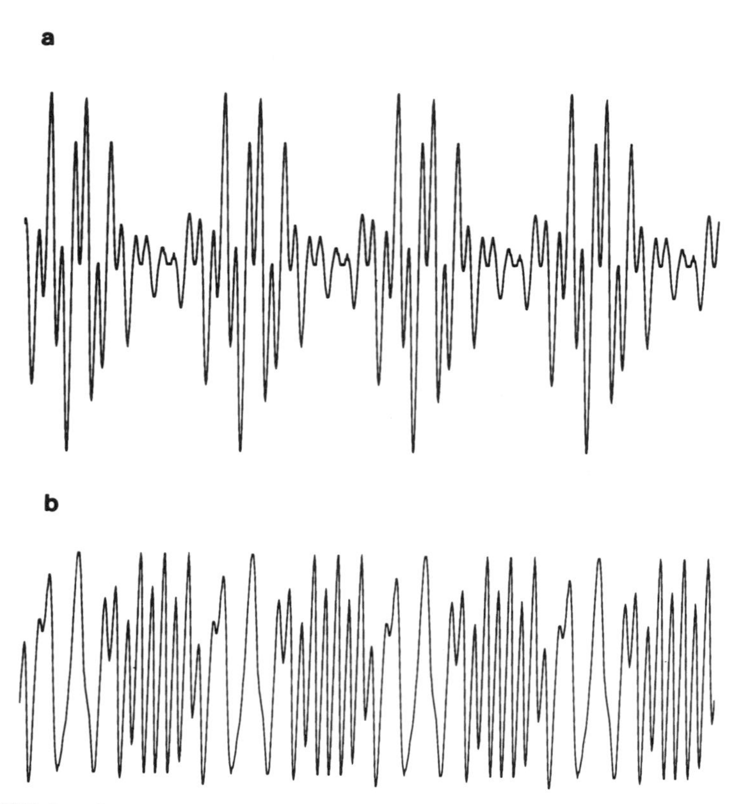

FIGURE 6 Output of a simulated auditory filter centered on 1000 Hz, in response to a complex consisting of two components at 500 and 1500 Hz. In the top part of the figure (a), the tones are amplitude modulated coherently at a rate of 100 Hz, whereas in the bottom part (b) they are modulated incoherently (180° modulator phase disparity). Note that, in (b), each component appears in turn. The stimuli were identical to those used by Bregman et al. (1985).

Finally, it is worth considering the role of AM phase disparities in the perceptual segregation of speech sounds. For PPAs, the data on concurrent segregation follow those for detection, in that the usefulness of the cue decreases with increasing *F*0: Summerfield and Assmann (1991) found that listeners could use PPAs to identify pairs of vowels with a common *F*0 of 50 Hz, but could not do so when the *F*0 was raised to 100 Hz. Surprisingly, the evidence for the use of differences in the phase of sinusoidal AM applied to competing speech sounds is less good. Summerfield and Culling (1992) measured the threshold at which a target vowel could be identified when it was presented against a masker vowel modulated either in phase or out of phase with the target, where both vowels were modulated at a rate of 2.5 Hz. Thresholds were lower in the out-of-phase condition, but this difference could easily be accounted for by the fact that the out-of-phase AM led to a momentarily advantageous "target to masker ratio" during part of the stimulus.

C. AM Phase and Onset Asynchrony

The previous discussion assumed that listeners were processing *ongoing* differences in AM phase, either to discriminate between sounds or to perceptually segregate them. However, it is important to bear in mind that, when two components are nominally turned on at the same time, an AM phase disparity will actually cause an effective onset disparity between them; one tone may be at a subaudible level at the beginning of the stimulus, during which time the other may be at its full amplitude. Even though the use of such cues may be reduced by turning the stimuli on and off fairly slowly, there is no evidence on how effective this precaution is or what rise time is sufficient to eliminate the use of onset cues. As we have shown, listeners are exquisitely sensitive to onset disparities between components and so, especially in detection experiments, the possible role of these onset disparities must be taken into account. For example, Strickland et al. (1989) found thresholds of about 50° for the detection of AM disparities between tones with frequencies of 1 and 2 kHz: As the modulation frequency was 8 Hz, this corresponds to a threshold of about 17 ms, well above that required to detect (abrupt) onset asynchronies between components of a complex sound (Zera & Green, 1993).

As far as we know, only one study has explicitly investigated the role of onset asynchronies in the processing of AM phase disparities. Carlyon (1994b) measured the detection of PPAs between two groups of frequency components that had (identical) *F*0s of either 40 or 80 Hz. Concerned that listeners might have been detecting the PPA by using onset asynchronies, he included a condition in which one of the groups was always turned on 75 ms before the other one and turned off 75 ms after it. He reasoned that, because an onset (and offset) difference was present in both the standard (no PPA) and signal (PPA) interval of each trial, detection in this new condition would be based on "ongoing" PPAs. His results showed that performance was impaired by this manipulation but that, at least for large asynchronies, it was still above chance. He therefore concluded that onset disparities might aid the detection of PPAs, but that they were not essential for it. The general applicability of this conclusion to paradigms using slower modulation rates (cf. Strickland et al., 1989; Yost and Sheft, 1989) and to those investigating supratheshold aspects of auditory grouping remains to be determined.

D. Summary

Ongoing differences in amplitude trajectory may aid sound segregation. Evidence from detection, "number of sources," and speech segregation experiments indicates that the usefulness of this cue deteriorates as the frequency of AM increases above about 50 Hz. However, care must be taken to distinguish what seems to be processing based on ongoing aspects of the stimulus from that which relies solely on temporal disparities at onset.

VII. FREQUENCY MODULATION

When the *F*0 of a periodic sound changes, all of its harmonics change frequency coherently; that is, in the same direction at the same time. One might expect listeners to use this cue to fuse the coherently frequency modulated (FM) components of one sound and to separate them from the components of other sounds that may be undergoing a different pattern of frequency change.

A. Detection of Across-Frequency Differences in FM Phase

There are a number of ways in which listeners could detect differences in the phase of FM applied to different components and use it to segregate them. One of these is trivial: For a harmonic sound, modulating one component out of phase with the rest causes it to become mistuned, and as we have discussed, listeners are very sensitive to mistuning. Therefore, when investigating the role of FM coherence, it is important to control for this "mistuning" cue. The question is: Can listeners detect FM incoherence in ways other than by the mistuning that it causes?

A number of authors have suggested that listeners could detect FM phase differences via deviations from a common pattern of movement in areas of excitation along the basilar membrane (Bregman, 1990, pp. 250–251; Cohen & Chen. 1992). However, Carlyon (1991) has argued that listeners cannot use such an across-frequency mechanism, at least to detect differences in the phase of FM applied to resolved frequency components. He required listeners to discriminate between a complex tone consisting of resolved components that underwent a coherent sinusoidal modulation and one in which a single component was modulated out of phase. He restricted the usefulness of within-channel cues by adding a background noise and found that listeners could do the task when the complex was harmonic, but not when it was inharmonic; hence, listeners were unable to detect differences in FM phase when the "mistuning" cue was removed by making the target component mistuned in both intervals. In a further experiment, he imposed various phase delays on the modulation applied to one component of a frequency modulated harmonic complex and calculated the amount of mistuning resulting from each phase delay. He then compared sensitivity to these modulator phase delays to that to "equivalent" static mistunings, in which one component was mistuned but the FM remained in phase. His results showed that sensitivity to FM phase disparities could be explained solely in terms of that to the resulting mistuning. More recent experiments (Carlyon, 1992, 1994c) have replicated and extended these findings to a range of modulator rates and types. However, it remains possible that listeners could exploit FM coherence in ways, such as by using within-channel cues, that were not available to the listeners in Carlyon's experiments. Therefore, although his experiments show that the most obvious method for exploiting FM phase differences is not adopted by the auditory system, they do not demonstrate that there are no other ways in which the cue might be used.

B. Identifying Speech Sounds

An alternative approach is to use a task that more rigorously taps the use of FM incoherence in perceptual segregation, without being overly concerned at the mechanism(s) involved. One such experiment was performed by Gardner et al. (1989), who used a variation of the /ru/–/li/ paradigm described earlier to examine the perceptual effects of modulating the *F*0 of the second formant of a vowel out of phase with the modulation imposed on the other formants. In terms both of the number of sources listeners reported hearing and the integration of *F*2 into the phonetic categorization of the vowel, the authors found no effects of FM phase differences that could not be accounted for by the mistuning they caused.

An important finding concerning the segregation of stimuli with overlapping spectra was reported by McAdams (1989). He asked listeners to judge the relative prominence of a vowel presented in a background of two other vowels, under conditions where the target and background vowels were modulated coherently or incoherently. In both conditions, the *F*0s of the constituent vowels differed by 30%, well beyond the separation at which mistuning creates its maximum effect. Therefore, any changes in mistuning caused by modulating the vowels incoherently should not have markedly affected prominence ratings. The results showed no difference in prominence ratings between coherent and incoherent FM. Interestingly, however, vowels modulated against a steady background *were* judged to be more prominent than when all three vowels were unmodulated. Another study showing an effect of the presence, but not the coherence, of FM was performed by Summerfield and Culling (1992), who investigated the detection and identification of masked vowels. They showed that listeners could not use different patterns of FM to help identify vowels synthesized with inharmonic frequency components, but that, when the masker was unmodulated, imposing modulation on the target did improve performance. However, it is not clear that the effects reported by McAdams and by Summerfield and Culling represent an effect of FM on the fusion of the components of a vowel, rather than simply an increase in the prominence of each individual component. In this respect, it is worth mentioning Gardner et al.'s (1989) finding that, when one formant of a vowel was mistuned from the others, imposing coherent FM on the entire vowel actually *increased* the number of sound sources heard.

C. Pitch Perception

Darwin, Ciocca, and Sandell (1994) have recently obtained evidence that the presence of FM can help to group together frequency components for the purpose of calculating pitch. They showed that a mistuned component of an otherwise harmonic complex could be mistuned by greater amounts and still contribute to the pitch of the complex if all the components were coherently frequency modulated than if they were not frequency modulated. Unfortunately this paradigm allows no clear predictions to be made about the effect of modulating one component out of phase with the others, because of the resulting time-varying mistuning. Nevertheless, the study provides the strongest evidence so far that the presence of FM increases *fusion,* rather than the prominence (McAdams, 1989), identifiability (Q. Summerfield & Culling, 1992), or naturalness of sounds (Chowning, 1980).

D. Summary

There is some evidence that the *presence* of FM aids the fusion of frequency components and can increase both the perceived prominence and the identifiability of a masked vowel. However, the results of a wide range of studies point against an independent role for FM phase differences in the perceptual segregation of concurrent sounds. Given that this "FM incoherence" seems such a plausible cue for segregation, it is worth considering why listeners do not use it.

First, as Summerfield (Summerfield, 1992; Summerfield & Culling, 1992) has pointed out, a mechanism specific to the detection of differences in FM phase would be computationally expensive. He notes that most sounds that one would want to segregate on this basis are harmonic and that, as FM incoherence always leads to inharmonicity (for which there already exists a mechanism), it would be ecologically disadvantageous to develop a separate mechanism that provides no new information. Second, if the existence of any FM (whether coherent or not) enhances the fusion of the components of a complex tone, relative to the case with no FM (Darwin et al., 1994), then there are two reasons why, by combining this sensitivity with that to mistuning, listeners could make very reliable "fusion vs. segregation" judgments. First, out-of-phase FM would not normally result in fusion because, for harmonic sounds, its effect would be counteracted by the resulting mistuning. Second, in-phase FM would provide further evidence that harmonically related components came from the same source. This is because unmodulated components from different sounds can, occasionally, be harmonically related by chance, but such "accidental" harmonicity is unlikely to survive FM. Thus, by a combined sensitivity to mistuning and modulation, listeners could fuse coherently modulated harmonic components and separate those that are modulated out of phase. We should perhaps await further evidence that FM does indeed promote fusion before concluding that listeners use this scheme, but it does have the advantage of being computationally more efficient than one based on correlating the movements of peaks in the excitation pattern. Finally, it is worth reiterating Carlyon's (1994c) point that a "peak correlation" mechanism for detecting FM phase disparities could produce the "wrong answer" when the peaks reflect maxima in the spectral envelope (e.g., formants), rather than in the spectral fine structure (e.g., harmonics). Unlike harmonics, formants of the same speech sound often change frequency in opposite directions ("incoherently"), even when they come from the same source. Hence, to work effectively, a correlation mechanism would have to be able to take into account whether a given peak was due to a harmonic or to a formant.

VIII. LATERALIZATION

There is a strong a priori case for believing that the components of a sound should be perceptually segregated according to their lateral position—all the components of a single sound source tend to come from the same place. In addition, it is not difficult to see how cues to lateral position could be incorporated into a perceptual grouping mechanism. For example, Jeffress's well-known model for spatially coding interaural time delays (Jeffress,

1948) provides a simple basis for grouping by interaural delay. However, although the early literature on selective attention emphasized the importance of different lateral positions for allowing listeners to ignore irrelevant voices (Cherry, 1953), evidence for a contribution of truly binaural mechanisms to perceptual grouping is not striking. For instance, in selective attention, the advantage of having two voices coming from different directions is due as much to the shadowing effect of the head as it is to binaural interaction (Plomp, 1976): A voice to one side of the head will be significantly attenuated at the more distant ear, and listeners can improve their processing of the other voice simply by attending to that ear. In addition, in the preceding sections we have come across many instances where a difference in lateral position of one or more frequency components has very little effect on the calculation of either the pitch or phonetic category of a vowel.

A. Hearing Out a Single Tone

There is some evidence that a difference in interaural timing imposed on one component of a complex can cause it to be heard out as a sound separate from the others. Kubovy, Cutting, and McGuire (1974) mixed together eight sinusoids whose frequencies corresponded to the those of the diatonic scale. When played diotically and continuously, it is very difficult to hear out any of the sinusoids. However, if the phase of one of the sinusoids is shifted between the two ears, that component "pops out" in a different location, and a tune can be played by sequentially changing the phases of different components.

B. Pitch Perception

Very little segregation by ear occurs in the calculation of the pitch of a complex tone. If two consecutive harmonics from two different fundamentals are presented simultaneously, subjects are virtually no better at identifying the two fundamentals when the harmonics are appropriately segregated by ear than when each ear receives one harmonic from each fundamental (Beerends & Houtsma, 1986). A similar conclusion can be drawn from the data shown in Figure 1 and in Darwin and Ciocca (1992), where a single mistuned component makes almost as great a contribution to the pitch of a complex tone when it is in the opposite ear as when it is in the same ear as the other harmonics. It seems that, although an initial analysis of the lateralization of individual components may occur at an early stage of processing, the pitch mechanism does not use this information. A scheme consistent with much of the evidence reviewed here is that components are first grouped together by the pitch mechanism irrespective of their initial lateralizations and that the initial measures obtained from the components in each perceptual group are then combined in the calculation of that group's laterality.

C. Identifying Speech Sounds

It has been known since the early 1950s (Fletcher, 1953) that subjects are still able to identify complex sounds such as speech or instrumental music when different parts of the spectrum are led to opposite ears. The fusion is convincing, with naive subjects being unable to identify which ear is receiving the low or the high part of the spectrum (Broadbent, 1955). A similar phenomenon has been obtained by temporally splitting the third formant of a stop consonant into a transition and a steady state. The isolated third-formant transition is played to one ear and the rest of the syllable to the other. In the resulting percept, subjects hear the third-formant transition as a separate "chirp" while also incorporating it into the phonetic percept as a cue to the place of articulation of the stop consonant (Rand, 1974; Liberman, Isenberg, & Rakerd, 1981).

Lateralization effects in grouping may be more substantial when the experimental paradigm allows the subject to attend *over an extended time* to sound from one direction rather than another. For example, Deutsch (1979) found that it was much harder to identify a tune when notes alternated between the ears, than when they were all presented to the same ear. In speech perception, a harmonic can be more effectively segregated from a vowel using a differential interaural time difference if the harmonic is preceded by a short string of similar tones with the same lateralization (Hukin & Darwin, in press).

IX. THE NATURE OF AUDITORY GROUPING

A. Grouping Is Not All or None

A common theme running through this chapter is the observation that, whatever the cue, its influence on auditory grouping depends on the task. For example, the top row of Table 1, which summarizes many of the experiments reviewed here, shows that a smaller mistuning of a low-numbered harmonic is required for it to be heard out from a complex (thereby increasing the number of perceived sound sources; column three, faint type) than for it to be excluded from the phonetic categorization of a vowel (column five, faint type). Although some caution should be exercised when comparing the results of experiments using different stimuli, it is clear

TABLE 1 Summary of the Experiments Described in the Text

Cue	Detection	# Sources	Pitch	Vowel	Sep. Speech	Lateralize
Harmonicity	L: 2–3 %[a] H: <1% (beats), or 2–3% (no beats) **LvL: 2%** **LvH: 10%**	L: 2% H: >5% **L: 2%** **H: 10%**	L: 3–8%	L: 8% **Up to 60%**	L: 2% (beats) H: 10%	L: 3%
Onset–Offset	<2 ms (order discrimination 15–30 ms) **< 2 ms**	35 ms **20–40 ms**	L: 80–300 ms	L: 40 ms **100–300 ms**		L: 40 ms **H: > 250 ms (or No)**
AM Phase	SAM: 30–60° **PPA: 2 ms**	SAM: Yes (only 180° studied)			**SAM: No** **PPA: 5–10 ms**	
FM Phase	L: No (in absence of harmonicity cues)	No		No	No	
FM Presence	L: 1% **L: 2%** **H: 10%**	No	L: Yes[b]	No	**Yes**	

Note: Each cell of the table shows, approximately, the minimum value of a cue for a listener to segregate a single harmonic (faint type) or a group of harmonics (bold type) from a complex sound. Each row shows the data for a single cue; each column shows one way in which the segregation is measured. Abbreviations: L = low, or resolved harmonics; H = high, or unresolved harmonics; SAM = sinusoidal AM, PPA = pitch-pulse asynchrony, Vowel = integration of component(s) into phonetic identity of vowel, Sep. Speech = identifying concurrent speech sounds. Where "beats" is appended to an entry, segregation or detection was based on beating between adjacent harmonics.

[a]Increases at very high Fos (Demany *et al.*, *1991*).

[b]Evidence that FM promotes fusion rather than segregation.

from the table that grouping is not an "all-or-none" process and that there are circumstances where the listener can, for example, hear out a component from the rest of the complex, yet still integrate it into the percept of that complex's pitch or phonetic quality. Why should this be?

The fact that grouping is graded may reflect the ambiguous nature of the concept of a sound source. Just as most visual objects can be regarded as consisting of a hierarchy of parts (the nail on the phalanx of the finger of the hand of the arm of the body), so sounds can be thought of as being hierarchically organized in terms both of production and perception (the scratch in the attack of the bottom note of the chord of the violin part at the beginning of the Kreutzer Sonata). The primitive grouping cues that can usefully segregate sounds at the bottom of the hierarchy (where our focus has been in this chapter) will have to be ignored to allow higher order properties to be extracted. For example, the sound of a bell has multiple pitches but a single timbre: Harmonically related subsets of a bell's components contribute to the different pitches heard within the overall sound. Therefore, to determine the timbre of the bell, we have to group together components that we must have perceptually segregated to calculate the bell note's multiple pitches. Similar arguments apply to the perception of many complex sounds, such as musical chords (with their multiple pitches) or speech (which is produced by more than one type of independent sound source, voicing and frication).

B. Understanding Speech in the Absence of Primitive Grouping Cues

The research reviewed in this chapter has, we believe, demonstrated the importance of primitive grouping cues for the perception of a wide range of speech and nonspeech sounds. However, it is possible to understand speech even in the absence of such cues. A striking demonstration comes from experiments on the perception of "sine-wave" speech (Bailey, Summerfield, & Dorman, 1977; Remez, Rubin, Pisoni, & Carrell, 1981). Here, frequency-modulated sine waves follow the formant frequency tracks that are used in conventional synthesis to specify the vocal tract resonant frequencies. On first hearing, many people fail to hear these sounds as speech, experiencing instead a number of independent whistles. Nevertheless, it is relatively easy, particularly when exposed to sentence-length utterances or when one's attention is drawn to speech, to hear a distorted voice speaking. The voice quality is of course bizarre, but much of the phonetic message can be understood. These phenomena have led to the claim that speech perception, unlike the perception of other sounds, does not utilize the results of primitive grouping mechanisms (Remez, 1987; Remez, Rubin, Berns, Pardo, & Lang, 1994).

At this point it is worth drawing a distinction between the two sides of the grouping coin. *Segregation* refers to the phenomenon whereby the listener is presented with a mixture of sounds and succeeds in assigning the frequency components in the mixture to the appropriate sources. In contrast, the experiments with sine-wave speech exploit the ability of the listener to *fuse* different parts of the frequency spectrum: Because, in these experiments, the sine-wave speech is the only sound present, perceptual segregation is not necessary and listeners can identify the speech sound by "combining everything." Under these circumstances, it is likely that listeners use their "top-down" knowledge about speech to interpret the rather weird input as a spoken message. However, the fact that listeners can sometimes understand speech in the absence of primitive grouping cues does not mean, of course, that such cues are never used, any more than our ability to interpret line drawings demonstrates that we never use shading in visual object perception. Indeed, there is evidence that applying one primitive cue, coherent AM, to the formants of sine-wave speech both improves its comprehension and makes it sound more natural (Carrell & Opie, 1992), in much the same way as adding shading improves the perception of a line drawing. Thus, although it is clear, as has been argued elsewhere (Darwin, 1981), that phonetic mechanisms must have some hand in selecting which elements make up a particular speech sound, it is also clear (Darwin, 1991) that primitive grouping mechanisms (such as those concerned with harmonicity and onset asynchrony) guide this selection process. In particular, an insight into how these mechanisms work and the nature of their input is essential for an understanding of how we interpret an auditory world in which, rather inconveniently, more than one sound is present at a time.

References

Assmann, P. F., & Summerfield, A. Q. (1990). Modelling the perception of concurrent vowels: Vowels with different fundamental frequencies. *Journal of the Acoustical Society of America, 88,* 680–697.

Assmann, P. F., & Summerfield, A. Q. (1994). The contribution of waveform interactions to the perception of concurrent vowels. *Journal of the Acoustical Society of America, 95,* 471–484.

Bailey, P. J., Summerfield, Q., & Dorman, M. (1977). On the identification of sine-wave analogues of certain speech sounds. Report no: SR-51/52. Haskins Laboratories.

Beerends, J. G., & Houtsma, A. J. M. (1986). Pitch identification of simultaneous dichotic two-tone complexes. *Journal of the Acoustical Society of America, 80,* 1048–1055.

Bregman, A. S. (1990). *Auditory scene analysis: The perceptual organisation of sound.* Cambridge, MA: Bradford Books, MIT Press.

Bregman, A. S., Abramson, J., Doehring, P., & Darwin, C. J. (1985). Spectral integration based on common amplitude modulation. *Perception and Psychophysics, 37,* 483–493.

Bregman, A. S., & Pinker, S. (1978). Auditory streaming and the building of timbre. *Canadian Journal of Psychology, 32,* 19–31.

Broadbent, D. E. (1955). A note on binaural fusion. *Quarterly Journal of Experimental Psychology, 7,* 46–47.

Broadbent, D. E., & Ladefoged, P. (1957). On the fusion of sounds reaching different sense organs. *Journal of the Acoustical Society of America, 29,* 708–710.

Brokx, J. P. L., & Nooteboom, S. G. (1982). Intonation and the perceptual separation of simultaneous voices. *Journal of Phonetics, 10,* 23–36.

Buell, T. N., & Hafter, E. R. (1991). Combination of binaural information across frequency bands. *Journal of the Acoustical Society of America, 90,* 1894–1900.

Carlyon, R. P. (1991). Discriminating between coherent and incoherent frequency modulation of complex tones. *Journal of the Acoustical Society of America, 89,* 329–340.

Carlyon, R. P. (1992). The psychophysics of concurrent sound segregation. *Philosophical Transactions of the Royal Society of London,* Series B, *336,* 347–355.

Carlyon, R. P. (1994a). Detecting mistuning in the presence of synchronous and asynchronous interfering sounds. *Journal of the Acoustical Society of America, 95,* 2622–2630.

Carlyon, R. P. (1994b). Detecting pitch-pulse asynchronies and differences in fundamental frequency. *Journal of the Acoustical Society of America, 95,* 968–979.

Carlyon, R. P. (1994c). Further evidence against an across-frequency mechanism specific to the detection of FM incoherence between resolved frequency components. *Journal of the Acoustical Society of America, 95,* 949–961.

Carlyon, R. P., & Shackleton, T. M. (1994). Comparing the fundamental frequencies of resolved and unresolved harmonics: evidence for two pitch mechanisms? *Journal of the Acoustical Society of America, 95,* 3541–3554.

Carrell, T. D., & Opie, J. M. (1992). The effect of amplitude comodulation on auditory object formation in sentence perception. *Perception and Psychophysics, 52,* 437–445.

Cherry, E. C. (1953). Some experiments on the recognition of speech, with one and two ears. *Journal of the Acoustical Society of America, 25,* 975–979.

Chowning, J. M. (1980). Computer synthesis of the singing voice. In *Sound generation in winds, strings, computers.* Stockholm: Royal Swedish Academy of Music.

Cohen, M. F., & Chen, X. (1992). Dynamic frequency change among stimulus components: Effects of coherence on detectability. *Journal of the Acoustical Society of America, 92,* 766–772.

Culling, J. E., & Darwin, C. J. (1993). Perceptual separation of simultaneous vowels: Within and across-formant grouping by F0. *Journal of the Acoustical Society of America, 93,* 3454–3467.

Culling, J. E., & Darwin, C. J. (1994). Perceptual and computational separation of simultaneous vowels: cues arising from low frequency beating. *Journal of the Acoustical Society of America, 95,* 1559–1569.

Cutting, J. E. (1976). Auditory and linguistic processes in speech perception: Inferences from six fusions in dichotic listening. *Psychological Review, 83,* 114–140.

Dannenbring, G. L., & Bregman, A. S. (1978). Streaming vs. fusion of sinusoidal components of complex tones. *Perception and Psychophysics, 24,* 369–376.

Darwin, C. J. (1981). Perceptual grouping of speech components differing in fundamental frequency and onset time. *Quarterly Journal of Experimental Psychology, 33A,* 185–208.

Darwin, C. J. (1984). Perceiving vowels in the presence of another sound: Constraints on formant perception. *Journal of the Acoustical Society of America, 76,* 1636–1647.

Darwin, C. J. (1991). The relationship between speech perception and the perception of other sounds. In I. G. Mattingly & M. G. Studdert-Kennedy (Eds.), *Modularity and the motor theory of speech perception* (pp. 239–259). Hillsdale, NJ: Lawrence Erlbaum.

Darwin, C. J. (1992). Listening to two things at once. In M. E. H. Schouten (Ed.), *The auditory processing of speech: From sounds to words* (pp. 133–147). Berlin: Mouton de Gruyter.

Darwin, C. J., & Ciocca, V. (1992). Grouping in pitch perception: Effects of onset asynchrony

and ear of presentation of a mistuned component. *Journal of the Acoustical Society of America, 91,* 3381–3390.
Darwin, C. J., Ciocca, V., & Sandell, G. R. (1994). Effects of frequency and amplitude modulation on the pitch of a complex tone with a mistuned harmonic. *Journal of the Acoustical Society of America, 95,* 2631–2636.
Darwin, C. J., & Gardner, R. B. (1986). Mistuning a harmonic of a vowel: Grouping and phase effects on vowel quality. *Journal of the Acoustical Society of America, 79,* 838–845.
Darwin, C. J., & Sutherland, N. S. (1984). Grouping frequency components of vowels: When is a harmonic not a harmonic? *Quarterly Journal of Experimental Psychology, 36A,* 193–208.
Demany, L., & Semal, C. (1988). Dichotic fusion of two tones one octave apart: Evidence for internal octave templates. *Journal of the Acoustical Society of America, 83,* 687–695.
Demany, L., Semal, C., & Carlyon, R. P. (1991). Perceptual limits of octave harmony and their origin. *Journal of the Acoustical Society of America, 90,* 3019–3027.
Deutsch, D. (1979). Binaural integration of melodic patterns. *Perception and Psychophysics, 25,* 399–405.
Duifhuis, H., Willems, L. F., & Sluyter, R. J. (1982). Measurement of pitch in speech: An implementation of Goldstein's theory of pitch perception. *Journal of the Acoustical Society of America, 71,* 1568–1580.
Fletcher, H. (1953). *Speech and hearing in communication.* New York: Van Nostrand.
Gardner, R. B., Gaskill, S. A., & Darwin, C. J. (1989). Perceptual grouping of formants with static and dynamic differences in fundamental frequency. *Journal of the Acoustical Society of America, 85,* 1329–1337.
Goldberg, J. M., & Brown, P. B. (1969). Response of binaural neurons of dog superior olivary complex to dichotic tonal stimuli: some physiological mechanisms of sound localization. *Journal of Neurophysiology, 32,* 613–636.
Goldstein, J. L. (1973). An optimum processor theory for the central formation of the pitch of complex tones. *Journal of the Acoustical Society of America, 54,* 1496–1516.
Green, D. M. (1971). Temporal auditory acuity. *Psychological Review, 78,* 540–551.
Green, D. M., & Dai, H. (1992). Temporal relations in profile comparisons. In Y. Cazals, L. Demany, & K. Horner (Eds.), *Auditory physiology and perception* (pp. 471–478). Oxford: Pergamon Press.
Hartmann, W. M., McAdams, S., & Smith, B. K. (1990). Hearing a mistuned harmonic in an otherwise periodic complex tone. *Journal of the Acoustical Society of America, 88,* 1712–1724.
Hill, N. J., & Darwin, C. J. (1993). Effects of onset asynchrony and of mistuning on the lateralization of a pure tone embedded in a harmonic complex. *Journal of the Acoustical Society of America, 93,* 2307–2308.
Hirsh, I. J. (1959). Auditory perception of temporal order. *Journal of the Acoustical Society of America, 31,* 759–767.
Hoekstra, A. (1979). Frequency discrimination and frequency analysis in hearing. Doctoral dissertation, Groningen University, the Netherlands.
Houtsma, A. J. M., & Smurzynski, J. (1990). Pitch identification and discrimination for complex tones with many harmonics. *Journal of the Acoustical Society of America, 87,* 304–310.
Hukin, R. W., & Darwin, C. J. (1995). Comparison of the effect of onset asynchrony on auditory grouping in pitch matching and vowel identification. *Perception and Psychophysics, 57,* 191–196.
Hukin, R. W., & Darwin, C. J. (in press). Effects of contralateral presentation and of interaural time differences in segregating a harmonic from a vowel. *Journal of the Acoustical Society of America.*

Jeffress, L. A. (1948). A place theory of sound localization. *Journal of Comparative and Physiological Psychology, 41,* 35–39.

Johnson, D. H. (1980). The relationship between spike rate and synchrony in responses of auditory-nerve fibers to single tones. *Journal of the Acoustical Society of America, 68,* 1115–1122.

Kiang, N. Y. S., Watanabe, T., Thomas, E. C., & Clark, L. F. (1965). *Discharge patterns of single fibers in the cat's auditory nerve.* Cambridge, MA: MIT Press.

Kubovy, M., Cutting, J. E., & McGuire, R. M. (1974). Hearing with the third ear: Dichotic perception of a melody without monaural familiarity cues. *Science, 186,* 272–274.

Liberman, A. M., Isenberg, D., & Rakerd, B. (1981). Duplex perception of cues for stop consonants. *Perception and Psychophysics, 30,* 133–143.

McAdams, S. (1989). Segregation of concurrent sounds. I. Effects of frequency modulation coherence. *Journal of the Acoustical Society of America, 86,* 2148–2159.

McFadden, D., & Pasanen, E. G. (1976). Lateralization at high frequencies based on interaural time differences. *Journal of the Acoustical Society of America, 59,* 634–639.

Moore, B. C. J. (1973). Frequency difference limens for short-duration tones. *Journal of the Acoustical Society of America, 54,* 610–619.

Moore, B. C. J., Glasberg, B. R., & Peters, R. W. (1985a). Relative dominance of individual partials in determining the pitch of complex tones. *Journal of the Acoustical Society of America, 77,* 1853–1860.

Moore, B. C. J., Glasberg, B. R., & Peters, R. W. (1986). Thresholds for hearing mistuned partials as separate tones in harmonic complexes. *Journal of the Acoustical Society of America, 80,* 479–483.

Moore, B. C. J., & Ohgushi, K. (1993). Audibility of partials in inharmonic complex tones. *Journal of the Acoustical Society of America, 93,* 452–461.

Moore, B. C. J., Peters, R. W., & Glasberg, B. R. (1985b). Thresholds for the detection of inharmonicity in complex tones. *Journal of the Acoustical Society of America, 77,* 1861–1868.

Pastore, R. E., Harris, L. B. & Kaplan, J. K. (1982). Temporal order identification: Some parameter dependencies. *Journal of the Acoustical Society of America, 71,* 430–436.

Plomp, R. (1964). The ear as a frequency analyzer. *Journal of the Acoustical Society of America, 36,* 1628–1636.

Plomp, R. (1976). Binaural and monaural speech intelligibility of connected discourse in reverberation as a function of the direction of a single competing sound source (speech or noise). *Acustica, 34,* 200–211.

Plomp, R., & Mimpen, A. M. (1968). The ear as a frequency analyzer II. *Journal of the Acoustical Society of America, 43,* 764–767.

Rand, T. C. (1974). Dichotic release from masking of speech. *Journal of the Acoustical Society of America, 55,* 678–680.

Rasch, R. (1981). Aspects of the perception and performance of polyphonic music. Ph.D. thesis, Drukkerij Elinkwijk BV.

Remez, R. E. (1987). Units of organization and analysis in the perception of speech. In M. E. H. Schouten (Ed.), *The psychophysics of speech perception* (pp. 419–432). Dordrecht: Martinus Nijhoff.

Remez, R. E., Rubin, P. E., Berns, S. M., Pardo, J. S., & Lang, J. M. (1994). On the perceptual organization of speech. *Psychological Review, 101,* 129–156.

Remez, R. E., Rubin, P. E., Pisoni, D. B., & Carrell, T. D. (1981). Speech perception without traditional speech cues. *Science, 212,* 947–950.

Roberts, B. & Bregman, A. S. (1991). Effects of the pattern of spectral spacing on the perceptual fusion of harmonics. *Journal of the Acoustical Society of America, 90,* 3050–3060.

Scheffers, M. T. (1983). Sifting vowels: Auditory pitch analysis and sound segregation. Ph.D. thesis, Groningen University, the Netherlands.

Shackleton, T. M., & Carlyon, R. P. (1994). The role of resolved and unresolved harmonics in pitch perception and frequency modulation discrimination. *Journal of the Acoustical Society of America, 95,* 3529–3540.

Smith, R. L. (1979). Adaptation, saturation, and physiological masking in single auditory nerve fibers. *Journal of the Acoustical Society of America, 65,* 166–178.

Stellmack, M. A., & Dye, R. H. (1993). The combination of interaural information across frequencies: The effects of number and spacing of components, onset asynchrony, and harmonicity. *Journal of the Acoustical Society of America, 93,* 2933–2947.

Strickland, E. A., Viemeister, N. F., Fantini, D. A., & Garrison, M. A. (1989). Within- versus cross-channel mechanisms in detection of envelope phase disparity. *Journal of the Acoustical Society of America, 86,* 2160–2166.

Summerfield, Q., & Assmann, P. F. (1991). Perception of concurrent vowels: Effects of pitch-pulse asynchrony and harmonic misalignment. *Journal of the Acoustical Society of America, 89,* 1364–1377.

Summerfield, Q. (1982). Differences between spectral dependencies in auditory and phonetic temporal processing: Relevance to the perception of voicing in initial stops. *Journal of the Acoustical Society of America, 72,* 51–61.

Summerfield, Q. (1992). Roles of harmonicity and coherent frequency modulation in auditory grouping. In M. E. H. Schouten (Ed.), *The auditory processing of speech: From sounds to words* (pp. 157–165). Berlin: Mouton de Gruyter.

Summerfield, Q., & Culling, J. (1992). Auditory segregation of competing voices: absence of effects of FM or AM coherence. *Philosophical Transactions of the Royal Society of London,* Series B, *336,* 357–366.

Trahiotis, C., & Bernstein, L. R. (1990). Detectability of interaural delays over select spectral regions: effects of flanking noise. *Journal of the Acoustical Society of America, 87,* 810–813.

Trahiotis, C., & Stern, R. M. (1989). Lateralization of bands of noise: Effects of bandwidth and differences of interaural time and phase. *Journal of the Acoustical Society of America, 86,* 1285–1293.

Vos, J. (1993). Perceptual separation of simultaneous complex tones: The effect of slightly asynchronous onsets. TNO-report no: IZF 1993 B-5. TNO Institute for Perception.

Wakefield, G. H., & Edwards, B. (1987). Discrimination of envelope phase disparity. *Journal of the Acoustical Society of America, 82,* S41.

Woods, W. A., & Colburn, S. (1992). Test of a model of auditory object formation using intensity and interaural time difference discriminations. *Journal of the Acoustical Society of America, 91,* 2894–2902.

Yin, T. C. T., & Chan, J. C. K. (1990). Interaural time sensitivity in the medial superior olive of the cat. *Journal of Neurophysiology, 64,* 465–488.

Yost, W. A., & Sheft, S. (1989). Across-critical-band processing of amplitude-modulated tones. *Journal of the Acoustical Society of America, 85,* 848–857.

Zera, J., & Green, D. M. (1993). Detecting temporal onset and offset asynchrony in multicomponent complexes. *Journal of the Acoustical Society of America, 93,* 1038–1052.

Zurek, P. M. (1985). Spectral dominance in sensitivity to interaural delay for broadband stimuli. *Journal of the Acoustical Society of America, 78,* S18.

CHAPTER 12

Timbre Perception and Auditory Object Identification

Stephen Handel

I. INTRODUCTION

Birds, water drips, wind, instruments, doors slamming, friends walking on wood or vinyl, objects breaking or bouncing—these are but a few of the events and objects that we can identify by sound alone. Jenkins (1984) points out that the world is eventlike, sound is produced by action and movement. Our ability to identify things by listening must be based on the information provided by their acoustic properties, and these properties are the result of the production process.

The basic problem is contextual variation. If the sound waves produced by each event were always identical, then identification would reduce to simple memorization (i.e., one sound wave to one object). But, the sound waves will not be physically identical at different times due to variations in the production process. Moreover, the sound waves generated by simultaneous events intermix so that the properties of each source are obscured. The listener must be able to identify a specific source in spite of the naturally occurring variation within the mixture.

What can we say in general about the auditory world? First, acoustic properties adhere to objects; the properties belong to and at the same time characterize the source. Second, these properties evolve over time. The

Hearing

changes typically are slow, continuous, and regular so that it is possible to track a sound over time. Third, the auditory world is transparent and linear. Sound waves combine together in an additive fashion to create the wave that reaches the listener; the waves do not occlude each other. This additivity allows for a hierarchical world of sound. Consider a violin. Sound waves are generated mainly by the vibrations of the top plate, the bottom plate, and the air within the violin body. Each of the resulting sound waves combines to represent the object (i.e., violin). The violin in turn may be part of an orchestra, and the orchestra in turn may be part of a larger scene that can include singers and environmental sounds. In a hierarchy of sound properties, there are local properties due to individual vibrations, as well as emergent properties due to the combination and interaction of local properties.

We will use the term *timbre* to refer to the perceptual qualities of objects and events; that is, "what it sounds like." Traditionally, timbre has been thought of as related to one acoustically measurable property such that each note of an instrument or each spoken sound of one voice would be characterized by a single value of that property. The traditional definition is by exclusion: The quality of a sound by which a listener can tell that two sounds of the same loudness and pitch are dissimilar (ANSI, 1973). Although this definition may tell us what timbre is not, it does not tell us what timbre is. We will make the argument here that, due to the interactive nature of sound production, there are many stable and time-varying acoustic properties. It is unlikely that any one property or combination of properties uniquely determines timbre. The sense of timbre comes from the emergent, interactive properties of the vibration pattern.

A basic issue is the level of explanation. One possibility is that timbre is perceived in terms of the actions required to generate the event. Thus, we might perceive speech in terms of the movements of the articulators, and we might perceive objects in terms of the physical action (e.g., rolling versus bouncing, bowing versus plucking). The physical properties of the source (e.g., size, shape, density, damping) and the type of action are perceived directly (Gibson, 1966; Balzano, 1986; Gaver, 1993). The perception of the production invariances would allow us to hear the same object in spite of large changes in the acoustic signal. Another possibility is that timbre is perceived simply in terms of the acoustic properties and that the connection between the acoustic properties and the object is learned by experience. In this view, the acoustic properties are used to figure out what event was most likely to have produced that sound. The useful acoustic information depends in a lawful way on environmental properties and processes that generate the acoustic wave. It is this learned schema that allows us to hear the same source in different contexts. Most probably, identification results both from the apprehension of acoustical invariants and from cognitive inferences.

II. SOUND PRODUCTION

We will consider the process of sound production to discover what there is to hear. One model for sound production is based on two possibly interactive components, the source and the filter. The basic notion is that the source is excited by energy to generate a vibration pattern composed of separate sinusoidal components or vibration modes. This pattern is imposed on the filter, which acts to modify the relative amplitudes of the components of the source input. These modified vibration components radiated from the filter are propagated to the listener. Thus the acoustic properties underlying timbre arise from the physical processes (e.g., banging, blowing, scraping) that excite the source and the physical material and shape (moist mouth tissue, wood plates, metal bars, conical pipes) of the objects that create the filter.

A. Source–Filter Model

1. Vibration Modes

Each vibration mode can be characterized by its resonant frequency and by its damping or quality factor. The resonant frequency of each mode is the frequency at which the amplitude of vibration is maximum. For some objects (a violin string in tension) the modes occur at integer frequency ratios determined by the tension and mass of the strings, but for other objects (a wooden plate) the modes occur at irregular ratios determined by the shape and physical material. Each excitation normally will excite a large number of these modes.

The damping, or inversely the quality factor (Q), of each vibration mode is a measure of its sharpness of tuning and its temporal response. For a heavily damped vibration mode (i.e., low Q), the amplitude of vibration is just slightly higher at the resonant frequency; the amplitude is relatively constant across frequency. Moreover, the temporal pattern of the amplitude of vibration of the mode is closely tied to that of the driving vibrator: If the driving vibrator increases or decreases its amplitude rapidly, then so will the amplitude of the mode. For a lightly damped mode (i.e., high Q), the amplitude of vibration is high at the resonant frequency, but low at surrounding frequencies. Here, the temporal pattern of the vibration mode changes more slowly than the temporal pattern of the driving vibrator.

2. Coupling between Source and Filter

It is now possible to consider some complexities of the source–filter model and thereby list some of the many acoustic properties that may underlie timbre and the ability to identify events. It is the intricacy of the variation in the sound that makes real instruments so wonderfully sounding, but this same intricacy creates the difficulty in understanding the ways listeners pick up the nature of the sound event.

When energy is initially applied to the source it generates a complex, time-varying vibration pattern that can be analyzed into the component vibration modes. Each mode can be characterized by its resonant frequency and damping. For some sources (e.g., a vibrating string, a wooden plate) there are discrete modes. For other kinds of sources, the vibration modes are continuous. For example, a drumstick hitting a drumhead, or the turbulent air rushing past the vocal folds when whispering is assumed to excite the filter (e.g., drum body, vocal tract) with a continuous range of frequencies. If the damping of all the modes is roughly equal, then the time until each mode reaches its maximum amplitude will be similar. If the damping is different, however, the modes will have asynchronous temporal patterns. Additionally, the source vibration may have an initial period of noise due to the time and energy it takes to get the vibration started. For example, a stable clarinet or trumpet mouthpiece source vibration requires feedback due to pressures pulses returning from the end of the instrument. Until this happens, there may be turbulence and a squeaky initiation sound.

The filter may contain a multitude of vibration modes, usually at nonharmonic frequencies, each with a different quality factor. Each time-varying vibration mode of the source can excite one or more damped modes of the filter so that the acoustic output is the product of the time-varying vibrations from each mode. Each note of an instrument, or each speech sound, engages different sets of source and filter vibration modes so that we should not expect a unique "signature" or acoustical property that can characterize an instrument, voice, or event across its typical range. The changing source filter coupling precludes a single acoustic correlate of timbre.

The change in acoustic structure across notes can be seen illustrated in Figure 1, which portrays violin scale notes across two octaves. For each note, the amplitude by time representation of each harmonic is shown. The amplitude of the fundamental harmonic (the leftmost peak for each note) varies dramatically. It is practically zero for G_3 (196 Hz), increases up to C_4 (256 Hz), decreases, and then reaches its maximum at A_4 (440 Hz) and C_5 (512 Hz).

3. Changing the Sound

It is possible to vary the strength of vibration modes by several mechanisms. For the source, the method and the strength of excitation can have a profound effect on the relative amplitudes of the harmonics. For a violin string, the maximum amplitude of the *n*th vibration mode is roughly $1/n^2$ of the amplitude of the first (lowest frequency) vibration mode for pluck-

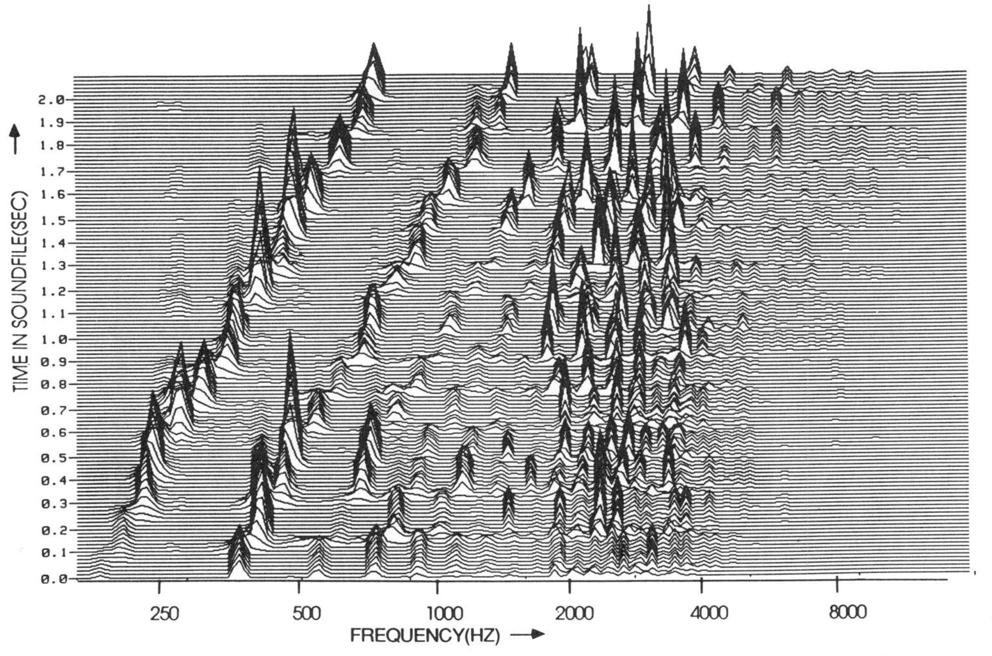

FIGURE 1 A representation of the harmonics of a violin playing the diatonic scale from G_3 (196 Hz) to G_5 (784 Hz). The relative amplitude of each harmonic is shown by the height of the peak. (From J. C. Brown, 1991, by permission.)

ing, but $1/n$ for bowing. Typically, stronger excitations generate relatively more high-frequency energy. For wind instruments, in which airflow is controlled by a reed or the performer's lips (e.g., clarinet, trumpet), as the blowing pressure increases, the amplitudes of the harmonics within the mouthpiece increase in a nonlinear manner. If the amplitude of the fundamental increases 3-fold, then the amplitudes of the second and third harmonics may increase 9-fold and 27-fold. Finally, the performer can vary the source by introducing vibrato. The frequency modulation simultaneously creates amplitude modulation of each source mode. Without the complex frequency and amplitude modulation due to vibrato, it is difficult to create a convincing simulation of a voice (Klatt & Klatt, 1990).

A performer can also change the filter characteristics. A trumpeter can vary the tubing and flare filter (as well as the radiation characteristics) by inserting a mute, and a singer can vary the vocal tract filter by changing the shape of the mouth. Singers often create a vibration mode around 2500 Hz (termed the *singers' formant*), which makes it easier for them to be heard over the sound of an orchestra (Sundberg, 1977).

B. Acoustic Structure

Listeners could use many acoustical properties to identify events. Spectral, transient, and noise cues typically used for single sounds will be presented first, followed by transition and rhythm cues based on sequences of sounds.

1. Frequency Spectrum

The classic cue is spectral shape, the relative amplitude of each partial. The spectral shape can be characterized by using the relative amplitudes of the partials to derive various statistical measures: (1) the central tendency (i.e., centroid); (2) the overall power obtained by summing the squared amplitude of each component; (3) the "slope" of the amplitudes, indicating the rate at which the amplitudes increase to the peak value and then fall off at higher frequencies; (4) the power spectrum across a set of critical bands or successive frequency regions (e.g., $\frac{1}{3}$ octave bands); (5) the variance of the amplitudes, which gives a measure of the "jaggedness" of the spectrum, useful when the source spectrum is continuous (an impact or noisy turbulence).

An alternative characterization is by means of the formant frequencies. Formants are spectral prominences (i.e., consistent high amplitude frequency ranges) created by one or more resonances in the filter. The formants of an instrument can be identified after averaging the frequency spectra across the playing range. In Figure 1, there appears to be a formant around 3000 Hz because there is an amplitude peak at that frequency for every note.

It is easy to understand how the frequency spectrum can specify an event that has but one sound (e.g., a church bell, a slamming door, a bouncing ball) or how the spectra can be used to categorize different instruments playing one identical note. It is more difficult to understand how the spectra can characterize an object or event that produces a range of sounds. The spectra will change from note to note as different vibration modes are stimulated (as shown in Figure 1). Therefore, it is impossible to claim that a particular pattern of amplitudes (i.e., spectral shapes) is the cause of identification or timbre because that pattern changes across notes.

2. Onset and Offset Transients

A second cue is the onset (attack) and offset (decay) of individual harmonics or different frequency ranges for nonharmonic objects. Differences in the attack and decay of harmonics for "string" instruments are due to variations in the method of excitation (plucked or bowed) and variations in the damping of the various vibration modes of the source and filter, and the differences for "wind" instruments (e.g., clarinet, trumpet) are due to linear and nonlinear feedback between the modes of the mouthpiece and sound body. On the whole, researchers have utilized qualitative descriptions to describe the patterning among the harmonics. For example, one description is based on the degree of synchrony of the attack transients.

The transients can provide a rich source of cues. Nonetheless, the same cautions that apply to spectral shape are relevant here. The onset and offset transients can be used to distinguish among single-sound objects or among one fixed set of instrumental sounds. But, for a given instrument or voice the pattern of the transients will vary across sounds. Typical changes in the onset and offset transients across notes can be seen in Figure 2, portraying piano notes across one octave. The shapes of the onset and decay of the fundamental vary across the octave.

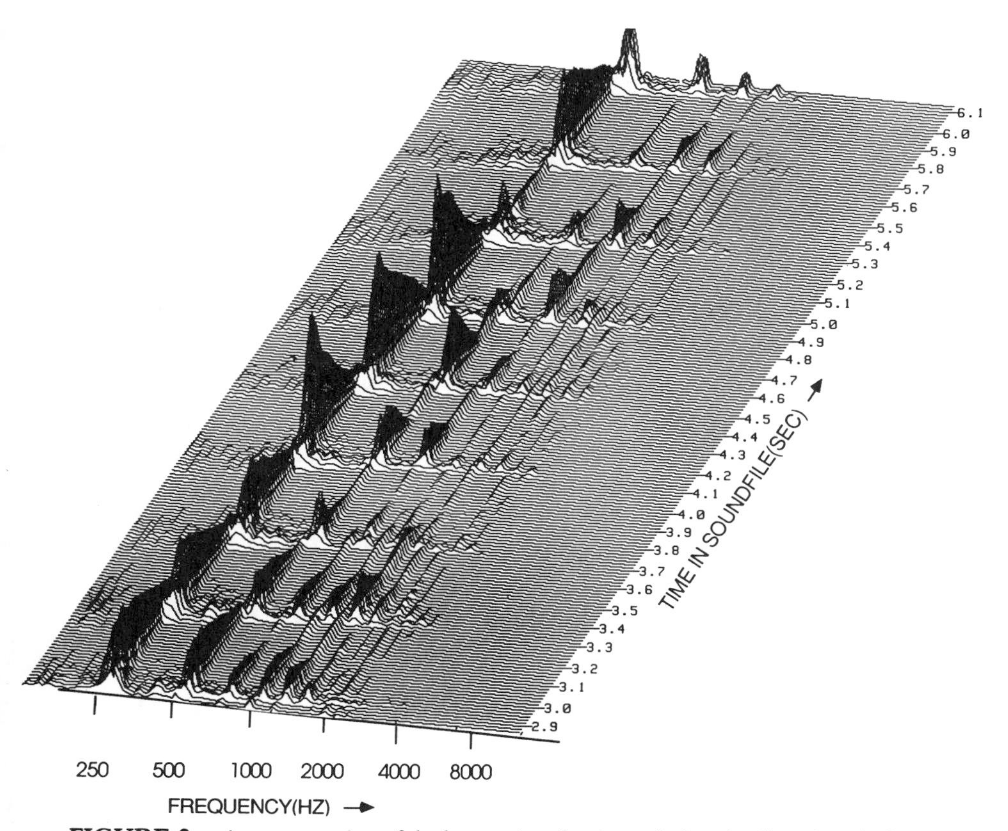

FIGURE 2 A representation of the harmonics of a piano playing the diatonic scale from C_4 (262 Hz) to C_5 (593 Hz). The onset and offset transients are shown by the rate of increase and decrease of the amplitude of each harmonic. (From J. C. Brown, 1991, by permission.)

3. Noise

A third cue is noise, nontonal sounds usually with broad frequency regions of high energy. Noise can occur when excitation energy is first applied to the source. For example, hitting a triangle creates a characteristic metallic sound, and bowing a violin creates an initial high frequency scratch before the bowing stabilizes. In addition, noise can be continuous. Examples include the breathy sound of the flute created by blowing across the mouthpiece, the hissing of a pipe, and the gurgling of flowing liquids. In fact, Gaver (1993) argues that inharmonic noise is characteristic of most natural events.

4. Transitions between Sounds

Another cue is the nature of the transition between one sound and the next one. During the transition, the decay and attack of successive sounds often overlap. The overlap may hinder identification by masking the transient cues or it may improve identification by creating a unique acoustic pattern not heard in discrete sounds. In speech, mechanical constraints on the speed of the articulators force speakers to overlap the articulation for successive sounds. Although it is not clear how coarticulation may be used to identify speakers, it may be significant.

5. Timing and Rhythm

Time provides a framework for auditory events where the onset and offset of sounds define those events. One temporal quality is whether the sound is roughly continuous (e.g., duct noise), oscillates in intensity (e.g., hand sawing), or is a series of discrete units (e.g., hammering, clapping, walking). Another temporal quality is the rhythm or timing between discrete sounds. Some physical systems are defined by damped rhythms in which successive sounds are progressively closer together in time (bouncing balls).

6. Overview

What does all of this mean? I think there are two important points.

1. The production of sound yields a large number of acoustic properties that can determine timbre and identification. The method of excitation and the physical properties of the source and filter will simultaneously affect the spectral shape, the transient, and the type of spectrum (harmonic or inharmonic). Although the discussion here has proceeded as if the source and filter are independent, that is not necessarily the case. For the clarinet and trumpet, pressure feedback from the filter acts to control the reed or lip

vibration of the source. The interlocking in production makes the acoustic cues somewhat related and correlated.

2. The cues to identification and timbre are not invariant and vary across notes, durations, intensities, and tempos. On top of these changes that usually reflect mechanical processes, performers may vary the sound by means of the excitation technique, intonation, and musical or linguistic emphasis. There are no "pure" cues. Every event exists in a context, and the acoustic properties depend on that context.

C. Perceptual Consequences

The perceptual consequences of the multiplicity of cues created by the sound production process are varied.

1. No predominant cue uniquely determines identification and timbre. Any single cue will provide some level of identification performance, and combinations of cues usually will produce better performance than a single one. Moreover, the effectiveness of any cue will vary across contexts. This makes comparisons of identification performance problematic. The ability to identify a trombone as opposed to a trumpet or a chain saw as opposed to a lawn mower will depend on the experimental conditions.

2. The interactive nature of sound production makes the acoustic cues correlated to some degree across a broad time span. Within limits, the cues are both redundant and substitutable. They are redundant because one part of the sound can be masked or removed and yet identification is still possible. They are substitutable because one part of the sound can be replaced by a different part with little loss. These two features make identification robust, enabling listeners to be accurate in noisy environments, or if they attend to different parts of the sound.

3. Listeners should make use of whatever cue(s) lead to best performance for a specific set of events. The cues selected will evolve as listeners become expert with the alternative sounds. Although it might appear that listeners should use acoustic cues that are invariant, there is little evidence that they utilize, or even can identify, the most stable ones.

III. EXPERIMENTAL RESULTS ON TIMBRE AND OBJECT PERCEPTION

The results will be organized in terms of instruments, voices, and environmental sounds. Within each section, results coming from various methodologies, including identification accuracy, similarity judgments, and verbal ratings will be intermixed.

A. Instruments

This section will start with research using single instruments playing one note, move to single instruments playing a short melody or a series of notes, and finally progress to two instruments simultaneously playing one note or a short melody.

1. Research Based on One Note

Grey (1977) investigated the perception of musical instrument timbre utilizing 12 instruments and 16 notes. Each note was approximately 311 Hz (E♭ above middle C). The notes were recorded and then reproduced by computer. The computer reproduction equalized fundamental frequency, intensity, and duration. It also simplified each note by approximating the amplitude envelope of each partial by 6 to 8 linear segments, thereby eliminating small scale variations in amplitude. The frequency variation was maintained including the initial noise with nonharmonic low-amplitude energy. The duration of each note was quite short, about 0.3 to 0.4 s. Twenty experienced musicians judged the similarity between every pair of notes.

The averaged similarity judgments represented the perceptual distance between any pair of notes. A multidimensional scaling procedure found the best configuration of the instrumental notes so that the calculated distances in the configuration were closest to the actual similarity judgments. (In general, multidimensional scaling is very good at identifying dimensions, but the placement of individual instruments at particular points is subject to chance error in the perceptual judgments; see Handel, 1989, for a further discussion of multidimensional scaling.) This procedure suggested that listeners were using three dimensions to make their judgments. To provide a perceptual and acoustical rationale for each dimension, Grey examined the frequency–amplitude–time representations for each note and proposed three descriptive measures that seemed to capture the similarities and differences among the instruments along each dimension. The instruments are shown in Figure 3 according to their positions on dimensions I, II, and III.

Dimension I is a measure of spectral energy distribution. At one end, the French horn and muted bowed cello have a restricted frequency range, and most of the energy is located at low frequencies. At the other end, the muted trombone and oboes have a wide frequency range, and a significant amount of energy is located in the upper partials. Dimensions II and III reflect the temporal variation of the notes. One end of dimension II includes the woodwinds (saxophones, clarinet, English horn), which are characterized by upper harmonics that start, reach their maxima, and decay at the same times. The other end includes the strings, brass, flute, and bassoon, which are characterized by upper harmonics that have differing patterns of attack and decay. Subjectively, this aspect of timbre reflects a static (woodwinds) versus dynamic (brass, strings) quality. One end of dimension III includes instruments that have low-amplitude, high-frequency inharmonic energy in the initial stage of the attack segment (e.g., clarinets). Subjectively, these instruments have a buzzlike but softer attack. The other end includes instruments that have dominant lower harmonics and that seem to have a more explosive initial attack (e.g., trumpets).

The emergence of the temporal dimensions of timbre may be due to the very short durations of the notes. The computer representations shown in Figure 3 portray sounds that consist mainly of the attack and decay; the steady-state section is almost nonexistent. Grey (1977) states that longer (1 s) unprocessed notes yield different configurations, in which the temporal dimensions do not emerge.

Wessel and Krumhansl (reported in Krumhansl, 1989) produced 21 distinct sounds using a frequency modulation synthesizer. Most of the sounds simulated traditional instruments, but others were hybrids such as a guitarnet (a combination of a guitar and a clarinet). One purpose was to determine the degree to which perceptual dimensions common to all instruments could characterize the timbre of each instrument and the degree to which a unique dimension specific to each instrument was necessary. For each instrument, the magnitude of the unique dimension can be thought of as the degree to which its timbre differs from that of all other instruments (see Winsberg and Carroll, 1989, for the specifics of the multidimensional scaling procedure).

The three common dimensions, shown in Figure 4, are remarkably similar to those found by Grey (1977). The first dimension was related to the attack quality (i.e., explosiveness). The trumpet (TPT) and trombone (TBN) have slower attacks with inharmonic transients whereas the vibraphone (VBS) and harpsichord (HCD) have more rapid attacks. The second dimension seemed to correlate with the temporal evolution of the spectral components (i.e., synchrony in Grey's, 1977, terminology). The trumpet (TPT) and string (STG) with asynchronous envelopes are at one end of the dimension, and the clarinet (CNT) and oboe (OBO) are at the other end. The third dimension was related to the amount of energy in the higher frequencies (i.e., brightness). Based on the magnitude of the unique dimension, the three common dimensions provide an excellent description of timbre for some traditional and hybrid instruments (e.g., trumpet, trombone, oboe/harpsichord, trumpet/guitar) but a poor description for other traditional and hybrid instruments (e.g., harpsichord, clarinet, vibraphone/

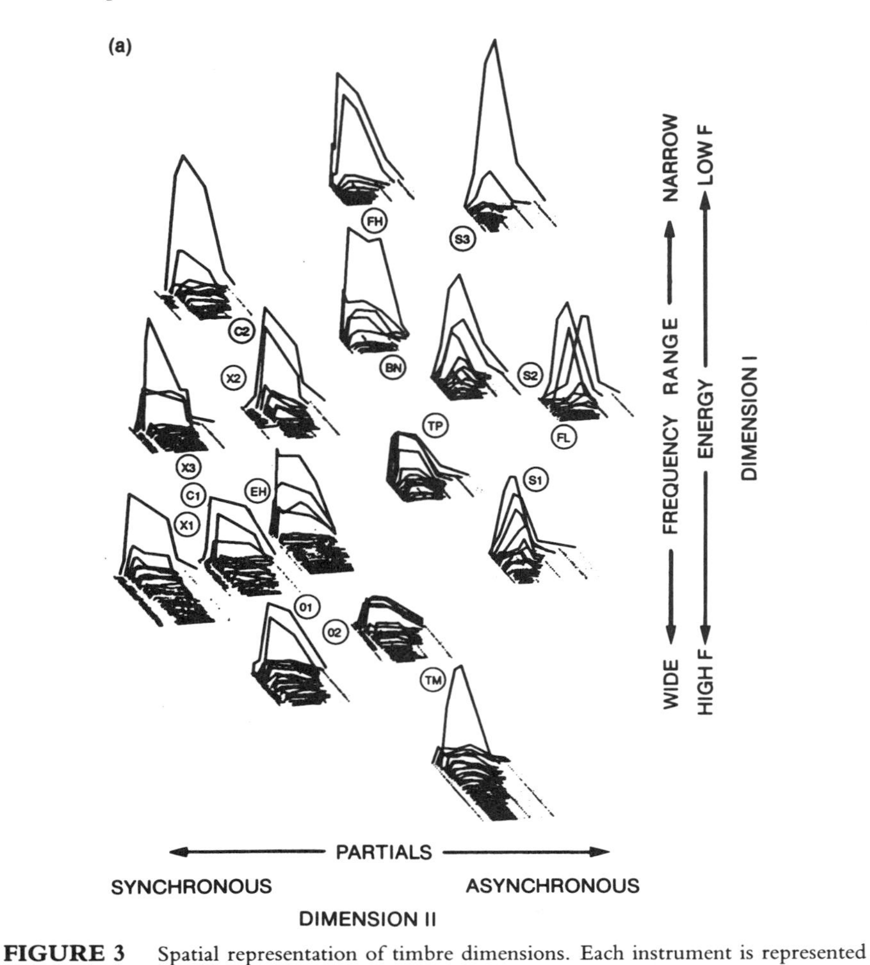

FIGURE 3 Spatial representation of timbre dimensions. Each instrument is represented by a simplified spectrogram. For each spectrogram, low frequencies occur at the top (upper left) and high frequencies occur at the bottom (lower right). Time is portrayed along the horizontal axis. Onset is at the left. The amplitude of each partial at each time point is represented by the vertical height. In (a), dimension I is combined with dimension II. In (b), dimension I is combined with dimensional III. Key: 01, 02 = oboes (different instruments and players); C1 = E♭ clarinet; C2 = bass clarinet; X1 = saxophone, medium loud; X2 = saxophone, soft; X3 = soprano saxophone; EH = English horn; FH = French horn; FL = flute; TM = muted trombone; TP = trumpet; BN = bassoon; Sl = cello, bowed near bridge producing nasal brittle sound (sul ponticello); S2 = cello, normal bowing; S3 = cello, bowing very lightly over the fingerboard producing a flutelike effect (sul tasto). (Adapted from Grey, 1977, by permission.)

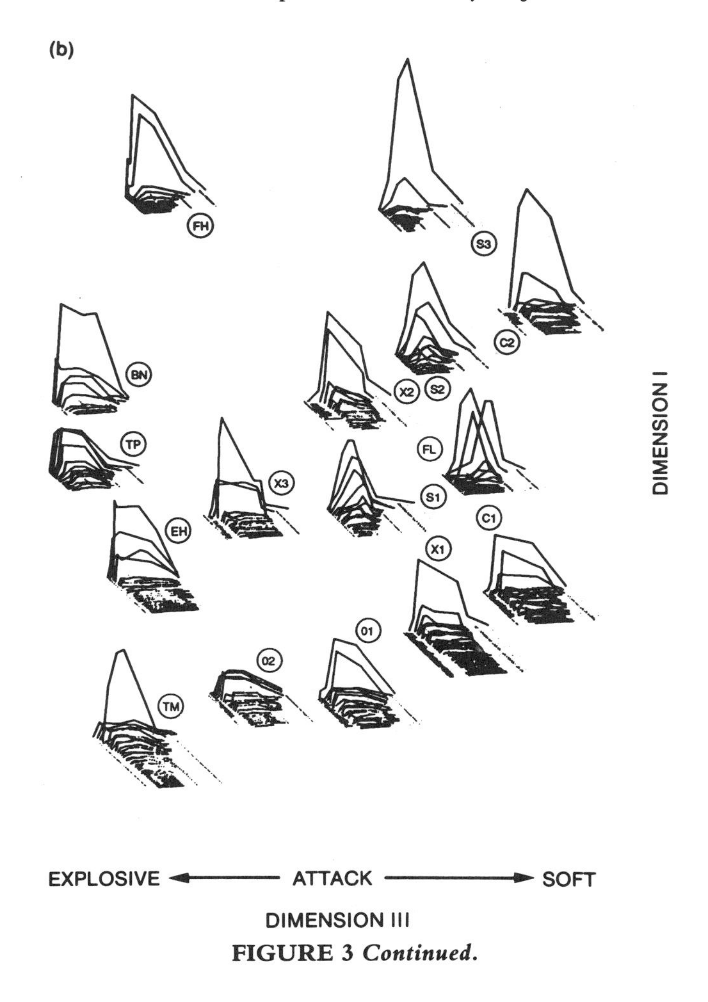

FIGURE 3 ***Continued.***

trombone, guitar/clarinet). For hybrids, the magnitude of the unique dimension was not based on those of its component instruments.

A more traditional experiment was done by Wedin and Goude (1972) using real instruments playing a sustained note of 440 Hz for about 3 s. All notes were roughly of equal loudness. The subjects either heard the entire note (3 s) or the middle steady-state segment of approximately 2 s. For both conditions, listeners judged the similarity between all pairs of notes.

There was no difference in judged similarity between whole notes and steady-state segments. Based on similarity judgments, the instruments could be placed in three groups, and each group could be characterized by a different spectral shape. In Figure 5, each group corresponds to one column,

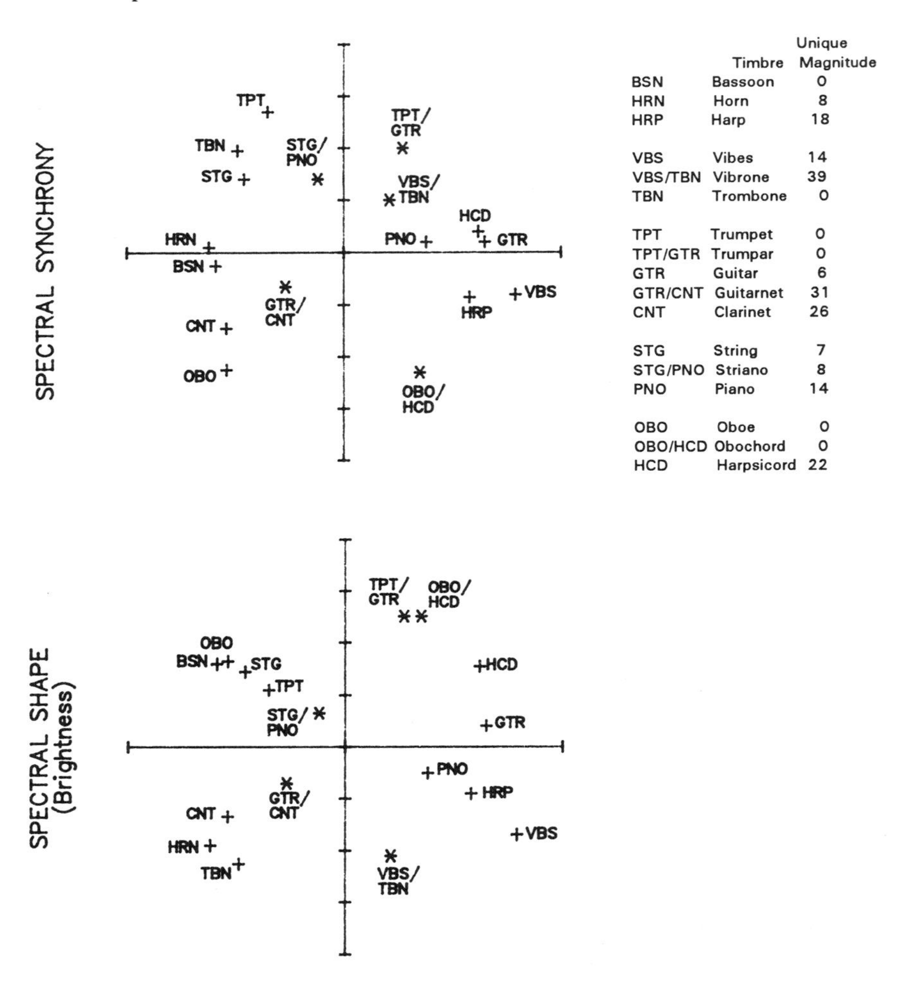

FIGURE 4 Spatial representations of the three timbre dimensions and the magnitude of the unique dimension for each instrument. A subset of 16 instruments including all common instruments and hybrids is shown. With the exception of the oboe/harpsicord hybrid, the other hybrids lie between their component instruments on all three dimensions. Traditional instruments are represented by pluses, and hybrids are represented by asterisks. (Adapted from Krumhansl, 1989.)

and for each instrument, the relative strength of each harmonic is shown. The leftmost column contains instruments characterized by a high-amplitude fundamental, along with high amplitudes of the middle harmonics; that is, harmonic richness. The middle column contains instruments

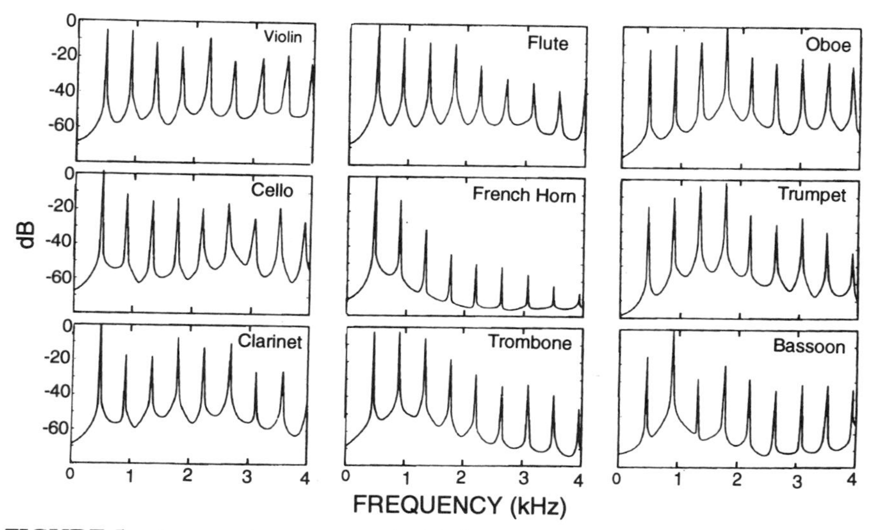

FIGURE 5 Spectra of the tones used by Wedin and Goude (1972). The columns represent the three different clusters of spectral shapes. (Adapted from Wedin & Goude, 1972, by permission.)

characterized by a high-amplitude fundamental along with decreasing amplitudes of the higher harmonics; that is, harmonic poorness. The rightmost column contains instruments characterized by a lower amplitude fundamental along with the highest amplitudes among the middle harmonics.

To summarize, the quality of timbre is multidimensional. In general, the important dimensions are related to the spectral shape and to the synchrony and rate of change of the attack and decay transients. But, variations in the presentation conditions can change the prominence of each one.

One way to confirm these results is to alter the signal in various ways and determine if the perceptual changes are predictable. One strategy is to construct synthesized hybrids. Grey and Gordon (1978) exchanged the relative maximum amplitudes of the harmonics between two instruments (i.e., the spectral shapes), but did not exchange the attack, and decay timings. The trumpet–trombone exchange is illustrated in Figure 6. Similarity judgments between pairs of tones were used to construct a three-dimensional perceptual space.

The results were compared to those obtained in a previous study using unmodified tones, as described earlier (Grey, 1977). In the original study, one of the dimensions of the perceptual space had been interpreted as relating to spectral shape. The pairs of tones that had exchanged spectral shapes did, in fact, exchange orders on this axis, confirming the original interpretation. Positions on the other dimensions, which had been interpreted in

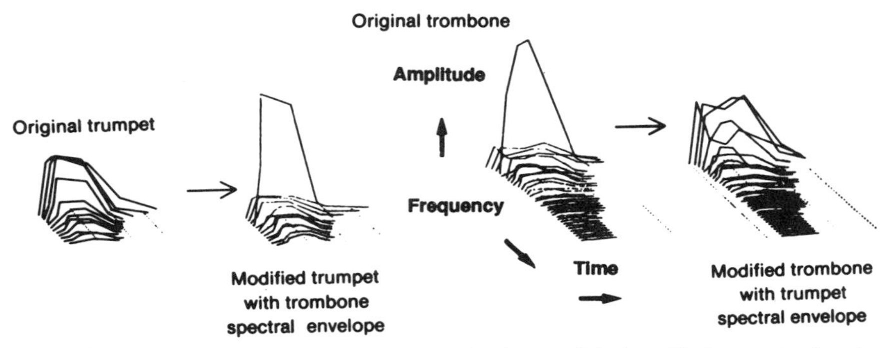

FIGURE 6 Each hybrid instrument retains the shape and timing of its harmonics, but the amplitudes are exchanged between the two original instruments. A trumpet and a trombone exchange is shown. (From Grey & Gordon 1978, by permission.)

terms of the temporal characteristics of the sounds, were, as expected, determined by the characteristics of the attack and decay of the sounds.

A second strategy is to isolate different segments of the acoustic signal. The amplitude–frequency–time graphs of various instruments shown in Figure 3 illustrate that each partial undergoes continuous change as it grows and decays. Throughout, even in the "steady state," each harmonic undergoes continuous small rapid amplitude and frequency changes.

In three simulations, Grey and Moorer (1977) progressively simplified instrument notes to discover the determinants of timbre. The first simplification smoothed out the amplitude modulations for each partial so that the amplitude envelope consisted of a series of six to eight linear segments (as in Grey, 1977). The second and third simplifications maintained the linear-segment amplitude envelope and added one additional simplification. The second eliminated the initial low-amplitude energy of the upper harmonics, correlated with the third dimension of the similarity judgment space described earlier. The third eliminated the frequency modulation of each partial so that all the small time-varying amplitude and frequency modulations that might be thought to give each instrument its unique richness were removed.

Surprisingly, the simplifications did not fundamentally change the timbre of any instrument, but instead resulted in changes perceived as slightly different articulations or playing styles. Subjects' judged that the tones with line-segment approximations of the amplitude envelope closely resembled the original tones. But, the additional elimination of the initial noise or the frequency modulation resulted in playing styles that were judged to be of poor quality. However, each simplification affected the various instruments differently. Clarinet sounds that contained low-amplitude, high-frequency noise were changed by the initial low amplitude energy simplification, whereas the soprano saxophone sounds that contained frequency modulation were changed by the elimination of the frequency variation.

Iverson and Krumhansl (1991) investigated timbre perception using entire tones, onsets only, and tones minus the onsets (i.e., steady state and decay). In each case, subjects judged the similarity among every pair of notes coming from 16 instruments. The similarity judgments were remarkably similar across the three contexts. The correlation between the judgments for the entire tones and onsets alone was 0.74, between the entire tones and tones minus onsets was 0.92, and that between onsets and tones minus onsets was 0.75. Thus the multidimensional solutions are quite similar. The configurations for the nonoverlapping onsets and tones minus onsets are superimposed in Figure 7.

The horizontal perceptual dimension is related to the rate of the amplitude change of each instrument. Percussive instruments like the tubular bells have fast onsets and progressive decays whereas blown instruments have slow onsets and remain at a steady amplitude before decaying due to the energy supplied by the player. The vertical dimension is related to the spectral shape. The French horn, tuba, and bassoon have much of their energy at lower frequencies but the muted trumpet has much of its energy at higher frequencies. Iverson and Krumhansl argue that subjects seem to be making judgments on the basis of acoustic properties that are found throughout the tone. The production constraints make the acoustic properties of the onset correlated with the acoustic properties of the remainder of the note.

What does this imply? First, the cues that determine timbre quality are interdependent because all are determined by the method of sound production and the physical construction of the instrument. For this reason, acoustic properties cut across the traditional segmentations of the sound (e.g., attack, steady state, decay) and can lead to the same perception using just part of the sound. This conclusion is supported by research investigating the identification of musical instruments using segments of a note (Berger, 1964; Saldanha & Corso, 1964; Wedin & Goude, 1972). From these experiments, it appears that the attack transient and part of the steady state or the true steady state alone can provide enough acoustic information for identification. The acoustic properties enabling source identification are found throughout the sound.

Second, the cues for timbre depend on context: the duration, intensity, and frequency of the notes, the set of comparison sounds, the task, and the experience of the subjects all determine the outcomes. At this point, *no* known acoustic invariants can be said to underlie timbre.

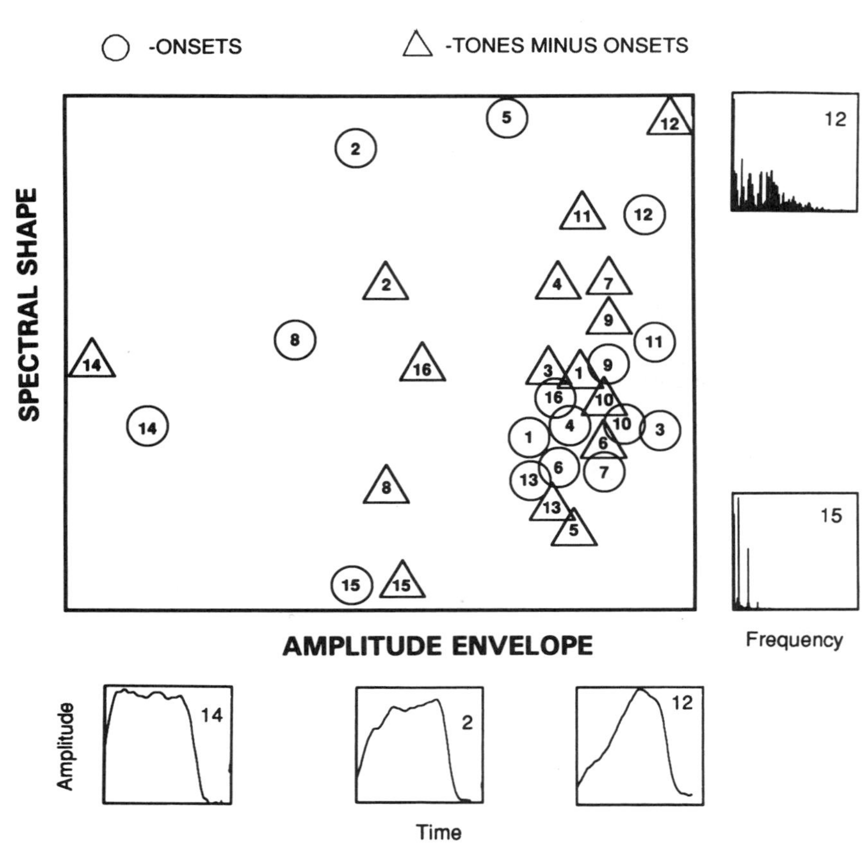

FIGURE 7 The two-dimensional perceptual configuration of the onsets (open circles) and the entire tones minus onsets (open triangles). Examples of individual tones are shown along the axes. The frequency scale ranges from zero to 22 kHz and the time scale ranges from zero to 186 ms. The vertical axes are amplitudes in arbitrary linear units. Key: 1 = bassoon, 2 = cello, 3 = clarinet, 4 = English horn, 5 = flute, 6 = French horn, 7 = oboe, 8 = piano, 9 = saxophone, 10 = trombone, 11 = trumpet, 12 = muted trumpet, 13 = tuba, 14 = tubular bells, 15 = vibraphone, 16 = violin. (Adapted from Iverson & Krumhansl, 1991, by permission.)

2. Sequences of Notes

It can be argued that single isolated notes do not provide a natural context for studying instrumental timbre. The simplest valid context may be a single instrument playing a short phrase, because the transients connecting notes might contain much of the acoustic information for identification. Kendall (1986) compared identification of a violin, clarinet, and trumpet using single notes and using seven- to ten-note folk song phrases. For both contexts, the experimental conditions included (1) the normal notes; (2) the steady states only, with silent gaps replacing the transients between notes; (3) the steady states only, connected together without transients or silent gaps; (4) the transients only, with silent gaps replacing the steady states; (5) simulated steady states generated by the repetition of a single cycle that eliminated all amplitude and frequency modulation.

The results for single notes mirrored other research. The transients alone (condition 4), the steady states alone (conditions 2 and 3), and the normal notes (condition 1) produced equivalent identification that was significantly better than that for single period simulated steady states (condition 5). In contrast, the results for the phrases indicated that the normal notes (condition 1) and steady states (conditions 2 and 3) were equivalent, and all three produced better identification than the equally difficult transients only (condition 4) and single period steady states (condition 5). Overall, performance was better using phrases than single notes.

There are two important points here. First, the time-varying steady-state component contributes to timbre in a way that the repetition of a single period does not. Second, the importance of any cue depends on context. For single tones, the transient, the steady state, and the transient plus steady state produce equivalent identification. For phrases, the steady state yields better identification than the transient. Thus, the attack transient becomes less useful for phrases than for individual notes.

3. Blends of Instruments

Krumhansl (1989) has speculated that there are levels of timbre, achieved by combining among natural events, instruments, and voices. Although the rules of acceptable instrumental blends have been discussed in orchestration handbooks (see Piston, 1955) little work has been done on the perception of the timbre of such blends. Kendall and Carterette (1991) studied the timbre resulting from two instruments played simultaneously. They used five instruments (clarinet, flute, oboe, saxophone, and trumpet), each played by a different musician, and constructed the ten possible pairs. The musical stimuli included instances in which both instruments played the same single note or the same melody and instances in which each instrument played a different single note (major third) or a different part of a harmonic melody. Although the outcomes differed slightly for each context, we will consider the combined results.

For all stimuli, subjects judged the similarity between two pairs of instruments, such as saxophone/clarinet (SC) with oboe/flute (OF). Multidimensional scaling techniques were used to generate the best fitting two-dimensional spatial representation, shown in Figure 8. Kendall and Carterette (1993) then interpreted the dimensions using verbal judgments of timbre. It appeared

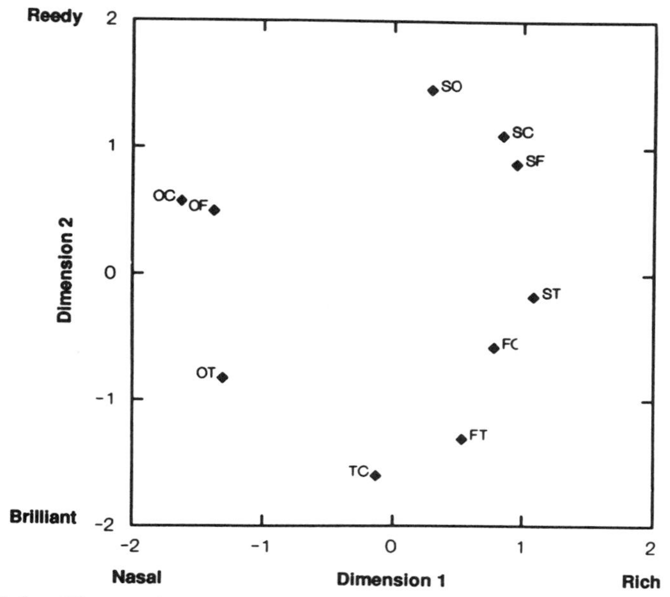

FIGURE 8 The two-dimensional configuration for the similarity among blends of two instruments. Key: C = clarinet, F = flute, O = oboe, S = Saxophone, T = trumpet. (Adapted from Kendall and Carterette, 1991, by The Regents of the University of California. Reprinted from *Music Perception*, Vol. 8, No. 4, p. 389.)

that dimension 1 represented nasal (negative) vs. rich (positive) and dimension 2 represented brilliant (negative) vs. reedy (positive). Dimension 1 was correlated with the amount of energy in the higher harmonics. The three "nasal" oboe pairs have a great deal of energy from the sixth to eighth harmonics whereas the remaining pairs typically have the maximum energy at the fundamental with progressively decreasing amounts at the higher harmonics.

It should be noted that a similarity space based on direct sound judgments is not simply related to a similarity space based on adjective judgments. For example, it is possible to rate each instrument pair along separate adjectives such as brilliant, edgy, mellow, nasal, and resonant, and then use the correlations in ratings to create a measure of similarity between pairs. This measure can be used to derive a configuration for the instrument pairs. If there was a equivalence between sound similarity and adjective similarity, the configurations ought to map onto each other. But, this is not usually the case. The adjectives are useful for conveying sound quality, but they are only tentative metaphors about timbre. The adjectives are easy to interpret in different ways, and the variability across subjects is often large; this may explain the lack of convergence.

4. Summary

Timbre is not correlated with a simple fixed acoustical property. It is inherently multidimensional and emergent, coming out of the interdependent but somewhat uncorrelated parts of the signal. Yet the sense of timbre pervades the entire signal. The attack and decay transient, the duration, the spectral shape, the inharmonic noise, and the amplitude and frequency modulation all contribute. However, the set of sounds that has been used in experiments is extraordinarily limited. No research has sampled notes across the playing range of one instrument and many studies have used unrepresentative timbres because the notes were at the extreme playing range of an instrument. This limitation has severely compromised any search for invariants.

B. Voices

The study of voice timbre and identification is different from that of instrument timbre and identification. There are millions of voices as opposed to the small set of instruments but the range of fundamental frequency of the vocal fold vibration is relatively limited compared to the playing range of nearly any instrument. Nonetheless, a single voice can produce a much greater variety of sounds than a single instrument.

We can still use the source–filter model to understand speech production (Fant, 1960). One excitation source is the airflow through the glottis as it opens and closes periodically. Another is noiselike, produced by forcing air past a constriction. The vocal tract is modeled by a succession of time-invariant filters that is assumed to be independent of the source excitation. Although this model is too simplified because there are several kinds of source-filter interactions (see Klatt & Klatt, 1990), it will serve as a useful guide.

1. Voice Identification

For the source, the possible acoustic cues for voice identification include the average fundamental frequency as well as the frequency range and contour of the vocal fold vibration. The typical fundamental frequency ranges for men, women, and children are 80–240 Hz, 140–450 Hz, and 170–600 Hz. For a normal speaker the fundamental frequency range in normal speech is roughly 2 to 1 while the range in singing is roughly 3 to 1. For the filter, the cues include the strengths, frequencies, and possibly the bandwidths of the formants. Speakers change the resonance modes and resulting formants by moving the articulators (particularly the tongue) to different positions.

Matsumoto, Hiki, Sone, and Nimura (1973) investigated the relative contribution of the vocal fold source and the vocal tract filter. Eight adult male speakers pronounced 0.5 segments of the Japanese vowel /a/ at three different fundamental frequencies (120 Hz, 140 Hz, and 160 Hz; a pure tone provided the reference frequency). Two sounds were presented successively, and the subjects judged whether the sounds were spoken by the same person. The measure of (dis)similarity was the percentage of times subjects judged the two voices as being different. The experimenters also measured the actual fundamental frequency, the spectral shape of the glottal source, the fluctuation of fundamental frequency, and the frequencies of the lowest three formants for each vowel.

The judgments among the 24 representations of the vowel /a/ (eight speakers at three different frequencies) were analyzed using multidimensional scaling, and then the derived perceptual attributes were related to the acoustic properties. The most important factor was the difference in fundamental frequency; it accounted for 50% of the variance in the perceptual distance between two representations. The next most important factor was the differences in formant frequencies; it accounted for an additional 25% of variance in the perceptual distance between vowels. However, if the fundamental frequency was held constant, the spectral shape and the fluctuation of the fundamental frequency predicted the perceived differences between two vowels better than the differences in the frequencies of the first three formants. Clearly, the set of sounds influences the relative importance of acoustic variables.

Kuwabara and Ohgushi (1987) studied the role of the formant frequencies in voice identification. They varied the center frequencies of the first three formants and found that shifts as small as 5% to 8% could severely hamper the ability to identify a known voice. Upward shifts of the second formant and downward shifts of the third formant created the greatest impairment. Thus, the frequency range roughly between 1000 Hz and 2500 Hz contains the most important information for voice identification.

Using a different approach to investigate the relative contributions of the source and filter, Miller (1964) created hybrid voices by pairing the vocal cord vibration pattern from one speaker with the vocal tract resonances from another speaker. In one experiment, vocal tract resonances were exchanged between two speakers. The word *hod* was used, and both speakers had the same fundamental frequency. Listeners judged that each hybrid more closely resembled the speaker whose vocal tract was represented. Matsumoto et al. (1973) constructed Japanese vowel hybrids that pitted the spectral shape of the source against the frequencies of the first three formants. The results were similar to those of Miller (1964): hybrids resembled voices with the same formant frequencies.

Several studies have investigated voice recognition using spoken phrases. For example, van Dommelen (1990) investigated the importance of the amplitude of the fundamental frequency, the fundamental frequency contour, and the speech rhythm (the durations of individual syllables). Five speakers were asked to imagine speaking the nine syllable sentence, "The seminar was truly a farce," and then to speak nine "ma" syllables using the same prosodic pattern. Several experimental manipulations were used to eliminate or equalize the three factors. In the most restrictive condition, three different initial fundamental frequencies were used but the durations of all the syllables were equalized and the fundamental frequency contours were equalized, so that they were linear decreasing functions of the fundamental frequency (286–239 Hz, 246–205 Hz, and 225–188 Hz). The listeners included the five speakers and six acquaintances, and their task was to judge which of the five speakers produced the synthesized utterance. In this case, the listeners judgments could have been based on the match between the speaker's own fundamental frequency and the average fundamental frequency of the declining contour as well as the match between the formant frequencies (i.e., spectral content) of each speaker and the synthesized version.

Fundamental frequency was the critical cue to speaker identification. The fundamental frequency contour and rhythm were of secondary importance, although identification was better using the original speech rhythm than using one constant duration for each syllable. Variable frequency contours or intonations and durational contrasts are probably inconsistent features for speaker identity because they often have semantic and emotional conotations. Finally, spectral cues were used by listeners to identify the speakers, particularly in instances where differences in fundamental frequency were small.

In a similar experiment, R. Brown (1981), starting with a control voice, varied the fundamental frequency by ±20%, the frequency of the first three formants by ±15%, and the formant bandwidths by ±50 Hz. Eight experimental voices were created from the three binary variables and listeners judged the similarity of the control voice to each one. All three variables affected similarity, demonstrating once again that voice quality is influenced by several acoustical cues.

In summary, fundamental frequency, formant frequencies, and formant bandwidths all affect judgment of voice quality and voice recognition. These cues appear analogous to the spectral amplitudes that partially characterize instrument timbre. R. Brown (1981) argues that these attributes are intrinsic to the speakers physiology (e.g., length, shape, and damping of the larynx) and should tend to be invariant over a wide range of contexts. However, other cues are utilized by listeners that are not tied to physiology. Van Dommelen (1990) found that the fundamental frequency contour and the syllable duration affected identity judgment, and R. Brown (1981) found that speaking rate affected similarity judgment. Moreover, Van Dommelen found that the use of acoustic cues was speaker dependent. One cue (e.g., fundamental frequency) was used to recognize one set of speakers but a different cue was used for a second set either because the speakers did not vary along the first cue or because there was a great deal of variation along the second.

The conclusion that there is no fixed set of invariant cues employed in all situations is supported by results using backward speech that maintains the long-term spectral information but alters dynamic cues (Van Lancker, Kreiman, & Emmorey, 1985) and by results using speech speeded up 33% or slowed down 33% without changing frequency (Van Lancker, Kreiman, & Wickens, 1985). For both conditions, the effects were not the same for all voices. For one example using speech rate, listeners recognized Jack Benny 90% of the time for all rates. However, listeners recognized John F. Kennedy 88% of the time at the normal speaking rates, 74% of the time at the faster rate, but only 17% of the time at the slower rate. The authors argue that the voice signal has a set of possible cues from which the listener can select a subset useful for recognizing a particular voice. The cues employed for one voice may be useless and inappropriate for another. Some cues will be distorted by being played backward and other ones will be distorted by changing rates. In all probability, listeners use whatever acoustical feature affords the best performance in a context (this is the same result found for instruments). However, it is dangerous to reify these features as being responsible for voice identification in real situations. It is quite possible that listeners use prototypes or schemas (Kreiman and Papcun, 1991) instead of features.

2. Voice Quality

Voice quality is the overall auditory coloring to which vocal fold (laryngeal) and vocal tract (superlaryngeal) factors contribute (Laver and Hanson, 1981). Before describing research on laryngeal effects, it is worthwhile to note briefly two ways in which the vocal tract can affect voice quality. First, to lower formant frequencies, the vocal tract can be lengthened by protruding the lips or by lowering the larynx. Second, to create a nasal formant (between 200 Hz and 300 Hz for men), the shape of the vocal tract can be changed by opening the velum and coupling the nasal cavity to the rest of the vocal tract. In both cases, there is a general acoustic effect beyond that required to produce a single sound. Thus, to describe a voice as nasal implies that the speaker is opening the nasal cavity for phonemes that normally do not have a nasal formant or that the voice sounds as if that is happening.

Ladefoged (1973) and Stevens (1977) have argued that due to the physiology of the larynx, speech sounds can be produced in only a few ways. Laver and Hanson (1981) defined six types of phonation: normal, vocal fry, falsetto, breathy voice, harsh, and whisper. Childers and Lee (1991) and Klatt and Klatt (1990) have attempted to characterize the first four types. Their research strategy was identical: first, use natural speech to find acoustical features that differentiate among the four types; second, synthesize different utterances by varying the magnitudes of these features; third, use listeners to judge the naturalness of the synthesized utterances in terms of phonation types.

A representation of the vocal fold movement, volume velocity waveforms of the air flow, and the source spectra for normal and breathy speech is shown in Figure 9. The motion of the vocal folds affects the airflow and that in turn simultaneously affects the amplitude of the fundamental, the spectral slope, and the amount of aspiration noise. Compared to normal (modal speech), breathy voices have a more gradual opening and closing of the glottis and usually the posterior opening of the glottis does not completely shut off the airflow. The open glottis introduces turbulent noise, particularly at frequencies above 2000 Hz that tends to replace weaker high frequency harmonics. In addition, the continuous airflow increases the amplitude of the fundamental frequency and increases the fall-off of the higher

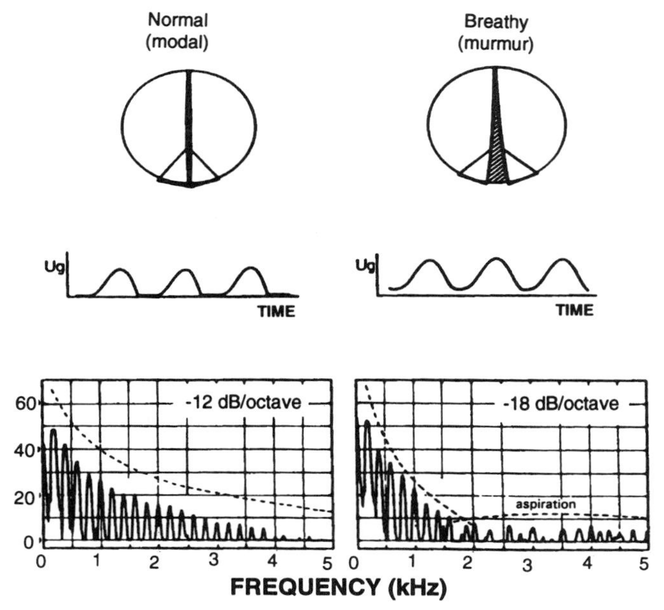

FIGURE 9 Representations of normal (modal) and breathy phonation. The top panel illustrates the closing of the glottis. The middle panel represents the airflow through the glottis across several periods. The bottom panel represents the spectra of the voice source. For breathy speech, the higher frequency harmonics fall off in intensity at a greater rate and are replaced by aspiration noise. (Adapted from Klatt & Klatt, 1990, by permission.)

frequency harmonics (−18 dB/octave). Thus, the acoustical cues to each type of phonation are interrelated due to the mechanics of the larynx.

Klatt and Klatt (1990) constructed a set of vowels to investigate the perception of breathiness. In some cases, only one acoustic feature was varied (e.g., the fundamental component could be increased in level by 10 dB). In other instances, two or three features were changed in correlated ways (e.g., spectral slope and aspiration noise). Increasing the level of the fundamental alone or changing the spectral slope alone did not affect perceived breathiness. Klatt and Klatt (1990) argue that each feature did not affect the judgment because, when either of these features is changed individually, it is a cue for another kind of speech variation and not for breathiness. The only single cue that affected perceived breathiness was level of aspiration noise. But, even here, context was important because the increase in noise necessary to influence the judgment was less if the other two features normally associated with breathiness were present.

In general, judgment of timbre for either instruments, voices, or natural events (discussed later) depends on rather complex cue interactions due to production constraints. Each cue is interpreted with respect to the values of the other cues and their use is situation dependent. These interactions make the synthesis of voices and instruments very difficult because the acoustic properties change in nonmonotonic ways across speaking and playing styles and across phonemes and notes.

C. Natural Events

Gaver (1993) suggests three main classes of natural events and resulting sounds: (1) vibrations of solids (e.g., scraping, rolling); (2) motions of gases (e.g., exploding balloons, wind); (3) impacts involving liquids (e.g., splashing, pouring). Unfortunately, research up to the present has been scattered across diverse events so that it is impossible to organize by type of event. Instead, we will consider first events composed of a single sound and then consider events composed of a series of sounds, where the timing and spectral changes among sounds may be the cue for identification.

1. Single Sounds

Halpern, Blake, and Hillenbrand (1986) investigated what makes a sound unpleasant. They asked listeners to judge the unpleasantness of a variety of sounds. A garden tool scraped across a piece of slate was the worst followed by rubbing two pieces of styrofoam together. They then manipulated the signal in two ways. First, the signal was split into frequencies below 2000 Hz and above 2000 Hz. Surprisingly, the low frequency components produced the unpleasant perception; the high frequencies alone do not sound unpleasant. Second, the overall amplitude envelope was simplified so that small amplitude variations were eliminated (much like in the Grey and Moorer, 1977, experiment). This did not affect the judged unpleasantness.

To study auditory perception of texture, Lederman (1979) had subjects listen to the sound produced when the experimenter ran her hand across grooved aluminum plates in which the grooves had different widths or were separated by different distances. The experimenter exerted either a small or a large amount of force as her hand crossed the plate. Incidental observations by the author suggested that as the width of the grooves increased making the surface rougher, the pitch generated by movement across the grooves decreased and the loudness, particularly at higher force levels, increased. Subjects could use the sound to judge texture although accuracy was far below that achieved by touch. Some subjects consistently used pitch or loudness but others switched to loudness at higher force levels, reflecting the discriminability of the cues. All subjects abandoned auditory cues when they could touch the plates themselves.

To study perceived mallet hardness, Freed (1990) used six mallets: In decreasing hardness, they were made of metal, wood, rubber, cloth-covered wood, felt, and felt-covered rubber. The mallets were used to strike four different sized aluminum cooking pans. Subjects listened to the sound of each strike, and judged the hardness of each mallet on a 10-point scale.

Listeners could judge hardness accurately. The hardness rating decreased monotonically across the six mallets. The average ratings for the hardest metal mallet and the softest felt-covered rubber mallet were approximately 9 and 3, respectively. Listeners judge the mallets as being harder when the sound energy is concentrated at higher frequencies and when the energy concentration shifts to lower frequencies over time (preliminary research had indicated that the first 300 ms of the sound was critical for identification, so the analyses were limited to that range). This can be seen in Figure 10(a): (1) The centroid of the frequency spectrum is greater for the harder mallets at nearly all time points; and (2) across time, the centroid shifts to lower frequencies for harder mallets but is relatively stable for softer ones. In addition, listeners judge mallets as being harder when the energy level is greater and the energy fall-off (i.e., slope) across time is higher. This can be seen in Figure 10(b): For the harder mallets, (1) the spectral level is higher, particularly at the onset of the sound and (2) across time the decrease in energy is also greater.

To study perceived walking, Li, Logan, and Pastore (1991) asked eight males and eight females to walk naturally across a hardwood stage. Typically, each walker used his or her own shoes constructed with leather soles and solid, synthetic heels, although three males did use the same shoe. Subjects listened to a recording of four steps and simply judged whether the walker was male or female. The probability of judging a male walker

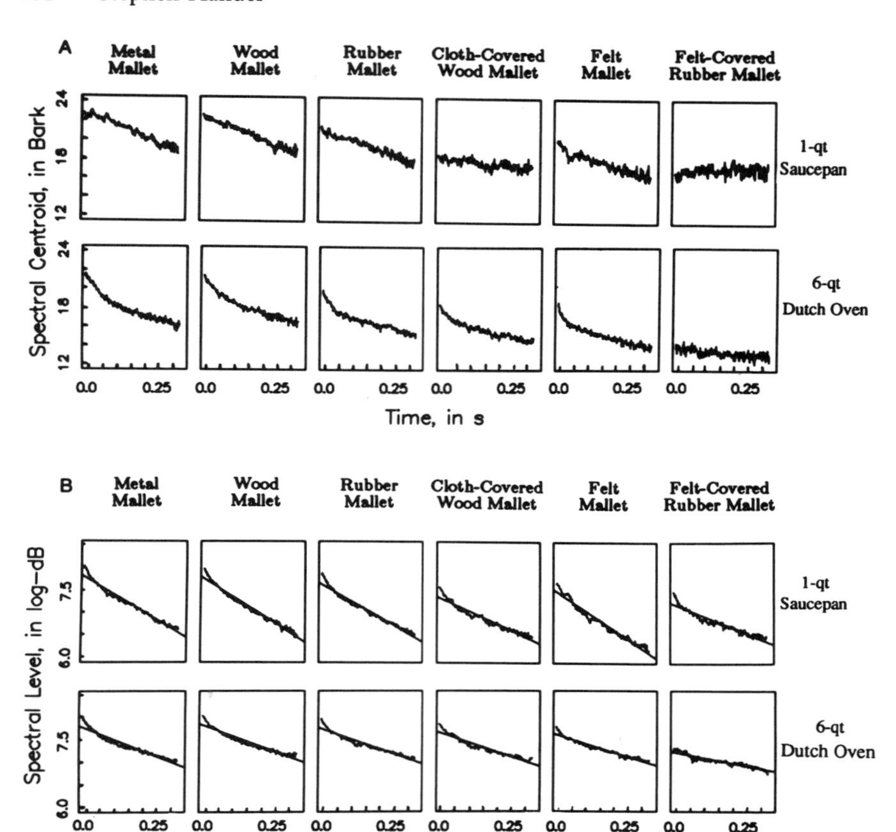

Time, in s

FIGURE 10 The spectral centroid and overall level as a function of time for the six mallets striking the smallest and largest aluminum pans. The best fitting regression line is superimposed in B. An explanation of the bark scale can be found in Moore, Chapter 6, this volume. (Adapted from Freed, 1990, by permission.)

correctly was 69% and the probability of judging a female walker correctly was 74%.

For the acoustical analysis, Li et al. (1991) averaged the spectra of the 12 steps for each of the walkers (see Figure 11). "Male" judgments were correlated with (1) a lower spectral centroid, (2) higher values of asymmetry (skewness) and peakiness (kurtosis), and (3) smaller amounts of high frequency energy. "Female" judgments were correlated with (1) a higher spectral centroid and (2) greater amounts of high frequency energy. The differences in the acoustic waves were attributed to gender differences in height and weight, which jointly affect the spectral centroid, and also shoe size.

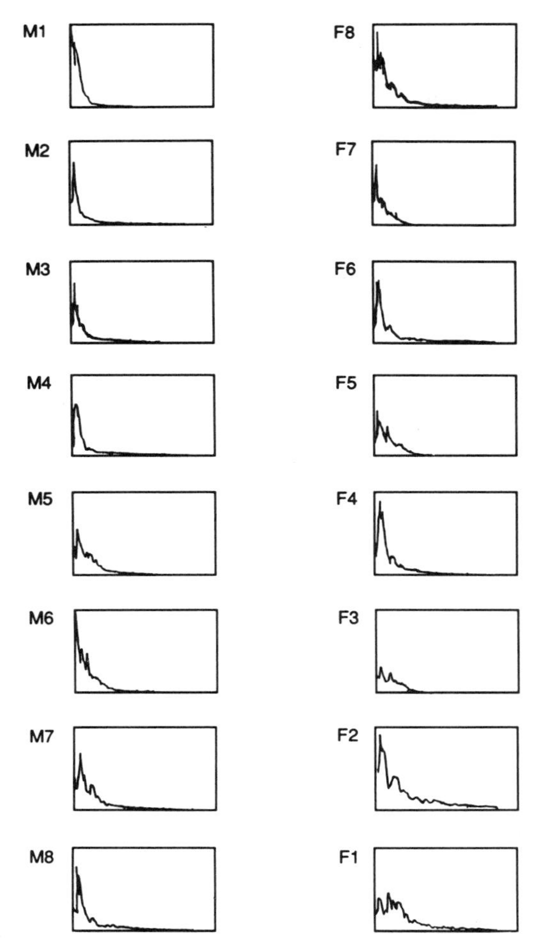

FIGURE 11 The mean spectra for the eight male and eight female walkers. The spectra of each of the 12 steps were calculated separately and then averaged to yield the mean spectrum for each walker. The order M1–M4, F8, M5–M7, F7, M8, F6–F1 represents the ranking of the probability of "male" judgments. (Adapted from Li et al., 1991, by permission.)

[Surprisingly, the pace (steps per second) was the same for males and females although listeners incorrectly judged faster rates as being female.]

The last study concerns clapping (Repp, 1987). Clapping can be used as a communicative signal, a sign of approval, or "notes" in music. Repp's fundamental question was whether the acoustical signal contained information concerning configuration and gender.

Repp clapped in eight different ways ranging from a flat, parallel mode in which the two hands perfectly overlapped to a perpendicular mode in which the fingers were cupped so that contact was palm to palm. The palm-to-palm resonance seemed to create a low-frequency peak, whereas the fingers-to-palm resonance produced a mid-frequency peak. Overall, the variation in hand configurations accounted for 50% of the spectral variability among the claps. The subjects perceptual task was to identify palm-to-palm clapping, finger-to-palm clapping, and one intermediate type of clapping. They were able to distinguish among the prototypical claps produced by Repp based on the presence of low-frequency peaks. They were not able to distinguish among the modes for the claps produced by a group of different subjects. The listeners used mainly clapping rate to infer hand position; they judged slower rates to have come from palm-to-palm clapping, but hand position was actually unrelated to rate. Due to another incorrect stereotype, listeners were unable to judge gender. They expected males to clap more slowly, produce louder claps, and produce claps with stronger low-frequency resonances. But, although males did have bigger hands there were no differences between males and females in the rate, intensity, or frequency spectra of claps. Perhaps when there is a high degree of variability among individual sounds as in walking or clapping, listeners move to information that is tied less to the production of each sound itself, such as the patterning among the sounds (even if it is an incorrect stereotype as in rate of clapping and walking).

2. Sequences of Sounds

To study the similarity among natural sounds, Vanderveer (1979) recorded events like hammering, finger snapping, and crumpling paper, and judges grouped together similar sounds. The most important sound quality was the gross temporal patterning. Hammering and knocking were grouped because of their rhythmic, repetitive sound. Sawing, filing (wood), and shaking a pin box were grouped because of their continuous, rough sound. Jingling keys and jingling coins were similar because of their irregular metallic sound. Two drinking glasses that clinked together several times and a spoon that clinked against a china teacup were similar because of their repeated ringing tones that soften in intensity. In these events, the rhythm, continuity, and temporal envelope of the sounds have the major influence on perceived similarity. Frequency spectra were of minor importance, probably because the sounds lacked tonal qualities. (Frequency spectra would characterize the material; e.g., crumpling metal vs. crumpling paper.)

Warren and Verbrugge (1984) demonstrated that the time-varying spectral patterning provides the information needed to distinguish between bouncing objects and breaking objects. Actually, each piece of a broken object bounces; the distinction between bouncing and breaking is between one bouncing object and many bouncing pieces. When one object (or one piece of an object) bounces, each brief impact excites all possible resonance modes of that object so that the sound of each impact is relatively constant. The impacts are "damped quasi-periodic": The impacts decrease progressively in both intensity and the interval between successive impacts. The acoustic information for bouncing is therefore one damped quasi-periodic sequence, such that each successive sound is similar. The acoustic information for breaking is overlapping damped quasi-periodic sequences, where each sequence (representing one piece) has a different frequency spectrum and a different damped quasi-periodic pattern. Successive sounds alternate irregularly, because each represents the impact of a different piece.

Warren and Verbrugge (1984) constructed several simulations:

1. Using a single damped quasi-periodic sequence as in bouncing, each successive impact was given a different sound, simulating the alternation of breaking parts. This destroyed the perception of bouncing: The identity of the sound impacts over time signifies the unity of a single object bouncing.

2. Again using the single bouncing sequence, each impact was given the same sound, created by the summation of the spectra of all broken pieces. This did lead to the perception of bouncing: The identity of impacts, even if they represent multiple pieces, signifies bouncing.

3. Using the overlapping damped quasi-periodic sequences representing the pieces of a broken object, each impact was given the same spectrum, simulating bouncing. This destroyed the perception of breaking: Distinct spectral properties of successive sounds are necessary to signify the multiple pieces of broken objects.

4. Using a strict periodic sequence in which the impacts were equally timed destroyed the perception of bouncing. An accelerating rhythm of impacts is necessary.

Warren, Kim, and Hussey (1987) studied the perception of elasticity in bouncing balls. They defined elasticity (e) by how readily an object returns to its initial shape after deformation; it ranges in value between 0 and 1. For a ball dropped on a rigid surface, the acoustic cue to elasticity (bounciness) is the rhythm of the sound impacts. If $e = 0$, the ball does not rebound and there is a damped thud; if $e = 1$, the ball bounces forever and there is a sequence of identically timed, equally loud sounds; if e is between 0 and 1, there is a damped sequence and the impact sounds speed up and have decreasing amplitude.

Visual and auditory simulations of a bouncing object consisted of a white circle moving up and down, appearing to bounce (with a constant intensity, 80 ms, low-pitch auditory tone at each impact) against a white line at the bottom of the screen. One cue to elasticity is identical for vision and audition: e is equal to the ratio of successive periods (t_2/t_1), between successive rebounds. Warren et al. (1987) demonstrated that the judgment of "bounciness" was equivalent for vision and audition. For both modalities, subjects judged bounciness on the basis of the duration of a single period, which is partially correlated to both elasticity and drop height, and not on the ratio of successive periods, which is perfectly correlated to elasticity but not drop height. Thus, listeners use temporal information to judge bounciness, but they cannot use the relative period cue that specifies it uniquely in either vision or hearing.

3. Discussion

The acoustic signal can convey information about the source. Listeners make use of their tacit knowledge about the effects of the physical properties of objects on sound production to perceive that object from the acoustic signal. We can emphasize the timbre of the sound (i.e., the proximal stimulus at the ear) or the event (i.e., the distal stimulus at the source) but they are normally equivalent. Event similarity and sound similarity are alternative ways of describing the relation between two different events because events and actions that are similar will generate sounds that are similar.

IV. NEUROPSYCHOLOGICAL EVIDENCE

The perception of instruments, voices, and other natural events seems to be similar. Although there are no fixed, invariant sets of acoustical properties, the cues and strategies that listeners use are equivalent in general form (see Ballas, 1993 for a similar conclusion).

In spite of this commonality, there is rather strong evidence that the processing of speech and music is done in different parts of the cortex. Previously, it was thought that each cerebral hemisphere was specialized to process one domain, either speech or music. It is now clear that the perception and production of speech and music involve many parallel subprocesses and that each may be lateralized differently. For example, playing the piano involves visual processing of the score, motor control of both hands, and monitoring of the sound output. Sergent, Zuck, Terriah, and MacDonald (1992), using PET techniques, provide evidence that each function is localized in different parts of the cortical lobes and cerebellum.

Auditory agnosia refers to a perceptual deficit in which individuals can perceive changes in the basic acoustical properties of frequency, intensity, and so on, but can not recognize or identify a sound event. The sound is stripped of its meaning (however, these individuals can usually identify those events by means of another modality such as vision). There are many cases of the double dissociation between speech and other auditory domains. In these, one individual can understand and speak complex verbal material, but cannot identify musical melodies or identify environmental events such as airplanes or doors slamming. A different individual shows the reverse pattern of deficits; that person can not understand verbal material but can identify melodies and nonspeech events (Spreen, Benton, & Fincham, 1965; Yaqub, Gascon, Al-nosha, & Whitaker, 1988; Lambert, Eustache, Lechevalier, Roosa, & Viader, 1989). Some patients who are unable to discriminate voices, can discriminate environmental sounds (Van Lancker & Kreiman, 1987).

A second kind of double dissociation occurs between identification and discrimination. Identification tasks usually require the subject to pick out which picture represents the sound object. Discrimination tasks usually require the subject to judge if two sound segments are from the same event or different events. Eustache, Lechevalier, Viader, and Lambert (1990) report two cases in which there appears to be a dissociation between identification and discrimination in their nonspeech performance. Case 1 with a left temporoparietal lesion was unable to identify common tunes but was able to discriminate whether two tunes were the same or different in terms of one false note, rhythm, or tempo. Case 2 with a right frontal lesion was able to identify environmental sounds and familiar tunes, but was unable to determine whether two sounds or two tunes were the same or different (this outcome seems paradoxical because someone who can identify two tunes should at least be able to compare the labels). Van Lancker, Cummings, Kreiman, and Dobkin (1988) present equivalent results for voices. Three patients with damage to the right hemisphere parietal region performed normally on voice identification but at the chance level for voice discrimination whereas two patients with temporal lobe damage of either hemisphere showed the reverse pattern: chance identification but normal discrimination. In sum, these two outcomes seem to illustrate that identification and discrimination occur in different cortical domains.

There is a puzzle. Studies of the perception of verbal and nonverbal sounds for both identification and discrimination tasks yield similar outcomes. Yet, the neuropsychological results suggest that the verbal and nonverbal domains are located in different cortical locations and that the identification and discrimination of events are independent and seem to be located in different cortical hemispheres. The similarity in perception is probably due to the equivalence of the acoustic information in all domains created by the similar production mechanisms and to the fact that the verbal and nonverbal domains are located in adjacent cortical regions, utilizing equivalent cellular architecture.

V. CONCLUDING DISCUSSION

An event or object has a distinct timbre, but that timbre cannot be related simply to one acoustic feature or to a combination of them. In the same fashion, event identification cannot be attributed to one acoustical feature or combination. Timbre is an emergent property that is partly a function of the acoustical properties and partly a function of the perceptual processes. Listeners may use a specific feature such as fundamental frequency or use global gestaltlike properties, depending on the task, the sets of alternatives, and other aspects of the context. The problem is the constancy of the sense of timbre for a given source in spite of the large changes in the acoustical properties.

I would like to suggest that perceiving sound objects is much like perceiving faces. A face percept is emergent; a face is made up of features like eyes, ears, lips, nose, and complexion but its "look" is based on the spatial configuration of these features. Similarly, a sound percept is emergent, it is made up of features but its timbre is based partly on the temporal interaction of those features. Face perception demonstrates constancy across changes in profile, orientation, hairstyle, and across changes due to aging. Timbre demonstrates constancy across changes in frequency, intensity, and so on. Finally, faces and sound objects show double dissociation of identification and discrimination (see Farah, 1990, for a discussion of visual agnosia).

Research on face perception may suggest new possibilities for investigating timbre perception. For example, there is both experimental and neurological evidence (reviewed by Bruce, 1990) that the perception of facial expressions and characteristics (e.g., a moustache) can occur in parallel with recognition. It may be that the perception of the timbre and the recognition of the sound event occur in parallel. As a second example, faces tend to be stored as composites when the instances are similar, but as prototypes when they differ greatly (Watt, 1988). It may be that there are multiple representations of the timbre of one instrument due to the very different sounds along the playing range.

Even though vision and audition traditionally are said to be fundamentally different kinds of senses (but see Handel, 1988), the similarity between face and sound object perception is a strong argument that perception in general is determined by the interaction of dynamic features. Natural events in all perceptual domains provide enormously complicated information in a formal mathematical sense. Paradoxically, this complexity may yield the stable unambiguous percept.

Acknowledgments

I would like to thank Judith C. Brown, Daniel J. Freed, Paul Iverson, Carol L. Krumhansl, Richard E. Pastore, and Cheryl B. Travis for help in preparing the figures and William M. Hartman and Brian C. J. Moore, in particular, for making significant improvements to this chapter.

References

American National Standards Institute (1973). *Psychoacoustical terminology.* S3.20. New York: American National Standards Institute.

Ballas, J. A. (1993). Common factors in the identification of an assortment of brief everyday sounds. *Journal of Experimental Psychology: Human perception and performance, 19,* 250–267.

Balzano, G. J. (1986). What are musical pitch and timbre? *Music Perception, 3,* 297–314.

Berger, K. W. (1964). Some factors in the recognition of timbre. *Journal of the Acoustical Society of America, 36,* 1888–1891.

Brown, J. C. (1991). Calculation of a constant Q spectral transform. *Journal of the Acoustical Society of America, 89,* 425–434.

Brown, R. (1981). An experimental study of the relative importance of acoustic parameters for auditory speaker recognition. *Language and Speech, 24,* 295–310.

Bruce, V. (1990). Perceiving and recognising faces. *Mind & Language 5,* 342–364.

Childers, D. G., & Lee, C. K. (1991). Voice quality factors: Analysis, synthesis, and perception. *Journal of the Acoustical Society of America, 90,* 2394–2410.

van Dommelen, W. A. (1990). Acoustic parameters in human speaker recognition. *Language and Speech, 33,* 259–272.

Eustache, F., Lechevalier, B., Viader, F., & Lambert, J. (1990). Identification and discrimination disorders in auditory perception: A report on two cases. *Neuropsychologia, 28,* 257–270.

Fant, G. (1960). *Acoustical theory of speech production.* The Hague: Mouton.

Farah, M. J. (1990). *Visual agnosia.* Cambridge, MA: MIT Press.

Freed, D. J. (1990). Auditory correlates of perceived mallet hardness for a set of recorded percussive sound events. *Journal of the Acoustical Society of America, 87,* 311–322.

Gaver, W. W. (1993). What in the world do we hear? A ecological approach to auditory event perception. *Ecological Psychology, 5,* 1–29.

Gibson, J. J. (1966). *The senses considered as perceptual systems.* Boston: Houghton Mifflin.

Grey, J. M. (1977). Timbre discrimination in musical patterns. *Journal of the Acoustical Society of America, 64,* 457–472.

Grey, J. M., & Gordon, J. W. (1978). Perceptual effects of spectral modifications on musical timbres. *Journal of the Acoustical Society of America, 63,* 1493–1500.

Grey, J. M., & Moorer, J. A. (1977). Perceptual evaluations of synthesized musical instrument tones. *Journal of the Acoustical Society of America, 62,* 454–462.

Halpern, D. L., Blake, R., & Hillenbrand, J. (1986). Psychoacoustics of a chilling sound. *Perception & Psychophysics, 39,* 77–80.

Handel, S. (1988). Space is to time as vision is to audition: Seductive but misleading. *Journal of Experimental Psychology: Human Perception and Performance, 14,* 315–317.

Handel, S. (1989). *Listening: An introduction to the perception of auditory events.* Cambridge, MA: MIT Press.

Iverson, P., & Krumhansl, C. L. (1991). Measuring the similarity of musical timbre. *Journal of the Acoustical Society of America, 89,* 1988.

Jenkins, J. J. (1984). Acoustic information for places, objects, and events. In W. H. Warren and R. E. Shaw (Eds.), *Persistence and change: Proceedings of the first international conference on event perception* (pp. 115–138). Hillsdale, NJ: Lawrence Erlbaum.

Kendall, R. A. (1986). The role of acoustic signal partitions in listener categorization of musical phrases. *Music Perception, 4,* 185–214.

Kendall, R. A., & Carterette, E. C. (1991). Perceptual scaling of simultaneous instrument timbres. *Music Perception, 8,* 369–404.

Kendall, R. A., & Carterette, E. C. (1993). Verbal attributions of simultaneous wind instrument timbres. II. Adjectives induced from Piston's *Orchestration. Music Perception, 10,* 469–502.

Klatt, D. H., & Klatt, L. C. (1990). Analysis, synthesis, and perception of voice quality variations among female and male talkers. *Journal of the Acoustical Society of America, 87,* 820–857.

Kreiman, J., & Papcun, G. (1991). Comparing discrimination and recognition of unfamiliar voices. *Speech Communication, 10,* 265–275.

Krumhansl, C. L. (1989). Why is musical timbre so hard to understand? In S. Nielzen and O. Olsson (Eds.), *Structure and perception of electroacoustic sound and music* (pp. 43–53). Amsterdam: Elsevier (Exerpta Medica 846).

Kuwabara, H., & Ohgushi, K. (1987). Contributions of vocal tract resonant frequencies and bandwidths to the personal perception of speech. *Acustica, 63,* 120–128.

Ladefoged, P. (1973). The features of the larynx. *Journal of Phonetics, 1,* 73–83.

Lambert, J., Eustache, F., Lechevalier, B., Roosa, Y. & Viader, F. (1989). Auditory agnosia with relative sparing of speech perception. *Cortex, 25,* 71–82.

Laver, J., & Hanson, R. J. (1981). Describing the normal voice. In J. K. Darby (Ed.), *Speech evaluation in psychiatry* (pp 57–78). New York: Grune and Stratton.

Lederman, S. J. (1979). Auditory texture perception. *Perception, 8,* 93–103.

Li, X., Logan, R. J., & Pastore, R. E. (1991). Perception of acoustic source characteristics: Walking sounds. *Journal of the Acoustical Society of America, 90,* 3036–3049.

Matsumoto, H., Hiki, S., Sone, T., & Nimura, T. (1973). Multidimensional representation of personal quality of vowels and its acoustical correlates. *IEEE Transactions on Audio and Electroacoustics, AU-21,* 428–436.

Miller, J. E. (1964). Decapitation and recapitation: A study of voice quality. *Journal of the Acoustical Society of America, 36,* 2002 (*A*).

Piston, W. (1955). *Orchestration.* New York: W. W. Norton.

Repp, B. H. (1987). The sound of two hands clapping: An exploratory study. *Journal of the Acoustical Society of America, 81,* 1100–1109.

Saldanha, E. L., & Corso, J. F. (1964). Timbre cues and the identification of musical instruments. *Journal of the Acoustical Society of America, 36,* 2021–2028.

Sergent, J., Zuck, E., Terriah, S. & MacDonald, B. (1992). Distributed neural network underlying musical sight-reading and keyboard performance. *Science, 257,* 106–109.

Spreen, O., Benton, A. L. & Fincham, R. W. (1965). Auditory agnosia without aphasia. *Archives of Neurology, 13,* 84–92.

Stevens, K. N. (1977). Physics of larynx behavior and larynx modes. *Phonetica, 34,* 264–279.

Sundberg, J. (1977). The acoustics of the singing voice. *Scientific American, 236,* 82–91.

Vanderveer, N. J. (1979). Ecological acoustics: Human perception of environmental sounds. *Dissertation Abstracts International, 40,* 4543B (University Microfilms No. 80-04-002).

Van Lancker, D. R., Cummings, J. L., Kreiman, J., & Dobkin, B. H. (1988). Phonagnosia: A dissociation between familiar and unfamiliar voices. *Cortex, 24,* 195–209.

Van Lancker, D., Kreiman, J., & Wickens, T. D. (1985). Familiar voice recognition: Patterns and parameters. Part II: Recognition of rate altered voices. *Journal of Phonetics,* 13, 39–52.

Van Lancker, D., & Kreiman, J. (1987). Voice discrimination and recognition are separate abilities. *Neuropsychologica, 25,* 829–834.

Van Lancker, D., Kreiman, J., & Emmorey, K. (1985). Familiar voice recognition: Patterns and parameters. Part I. Recognition of backward voices. *Journal of Phonetics, 13,* 19–38.

Warren, W. H., Kim, E. E., & Husney, R. (1987). The way the ball bounces: Visual and auditory perception of elasticity and control of the bounce pass. *Perception, 16,* 309–336.

Warren, W. H., Jr., & Verbrugge, R. R. (1984). Auditory perception of breaking and bouncing events. *Journal of Experimental Psychology: Human Perception and Performance, 10,* 704–712.

Watt, R. J. (1988). *Visual processing: Computational psychophysical and cognitive research.* Hillsdale, NJ: Lawrence Erlbaum.

Wedin, L., & Goude, G. (1972). Dimension analysis of the perception of instrumental timbre. *Scandanavian Journal of Psychology, 13,* 228–240.

Winsberg, S., & Carroll, J. D. (1989). A quasi-nonmetric method for multidimensional scaling via an extended Euclidean model. *Psychometrika, 54,* 217–229.

Yaqub, B. A., Gascon, G. G., Al-nosha, M., & Whitaker, H. (1988). Pure word deafness (acquired verbal auditory agnosia) in an Arabic speaking patient. *Brain, 111,* 457–466.

Index

A

B

Q

R

S